NATO ASI Series

Advanced Science Institutes Series

A series presenting the results of activities sponsored by the NATO Science Committee, which aims at the dissemination of advanced scientific and technological knowledge, with a view to strengthening links between scientific communities.

The Series is published by an international board of publishers in conjunction with the NATO Scientific Affairs Division

A	Life Sciences	Plenum Publishing Corporation
B	Physics	London and New York
C	Mathematical and Physical Sciences	Kluwer Academic Publishers Dordrecht, Boston and London
D	Behavioural and Social Sciences	
E	Applied Sciences	
F	Computer and Systems Sciences	Springer-Verlag Berlin Heidelberg New York
G	Ecological Sciences	London Paris Tokyo Hong Kong
H	Cell Biology	Barcelona

The ASI Series Books Published as a Result of
Activities of the Special Programme on
SENSORY SYSTEMS FOR ROBOTIC CONTROL

This book contains the proceedings of a NATO Advanced Research Workshop held within the activities of the NATO Special Programme on Sensory Systems for Robotic Control, running from 1983 to 1988 under the auspices of the NATO Science Committee.

The books published so far as a result of the activities of the Special Programme are:

Vol. F25: Pyramidal Systems for Computer Vision. Edited by V. Cantoni and S. Levialdi. 1986.

Vol. F29: Languages for Sensor-Based Control in Robotics. Edited by U. Rembold and K. Hörmann. 1987.

Vol. F33: Machine Intelligence and Knowledge Engineering for Robotic Applications. Edited by A. K. C. Wong and A. Pugh. 1987.

Vol. F42: Real-Time Object Measurement and Classification. Edited by A. K. Jain. 1988.

Vol. F43: Sensors and Sensory Systems for Advanced Robots. Edited by P. Dario. 1988.

Vol. F44: Signal Processing and Pattern Recognition in Nondestructive Evaluation of Materials. Edited by C. H. Chen. 1988.

Vol. F45: Syntactic and Structural Pattern Recognition. Edited by G. Ferraté, T. Pavlidis, A. Sanfeliu and H. Bunke. 1988.

Vol. F50: CAD Based Programming for Sensory Robots. Edited by B. Ravani. 1988.

Vol. F52: Sensor Devices and Systems for Robotics. Edited by A. Casals. 1989.

Vol. F57: Kinematic and Dynamic Issues in Sensor Based Control. Edited by G. E. Taylor. 1990.

Vol. F58: Highly Redundant Sensing in Robotic Systems. Edited by J. T. Tou and J. G. Balchen. 1990.

Vol. F63: Traditional and Non-Traditional Robotic Sensors. Edited by T. C. Henderson. 1990.

Vol. F64: Sensory Robotics for the Handling of Limp Materials. Edited by P. M. Taylor. 1990.

Vol. F65: Mapping and Spatial Modelling for Navigation. Edited by L. F. Pau. 1990.

Vol. F66: Sensor-Based Robots: Algorithms and Architectures. Edited by C. S. G. Lee. 1990.

Traditional and Non-Traditional Robotic Sensors

Edited by

Thomas C. Henderson

Department of Computer Science
University of Utah
3190 Merrill Engineering Building
Salt Lake City, Utah 84112, USA

Springer-Verlag Berlin Heidelberg New York
London Paris Tokyo Hong Kong Barcelona
Published in cooperation with NATO Scientific Affairs Division

Proceedings of the NATO Advanced Research Workshop on Traditional and Non-Traditional Robotic Sensors, held in Maratea, Italy, August 28 – September 2, 1989.

Library of Congress Cataloging-in-Publication Data
NATO Advanced Research Workshop on Traditional and Non-Traditional Robotic Sensors (1989: Maratea, Italy)
Traditional and non-traditional robotic sensors/edited by Thomas C. Henderson. p. cm.—(NATO ASI series. Series F, Computer and system sciences; vol. 63) "Proceedings of the NATO Advanced Research Workshop on Traditional and Non-Traditional Robotic Sensors, held in Maratea, Italy, August 28–September 2, 1989"— T.p. verso.

ISBN-13: 978-3-642-75986-4   e-ISBN-13: 978-3-642-75984-0
DOI: 10.1007/978-3-642-75984-0

1. Robotics—Congresses. 2. Detectors—Congresses. I. Henderson, Thomas C. II. Title. III. Series: NATO ASI series. Series F, Computer and system sciences; vol. 63. TJ210.3.N377   1989   629.8'92—dc20   90-46060

© Springer-Verlag Berlin Heidelberg 1990
Softcover reprint of the hardcover 1st edition 1990

2145/3140-543210 – Printed on acid-free-paper

Preface

This book contains the written record of the NATO Advanced Research Workshop on Traditional and Non-Traditional Robotic Sensors held in the Hotel Villa del Mare, Maratea, Italy, August 28 – September 1, 1989. This workshop was organized under the auspicies of the NATO Special Program on Sensory Systems for Robotic Control. Professor Frans Groen from the University of Amsterdam and Dr. Gert Hirzinger from the German Aerospace Research Establishment (DLR) served as members of the organizing committee for this workshop.

Research in the area of robotic sensors is necessary in order to support a wide range of applications, including: industrial automation, space robotics, image analysis, microelectronics, and intelligent sensors. This workshop focused on the role of traditional and non-traditional sensors in robotics. In particular, the following three topics were explored:

– Sensor development and technology,

– Multisensor integration techniques,

– Application area requirements which motivate sensor development directions.

This workshop brought together experts from NATO countries to discuss recent developments in these three areas.

Many new directions (or new directions on old problems) were proposed. Existing sensors should be pushed into new application domains such as medical robotics and space robotics. Much work is required on the development of silicon sensors and model-based sensors. The integration of olfactory sensors seems a useful step. Error recovery for sensors, as well as maintenance and prognosis of sensors, needs to be addressed. It would be very useful to have adequate sensor simulations for a wide range of sensors. It appears that internal sensors will switch from strain gauges to induction-based sensors; this needs to be developed. Inertial sensors need to be incorporated in the sensor-based control loop for robotic systems. In the context of active sensing, it is necessary to define the relation between quantitative versus qualitative sensing and sequencing and integration of sensing with the given task. Behavioral specifiactions seem to offer an approach to this problem.

One major problem that arose was the definition of tactile sensors; just what does a tactile sensor measure? It seems necessary to distinguish at a high level between the cutaneous and kinesthetic properties of tactile sensors. Another concern is the field performance of commercially available sensors, as well as the requirement of a meaningful set of specifications for a sensor. It seems that it may be necessary to retrofit sensors to existing robots, and this may pose some problems. Another major concern is the fact that the development of sensor systems is tied to commercial development, which puts a premium on industrial needs rather than general research needs. It might also be useful if there existed an integrated testing lab for sensors and associated software systems. Of immediate concern is the lack of fast sensor and controller interfaces. Furthermore, many participants felt the need for torque sensors at robot joints. Finally, safety issues with sensors have yet to be dealt with adequately.

The papers in this volume constitute a cross-section of the current research in robotic sensors and motivate the new directions and problems summarized above. We were quite fortunate to obtain such an excellent group of active researchers to participate and contribute to this workshop. In addition, another major goal of this NATO program was met, in that the group mixed very well socially, and we all made new acquaintances from the NATO research community. This kind of meeting provides a wonderful opportunity for the detailed exchange of ideas which, in turn, motivates further research cooperation among the participants.

Finally, I would like to thank all the people who helped make the meeting possible and who saw to it that it ran smoothly: the staff and management of the Hotel Villa del Mare, and Vicky Hawken from the University of Utah who handled the administrative details of the meeting with wit and charm.

Salt Lake City   Thomas C. Henderson
March 1990

Table of Contents

Fast Sensory Control of Robot Manipulators

F. Dessen and J. G. Balchen
Norwegian Inst. of Technology, Div. of Engineering Cybernetics
N-7034 Trondheim, Norway

Abstract: A positional deviation sensor for contact-free physical guidance of a manipulator is described. The manipulator is set to track a motion marker close up. Factors limiting and means for improving the performance are pointed out. The result is a system that can be used for real-time training of spray-painting robots. The means are easily extened to general sensory control.

1. INTRODUCTION

This contribution presents a new system for contact-free positional sensing of a tool or teach-handle moved freely by the operator (Balchen, 1984; Dessen, 1987; 1988). It consists of a small-range positional deviation sensor (PDS) mounted at the tip of the manipulator. A motivation for employing the positional deviation sensor is given by considering the programming of a paint spraying robot. During the training session, the operator holds the spray-gun in his own hands, and carries out the task in his usual manner. At the same time, the robot manipulator follows the tool slavishly, without any contact. The situation is shown in Fig. 1. The manipulator is able to track the tool because of feedback from the PDS, in a similar manner to using a force sensing handle for the same purpose.

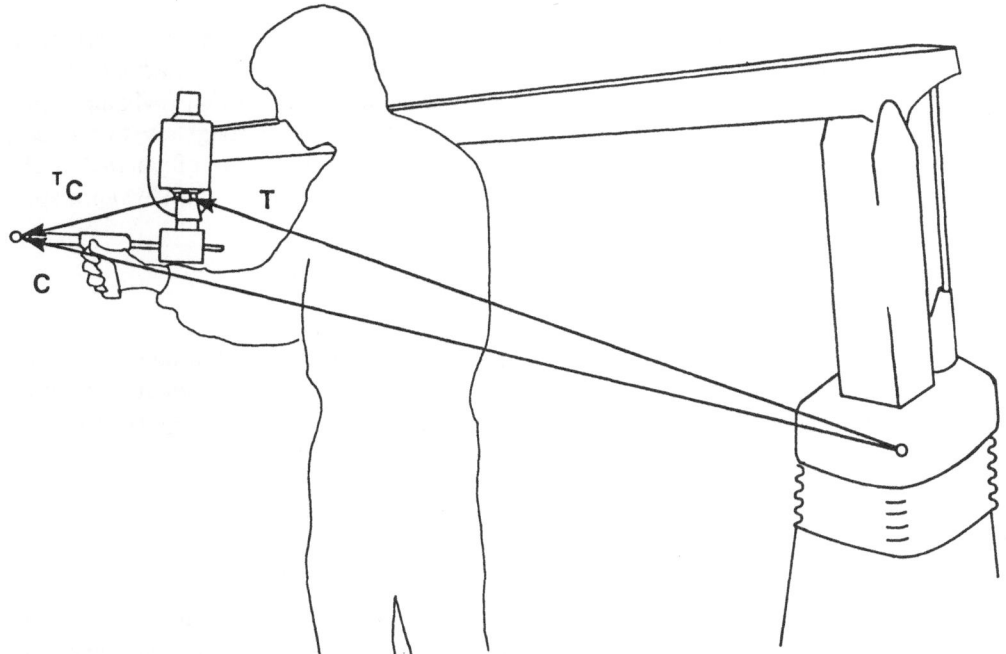

Figure 1. Manual lead-through programming

NATO ASI Series, Vol. F 63
Traditional and Non-Traditional Robotic Sensors
Edited by T. C. Henderson
© Springer-Verlag Berlin Heidelberg 1990

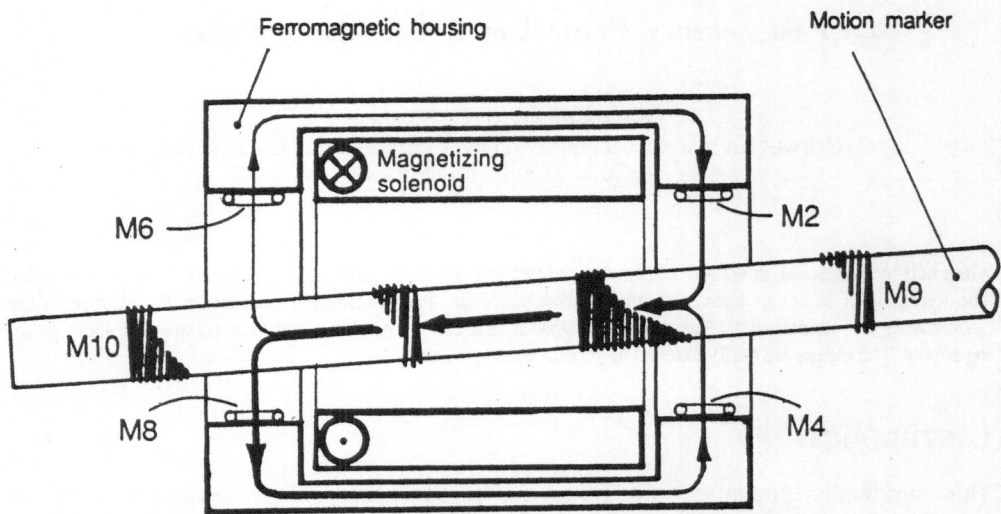

Figure 2. Measurement sensor. Flux through each solenoid varies according to motion.

The PDS consists of two complementary parts; one fixed to the tool, and the other to the wrist of the manipulator. These two parts are made of ferromagnetic material, Fig. 2, and are magnetized in opposite axial directions by means of a solenoid mounted inside the larger, hollow part. This creates a radial magnetic field between them which varies according to their relative position. By measuring the magnetic field at selected points, relative motion is monitored.

Details on the system are presented in (Dessen, 1988). Presented below is a discussion of problems concerning high-speed tracking of the tool marker. First, tracking performance is discussed in terms of a simple 1# (degree of freedom) system, reviewing basic servo system concepts. Means for improving the performance are sought in two directions, the first one being to increase the system bandwidth by means of internal feedback. The second involves optimizing the use of the sensor's range by introducing a special coordinating controller

2. TRACKING SYSTEM

The success of the complete training system relies on the presence of a control system which makes the manipulator follow the tool closely at all times. Especially, real-time training of paint spraying robots requires outstanding velocity and accelerational capabilities. The purpose of this section is to outline a coordinating control structure and to point out factors that limit its performance.

2.1. Servo coordination

Very often, playback control systems work in servo coordinates. This means that any motion reference is converted into a vector of servo references. A corresponding vector e of control errors is used to compute the required action. With the PDS, no servo reference is available. It is easy however to produce the control error in servo coordinates directly from the deviation sensor data, vector $^M d$. The relationship is usually represented by the first-order approximation

$$^M\mathbf{d} = \mathbf{J}_{eM}\mathbf{e} \tag{2.1}$$

where $\mathbf{J}_{eM}$ is the Jacobian matrix $\partial^M\mathbf{d}\,/\,\partial\mathbf{e}$. If it is nonsingular, the control error may be computed by

$$\mathbf{e} = \mathbf{J}_{Me}{}^M\mathbf{d} \qquad \text{where} \qquad \mathbf{J}_{Me} = \mathbf{J}_{eM}^{-1} \tag{2.2}$$

Often, $\mathbf{J}_{Me}$ will not be computed explicitely. Instead $^M\mathbf{d}$ will go through a sequence of intermediate differential tranformations until finally $\mathbf{e}$ is obtained.

Once $\mathbf{e}$ is given, controllers working in joint coordinates can be applied. The usual approach at this level is to employ separate controllers for each servo, neglecting possible coupling between the actuators. In many cases, this assumption is reasonable though not completely true. In the present case, coupling will be neglected by assuming it to be taken care of by internal control. More precisely, internal speed control of N servos is assumed with a resulting transfer matrix

$$\mathbf{G}(s) = \mathrm{diag}[g_1(s),\, g_2(s),\, ...,\, g_N(s)] \tag{2.3}$$

which relates actual speed and speed reference by

$$\dot{q}_i(s) = g_i(s)\,\dot{q}_{i0}(s)$$

Seen from an added positional controller, the process transfer matrix is

$$\mathbf{H}(s) = \tfrac{1}{s}\,\mathbf{G}(s) = \mathrm{diag}[\tfrac{1}{s}\,g_1(s),\, ...\, ,\, \tfrac{1}{s}\,g_N(s)] \tag{2.4}$$

In practice $\mathbf{H}(s)$ will include several off-diagonal coupling terms. However, these are assumed to be taken care of by the speed controller.

The outlined structure may be considered as a hierarchical control system where the higher level is a coordinating positional controller whereas the lower level performs decoupling speed control. This partitioning will be used throughout since it is believed to give a better understanding of the control problem at hand.

2.2. Performance

It is of interest to have a rough idea of the expected tracking performance. For this purpose, a simple 1# positional control system will be considered. Conforming to (2.4), the process transfer function is written

$$h(s) = \tfrac{1}{s}\,g(s) \tag{2.5}$$

The control system will be analyzed in terms of its open-loop transfer function, which is assumed to be stable. At first, a proportional controller with gain k_p will be applied. It follows from the Bode-Nyquist stability criterion that the closed-loop system will be stable whenever

$$k_p < \alpha\omega_\varphi = \omega_x \tag{2.6}$$

4

where ω_φ is the $-180°$ phase shift frequency for $h(j\omega)$, and $\alpha = |g(j\omega_\varphi)|^{-1}$

In order to obtain simple expressions for the performance, controllers are assumed to be designed using the Ziegler-Nichols method, where control parameters are based on the values of ω_x and ω_φ. For a proportional (P) controller, the method yields

$$h_r(s) = k_p = 0.5\,\omega_x \tag{2.7}$$

The resulting open-loop transfer function becomes

$$h_0(s) = h_r(s)h(s) \tag{2.8}$$

Tracking performance will be studied in terms of the closed-loop error transfer function

$$N(s) = e(s)/q_0(s) = \frac{1}{1 + h_0(s)} \tag{2.9}$$

since the relationship between reference and control error is

$$e(s) = q_0(s) - q(s) = q_0(s) - h_0(s)e(s) \tag{2.10}$$

Assuming that the control error amplitude is restricted by $|e(j\omega)| < E$ for all real ω, the reference is restricted by

$$|q_0(j\omega)| < E\,|N(j\omega)|^{-1} \tag{2.11}$$

Hence, the allowed tool motion amplitude is closely related to the allowed positional deviation, and may for any single frequency ω be obtained almost directly from $N(j\omega)$.

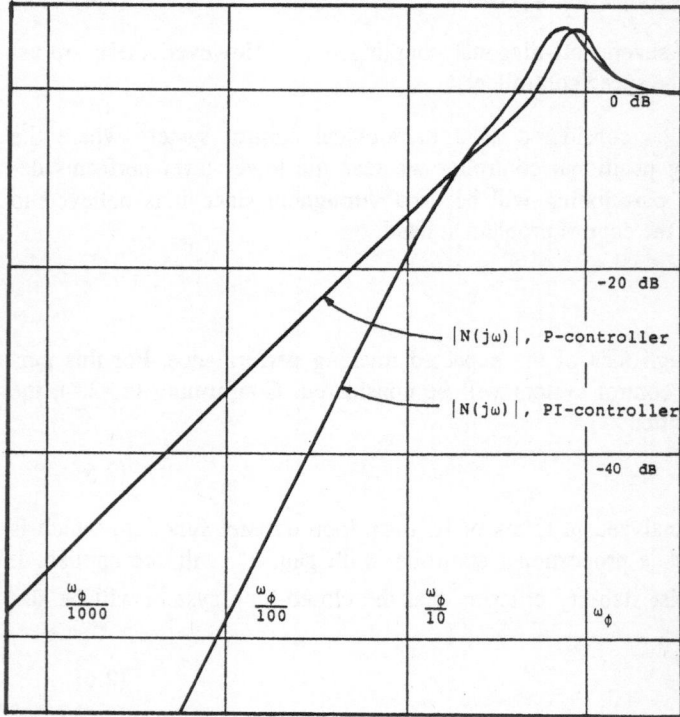

Figure 3. Amplitudes of typical closed-loop error transfer functions for P and PI type servos. Low amplitude means good performance.

By studying the asymptotical behaviour of $N(s)$ as s approaches zero, it is seen that a maximum steady-state velocity of

$$v_P = \lim_{s \to 0} \frac{sE}{N(s)} = E\,k_p = 0.5\,E\,\omega_x = 0.5\,E\,\alpha\,\omega_\varphi \qquad (2.12)$$

is obtained. Since the range of the PDS of order ± 1cm only, the resulting speed will often be low. A proportional plus integral (PI) controller, where

$$h_r(s) = k_p\,\frac{1 + T_i\,s}{T_i\,s} \qquad (2.13)$$

will by the Ziegler-Nichols method be assigned

$$k_p = 0.45\,\omega_x \;\; ; \;\; T_i = 5/\omega_\varphi = 5\,\alpha/\omega_x \qquad (2.14)$$

This time, the asymptotical behaviour of $N(s)$ implies unbound velocity with a maximum steady state acceleration

$$a_{PI} = \lim_{s \to 0} \frac{s^2 E}{N(s)} = E\,\frac{k_p}{T_i} = \frac{0.45}{5\,\alpha}\,E\,\omega_x^2 = 0.09\,E\,\alpha\,\omega_\varphi^2 \qquad (2.15)$$

Compared to the use of a proportional controller, this is an improvement. However, the servo may still react slowly.

Performance measures v_P and a_{PI} are used often in place of function $N(s)$ since they give an immediate understanding of important properties of the system in question. Whenever v_P is used, it is understood that the underlying closed-loop error transfer function has the slope $+20$ dB/decade at low frequencies. Whenever a_{PI} is used, this slope is understood to be $+40$ dB/decade. For both P and PI control, sample Bode plots of resulting error transfer functions are shown in Fig. 3.

As seen from (2.12) and (2.15), the performance depends on the deviation sensor by E and on $g(s)$ by α and ω_φ. Trivially, E may be increased by enlarging the positional deviation sensor. A second approach is to coordinate the servos so that maximum use is made of the existing sensor range. This is considered in § 6. Due to the second order dependence on ω_φ in (2.15), means for improving the lower-level control system are of special interest. Such means are discussed in §§ 7 and 8.

2.3. Force sensing handles

During the initial experiments, it was of interest to compare the PDS with a force sensing device. For a moment, the space between the two parts of the sensor was packed with rubber foam in order to obtain mechanical contact with a reasonable stiffness. The result was a more stable system, however the sensor workspace was reduced because of the foam, and now considerable force had to be applied in order to make the manipulator move.

Consider now the possibility of making use of the improved stability to increase the controller gains. This again will increase the performance of the system. A quick stability analysis can be made by assuming a certain stiffness k_s between the two parts of the sensor,and that the human operator represents a stiffness k_h. Defining the position of the operator and the manipulator as q_0 and q respectively, the position of the tool marker will be found as

$$q_m = \frac{k_s q + k_h q_0}{k_s + k_h} \tag{2.16}$$

The control error is defined as $e = q_0 - q$ whereas the measured control error is

$$e_m = q_m - q = \frac{k_h}{k_h + k_s} e \tag{2.17}$$

This effect results in a corresponding damping of the original feedback gain, which now can be increased to

$$k_p = 0.5 \, \omega_x \frac{k_h + k_s}{k_h} \tag{2.18}$$

according to (2.7) or a similar value corresponding to (2.14).

The problem is that the stiffness of the human muscular system varies according to the type of motion required, and will often increase during swift and precise motion. In that case the damping given by (2.17) will decrease, and if k_p is designed for a lower k_h than the actual value, the system may become instable. Safe design requires a large k_h to be inserted into (2.18), and as k_h increases, k_p approaches the value given by (2.7), which is the only one that is perfectly safe.

A more detailed analysis of the situation is presented in (Hirzinger, 1982). However, the above considerations provide a simple link between the stability problems encountered with a positional deviation sensor and with a force sensing handle. Moreover, simple stability criteria have been obtained for both cases.

3. INCREASING THE BANDWIDTH

Above, simple expressions were developed which relate tracking performance and process bandwidth. Assumed controllers were of the P and the PI type, subject to a process with internal speed control. What remains is to investigate how different internal controllers affect the performance. Dynamic models for the hydraulic servos of the TR-400 are developed in (Dessen, 1988). Neglecting possible coupling between the different axes,

$$g(s) = \frac{1}{1 + 2\zeta(s/\omega_0) + (s/\omega_0)^2} \tag{3.1}$$

is a good representation of a single servo, as seen from the higher level.

3.1. Internal feedback

The bandwidth of the lower-level system can be increased by introducing feedback from torque and angular velocity. These quantities may be obtained either by measurements or by state estimation. When considering a single motor, the internal controller will take the form

$$u = -g_\tau \tau + g_v(\dot{q}_0 - \dot{q})$$ (3.2)

It follows from the theory of linear, modal control that with this feedback structure any desired values for ω_0 and ζ can be obyained. However, this will not be the case when applied to the real system. An upper bound for what can be obtained in practice is found by looking at some of the dynamic effects which were neglected while developing the model. The characteristic frequencies for three of these are

Link 3 resonance: $\omega_{pf} \approx 10^3 \text{rad/s}$

Spool valve dynamics: $\omega_v \approx 10^3 \text{rad/s}$

Finite system sample rate: $\omega_s \approx 10^3 \sim 10^4 \text{rad/s}$

However, unexpected effects may appear which will narrow this bound.

3.2. Experiments

Means for improving the dynamics of single and multi-axis trackers have been investigated experimentally with a TR-400 manipulator. The experiments were carried out using a specially designed control system run on a single TMS 32020 signal processor board which fits into the extention bus of an IBM AT. Two adjoint input/output boards are connected to the I/0-bus of the TMS 32020. The first board serves as an interface to the positional deviation sensor, and allows simultaneous sampling of the measurement vector at a rate of 2 per ms. The data are preprocessed and fused by means of analog computation. The second I/O board contains the interface to the TR-400 manipulator. The aquisition of joint displacement values is restricted to 1 element per ms, which for the manipulator at hand implies a 5 ms cycle for updating the Jacobian matrix described in §2. However, the setting of the control action vector u, and sampling of the hydraulic pressure of servos 1, 2 and 3 can be carried out simultaneously. Here, a cycle time of 0.5 ms is easily obtained. The experiments were carried out with an effective sample time of 5 ms. Conversion of the positional deviation measurements into the equivalent joint-space control error vector e was done using the sequential scheme mentioned in § 2.1. In addition to vector e, pressure measurements for servos 1, 2 and 3 were available.

Below, experimentally obtained best-possible gains are presented for each of four different control structures;

1. Pure proportional control, cf (2.7)
2. PI-control, cf (2.13)
3. P-control and internal feedback, cf (3.2)
4. PI-control and internal feedback, cf (3.2)

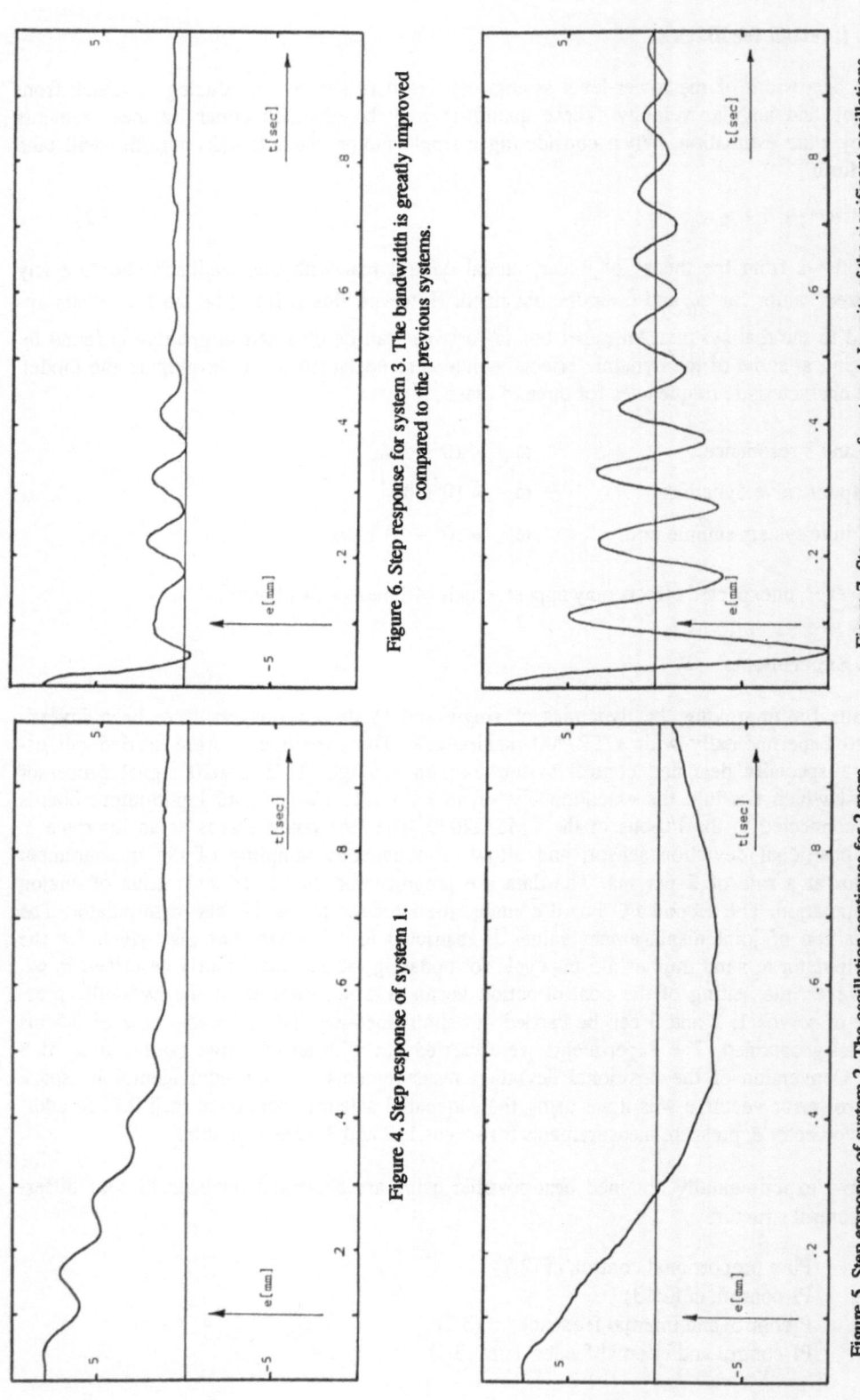

Figure 6. Step response for system 3. The bandwidth is greatly improved compared to the previous systems.

Figure 7. Step response of system 4. In spite of the significant oscillations, this is by far the best tracking system.

Figure 4. Step response of system 1.

Figure 5. Step ersponse of system 2. The oscillation continues for 2 more seconds. The tracking performance however is far better than for system 1.

applied to servo 3 of the TR-400. This servo has appeared to be the most difficult one to control, due to the elasticity of link 3. Found for Controller 1 is

$$k_p = 3.3 \text{ rad/s}$$

The same value applies to Controller 2 as well, which has the additional

$$T_i = 250 \text{ ms}$$

Applying (2.12) and (2.15), the corresponding performance measures

$$v_P = E k_p \approx 0.033 \text{ m/s} \quad ; \quad a_{PI} = E k_p/T_i \approx 0.13 \text{ m/s}^2$$

are found for Controllers 1 and 2 respectively. A far better performance is found when using nonzero internal feedback gains. The bandwidth of $g(s)$ is increased so that

$$k_p = 41 \text{ rad/s}$$

is possible for Controller 3. In addition, for the even better Controller 4 comes

$$T_i = 30 \text{ ms}$$

In these cases, the application of (2.12) and (2.15) respectively results in

$$v_P \approx 0.41 \text{ m/s} \quad ; \quad a_{PI} \approx 13.7 \text{ m/s}^2$$

To compare the performance of the different controllers, the step response for each is shown in Figs. 4 through 7. These give a good indication of the resulting bandwidths, but are in many ways unfair since the system is not designed to cope with sudden changes in the tool position.

4. OPTIMAL COORDINATION

In the 1# system treated in § 2, the control error magnitude was restricted by a positive constant E. When several degrees of freedom are to be coordinated, the situation becomes more complicated. Maximum control error is then restricted by the multi-dimensional play between the two parts of the sensor. In order to make full use of the resulting 6-dimensional volume, a taylored coordinating controller will be required.

4.1. Parametric optimization

To simplify the design of the higher-level control system, a quadratic performance index can be used to reflect the limitations given by the sensor. In terms of deviation vector M_d, this index can be given in general matrix notation by

$$L_1 = M_d^T Q' M_d \tag{4.1}$$

However, since the controllers in § 2 work in servo coordinates, it is convenient to express L_1 in terms of e using (2.2);

$$L_1 = e^T J_{eM}^T Q' J_{eM} e = e^T Q e \tag{4.2}$$

A similar performance index may be used to express the effort made by the servos,

$$L_2 = \mathbf{u}^T \mathbf{P} \mathbf{u} \tag{4.3}$$

A controller which minimizes the performance index

$$L = E(\mathbf{e}^T \mathbf{Q} \mathbf{e} + \mathbf{u}^T \mathbf{P} \mathbf{u}) \tag{4.4}$$

is then sought for the multivariable process

$$\mathbf{q}(s) = \mathbf{H}(s) \mathbf{u}(s) \tag{4.5}$$

Here $\mathbf{e} = \mathbf{q}_0 - \mathbf{q}$ and $E(\cdot)$ denotes stochastic expectation. $\mathbf{H}(s)$ is given by (2.4). The solution to this problem is readily obtained obtained from the theory of linear-quadratic optimal control which requires feedback from the complete underlying state vector of (4.5). However, it is desirable at present to retain the assumption of hierarchical control made in § 2, and thus only take feedback from the output vector $\mathbf{q}$, assuming the structure of the controller to be

$$\mathbf{u} = \mathbf{S}\mathbf{e} = \mathbf{S}(\mathbf{q}_0 - \mathbf{q}) \tag{4.6}$$

Optimal values for the elements of $\mathbf{S}$ when applied to (4.5) can be found by parametric optimization. Inserting (4.6) in (4.4), results in

$$L = E(\mathbf{e}^T [\mathbf{Q} + \mathbf{S}^T \mathbf{P} \mathbf{S}] \mathbf{e}) \tag{4.7}$$

Laplace transforming, and inserting $\mathbf{e}(s) = \mathbf{N}(s) \mathbf{q}_0(s)$ further results in

$$L \sim \int_{-j\infty}^{j\infty} (\mathbf{q}_0^*(s)\mathbf{N}^*(s)[\mathbf{Q} + \mathbf{S}^T \mathbf{P} \mathbf{S}]\mathbf{N}(s)\mathbf{q}_0(s))ds \tag{4.8}$$

where * denotes the conjugate transpose. Once a scheme is available for the computation of L, iteration may be applied to obtain the optimal $\mathbf{S}$. It is characteristic that the solution not only depends on the performance index and the process itself, but also on the motion reference $\mathbf{q}_0$ applied. Instead of applying (4.8) to obtain L, an estimate may be found from measurements of the true $\mathbf{e}(t)$ and $\mathbf{u}(t)$, applied to (4.4). This results in an experimenting, adaptive controller where the same iterational scheme as above can be used for real-time adjustments of $\mathbf{S}$.

4.2. Stability

If the state-space representation of (2.4) is known, closed-loop stability may be checked by eigenvalue computation. For instance if, as in § 3,

$$g_i(s) = \frac{1}{1 + 2\zeta_i(s/\omega_i) + (s/\omega_i)^2} \qquad , \quad i \in \{1, N\} \tag{4.9}$$

the state-space representation of $\mathbf{H}(s)$ may be found in the form

$$\dot{\mathbf{x}} = \mathbf{A}\mathbf{x} + \mathbf{B}\mathbf{u} \quad ; \quad \mathbf{q} = \mathbf{D}\mathbf{x} \tag{4.10}$$

and the controller corresponding to (4.6) as $\mathbf{G} = [\, \mathbf{0} \quad \mathbf{0} \quad -\mathbf{S}\,]$, resulting in the closed-loop system matrix $\mathbf{A} + \mathbf{B}\mathbf{G}$ with which the system is stable if all its $3N$ eigenvalues are in the left half of the complex plane.

Often, $H(s)$ is more complicated than (4.9) suggests. For this reason, it may be convenient to be able to check stability by other means. This may be done by applying the generalized Nyquist stability criterion (MacFarlane, 1977). As with the standard Nyquist criterion, this is based on studying open-loop system transfer functions as complex frequency s traverses the standard Nyquist D-contour (clockwise encirclement of the right half of the complex plane).

Our system has the open-loop, stable transfer matrix

$$H_0(s) = H(s)S \tag{4.11}$$

Letting s traverse the Nyquist contour D, the system is closed-loop stable if and only if the net sum of encirclements of the critical point $(-1+j0)$ by the loci of the N eigenvalues $\{\lambda_i(s)\}$ of $H_0(s)$ is zero.

The problem of solving for the eigenvalues of H_0 is simpler than for the complete state-space system matrix $A + BG$. On the other hand, the eigenvalues of $H_0(s)$ must in principle be found for every $s \in D$. At least, their magnitudes at each crossing of the real axis must be known. In any case, due to the complexity of the eigenvalue problem, it is difficult to sort out the set of coordinating matrices S which result in a stable system.

Such results may however be found for the special case where

$$H(s) = I\, h(s) \tag{4.12}$$

which in the present case means that all $g_i(s)$ in (2.4) are indentical. Then the open-loop transfer matrix is

$$H_0(s) = S\, h(s) \tag{4.13}$$

so that, if $\{\sigma_i\}$ are the eigenvalues of S,

$$\{\lambda_i(s)\} = \{\sigma_i\, h(s)\} \tag{4.14}$$

are the eigenvalues of $H_0(s)$. Since the stability criterion is in terms of encirclements of the critical point $(-1+j0)$ by each $\sigma_i\, h(s)$, it may alternatively be stated in terms of the critical points $\{-\rho_i\}$ and $h(s)$, where ρ_i is the reciprocal of σ_i. The following corollary is thus obtained:

Coordination of identical servos: *If $H(s) = I\, h(s)$, the closed system is stable if and only if the net sum of encirclements of the critical points $\{-\rho_i\}$ by $h(s)$ as s traverses D is zero.*

An example is given in Fig. 8. As can be seen, stability requires that all $-\rho_i$ are placed left of the curve traced out by the transfer function. In the reciprocal plane, Fig. 9, corresponding restrictions on the eigenvalues of $-S$ are obtained. Note that these results are due to the simplicity of (4.14), and may not be extended to the general case. The above

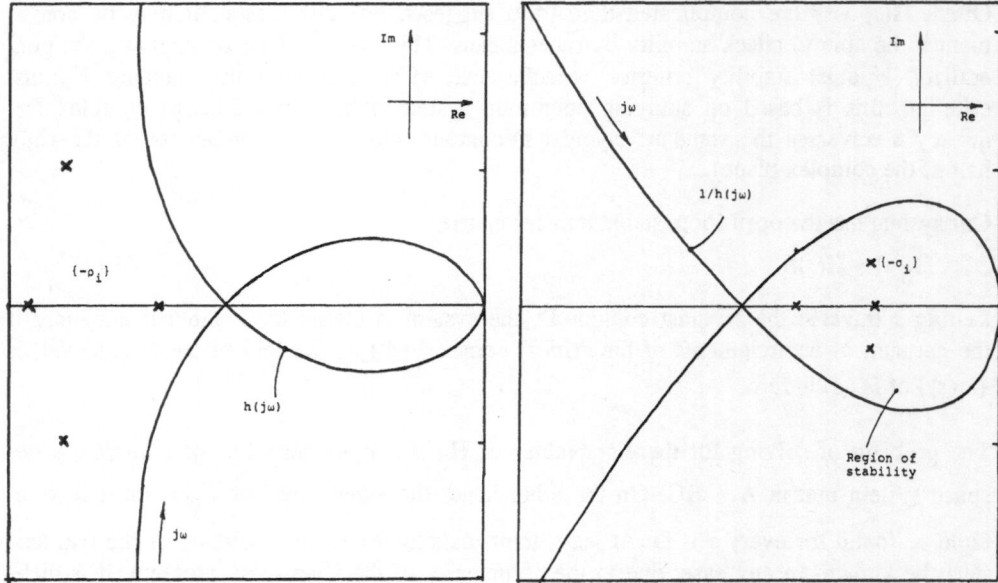

Figure 8. A stable system.   Figure 9. The resiprocal of the plot in Fig. 8.

results can however be used to give some idea of what can be gained by coordination.

4.3. Example

As an illustration, a 2# system corresponding to servos 3 and 4 of the TR 400 manipulator will be made to track translational and rotational motion as shown in Fig. 10. Follow-

Figure 10. Simple optimization problem. A good controller will make servo 2 assist servo 1.

ing § 4.1, the corresponding performance index L_1 can be written

$$L_1 = e^T Q e = e^T J_{eM}^T Q' J_{eM} e \qquad (4.15)$$

where

$$Q' = \begin{bmatrix} E_1^2 & 0 \\ 0 & E_2^2 \end{bmatrix} \quad ; \quad J_{eM} = \begin{bmatrix} a_3 + a_4 & a_4 \\ 1 & 1 \end{bmatrix} \qquad (4.16)$$

For the computations, parameter values

$$E_1 = 1\text{cm} \quad ; \quad E_2 = 0.3 \text{ rad}$$

$$a_3 = 1.6 \text{ m} \quad ; \quad a_4 = 0.25 \text{ m}$$

are taken, and the dynamics of the two servos is assumed to be represented by

$$\mathbf{H}(s) = \mathbf{I}\, h(s) \quad , \quad h(s) = \frac{1}{s[1 + (s/\omega_0) + (s/\omega_0)^2]} \tag{4.17}$$

The scope is now to investigate how the coordinating feedback matrix $\mathbf{S}$, as compared to the diagonal feedback described in § 2.1, can alter the tracking performance. The results will be visualized in terms of the two-by-two error transfer matrix

$$\mathbf{N}(s) = \{\frac{e_i}{y_{j0}}\} \tag{4.18}$$

where e_1, e_2, y_{10} and y_{20} respectively denotes translational and orientational deviation and tool marker motion. In matrix state-space notation, the controlled system becomes

$$\dot{x} = \mathbf{A}x + \mathbf{B}y_0 \quad ; \quad e = \mathbf{D}x + \mathbf{E}y_0 \tag{4.19}$$

where

$$\mathbf{A} = \begin{bmatrix} 0 & 0 & -\mathbf{S} \\ \Omega^2 & -2\mathbf{Z}\Omega & -\Omega^2 \\ 0 & \mathbf{I} & 0 \end{bmatrix} \quad ; \quad \mathbf{B} = \begin{bmatrix} \mathbf{S}\mathbf{J}_{eM}^{-1} \\ 0 \\ 0 \end{bmatrix}$$

$$\mathbf{D} = \begin{bmatrix} 0 & 0 & -\mathbf{J}_{eM} \end{bmatrix} \quad ; \quad \mathbf{E} = \begin{bmatrix} \mathbf{I} \end{bmatrix}$$

$$2\mathbf{Z} = \mathbf{I} \quad ; \quad \Omega = \omega_0 \mathbf{I}$$

and $\mathbf{N}(s)$ is computed using

$$\mathbf{N}(s) = \mathbf{D}(s\mathbf{I}-\mathbf{A})^{-1}\mathbf{B} + \mathbf{E} \tag{4.20}$$

Rather than computing matrix $\mathbf{S}$ as described in § 4.1, the feedback is established by heuristic minimization, subject to parameters c_0 and c_1 of

$$\mathbf{S} = c_0\omega_0 \begin{bmatrix} 1 & 0 \\ c_1 & 1 \end{bmatrix} \tag{4.21}$$

The reason for prescribing this specific feedback structure is, apart from the fact that it will give good results, that the necessary and sufficient conditions for stability simply are

$$0 < c_0 < 1$$

This follows from the corollary presented in § 4.2 by noting that both critical points corresponding to matrix $\mathbf{S}$ are

$$-\rho_i = -1/c_0\omega_0$$

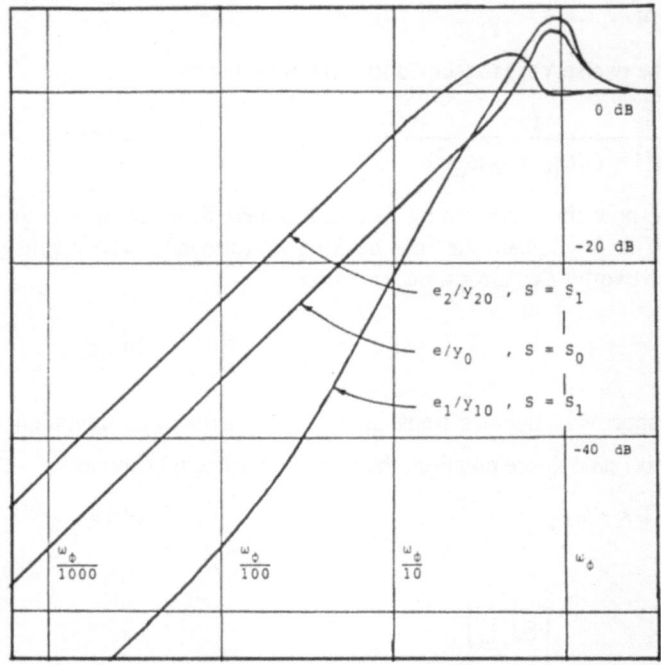

Figure 11. Amplitudes of the closed-loop transfer functions given by (4.22) and (4.23).

and that no encirclements of these are made by $h(s)$ as long as

$$-\infty < -\rho_i < -1/\omega_0$$

The result is

$$S_1 = \begin{bmatrix} 1/3 & 0 \\ 2/m & 1/3 \end{bmatrix} \omega_0 \;;\; c_0 = 1/3 \;;\; c_1 = 6\,\mathrm{m}^{-1} \tag{4.22}$$

To compare, application of the Ziegler-Nichols method as in § 2.2 results in

$$S_0 = \begin{bmatrix} 1/2 & 0 \\ 0 & 1/2 \end{bmatrix} \omega_0 \;;\; c_0 = 1/2 \;;\; c_1 = 0 \tag{4.23}$$

Amplitude-frequency plots of (4.20) with S given by (4.22) and (4.23) are shown in Fig. 11. It is seen that the use of (4.22) rather than (4.23) results in a trade-off between translational and rotational tracking performance. From the asymptotical behaviour of the plots at low frequencies, it is seen that for

$$S = S_0 \;:\; \frac{e_1}{y_{10}}(s) = \frac{e_2}{y_{20}}(s) = s/15\;\mathrm{sec}^{-1}$$

$$S = S_1 \;:\; \frac{e_1}{y_{10}}(s) = s/160\;\mathrm{sec}^{-1} \;;\; \frac{e_2}{y_{20}}(s) = s/5\;\mathrm{sec}^{-1}$$

The design of the PDS restricts the maximum translational and rotational deviation to

$$E_1 \approx 1\;\mathrm{cm} \;;\; E_2 \approx 0.3\;\mathrm{rad}$$

which results in the maximum allowed steady-state velocities

$$S = S_0 : v_0 = 15 \text{ cm/s} \quad ; \quad w_0 = 4.5 \text{ rad/s}$$
$$S = S_1 : v_1 = 160 \text{ cm/s} \quad ; \quad w_1 = 1.5 \text{ rad/s}$$

at $\omega_0 = 30$ rad/s. Thus, feedback matrix S_1 results in a better *performance distribution*.

5. CONCLUSION

The results in § 3 show that the performance of the tool tracking system is greatly improved by the use of internal feedback. For Controllers 1 and 2, the obtained parameters reflect a trade-off between stable behaviour and swift tracking. For structures 3 and 4 however, a ceiling was reached while adjusting the velocity gain. This was due to the appearance of significant noise rather than instability. Even so, the performance of the system when using all the available measurements was outstanding, and is believed to be sufficient for the real-time teaching of high-speed paint-spraying robots. It is characteristic throughout that a tracking controller which is optimal from the point of view of a human operator turns out oscillatoric. Throughout the plots, an additional oscillatoric mode at approximately 60 rad/sec is observed which is an obvious obstacle to further improvement. The crossing of this barrier will be the scope of future experiments. It is believed that this will require better noise protection, and a more detailed look at the structural elasticities of the manipulator.

Optimal coordination was considered in § 4. In the given example, a non-zero displacement a_4 between the sensor and the manipulator wrist was assumed, which enables the wrist servo to assist in the compensation of translational deviation. The experimental system described in § 3 does not however provide this displacement, so no attempt was made to obtain practical results. Nevertheless, the principle of optimal coordination seems promising, and is believed to be of interest in connection with many applications in addition to direct teaching.

Internal feedback was considered separate from the discussion of optimal coordination. Using both approaces, a very-high-performance system will result, which still can have a good stability margin.

REFERENCES

Balchen, J.G. (1984). *Norwegian Patent Application 870419.*

Dessen, F. , J.G.Balchen. (1987). A Six Degrees of Freedom Positional Deviation Sensor for the Teaching of Robots. *Proc. of the NATO ARW on Sensor Devices and Systems*, Spain, October 13-16.

Dessen, F. (1988). *Direct Training of Robots using a Positional Deviation Sensor.* Dr. ing. dissertation, Norwegian Inst. of Tech.

Hirzinger, G. (1982). Robot Teaching via Force Torque Sensors. *6th Eur. Meeting on Cybernetics and Systems Research EMCSR 82,* April 13-16.

MacFarlane, A.G.J. , I. Postlethwaite. (1977). Generalized Nyquist Stability Criterion and Multivariable Root Loci. *Int. J. of Control*, 25, 81-127.

FORCE/TORQUE AND TACTILE SENSORS
FOR SENSOR-BASED MANIPULATOR CONTROL

H. Van Brussel, H. Beliën, Bao Chao-Ying
Katholieke Universiteit Leuven
Department of Mechanical Engineering
Celestijnenlaan 300B
B-3030 Leuven, Belgium

Abstract

The autonomy of manipulators, in space as well as in industrial environments can be dramatically enhanced by the use of
force/torque and tactile sensors.

In a first part the development and future use of a six-component force/torque sensor for the Hermes Robot Arm (HERA) Basic
End-Effector (BEE) is discussed.

Further, a multifunctional gripper system based on tactile
sensors is described. The basic transducing element of the
sensor is a sheet of pressure-sensitive polymer. Tactile image
processing algorithms for slip detection, object position estimation and object recognition are described.

1. Introduction

The HERA is a symmetric six-degrees-of-freedom manipulator arm
with an anthropomorphic configuration and an overall length of
11.2 meter. It is designed to perform following operational
functions: capture, berthing, release, inspection, insertion
and retraction, transfer, placement, actuation, tool operation, EVA-support. It can be operated in the following modes:
automatic mode, (tele)-operator-controlled mode, single-joint
mode.

Several of the above mentioned functions require the use of
closed loop control strategies, based on active force
feedback [1]. This requires the presence of a multi-component
force/torque sensor imbedded in the HERA BEE (fig.1).

NATO ASI Series, Vol. F 63
Traditional and Non-Traditional Robotic Sensors
Edited by T. C. Henderson
© Springer-Verlag Berlin Heidelberg 1990

Active force feedback seems to be an appropriate control mode for all "compliant motion" functions. These are functions where the manipulator is in direct contact with its environment (e.g. insertion). Good results have been obtained in industrial environments with force-around-position control loops [2,3]. These schemes also seem applicable to space manipulators [1]. The main difference with respect to industrial manipulators is the back effect of the contact forces on the position loops which has to be taken into account in space manipulators [1], but can be neglected in industrial robots.

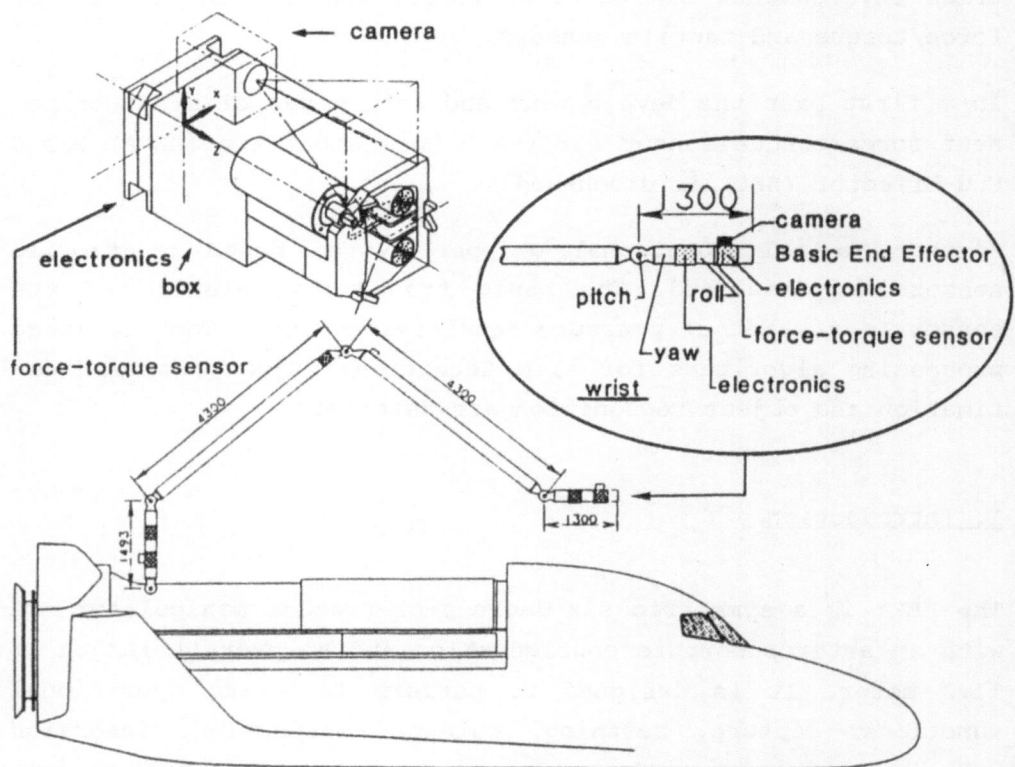

Fig.1. General layout of HERA-arm on HERMES.

At KULeuven, a prototype six-component force/torque sensor for the HERA has been developed and tested. Some particular design features are outlined hereafter.

2. The HERA Force Torque Sensor (FTS)

Designing force sensors stands for making compromises. Indeed, many compromises have to be made in order to overcome the numerous restrictions which may be imposed on the design. In the case of the HERA FTS, both the functional requirements as well the imposed restrictions were very severe :

- nominal F/T ranges   : forces : 200 N
                                torques : 150 Nm
- resolution   : 1/1000
- accuracy   : 0.5% F.S.
- minimal stiffness   : translation : 100 kN/m
                                rotation   : 200 kN/rad
- overload   : 500%
- operational temp.   : -50 to +70°C
- max. weight (incl.   : 2.5 kg
  electronics)
- max. outer dimensions   : 220 x 35 mm
- 4 corner mountings, with provisions for central passage of electrical cables.

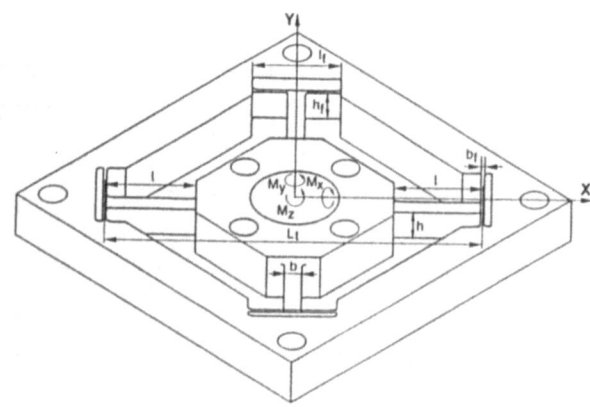

Fig. 2. Mechanical layout of HERA BEE six component force/torque sensor.

Relying upon a decade of expertise in the field of force torque sensor design at the KU Leuven [4], a maltese cross sensor was proposed mainly because of its compact structure and its excellent decoupling behaviour (cfr. fig. 2).

This sensor consists of four cantilever beams, at one end rigidly connected to a central block (the central hole allowing the passage of a wire bundle) and at the other end suspended to the outer frame by two flexures. These flexures create 4 DOF boundary conditions with respect to beams providing the decoupling mentioned above. Indeed, this configuration results in a simple load situation at the outer end of the beams containing only two forces, a horizontal and a vertical one.

$$F_h = \frac{F_x \text{ or } F_y}{2} + \frac{Mz}{2.l_t} \qquad F_v = \frac{Fz}{4} + \frac{M_x \text{ or } M_y}{l_t} \qquad (1)$$

Using a CAE package called SENCAD [5], both the sensor geometry as well as the resulting stresses and strains were calculated. SENCAD is an interactive software package, specially developed for the design of multi-component F/T sensors. SENCAD used following starting data :
- sensor material : Al7075 T7351
    - $\sigma 0.2 = 395 \quad N/mm^2$
    - $\sigma_b = 475 \quad N/mm^2$
    - $E_{MOD} = 72000 \ N/mm^2$
    - estimated linear deformation range : app. 150 N/mm^2
- strain gage : Micro-Measurement type WK 13/125/AD350
    - ϵmax : app. 1.5%
    - accuracy : < 0.5%
    - temp. range : -269 ^oC + 290 ^oC
    - max. supply voltage (full bridge) : 10 Volt
- max. gain of the signal amplifier unit : $G_{max} = 2000$
  (based upon the desired accuracy in the operational temperature range)
- max. outer dimension $\qquad L_t$: 200mm
  min. width of the central block $\quad L_o$: 110mm
- the nominal force/torque ranges : 200 N/150 Nm
- design criterion : $\epsilon_{F_v} = \epsilon_{F_h}$

In a first run, the beam dimensions b and h, the sensor stiffness and the maximum detectable strains are calculated assuming perfect 4 DOF boundary conditions at the end of the sensor beams. The results of this calculation were :

b = 10 mm	$\epsilon_{Fx(y)}$ = 221 μstrain	$k_{x,y}$ =	6.7 kN/mm
h = 169 mm	ϵ_{Fz} = 66 μstrain	k_z =	38.1 kN/mm
l = 45 mm	$\epsilon_{Mx(y)}$ = 976 μstrain	$k_{\alpha x,\alpha y}$ = 1830	kN/rad
	ϵ_{Mz} = 821 μstrain	$k_{\alpha z}$ = 1370	kN/rad

At this stage the most challenging problem with respect to the HERA sensor became visible. Indeed, an extreme sensitivity mismatch can be noticed with respect to the individual F/T components. This fenomenon is directly related to the particular design of the HERA BEE, where a long electronic control box is positioned between the sensor and the tool flange, resulting in an odd specification of nominal forces versus nominal torques (cfr. fig. 1).

Furthermore, one can notice a slightly insufficient rotation stiffness suggesting a further decrease of the beam length.

However, as the electronics had to be integrated inside the sensor, a decrease of its internal space was rejected.

Having to accept results mentioned above, a second calculation step was started in order to determine the flexure dimensions and its influence on the sensor sensitivity and stiffness.

direction	nominal strain	calculated stiffness	sensor deflections
x,y	179 μstrain	6.7 kN/mm	0.03 mm
z	65 μstrain	38.1 kN/mm	0.01 mm
α_x,α_y	954 μstrain	1830 kN/rad	0.1 mm*
α_z	810 μstrain	1370 kN/rad	0.15 mm*

*at r= 140 mm

The sensor deflections are important with respect to the design of the overload protections. The demands for large overload capacity and high sensor sensitivity are very difficult to realize under the condition that no plastic deformation may occur. Therefore mechanical stops limit the displacement of the outer frame with respect to the central block when 125% of the nominal load is exceeded. To ensure a proper sensor operation, all mechanical parts, especially the mechanical stops have to be machined at very close tolerances.

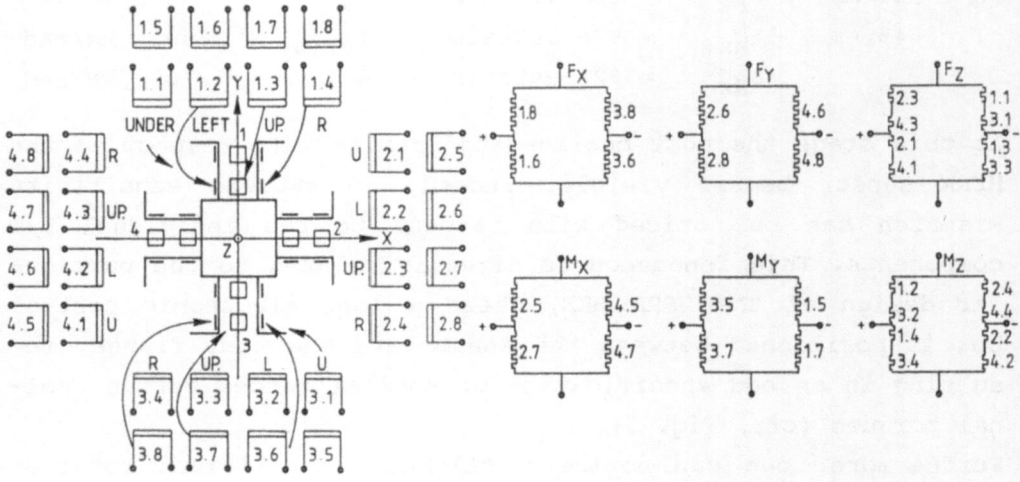

Fig. 3. Placement of strain gauges for achieving a decoupled sensor coupling matrix.

In order to overcome the problem of non-matching sensitivity to some extent, a special strain gage configuration has been provided (fig. 3). Full bridges with four strain gauges are used for F_x, F_y, M_x, M_y and eight gauges for F_z and M_z, resulting in a total number of 32 gauges. The gauges detecting the strain due to the forces are placed at the inner side, while those detecting the strain due to the torques are placed at the outer side. The loss of sensitivity due to the non optimum position of the latter gauges may be neglected.
The approximate signal equations are :

$$- S_{1,2} \approx G_1 \cdot k \cdot V_s \cdot \frac{F_{x(y)}}{2} \cdot \frac{1}{E_{mod}} \cdot \frac{6}{h \cdot b^2} \qquad (2)$$

$$- S_3 \approx G_3 . k . V_s . \underline{F_z} . \underline{\frac{1}{E_{mod}}} . \underline{\frac{6}{b.h^2}} \qquad (3)$$

$$- S_{4,5} \approx G_4 . k . V_s . \underline{\frac{M_{x(y)}}{lt}} . \underline{\frac{1}{E_{mod}}} . \underline{\frac{6}{b.h^2}} \qquad (4)$$

$$- S_6 \approx G_6 . k . V_s . \underline{\frac{M_z}{2.lt}} . \underline{\frac{1}{E_{mod}}} . \underline{\frac{6}{h.b^2}} \qquad (5)$$

Where :

S : output signal

G : amplifier gain

k : strain gage const

V_s: bridge supply volt

The amplifier gain can be optimized for each F/T component and even the supply voltage for the F_z and M_z bridges can be doubled because of their higher bridge resistance with respect to the other bridges. These advantages become clear when compared with another configuration, containing only 16 gages in 8 full bridges, 2 bridges for each beam, one measuring the horizontal and one measuring the vertical bending strain resulting in following equations :

$$- S_{1,2} \approx G.k.V_s . [\underline{\frac{F_{x(y)}}{2}} + \underline{\frac{M_z}{2.lt}}] . \underline{\frac{1}{E_{mod}}} . \underline{\frac{6}{h.b^2}} \qquad (6)$$

$$- S_{3,4} \approx G.k.V_s . [\underline{\frac{F_{x(y)}}{2}} - \underline{\frac{M_z}{2.lt}}] . \underline{\frac{1}{E_{mod}}} . \underline{\frac{6}{h.b^2}} \qquad (7)$$

$$- S_{5,6} \approx G.k.V_s . [\underline{\frac{F_z}{2}} + \underline{\frac{M_{x(y)}}{2.lt}}] . \underline{\frac{1}{E_{mod}}} . \underline{\frac{6}{b.h^2}} \qquad (8)$$

$$- S_{7,8} \approx G.k.V_s . [\underline{\frac{F_z}{2}} - \underline{\frac{M_{x(y)}}{2.lt}}] . \underline{\frac{1}{E_{mod}}} . \underline{\frac{6}{b.h^2}} \qquad (9)$$

In the this case, each F/T component has to be calculated using at least 2 bridge signals.This would result in a poor accuracy due to the very low degree of signal utilization with respect to the force components.

It is obvious that the formula's mentioned above do not take in account the cross-sensitivity effects which are always pre-

sent in these kind of sensors. Cross-sensitivity is mainly related to machining errors and misalignment of strain gauges.
Although for normal robotic applications, cross-sensitivity
effects (typically 3%) may be neglected, it is possible to
correct their influence by taking into account the complete
calibration (decoupling) matrix.
Using a static calibration procedure the 6x6 coupling matrix
A, relating the force vector F with the output signal vector
S, can be determined :

$$S = A.F. \qquad (10)$$

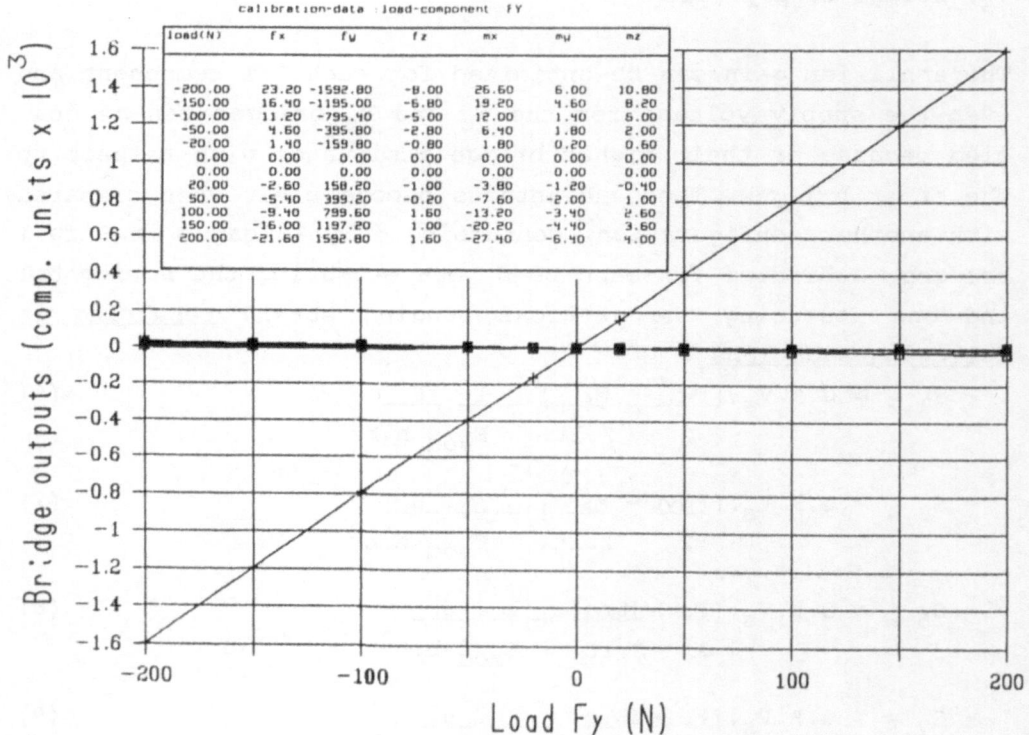

calibration-data : load-component FY

load(N)	fx	fy	fz	mx	my	mz
-200.00	23.20	-1592.80	-8.00	26.60	6.00	10.80
-150.00	16.40	-1195.00	-6.80	19.20	4.60	8.20
-100.00	11.20	-795.40	-5.00	12.00	1.40	6.00
-50.00	4.60	-395.80	-2.80	6.40	1.80	2.00
-20.00	1.40	-159.80	-0.80	1.80	0.20	0.60
0.00	0.00	0.00	0.00	0.00	0.00	0.00
0.00	0.00	0.00	0.00	0.00	0.00	0.00
20.00	-2.60	158.20	-1.00	-3.80	-1.20	-0.40
50.00	-5.40	399.20	-0.20	-7.60	-2.00	1.00
100.00	-9.40	799.60	1.60	-13.20	-3.40	2.60
150.00	-16.00	1197.20	1.00	-20.20	-4.40	3.60
200.00	-21.60	1592.80	1.60	-27.40	-6.40	4.00

Fig.4. Sensor outputs for pure force load F_y.

By applying one F/T component at a time over its full range,
the corresponding column of A can be calculated.
During the actual F/T measurements the signal vector S can be
multiplied by the inverse matrix A in order to compensate for

cross-sensitivity influences, hereby enhancing the overall sensor accuracy. However, due to the propagation of the calibration errors, the gain with respect to accuracy is limited. Figure 4 shows a typical calibration result indicating the different bridge outputs for a force Fy applied to the sensor. Figure 5 shows the finished sensor prototype with its built-in data processing electronics. The mechanical stops are not visible as they are hidden in the back plane.

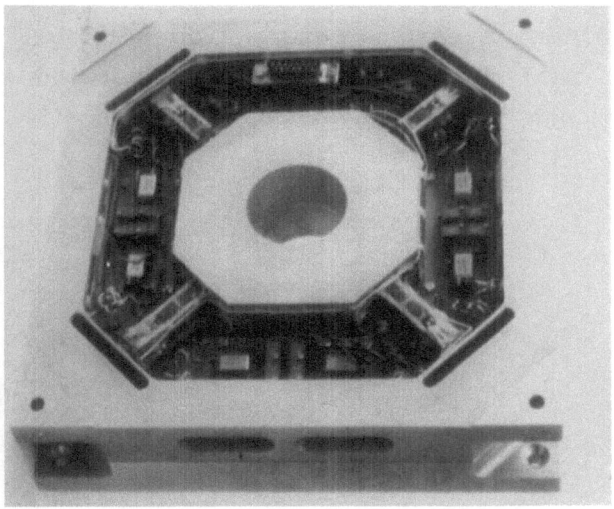

Fig.5. Photograph of finished sensor prototype with built-in electronics.

3. Controlled gripper system

The above described force/torque sensor can enhance the autonomy of manipulators performing compliant motion tasks, by applying active force-feedback. Additionally, the gripping capabilities of those manipulators can be made more intelligent by introducing tactile perception within the gripper.

At KULeuven some years ago, a high-resolution tactile sensor has been developed to be incorporated into a two-jaw gripper mechanism and aimed at slip-detection, object localisation and recognition [7].

The general layout of the sensor is outlined in fig. 6. It consists of a pressure-sensitive contact layer mounted on a specially laid-out printed circuit board consisting of row and column tracks. In this way the sensor surface is divided into a matrix of 16x16 "islands" (cells). By using a scanning mechanism, these islands are consecutively electrically isolated from their neighbours, thus allowing to measure the local electrical resistance. This latter is function of the pressure exerted on that cell. The digitalization module provides either binary pressure information per cell (through a comparator) or real digital information according to the analog output (through an A/D-convertor).

Fig.6. Layout of the tactile sensor built into a two-jax gripper (a); detail of scanning pcb (pressure sensitive layer removal).

The main features of this sensor include:
- matrix size: 16x16 cells (or 32x32);
- spatial resolution: 1.2 mm (or 0.6mm);
- allows detection of 256 pressure levels from $1N/cm^2$ to $50N/cm^2$ per cell. (The uncertainty level is 4 bits, leaving a real resolution of 16 distinct levels);
- a total acquisition time for 2 x 256 cells of 75ms;
- a wide operating temperature from -30°C to 100°C.

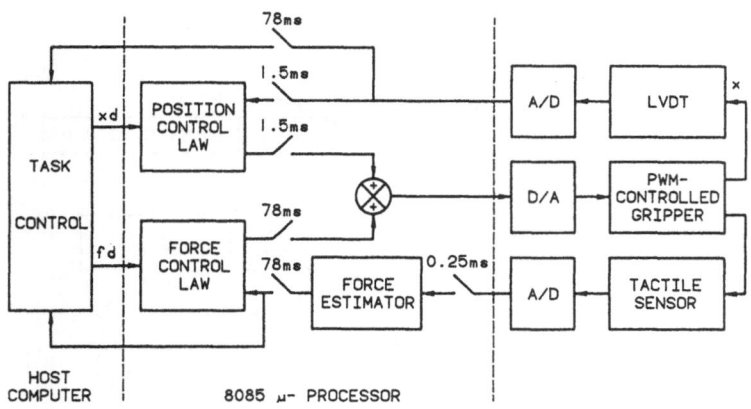

Fig.7. Block diagram of the gripper controller.

A detailed description can be found in [7]. A further development tested out recently is the implementation of the sensitive layer on an elastic printed circuit board, an interesting
feature when one wants to use the sensor on curved surfaces,
like the fingers of a dextrous hand.
Two sensors have been incorporated at the inside surfaces of
an off-the-shelf two jaw gripper (fig.6). The global control
scheme is illustrated in fig.7. It consists of a hybrid posi-
tion/force controller. As long as there is no contact force,
the system acts as a pure position controlled gripper. Control
of the gripper fingers is achieved by a pneumatic piston,
driven by a pulse width modulated pneumatic controller based
on fast-acting pneumatic valves and pressure transducers. An
LVDT displacement transducer provides the position feedback.
The tactile sensor acts as force transducer. The overall architecture of the sensory controlled gripper system is illustrated in fig.8.

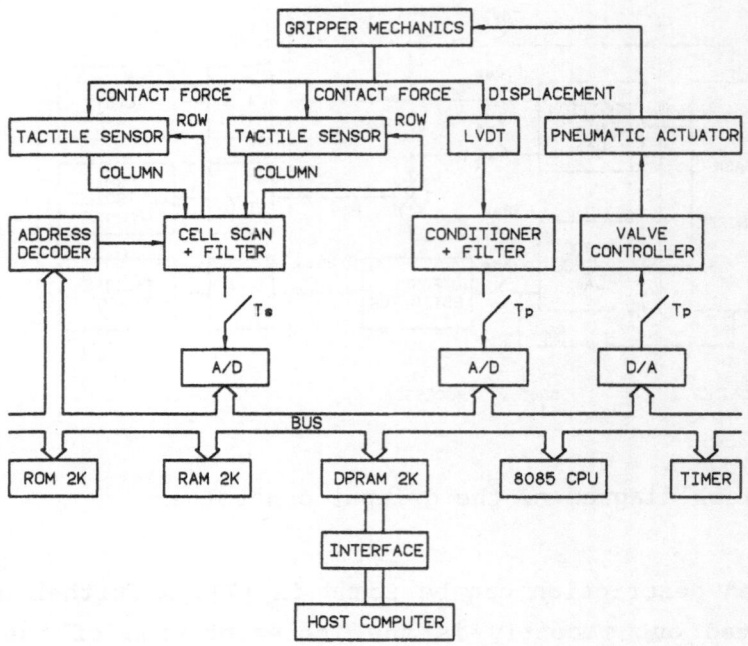

Fig.8. Overall architecture of the sensory controlled gripper system.

Slip detection

Slip detection is important in grasping unknown fragile objects. The here described sensors can only detect normal contact forces and no tangential forces. The key point is how to detect slip by only measuring normal forces. Detecting a shift of the gravity centre of the tactile image can only work when the sensor's active contact surface is not fully covered by the object. Moreover, image noise normally prevents detection of minute changes in the computed centre of gravity. Therefore, the solution adopted here detects changes in the contact area. For instance, the contact area reduces and the pressure value of the most loaded cell changes when slip occurs. By combining both features after proper weighing and by using a simple digital filter a very sensitive slip detection method could be worked out. The most noticeable advantage of this solution is that there is no limitation on the size of the

grasped object. Experiments have shown very satisfactory results.

Object location estimation
Compared to vision, the use of tactile sensing for identifying
object location is advantageous, because:

- much less data is required, reducing
image processing time;
- the measurements are direct, without distortion, shadows,
projection errors, etc.;
- no problems occur with obscured objects;
- it is much cheaper than vision;
- location and even recognition are combined with the grasping
function.

Important, prior to determining position and orientation of a
grasped object, is to start from noise-free tactile images.
This is obtained here by dynamic image comparison and proper
thresholding. Fig.9 shows the images before and after such
filtering. The object's position and orientation coordinates
are determined by calculating its centre of gravity and the
direction of its principal axes of inertia of the enhanced
tactile image:

$$x_c = \Sigma x_i/A \quad ; \quad y_c = \Sigma y_i/A \tag{11}$$

$$\theta = 0.5 \arctan \frac{-2\Sigma (x_i-x_c)(y_i-y_c)}{\Sigma(y_i-y_c)^2 - \Sigma(x_i-x_c)^2} \tag{12}$$

where x_i and y_i are resp. the column and row coordinates of an
active cell i; A is the number of active cells x_c, y_c is the
location of the centre of gravity θ is the angle between the
minor principal axis of inertia and the x-axis (fig.9).
As an example, a cylinder was grasped 20 times under three
typical orientations with respect to the x-axis: 0°, 45°, 90°.
Fig.10 shows typical tactile images; table 2 shows the relevant position data and the standard deviations, based on 20
measurements. It can be concluded that the obtained results

are very reliable: standard deviations on position coordinates of less than 0.5mm and of 0.6° on the angles, this being obtained with a sensor spatial resolution of only 1.2mm !

$$(x_c, y_c) = (x_c, y_c) \pm (\sigma_{xc}, \sigma_{yc}) \qquad \theta = \theta \pm \sigma_\theta$$

a	(8.50, 7.50) ± (0.12, 0.07)	0.0° ± 0.5°
b	(7.93, 8.04) ± (0.03, 0.2)	89.94° ± 0.4°
c	(8.16, 7.84) ± (0.2, 0.2)	44.81° ± 0.6°

Table 2. Position data and standard directions for cases a, b and c of fig.10.

Object recognition

An immediate further use of tactile sensors is extracting knowledge from the tactile image to define shape features of the grasped object and making decisions about the class the object belongs to, out of a finite number of classes.

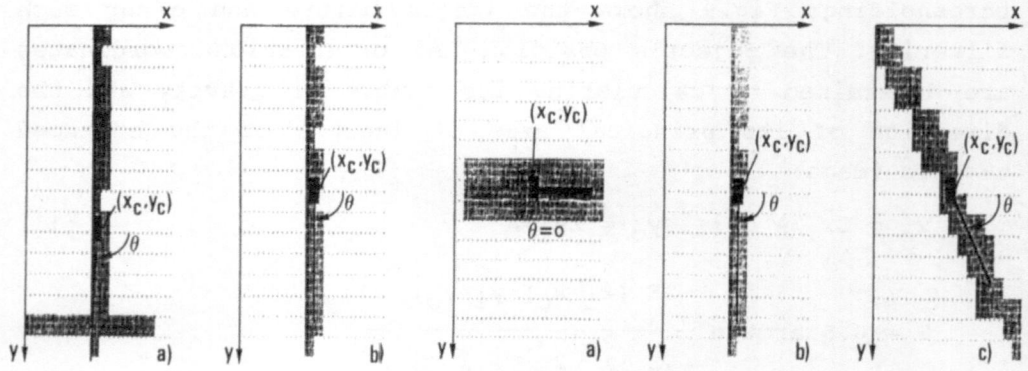

Fig.9. Tactile image before (a) and after (b) noise elimination.

Fig.10. Typical tactile image obtained by grasping a cylindrical object under 0° (a), 90° (b) and 45° (c).

Our tactile sensor can extract following features:

- contact force;
- contact area A;
- perimeter of contact area P;
- moment invariants of A;

- smoothness of contact area, defined by A/P;
- softness of an object.

A recognition programme was developed using the first four features mentioned above, together with the object thickness (measured by the LVDT).

Assume there are n classes of objects and each class has m features f_{ij} (j-th feature of i-th class). An nxm feature matrix F can be defined. In this matrix, row i contains the different features for object i, column j contains feature j for all the object classes. During a learning phase, with k sample measurements for each object class, the nxm expected value matrix $\bar{F}$ and the nxm standard deviation matrix $\Sigma = [\sigma_{ij}]$ can be derived using standard statistical techniques. For the recognition phase, a 1xm feature row vector $\underline{M} = [m_j]$ is defined for an object to be recognized. Then we compare the matrix $\underline{DF}$ defined by:

$$\underline{DF} = [df_{ij}] \quad = \bar{F} - [1....1]^T_{1xn} \underline{M} = [\bar{f}_{ij} - m_j] \qquad (13)$$

with the standard deviation matrix Σ, element per element.

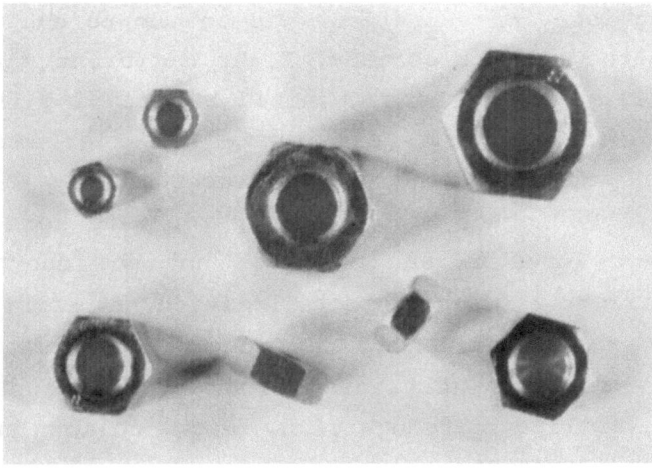

Fig.11. Set of nuts to be recognized by tactile gripper.

This results in a matrix P, with elements p_{ij} defined as follows:

$$p_{ij} = \begin{cases} 0 & \text{if } df_{ij} > 3\sigma_{ij} \\ 1 & \text{otherwise} \end{cases} \qquad (14)$$

Statistically, $p_{ij} = 0$ means that the probability that the j-the object feature belongs to object class i is less than 0.003. Contrarily, $p_{ij} = 1$ means that a large probability exists that the j-th object feature belongs to object class i. The recognition is finally made by means of a score vector S, defined by:

$$S = [s_i]_{nx1} = P.W \qquad (15)$$

where $\underline{W} = [w_j]_{mx1}$ is a weight vector, representing the relative importance of the different object features. W is determined according to the knowledge obtained in the learning phase. The object is said to belong to that object class i which yields the largest score S_m, thereby exceeding a certain recognition threshold T_r:

$$S_m = \max \{S_i \geq T_r\}, \ 1 \leq i \leq n \qquad (16)$$

This T_r can be determined by trial and error. When no S_i exceeds T_r, then the object does not belong to any class and cannot be recognized. An experiment was set up to evaluate the performance of the above algorithm. Eight different classes of nuts (see fig.11) were to be recognized by grasping them with the sensory based gripper. Per class, 50 experiments were performed. This resulted in a 100% recognition, without a single failure. Some observations are appropriate here. First, getting an exact image of an object is not so important for object recognition as for object location. A small image distortion does not influence the recognition result very much. Second, features obtained from other sensors may significantly facilitate recognition, e.g. object thickness information from the LVDT.

4. References

1. DE SCHUTTER J., 1984, Design techniques for robots, Internal report, KULeuven (Belgium), Dept. of Mech. Eng.
2. DE SCHUTTER J., VAN BRUSSEL H., 1988, Compliant Robot Motion, I. A Formalism for Specifying Compliant Motion Tasks, Int. J. of Rob. Res., Vol.7, No.4, pp.3-17.
3. DE SCHUTTER J., VAN BRUSSEL H., 1988, Compliant Robot Motion, II. A Control Approach Based on External Control Loops, Int. J. of Rob. Res., Vol.7, No.4, pp.18-33.
4. VAN BRUSSEL H., BELIEN H., THIELEMANS H., 1985, Force sensing for advanced robot control. In Proc. 5th Int. Conf. on Robot Vision and Sensory Control, ed. N.J. Zimmerman, Bedford (U.K.): IFS.
5. NEYRINCK F., 1988, Computer-aided design of force sensors (in Dutch), Thesis 88E1, KULeuven (Belgium), Dept. of Mech. Eng.
6. DE SCHUTTER J., 1986, Compliant robot motion: task formulation and control, Ph.D.Thesis, KULeuven (Belgium), Dept. of Mech. Eng.
7. VAN BRUSSEL H., BELIEN H., 1986, A high-resolution tactile sensor for part recognition, In Proc. 6th Int. Conf. on Robot Vision and Sensory Control, ed. M. Briot, Bedford (U.K.): IFS.

3D RANGE IMAGING SENSORS

D.J. Conrad

R.E. Sampson

Environmental Research Institute of Michigan (ERIM)

Image Processing Systems Division

P.O. Box 8618

Ann Arbor, Michigan 48107

USA

Abstract

Generalized robotic applications involving vision systems necessitate the development of real time recognition of three dimensional (3-D) surfaces. Range imaging systems collect 3-D coordinate data from object surfaces and can be useful in a wide variety of robotic applications, including shape acquisition, bin picking, assembly, inspection, and robot navigation. Range imaging sensors for such systems are unique imaging devices in that the image data points (pixels) explicitly represent scene surface geometry in a sampled form.

At least five different fundamental physical principles have been used to obtain range images: (1) laser radar, (2) triangulation, (3) interferometry, (4) lens focusing, and (5) tactile sensing. One of these techniques, active laser range sensing, described in detail in this paper, measures surface geometry directly and thus avoids the extensive computations necessary for reconstruction of an approximate range map from multiple camera views or inference from reflectance information. In fact, for many robot vision applications, active scanners represent the only viable method for obtaining the scene information necessary for real time operation. This paper provides a brief overview of range imaging techniques with an emphasis on laser-based active sensing methods and demonstrated sensors that are likely to be used for robot control.

NATO ASI Series, Vol. F 63
Traditional and Non-Traditional Robotic Sensors
Edited by T. C. Henderson
© Springer-Verlag Berlin Heidelberg 1990

1 Introduction

Fundamentally, 3D range imaging sensors measure distance. Historically these devices were point sensors and had their roots in applications involving manufacturing, surveying, and laboratory research where precise distance measurements are required. The advent of imaging technology has provided a means for extending range sensing to a significantly broader set of applications. One of the most promising opportunities for range sensing involves generalized robotic applications. To date, most robotic applications operate "blind" or with only limited sensing capabilities. Even a quick review of recent research in robotic/AI technology is sufficient to indicate that this area will evolve in a way that allows the robot to perceive and reason about it's environment. For reasons elaborated on in the paragraphs which follow, ranging sensors will play a major facilitating role in this evolution.

1.1 What is a Generalized Robotic Application?

Candidate tasks for robotic applications are typically performed by humans and involve high safety risk, mundane, or other undesirable work. Examples might include 1) bin picking on an assembly line, 2) hazardous material handling/inspection, and 3) complex and tedious assembly operations. A fundamental requirement for automating these, or similiar tasks, is the ability of the robot to sense and recognize the objects it is working with. Additionally, some applications require mobility, thus sensing for navigation and obstacle avoidance is also a necessity. Using a human being as a model, operating criteria for a generalized robotic sensing capability would include:

- Near range accuracy on the order of mm's.

- Coarse operating range on the order of 10's of m.

- Data formatted to support object recognition at near-real-time processing rates.

In order for the robot to be effective, its sensing and processing capabilities must operate at near-real -time rates. 3-D range image sensing, in an active mode, offers the best alternative for meeting the general operating criteria outlined above.

1.2 Why Range Sensing Rather Than "Conventional" Intensity-Based Sensing?

The physical design of robots has tended and continues to be anthropomorphic. Robot design based on the human model has proved to be convenient and sufficient for most all applications. Extending this design model to robotic sensing, however has proved to be totally inadequate when considering the criteria presented above. The human vision system is reflectance based and does not sense range directly, but instead uses a sophisticated algorithm for inferring distance based on a stereo sensor arrangement. Emulation of the stereo arrangement has proven to be effective for producing 3-D range imagery, however the processing required to accomplish this and support real-time object recognition is extensive. No present system can emulate the real-time parallel processing power of the human brain as applied to the object recognition problem.

In defining the 3-D object recognition problem, Besl [Besl 85] describes fundamental differences between conventional intensity-based imaging sensors and the relatively newer range-based devices. Intensity sensors directly sample and encode a 2-D spatial distribution function of scene illumination and surface reflectance information. Pixels in an intensity image contain brightness information and do not explicitly represent depth information. However, depth information may be inferred if certain geometrical and reflectance properties of the sensor/scene arrangement are either known or estimated. Alternatively, pixels in a range image directly encode distance information between the sensing plane and objects in the scene. Instead of being a sampled function of illumination and surface reflectance, range pixels more explicitly represent a sampled function of scene surface geometry. This surface geometry nature of range data is a direct measure of the 3-D shape of objects within the sensor's field of view. Therefore, the process of recognizing objects, based on their shape, should be less difficult in range imagery than in intensity images, due to the explicitness of the encoded information.

1.3 Range Sensing Techniques

Range sensing techniques have been generally classified as either active or passive. Active techniques are regarded as such because they project energy onto a scene to measure range. Passive techniques use natural scene illumination and estimations of scene reflectance and geometry to infer range. At least five different fundamental physical principles have been used to obtain range imagery: 1)Tactile sensing, 2)Lens focusing, 3)Triangulation, 4)Interferometry, and 5)Laser radar. This paper will briefly review techniques based on each of these principles. Consideration is given on the suitability of each technique for range sensing in the generalized robotic application domain. For a more complete review of range imaging sensors the interested reader is referred to [Besl 88].

2 Tactile Sensing

Contrary to what the name may imply, tactile sensing does not necessarily require direct contact with the surface. Techniques which fall in this category sense range information either by directly touching the surface, with a tactile position sensor, or by getting close to the surface and using a coupling effect of a proximity sensor. Stylus profilometers are an example of direct contact devices which have been used for measuring small surface irregularities, on the order of $0.1nm$ or 1 Angstrom. This technique requires scanning, or some form of relative motion, between the stylus and the object's surface. Thus, a major disadvantage in many cases, is that surfaces can be damaged by the sharp stylus as it rides across the surface. Surprisingly, and in spite of the surface damage, automobile manufacturers have used this approach to quantitatively sample the quality and consistency of spray paint application devices.

Allen [Allen 85] describes another tactile sensing approach which uses an array of touch/force sensitive elements. Both Harmon and Hillis describe [Harm 82,Hill 82] a variety of physical principles which can be used to construct such arrays. In this approach, the tactile sensing array is pushed into contact with an object and binary, or in some cases, grey value images are formed. The pixels in this imagery represent points on the object which are in contact with the sensor. In addition to measuring position, grey value tactile imagers quantify the force exerted on the tactile sensing elements, from which the range to the surface can be computed. While these sensors do directly acquire 3D surface coordinate data, limitations associated with direct contact and low resolution limit their utility for supporting object recognition and generalized robotic applications.

Besl [Besl 88] includes proximity based sensors in the tactile category and we will not deviate in this paper. Proximity sensors have been designed and built based on many non-contact, close-coupling techniques, including inductance, capacitance, ultrasonics, amplitude modulated LED's, fiber optics, mirror tunnels, and ring of light methods. To call these techniques proximity based implies a short stand off distance. Additionally, the techniques tend to be point based, thus scanning to cover an area has had to be mechanically accomplished, significantly reducing the imaging speed. Because of limitations in working volume, speed and accuracy these techniques do not readily support range sensing for object recognition. Instead, research using proximity methods has focused on obstacle avoidance sensing for robot navigation. Recent research [Kore 88] has shown that arrays of ultrasonic proximity sensors can be used to autonomously adjust a moving robot's path to avoid both static and dynamic obstacles.

3   Lens Focusing

Range sensing using this principle has been researched and discussed extensively in the literature [Horn 68,Tene 70,Jarv 76,Krot 86]. Range from focus techniques operate using the Gauss thin lens law. This law states that the reciprocal of the range to a point (z) is equal to the difference between the reciprocal of the focal length of the lens (f) and the reciprocal of the distance between the lens and focal plane (w):

$$\frac{1}{z} = \frac{1}{f} - \frac{1}{w} \tag{1}$$

Besl [Besl 88] shows the basic relationships for focusing and what it means for an imaged point at a depth z to be in focus at the detector. Some autofocusing mechanisms mechanically adjust w to maximize an electronic measure of the image focus, such as the high frequency energy in the video signal. Range images can be computed using such autofocus mechanisms. By sweeping w within some predetermined range, local focus measures can be computed to obtain the z value of each pixel. Such techniques are discussed and applied in [Jarv 76,Jarv 83,Mart 86].

Another method of obtaining range from focus is to note that the blurring of an out-of-focus point is a function of z and w. The total blur of an ideal point passing through an imaging system is characterized by the imaging system's point spread function. If the blur is circular, it is referred to as the circle of confusion [King 83], or blur circle. The radius of this circle is minimum when the detector is $w(z)$ away from the lens, where $w(z)$ is the focal distance of a point z from the lens. If a point blur is modeled as a 2-D Gaussian intensity distribution of diameter s for a fixed focal plane of distance w, the range to that point is:

$$z(s) = \frac{wf}{w - f \pm sF} \tag{2}$$

where $F = f/D$ is the f-number of the lens (D is the aperture diameter).

Pentland [Pent 85,Pent 87] experimented with ranging techniques using the "focal gradient" as characterized by local blur. Grossman performed similiar research [Gros 85,Gros 87] and determined that such systems yield accuracies of $\pm 1.25cm$ over a depth of field of $35cm$ at a standoff of 1 meter. Krotkov and Martin [Mart 86] implemented a range-from-focus system that maximizes a focus measure consisting of the absolute first differences on a scan line for a particular focus adjustment at each point in an image using 10×10 windows. They obtained an accuracy of 10% of the range in a depth of field of 1 meter at 1 meter standoff. While autofocus mechanisms in cameras act as point range sensors, [Gold 82,Dent 80] most commercially available units do not use focusing principles to determine range. Jarvis [Jarv 82] used the Canon "Sure-Shot" autofocusing mechanism (an active triangulation system using a narrow frequency modulated infrared beam and a small rotating lens) to obtain range images. The 64×64 images took 50 minutes to compute.

4   Triangulation

Triangulation is probably the oldest and most common method of measuring distance to remote points. Techniques which use this method rely on basic triangle trigonometry. Specifically, the Law of Sines provides that if the length of one side, along with two interior angles, of the same triangle, are known, then this is sufficient information for determining the lengths of the two remaining sides, along with the third angle.

Ranging systems which use triangulation may be either passive or active. The subsections which follow briefly discuss triangulation techniques from both categories.

4.1   Stereo

Stereo ranging has been an area of active research in computer vision for some time [Levi 73,Hann 74,Gana 75,Marr 76,Yak 78]. Stereo is a passive ranging technique which typically uses two cameras. Stereo techniques compute range by solving the geometry of a triangle with two known angles and the length of the line separating them (the camera baseline separation). The angle measurements result from the line and pixel coordinates of imaged points when expressed relative to the known geometry of each camera's boresight. The key problem in passive stereo is that of determining the relationship between the pixels of one camera image and the corresponding pixels of the other. This is referred to in the literature as the correspondence problem. The usual approach to this problem is to match "interest points" between the two images. These are generally points or areas of large intensity discontinuities. Once a correspondence between pixels is established, the disparity is computed and, hence, 3-D coordinates are recovered from the observed 2-D quantities (image line and pixel coordinates).

The human vision system is an excellent example of a passive stereo triangulation system. In this implementation, the matching, or correspondence problem, is significantly aided by the parallel processing efficiency of the brain and an acquired training of the way objects usually appear. A computer generalization of the human stereo matching technique has proven to be very difficult and computationally intensive. Thus, emulation of this approach for the generalized robotic control problem is not an optimum alternative.

4.2   Shape From Shading

Another passive approach substitutes one of the two cameras used in the stereo technique, with a broad area illumination source. This approach, called shape-from-shading, attempts to infer 3-D surface shape by estimating diffuse reflection properties of Lambertian surfaces. This work was first pioneered by Horn [Horn 70] and has since been continued by [Horn 75,Ikeu 81,Bolle 84,Pent 84]. Unlike stereo techniques which operate on interest points of high intensity discontinuities, shape from shading techniques attempt to recover range information from smooth surfaces. Shape from shading techniques generally compute a normalized reflectance map by dividing the original intensity image pixel values by the incident light intensity and an estimate of the diffuse reflection coefficient of the surface. The resulting intensity values represent the pointwise dot product of the unit surface normals and the unit vector from the surface in the direction of the light source. This relation is known as the image irradiance equation. The latter of these normals is known, and thus the computation of the surface normals reduces to computing the reverse of the dot product, with one of its component vectors known. This computation is intensive (when considered on an image basis) and inherently ambiguous, as demonstrated by Besl [Besl 88]. These two factors make the technique undesirable for generalized robotic applications.

4.3   Structured Light

The key problem in passive triangulation ranging systems is the correspondence problem described earlier. In active triangulation, this problem is reduced to a detection problem by replacing one of the cameras with a structured light source. By projecting a narrow beam of light into a scene, a very small point-like area of an object surface will be illuminated. Since the projection angle of the beam and the separation between the beam source and the camera's focal axis are known, range can be computed by solving the geometry of the triangle. Several illumination techniques have been used to validate this concept. These include a single scanning spot, patterns of spots, light lines, and patterns of alternating lines and dark spaces. For a thorough description of the various techniques, see Besl [Besl 88].

Structured light approaches have a demonstrated ability to significantly reduce the correspondence problem associated with passive triangulation techniques. Although the resolution and speed do not support real-time object recognition as well as Laser Radar (yet to be discussed), they do offer potential for coarser object recognition in generalized robotic applications.

5   Interferometry

Interferometric ranging techniques are based on light wave interference phenomena. Specifically, when two monochromatic (single wavelength) wave phenomena both occur in the same space at the same time, they interfere with each other. The simple trigonometric identity:

$$2cosAcosB = cos(A - B) + cos(A + B) \tag{3}$$

is fundamental to understanding interference phenomena. Besl [Besl 88] describes several techniques which use this principle. Some of these techniques have demonstrated utility at measuring surface height variations on the order of nanometers. Other approaches have been used for producing both relative and absolute depth maps of slowly varying surfaces. One interferometric technique which may have some generalized robotic utility is called moire interferometry. This technique has been used to produce range information with sufficient resolution and accuracy to support limited object recognition capabilities.

In moire interferometry, an interference pattern is created by inserting diffraction gratings into the optical path. In one particular implementation, called projection moire, a matched pair of gratings are used. One grating is placed in front of a focused light source, acting as a projector. The other grating is placed in front of a viewing camera. The projector grating causes a pattern of alternating bright and dark bands (fringes) of light to appear on the irradiated surface. This projected pattern of light is actually a spatially amplitude-modulated light signal. When the projected pattern falls on a smooth surface, the shape of the surface modulates the phase of the spatial light signal. In other words, the surface shape effectively deforms the projected light pattern. The modulation, or spatial frequency, of the projected light pattern is a direct function of the projection grating's pitch. Thus, by viewing the deformed pattern through a grating of identical pitch, the spatial light signal is demodulated, and interference fringes are created which carry information about the surface shape. The 2D camera image of these fringes may be processed to extract this surface shape information. The general problems with this technique include accurate calibration and automated analysis of the fringe images. The image analysis problem is compounded by the fact that the gray-level variations in the fringe imagery are a function of local contrast and local surface reflectance, as well as phase due to distance. In general, this technique suffers from the same problem as others previously discussed, that is the significant processing required to transform a 2-D spatial intensity image into a 3-D depth map. Object recognition to support generalized robotic applications is better facilitated by a sensing approach which directly produces 3-D surface coordinate data. Such a sensing technique is discussed in the next section.

6 Laser Radar

An imaging radar is defined as any sensor that emits time-varying signal energy, detects time-varing signal energy, and uses the difference in the emitted and received signals to determine range for each pixel in the image. For robotic vision applications, infrared lasers have been found to be useful as an imaging radar energy source. Besl [Besl 88] describes three different laser radar techniques, including time-of-flight/pulse detection, amplitude modulation, and frequency modulation. The Environmental Research Institute of Michigan (ERIM) has experience with each of these techniques and has actually built and delivered several laser radars (for various different applications). ERIM has demonstrated that an amplitude modulated approach is particularily well suited to generalized robotic applications [Zuk 83,Loug 85,Samp 84]. The following paragraphs discuss the operating principles of this approach.

Figure 1 illustrates a block diagram of a typical amplitude-modulated laser ranging sensor. Laser diodes typically can be modulated at frequencies up to $1.5GHz$. Examples presented in this paper will use a frequency of $720MHz$, which produces a half

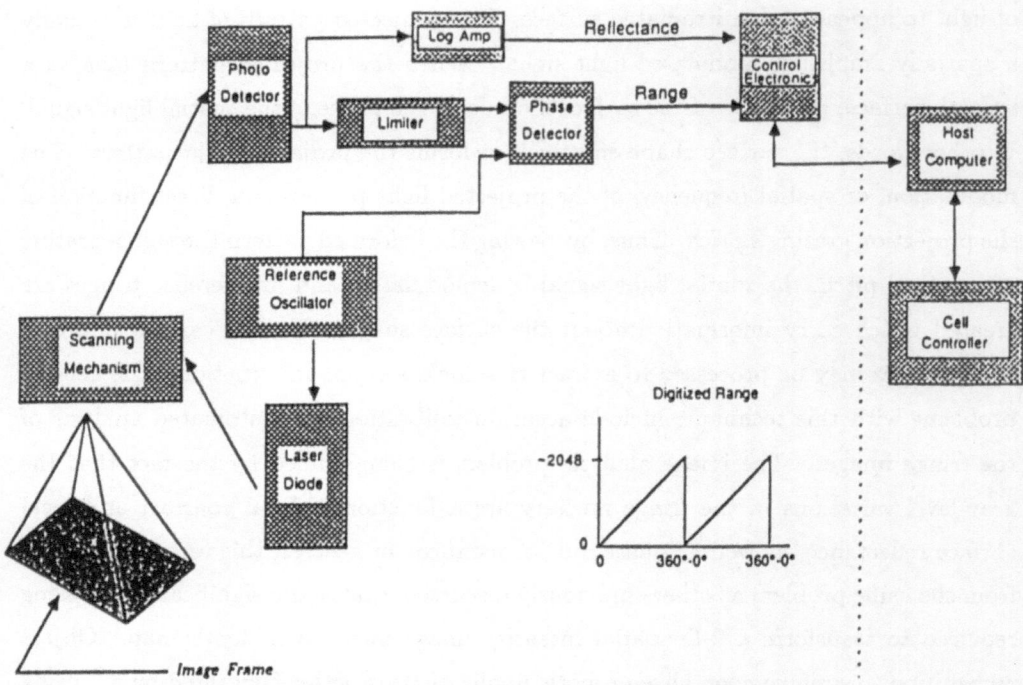

Figure 1. 3-D Laser Scanner, Simplified Block Diagram.

wavelength, in air, of 8.2 inches (the free space wavelength of a $720MHz$ waveform is 16.4 inches). The significance of the modulation frequency will be shown later.

Laser ranging sensors typically operate by transmitting a modulated infrared laser light beam to the object or scene of interest. The laser light is reflected, received by a photodetector, and electronically demodulated. The sinusoidal demodulated signal is then compared to the transmitted signal by using a digital phase detector. The phase difference is proportional to the range (Figure 2). In ERIM's most recent sensor, the range is quantized to a resolution of 11 bits, providing 2048 incremental levels of range. To acquire 3-D information from the entire scene, the narrow laser beam is mechanically scanned (in a raster format) across the total surface. The scanning is accomplished by using servo motor driven mirrors. Recent research is directed toward solid-state scanning techniques.

Since phase measurements become ambiguous once the phase difference exceeds 360° (a true 375° difference appears to be the same as a 15° difference), an ambiguity interval is defined [Zuk 83]. The ambiguity interval is the maximum range that allows the phase difference to go through only one complete cycle of 360°. The ambiguity interval, R_a, is mathematically defined:

$$R_a = \frac{c}{2f} \tag{4}$$

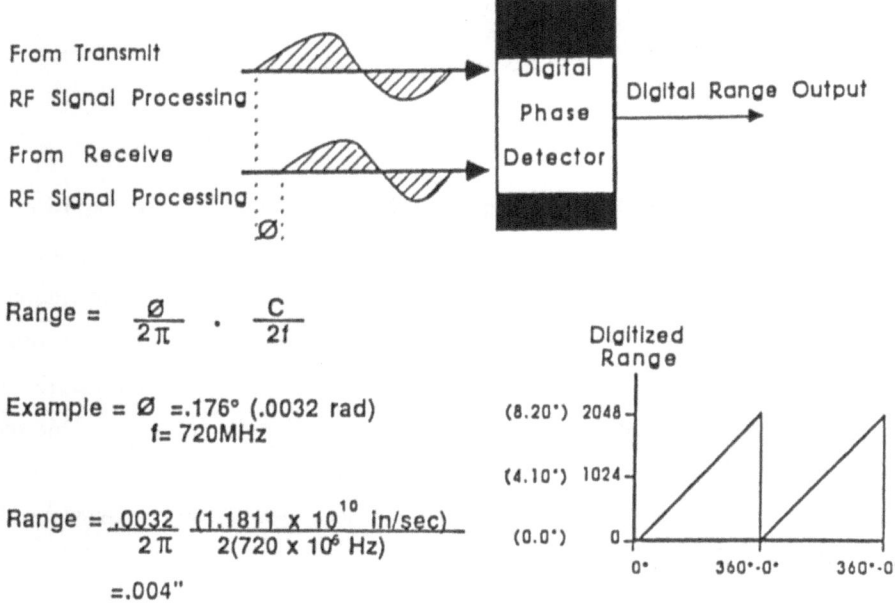

Figure 2. Phase Detector

where: f = modulation frequency

   c = speed of light

In imagery produced by amplitude modulated laser range sensing, each pixel value starts out as a phase difference. To determine the range to which this phase difference corresponds requires knowledge of the modulation frequency. The following formulation describes the range computation:

$$Range = \frac{\phi}{2\pi} \frac{c}{2f} \tag{5}$$

where: ϕ = phase shift

   c = speed of light

   f = modulation frequency

As previously described, when phase differences exceed 360°, range differences exceeding R_a become ambiguous. For an ambiguity interval of 8.2 inches, a true range of 9.7 inches will appear to the laser imager as 1.5 inches. With this in mind, the choice of modulation frequency is determined by the intended altitudinal (range) depth of field. Through proper choice of laser modulation frequency, signal processing electronics, laser beam diameter, and laser scanning mechanism, the ranging system can be tailored to meet most user requirements.

7 Summary

Generalized robotic applications require robust real-time object recognition capabilities. These object recognition capabilities are best met by range imaging sensors which directly produce 3-D, or depth map, imagery. Several of the ranging techniques briefly discussed in this paper, indirectly produce range imagery through transformation of 2-D intensity imagery. This transformation sometimes involves assumptions regarding the imaged scene and is almost always computationally intensive. One technique, Laser Radar, directly produces real-time 3-D range imagery and thus offers a promising alternative for supporting near-real-time object recognition for generalized robotic applications.

References

[Allen 85] ALLEN, P.K. 1985. Object Recognition Using Vision and Touch. Ph.D. Dissertation, Univ. of Penn.

[Besl 85] BESL, P.J., AND JAIN, R.C. 1985. Three Dimensional Object Recognition. *ACM Computing Surveys* **17**, 1(March), 75-145.

[Besl 88] BESL, P.J., 1988. Range Imaging Sensors. General Motors Research Laboratory. GMR-6090.

[Bolle 84] BOLLE, R.M., AND COOPER, D.B. 1984. Bayesian Recognition of Local 3-D Shape By Approximating Image Intensity Functions With Quadric Polynomials. *IEEE Trans. Pattern Anal. Machine Intell.* **PAMI-6**, 4(July), 418-429.

[Dent 80] DENTSMAN, H. 1980. State-of-the-Art Optics: Automated Image Focusing. *Industrial Photography*, July, 33-37.

[Gana 75] GANAPATHY, S. 1975. Reconstruction of Scenes Containing Polyhedra From Stereo Pairs of Views. Stanford AI Lab. Memo AIM-272. Stanford Univ., Stanford, CA.

[Gold 82] GOLDBERG, N. 1982. Inside Autofocus: How The Magic Works. *Popular Photography*, Feb., 77-83.

[Gros 87] GROSSMAN, P. 1987. Depth From Focus. *Pattern Recognition Letters* **5**, 1 (Jan), 63-69.

[Gros 85] GROSSMAN, P. 1985. Depth From Focus. Project Alvey Committee Meeting for Machine Vision, Univ. Sussex, Brighton, England.

[Hann 74] HANNAH, M.J. 1974. Computer Matching of Areas in Stereo Imagery. Ph.D. Dissertation, AIM 239. Computer Science Dept., Stanford Univ., Stanford, CA.

[Harm 82] HARMON, L.D. 1982 Automated Tactile Sensing. *Int'l. J. Robotics Research* **1**, 2(Summer), 3-32.

[Hill 82] HILLIS, W.D. 1982 A High Resolution Imaging Touch Sensor. *Int'l. J. Robotics Research* **1**, 2, 33-44.

[Horn 68] HORN, B.K.P. 1968 Focusing. MIT, Project MAC, AI Memo 160.

[Horn 70] HORN, B.K.P. 1970. Shape From Shading: A Method For Obtaining The Shape of a Smooth opaque Object From One View. MIT, Project MAC, MAC-TR-79.

[Horn 75] HORN, B.K.P. 1975. Obtaining Shape From Shading Information. In *The Psychology of Computer Vision*. (P.H. Winston, Ed.), McGraw-Hill, New York. pp 115-156.

[Ikeu 81] IKEUCHI, K AND HORN, B.K.P. 1981. Numerical WShape From Shading and Occluding Boundaries. *Artificial Intell.* **17**, 141-184.

[Jarv 76] JARVIS, R.A. 1976. Focus Optimization Criteria for Computer Image Processing. *Microscope* **24**, 2 (2nd Quarter), 163-180.

[Jarv 82] JARVIS, R.A. 1982. Computer Vision and Robotics Laboratory. *IEEE Computer* **15**, 6 (June), 9-23.

[Jarv 83] JARVIS, R.A. 1983. A Perspective on Range Finding Techniques for Computer Vision. *IEEE Trans. Pattern Analysis Mach. Intell.* **PAMI-5**, 2(Mar), 122-139.

[King 83] KINGSLAKE, R. 1983. Optical System Design, Academic Press, New York.

[Kore 88] KOREN, Y., AND BORENSTEIN, J. 1988. Obstacle Avoidance With Ultrasonic Sensors. *IEEE Journal of Robotics and Automation*, Vol RA-4, No.2, pp 213-218.

[Krot 86] KROTKOV, E.P. 1986. Focusing, Ph.D. Dissertation, U. Penn, Phila, Pa.

[Mart 86] KROTKOV, E. AND MARTIN, J.P. 1986. Range From Focus. *Proc. IEEE Int'l. Conf. on Robotics and Automation* (April 7-10, San Francisco, Ca.) IEEE-CS, pp 1093-1098.

[Levi 73] LEVINE, M.D., O'HANDLEY, D.A. YAGI, G.M. 1973. Computer Determination of Depth Maps. *Computer Graphics Image Processing* **2**, 131-150.

[Loug 85] LOUGHEED, R.M. AND TOMKO, L.M. 1985. Robot Guidance Using A Morphological Vision Algorithm. *Proc. of SPIE Conf. on Intell. Robots and Computers* September 1985.

[Marr 76] MARR, D. AND POGGIO, T. 1976. Cooperative Computation of Stereo Disparity. *Science* **194** 283-287.

[Pent 84] PENTLAND, A.P. 1984. Local Shading Analysis. *IEEE Trans. Pattern Analysis Mach. Intell.* **PAMI-6**, 2(March), 170-187.

[Pent 85] PENTLAND, A.P. 1985. A New Sense of Depth of Field. In *Proc. 9th Int'l. Joint Conf. Artificial Intell.* (Aug., Los Angeles, Ca.)

[Pent 87] PENTLAND, A.P. 1987. A New Sense of Depth of Field. *IEEE Trans. Pattern Analysis Mach. Intell.* **PAMI-9**, 4 (July), 523-531.

[Samp 84] SAMPSON, R.E., SWONGER, C.W., AND VANATTA, P.W. 1984. Real Time Three-Dimensional Image Processing For Robotic Applications. Environmental Research Institute of Michigan, Ann Arbor, Michigan. June 1984.

[Tene 70] TENENBAUM, J. 1970. Accomodation in Computer Vision. Ph.D. Dissertation, Stanford University, Ca.

[Yak 78] YAKIMOVSKY, Y. AND CUNNINGHAM, R. 1978. A System For Extracting 3D Measurements From a Stereo Pair of TV Cameras. *Computer Graphics Image Processing* **7**, 195-210.

[Zuk 83] ZUK, D.M. AND DELL'EVA, M.L. 1983. Intelligent Task Automation. Environmental Research Institute of Michigan, Ann Arbor, Michigan.

AN ALTERNATIVE ROBOTIC PROBOSCIS

MR J B C DAVIES

DEPARTMENT OF MECHANICAL ENGINEERING
HERIOT-WATT UNIVERSITY
EDINBURGH, UK

Abstract

A flexible, simple robotic proboscis has been developed based upon the differential extension of 3 extendable tubes. Pneumatic pressure is used to create the tube extension and control is achieved via a micro computer and analogue proportional control valves. Motion and flexibility is very organic and although power/weight ratio is high, absolute payload and stiffness are low.

Introduction

Industrial robots are rigid, reliable, programmable devices that have been very successfully applied in areas where sequential flexibility is important i.e. where individual sequences remain unchanged, but subsequent complete sequences may be different. Despite this success, alternative robotic devices having superior power/weight characteristics or improved flexibility have been constructed e.g. spine robot of Robotics AB of Moludal, Sweden and the "snake" robot of UMETANI in the Tokyo Institute of Technology. Both these robots are mechanically operated and consist of a series of planes connected by a central axis and pivoted by a series of control wires. Consequently, control of the robot motions is a complex problem involving a large number of control inputs. In contrast to this approach, the Department of Mechanical Engineering at Heriot-Watt University has developed an alternative robotic proboscis affectionately titled the "Elephant's Trunk". A simple tube, when subjected to internal pressure extends and, if connected to an identical, unpressurised tube, instead of extending linearly, adopts a curved shape under the influence of the non extending member. If 3 such tubes are connected as shown in Fig 1, by varying the internal pressure differentials, the proboscis can be made to adopt a wide range of curvatures and positions. To validate this proposition, a simple rig was constructed to operate by analogue proportional pneumatic valves controlled by a BBC computer.

A cubic framework was built and after some initial analysis and trial a suitable convoluted air hose was selected to form the basic members of the proboscis. For convoluted air hose of this design, for any increase in internal air pressure, the longitudinal expansion is markedly greater than the radial expansion.

Control System

The proboscis is controlled from a microcomputer connected via a standard input/output interface through D/A to converters to 3 analogue proportional pneumatic valves, Fig 3. Pressure values are sensed inside the control valves and a local control loop equates output pressure with an input signal. Control of the robot is entirely open loop, but attempts are being made to close the control loop. Work by Daniels at Oxford University on large compliant structures may be relevant to this problem albeit requiring a reference structure.

Currently the proboscis can be driven in 3 ways

a) Following a joystick or the cursor keys in a tele-operator condition
b) A teach mode where specific positions are recorded in terms of 3 input pressures for subsequent replay
c) Existing sequences can be replayed to perform previously defined motions.

Motion attributes may be associated with individual sequences e.g. pressure change rate, dwell cycle rate.

Calculation of Expected Configuration

Expected configuration (2 dimensional only)

From standard beam theory

$$\frac{\sigma}{y} = \frac{M}{I} = \frac{E}{R}$$

where σ = stress

y = position relative to the neutral axis

M = bending moment

I = 2nd Moment of area of section

E = youngs Modulus

R = Radius of curvature of neutral axis

For the purposes of this proboscis, we are primarily interested in the radius of curvature at the neutral axis.

At any instant, the proboscis may be considered as a single member having the cross section shown in Fig 3. We will further assume that the proboscis may be treated as in pure bending and so the radius of curvature is given by

$$R = \frac{EI}{M}$$

To evaluate the system, readings of the angle of the end face of the proboscis were taken for various internal pressures and for one and two tubes pressurised. From Fig 2 it can be seen that whilst all the graphs exhibit similar characteristics, as the number of pressurised tubes increased, so does the stiffness of the proboscis. It was also apparent that the equation was non-linear, suggesting that the product EI was not constant. This variation could be caused by dimensional variations in the overall size of the proboscis i.e. changes in a and b, or, as indicated by Fig 2, variations in the stiffness of the tube as the imposed strain increases. Measurements suggest that the variation/reduction in the overall tube diameters/pitch is small, but the tube construction as coil wound PVC extrusion, was highly likely to possess non-linear deformation characteristics. To reduce geometric effects, a calibration curve was generated by measuring only the linear extension of a single tube. This effectively provided a measure of the variation of E and was utilised in subsequent displacement calculations.

Current State and Limitations

During the initial construction stages, difficulties were experienced in attaching the end plugs to the extension tubes. This was finally solved. Photograph 1 shows the current state of the test rig. It operates reliably and utilises air pressures up to 1 bar. Fig 4 shows the operating surface of the proboscis, but repeatability and drift are problems intrinsically associated with the tube construction. Only spiral wound tubes are available off the shelf and this inevitably introduces a torsional mode into the movement of individual tubes. A concentric non-spiral tube would improve performance and enable the orientation of an end effector to be more easily predicted.

Stiffness of the proboscis is very low and, whilst its power/weight ratio is high, its natural frequencies are likely to limit future applications to low or fixed load carrying activities. Closing the feedback loop would significantly improve the operating characteristics as a load carrying robot.

Potential applications are listed in the Table below, but inspection tasks currently head the list, especially for nuclear applications.

Future Developments and Applications

The proboscis requires a tailor-made tube element, miniaturised electronics and an improved closed loop control system. A myriad of applications have been identified from film extra through differential pressure indicator to fibre optic carrying inspection device and funds are being sought to develop any one of the applications currently identified.

The project has been developed using funds from the Department of Mechanical Engineering of Heriot-Watt University.

Conclusions

The organic Elephant's Trunk proboscis is a simple reliable device based on the differential extension of 3 linked members. Application potential is very large but significant funding is required to produce a specific solution to a particular application.

GENERAL CONSTRUCTION
SECTION THROUGH TUBES

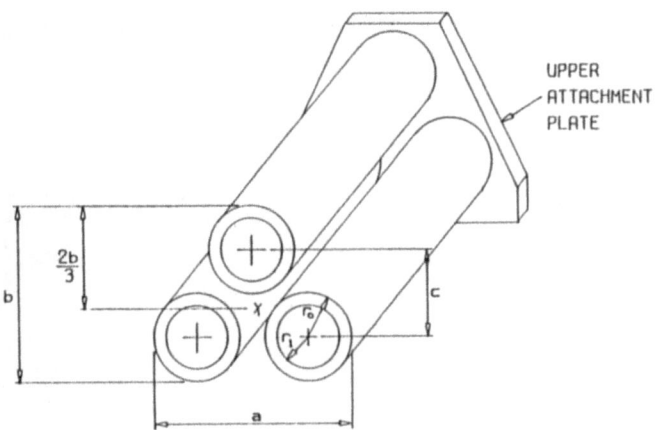

UPPER
ATTACHMENT
PLATE

FIG 1

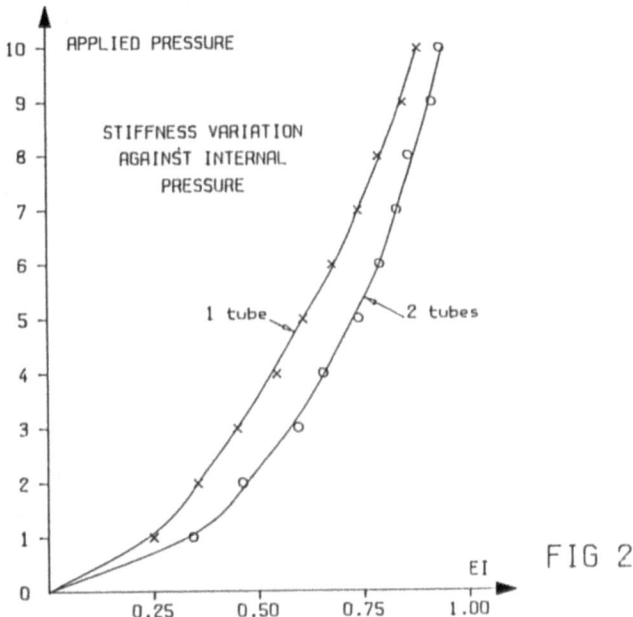

APPLIED PRESSURE

STIFFNESS VARIATION
AGAINST INTERNAL
PRESSURE

1 tube

2 tubes

EI

FIG 2

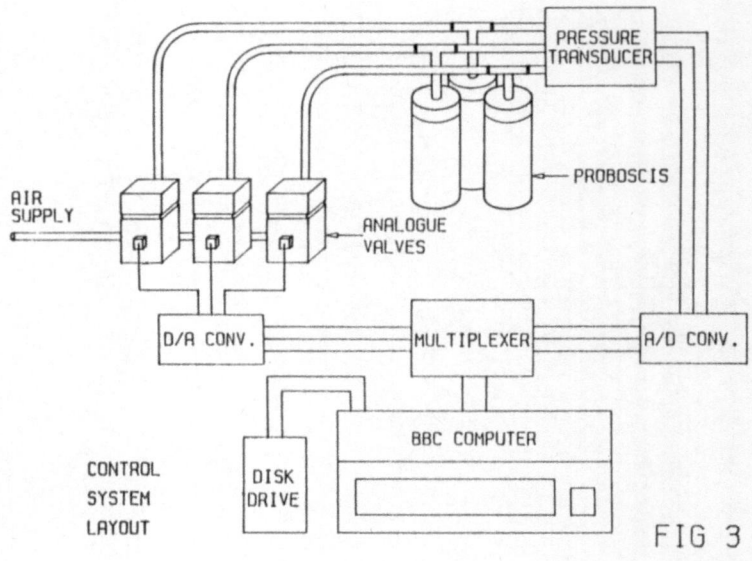

PRESSURE
TRANSDUCER

— PROBOSCIS

AIR
SUPPLY

ANALOGUE
VALVES

D/A CONV.

MULTIPLEXER

A/D CONV.

DISK
DRIVE

BBC COMPUTER

CONTROL
SYSTEM
LAYOUT

FIG 3

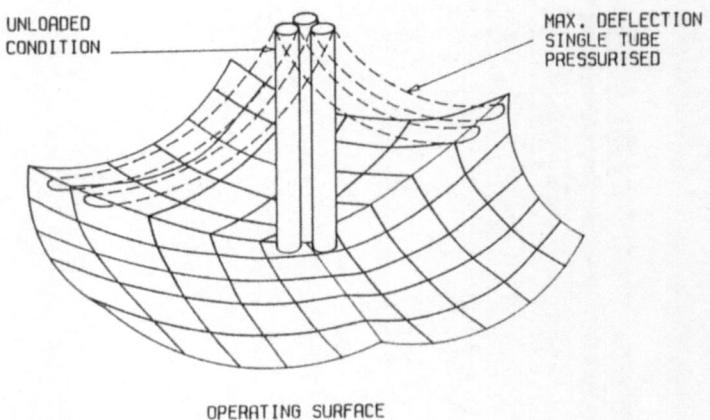

UNLOADED
CONDITION

MAX. DEFLECTION
SINGLE TUBE
PRESSURISED

OPERATING SURFACE

FIG 4

Table 1	ELEPHANT'S TRUNK - APPLICATIONS

1 Internal pipe inspection.
2 Security surveillance around corners.
3 Microphone boom.
4 Film effects - ET neck.
5 Brush/Sweeper/Vacuum device.
6 Snake - Burglar alarm. Heriot Hydra.
7 Underwater - pressure compatible/differential.
8 Outer space - higher power/weight ratio.
9 Fire hose direction control - remote.
10 Liquid/powder leading or unloading.
11 Aircraft airborne refuelling.
12 Tailor Mannequins - Shop windows.
13 Window cleaning on high buildings.
14 Car manoeuvring aid on bumper.
15 Sorting device into pigeon holes.
16 Painting round/cylindrical objects e.g. lamp post.
17 Grain handling Gr 10.
18 Seed sowing.
19 Apple/fruit picking.
20 Cement pouring.
21 Aircraft escape chute.
22 Medical - adjustable splint.
23 Adaptive chair shape.
24 Hazardous environment inspection.
25 Multipressure indicator.

Dynamic Robot Vision

Erik Granum and Henrik I. Christensen[*]
Aalborg University, Institute of Electronic Systems,
Laboratory of Image Analysis
Badehusvej 23, DK-9000 Aalborg, Denmark

Abstract

In computer vision efficient methods for detection and interpretation of motion of objects have been developed. As technology advances, the ambition to include this ability in robot vision systems appears more and more realistic. However, to become of practical use, real time performance (in some sense) is required, and the current possibilities for this are still limited.

Many different approaches to motion analysis have been proposed in the literature. Motion information may be derived from image analysis systems at different levels of the general scheme of image processing and interpretation. However, to achieve a result in terms of motion descriptions, most of these methods depend extensively on image preprocessing (and interpretation) or on integration into an image postprocessing (and interpretation) system.

A number of methods are reviewed and evaluated with regard to dependency on supplementary processing and with regard to current potential for real time application. Also we discuss their weaknesses due to problems of ambiguity and noise. However, one can take into account that real time operation also means continuous operation and thereby that a temporal context is provided. This allows concentration on changes most of which are predictable, and savings in computing as well as improved robustness to noise and ambiguities can be achieved.

In conclusion we find that high level token matching currently is one of the most promising approaches, and an experimental implementation is used to demonstrate a possible approach to motion analysis in real time.

---

[*]   This research has in part been sponsored by the Danish Technical Research Council, FTU grant 5.17.5.6.06

1 Introduction

The use of vision sensors may provide the control of a robot system with a great deal of important flexibility for many applications. It depends on how intensive the actual usage of vision sensors for robot control is. The use of vision for robot control may be categorised into 3 groups according to the use of temporal information; i.e., 1) static, 2) pseudo dynamic, and 3) dynamic robot vision. The three categories are outlined below.

1) The acquisition of a single image (pair) of the workspace before the robot action is to take place allows for the control to take the actual arrangement of objects into account. The information gathered this way may suffice to plan a complete and successful operation, provided that everything, not purposely manipulated by the robot, remains static.

2) As robot action or motion progresses additional images may be acquired. However, time intervals between image captures are relatively large and every image (pair) is analysed independently. The scene information thus acquired can be used to check whether the robot since the last image has moved to the position expected and whether the conditions of the scenario have changed in a way which calls for adjustments of the action plan.

3) Images are acquired so frequently relative to the highest possible velocity in the scene that all motion can be detected and monitored. This case opens up for quite a variety of applications including the use of vision as feedback sensor in a control loop.

Cases 1 and 2 are commonly used while 3 is a rarity. It requires a sort of real time processing of image sequences, correspondence analysis between these images and so on, for example motion analysis.

It is not obvious how this could best be done, and this paper discusses methods of real-time motion analysis for a dynamic robot vision system. To limit the scope of our discussion, we will mainly refer to the analysis of the perspective projection of 3-D motion onto the image plane, f(x,y), of a camera. The many other interesting problems relating to proper 3-D interpretation etc. are generally not included here. Monocular vision using a static camera is the typical reference. Also for the general discussions the illumination of the scene observed is assumed constant.

In the following Section 2 we distinguish between motion detection and motion interpretation as an introduction to principles of motion analysis in Section 3. A number of motion detection methods are reviewed in Section 4. Section 5 outlines a scheme for a model driven approach, which is demonstrated in Section 6. Section 7 provides a brief summary of the paper.

2 Detection and Interpretation of Motion

A distinction is made between *motion interpretation* and *motion detection*. Let motion detection designate the process of detecting time dependent changes between subsequent images of a time sequence. The combination of such changes with the information about the *spatial context* of the image(s) (from traditional image analysis and interpretation) provides the input to an actual process of motion interpretation, which in principle have the potential of evaluating a wider *temporal context*.

Indirectly this presents a conceptual classification of motion detection as a low level process and motion interpretation as a high level process. The mechanism for motion detection provides this primitive to be used for motion perception (motion interpretation), and other perceptual tasks such as *shape and structure from motion* [12,34].

However, as will appear from the following discussion, in some contexts of practical computer vision systems it may be beneficial to execute the motion detection process at a relatively high level of processing.

In spite of the limitations on scope given in the introduction we want to consider these two concepts of motion detection and analysis in the context of computer vision systems. For the discussions below the very simplified vision system models of Figure 1 are used. Model I exemplifies an approach where data, either raw or briefly prepocessed, are input to the motion detection process before the general spatial contexts of the image(s) are considered. Model II, on the other hand, illustrates approaches where not only segmentation but also some higher level image analysis and interpretation is required before motion detection can take place.

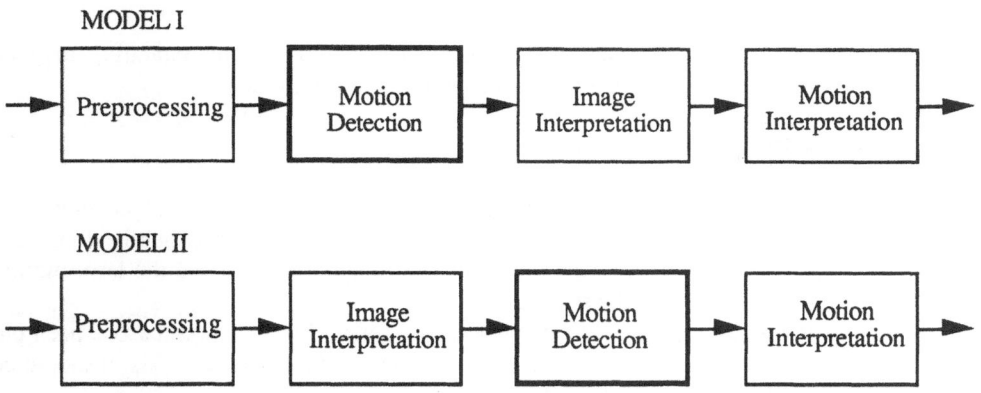

Figure 1 *Simplified models of vision systems for motion analysis. Model I illustrates motion detection on relatively "raw" data before or independent of the spatial image analysis. In Model II motion detection is dependent on results of some high level image analysis*

3. Motion Analysis

Let the following be a suggestion for a definition of this higher level process:

Motion analysis is examination in detail of the process of continual change in the physical position of objects. This is done by breaking down the process into its essential features in order to determine their relationships and thereby discover the meaning of the whole process.

Motion analysis may be approached in a "microscopic" and analytic fashion by assigning a velocity vector to each point in space and considering the *motion field* they form. Alternatively a more "macroscopic" approach may be adopted by considering space segmented into tokens, objects, etc., which are represented by points in space. The trajectories of these points are then the subjects of analysis. We will mainly confine our discussion to the latter approach.

More specifically we consider the Motion Analysis process (Figure 1) to have two basic functions:

1) to accumulate the sequence of trajectory samples from the motion estimation process and organise them in their temporal context.
2) to evaluate these trajectories in their temporal context such as to facilitate various purposes, e.g.
2a) short term prediction of motion in support of the whole motion analysis process itself,
2b) segmentation of the trajectories into structurally meaningful parts for description and recognition of motion patterns.

Depending on the application, further functions may apply, e.g.

3) to provide quantitative and/or qualitative descriptions of motion, i.e. functioning as a measuring instrument for various applications,
4) to predict long term motion behaviour of objects for the purpose of navigation and collision avoidance in a dynamic scene.

The very high level processes of knowledge based motion understanding may be considered separately or as a part of a wide concept of motion analysis.

3.1 Activities in Motion Analysis.

The term *motion analysis* may be found in the text of publications on the subject of motion in connection with computer vision. It is remarkable, though, that very few publications have this term in their title. And many which do could more appropriately use the term motion detection or motion estimation in the sense of Section 2.

This seems to characterise the state of the art. Most effort has for good reasons been put into the necessary prerequisite - the low level processing part of motion analysis. It should be appreciated, however, that much of the effort in connection with the flow-methods has quite some bearing on motion analysis in a wider sense, as it attempts to bring about a consistent analytical formalism capturing fundamentals of motion phenomena [34,5,33].

Motion analyses in the sense of looking at *motion patterns* of objects in a wider temporal context have mainly been published in connection with specific applications. One of the most obvious applications is analysis of human body motion [22,26,27]. Other examples are motion analysis (and understanding) of the left ventricular wall [31], analysis of cloud movements and the like (included in the survey of Martin and Aggarwal [23]) and motion patterns of water surface waves [20].

With a few exceptions, such as navigation systems for autonomous land vehicles [35], all computational motion analyses reported so far are retrospective motion analysis, i.e. using synthetic or recorded data.

3.2 Observing a Dynamic Scene

Let us consider, from a "macroscopic" point of view, the scene phenomena we may observe - here classified into 3 groups:

- static "objects"
- moving "objects"
- (sudden) events.

The analysis of **static objects** (i.e. all static regions of the image) is the usual image or scene analysis problem, which in principle has a separate and less time critical solution. For ambitious vision systems a parallel channel may analyse static phenomena in any required detail at a slower speed and relatively independent of the frame rate necessary for the analysis of dynamic phenomena. Regions of the image plane with static data can be singled out, and averaging over several frames for reduction of noise and so on is possible.

For the **moving objects** we expand on the previous breakdown of processing:

- motion detection (the PRESENT).
  MODEL I approach: is anything anywhere moving NOW, and if so what is it?
  MODEL II approach: given an object - is it moving NOW?

- motion estimation (the PRESENT).
  How is it moving NOW?

- motion interpretation (temporal context).
  How was the object moving? (the PAST).
  How will the object move? (the FUTURE).
  What is the 3-D structure?
  etc.

By accumulating over time the detections and estimations of "moving NOW" we gain information on how the motion was in the PAST and we have a basis for predicting how motion will be in the FUTURE.

Sethi and Jain [29] discuss many important aspects of motion analysis and analysis of dynamic scenes. They consider Jenkin's smoothness of velocity [21] or *smoothness of motion* and introduce the notion of *path coherence* on the assumption that for real-world dynamic scenes: *the motion of an object at any time cannot change abruptly.*

In the realm of this assumption the motion to be expected in the near future can be predicted with quite some confidence. For higher level vision this information is essential for understanding dynamic scenes and the predictions can be used for collision avoidance etc. For low level processes short term predictions can be utilized for dynamically directing computer resources to those areas of the image where motion signals are expected and the motion detection itself can be supported to improved robustness with respect to noise and ambiguities. For example in the case of matching the information of two consecutive frames, one might rather predict the content, i.e. the "few" changes, of the most recent frame and then match the predicted with the measured frame.

Cases where the above assumption does not hold we consider **sudden events**. In the extreme these are explosions, collisions etc. The distinction between motion and events is not clear cut, though, and may even depend on current status of the interpretation process. Consider an elastic ball bouncing on a hard surface. The first impact of the ball on the surface might well be considered an event. However, when the phenomena and its components are recognized, the trajectory of the ball is in principle predictable and may satisfy some extended notion of Sethi and Jain's path coherence [29].

3.3 Real Time Analysis

Qualitatively, real time processing means processing data as fast as they are generated, often with the purpose of achieving just a high level description of some information carried by the data. Hence all raw data can be disposed of right after analysis.

The speed of generation of data is very application dependent. In our case images should be sampled frequently enough to avoid those ambiguities in the motion detection that could be resolved by proper sampling. However, the standard rate of 25-30 frames per second is often an understood reference as it seems to be fairly compatible with human visual perception in many ways. Lower sampling rates might suffice in many applications.

Should we consider real time analysis today without a priori relying on very special hardware, the following practical principles are worth considering:

a) make the processing path up to the higher level motion analysis as short and direct as possible.

b) supplement with parallel concurrent processes as need may be needed for more sophisticated analysis.

c) break down the direct processing path into processing modules, which can be coupled asynchronously. This provides an automatic buffering capacity, as subsequent modules at peak load may process subsequent frames. A delay of several frames may well be acceptable in many applications. Human performance is characterised by a delay of some 100 ms.

d) real time operation implies a continuously running process which after initialisation can take advantage of the fact that most data are repeated frame by frame. Use feedback and prediction to control computer resources and algorithms by concentrating on changes and/or features relevant for the goal of the system.

In any case, successful real time systems of today are likely to rely on restricted and known scenarios, and thus to be dedicated systems for special applications.

4 Methods for Motion Detection

In connection with motion detection the term motion estimation is often used for the quantitative determination of the movement of objects between subsequent frames. Here the estimation generally will be considered as part of the detection process.

The approaches to motion detection may be grouped in many different ways, e.g. under Huang's three headings [17]: (i) methods of differentials (dffr), (ii) spectral methods (spct), and (iii) matching methods (mtch). Several examples of these major approaches are discussed below, and simple block diagrams are shown in Figure 2 for each example:

a.	(dffr)	Image differencing
b.	(dffr)	Optic flow (Image flow)
c.	(df/sp)	Spatio-temporal filtering
d.	(spct)	Fourier methods
e.	(mtch)	Correlation methods
f.	(mtch)	High level token matching
g.	(mtch)	Low level token matching.

4.1 Image Differencing

In this category the initial analysis is a subtraction of subsequent frames, i.e. rather than using raw images, the analysis is based on or supplemented by the difference images:

$D(x,y,n) = f(x,y,n) - f(x,y,n-1).$

The difference images support segmentation into static and moving regions, as only objects that have changed position between subsequent images will have noticeable non-zero pixel values. Some additional processing is required to single out the moving objects in a reasonably robust fashion. Motion detection based on difference images is theoretically not very attractive, but being simple and yet useful [20,39], it has a potential for fast (and possibly real time) systems for practical applications. Recently Wiklund & Granlund [37] have used the technique for detecting moving objects in traffic scenes.

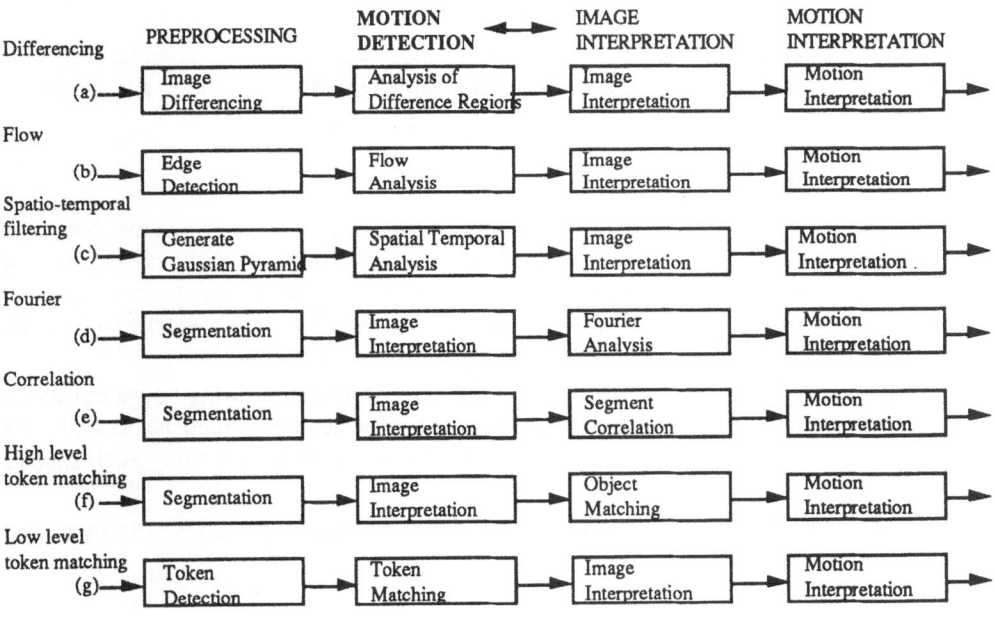

Figure 2   *Simplified block diagrams of motion analysis systems for each of 7 methods of motion detection*

4.2 Optic Flow (Image Flow)

These methods consider a time varying *motion field* of velocity vectors (spatio-temporal derivatives) associated with each point in space, or after projection, each point in the image plane. The perceived (directly computable) *optic flow (field)*, however, is unfortunately not the same as the motion field, unless at sharp edges and under various constraints like constant illumination etc. [15,33].

Much important research in this area on fundamentals of motion phenomena, including recovery of structure from motion, is going on [36,34,33,5]. But these methods are in general computationally demanding and very sensitive to noise - in particular when higher order differentials are included.

In its simpler version we can determine the optic flow from first order differentials using Horn and Schunck's equation [16] conditioned on constant illumination:

$$I_x v_x + I_y v_y = -I_t$$

I_x, I_y, and I_t are derivatives of the image intensity function with respect to x, y, and time, respectively. The v_x and v_y are velocity components. As this formula is only meaningful at edges, where the spatial derivatives are different from zero, the image sequence is pre-processed with an edge detector. At the edges the equation will provide only the motion component perpendicular to the edge, a limitation called *the aperture problem*. Additional analysis using spatial context is required to coordinate this description to the *true optic flow* for use at object or surface level, which is a task with its own complexity, and of a degree depending on the number of domain constraints used.

In spite of promises that special hardware is under way [6,34], this group of ambitious methods is likely to require more time to mature before they can demonstrate a potential in a real time context.

4.3   Spatio-Temporal Filtering

This approach is even more ambitious and computationally demanding than the above flow methods. Heeger [14] uses a family of twelve 3D Gabor filters which he convolves with Gaussian pyramid representations of each of the images in a sequence. By combining the output of the convolutions he is in principle able to estimate the true flow per pixel, i.e. the aperture problem is coped with by using a wider spatial and temporal context.

Heeger gives a spectral interpretation of the filtering process, showing how a translating texture transforms to a plane through the origin in the spatio-temporal frequency domain, and how the family of tuned filter functions "samples" that domain identifying such spectral structures.

It hardly needs mention, that however interesting this type of approach is, we may have to wait for real time operation.

4.4   Fourier Methods

These methods rely on some useful properties of the Fourier domain. Consider an image f(x,y) and a translated version g(x,y) = f(x+Δx, y+Δy). Let F(u,v) and G(u,v) denote the Fourier transform of *f(x,y)* and *g(x,y)*, respectively, then the two transforms are related by

$$G(u,v) = F(u,v) * e^{-j2\pi(u * \Delta x + v * \Delta y)}$$

Huang [17] shows how the translation of an object in the time domain can be computed from phase shifts in the frequency domains. Similarly the spectrum rotates with the object, and motion in depth (scale change) scales the spectrum in the opposite way.

However, all this relies on the facts that only objects with identical motion are present in the pair of images being analysed and that all background is uniform. This implies image segmentation prior to analysis, and if more objects move independently, these have to be detected and objects to be paired prior to Fourier analysis. Hence only motion estimation can be computed this way, and only reasonably conveniently if the objects have nice straight edges, which show up as spectral lines through the origin of the Fourier domain.

Fourier methods, as here presented, have apparently little to offer in general and even less in a real time context, but their basis is worth remembering as a spectral interpretation of motion phenomena can be useful. The more recent spatio-temporal filtering methods are examples hereof.

4.5 Correlation Methods

Segments of successive images are compared using correlation or a similar measure [2], e.g. a segment with a moving object from one image is cross-correlated with a relevant segment in the next image. The location with maximum correlation is taken as the position of the segment (object) in the latter image. These methods are sensitive to scaling and orientation, i.e. if the objects can rotate, the correlation must be repeated corresponding to each of the possible new orientations.

Preferably, but not necessarily, the objects should be segmented out prior to correlation. Ciccotelli et al. [10] compare the two conditions and their algorithm operates well with background data included, and they are optimistic with regard to the use of their system for special robotics applications. It is a question, though, whether these methods will find general use. Both inherent limitations of the methods as well as their computational requirements seem to be against them.

4.6 High Level Token Matching

High level primitives such as identified surfaces or entire objects with feature vectors from each image are used for motion detection. Fast and possibly simplified preprocessing is usually required to make this approach relevant for real time operation. If adequate preprocessing is provided, however, then motion detection and estimation can be made fast and using fairly sophisticated matching methods. Relatively few tokens (objects) per frame are generally to be expected, some bookkeeping per token is possible (scene model), thorough prediction in terms of token positions etc. for next frame can be computed, and occlusions may be taken into account.

Chow and Aggarwal [7] use a scene model and prediction for planar moving objects (no occlusions), Anderson [4] uses 1st-4th order moments for object features at 60 Hz frame rate, Granum and Munk [13] propose model and prediction coping with occlusions, Sethi and Jain [29] optimise for path coherence, and Christensen and Granum [9] improve the approach of [13] with fast probabilistic relaxation etc. in the matching process.

The last example involved testing in operation at 5 frames per second for 5 objects on a standard UNIX system with non-optimised prototype implementation. This and the other methods still rely on the fast preprocessing, which may well introduce limitations on the possible applications.

4.7 Low Level Token Matching

For this approach images are preprocessed only so that low level primitives (tokens) like corners, edges, peaks or ridges can be identified. Tokens of subsequent frames are matched using some measures of similarity including spatial distance. A primary problem is ambiguity

[32], particularly because of high numbers of possibly very similar tokens, so that it may not be obvious which primitives are to be matched (Figure 3). Noise and occlusions may make tokens disappear and reappear, all of which adds to ambiguities. This call for complex matching methods such as probabilistic relaxation, which may be expensive for high numbers of tokens.

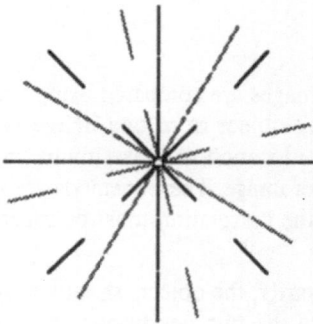

Figure 3 *Illustration of matching ambiguity*

Roach and Aggarwal [28] discuss problems of motion estimation from such tokens given that matching was correct, and Nagel [24,25] describes various examples of the approach. Crowley et al. [11] are very optimistic with a special approach, which involves tracking line segments and using a (simplified) Kalman filter supported prediction to maintain a coarse but dynamic image flow model. Reported experimental results are claimed to be very successful and special hardware for near video-rate processing is under way.

4.8 Summary of Motion Detection

From this presentation of the methods and the summary of Figure 2, it appears that all three differential methods and the low level token matching correspond to Model I of Figure 1. Optic flow and low level token matching requires some preprocessing to single out specific low level image features for the motion detection to concentrate on. In all four cases a large amount of data is to be processed for motion detection at frame rate. The motion information obtained, however, may be used to support the subsequent spatial image analysis and interpretation.

Optic flow and spatio-temporal filtering are currently not yet very attractive from the point of view of real time processing. In contrast, image differencing and low level token tracking have more of a potential of being engineered to meet the requirements of (restricted) real time systems.

The methods of Model II of Figure 1 rely more or less on images passing through a spatial analysis and some interpretation at frame rate, which is a considerable requirement for processing power. It may be easier though, to engineer special solutions to this analysis for real time systems of restricted performance. In any case, given "interpreted" images, the processing per motion detection can be drastically reduced and be made more robustly. This advantage is most relevant for the high level token matching. The two other methods basically return to raw data level again for the regions segmented out, and much number crunching still has to be carried out. Compared to high level token matching, they seem to have little extra to offer and this little is computationally expensive.

5 Model Driven Motion Analysis

In this section we would like to present what we consider the essence of the speculations in Section 3 - the scheme for *model driven motion analysis* (detection and estimation). The scheme is illustrated in Figure 4, and cover the following main functions:

1) dynamic "world" model
- accumulator of ("raw") motion data etc.
- data for predictor
- asynchronous coupling to higher level motion analysis etc.
An entry for every individual object is maintained with all available static and dynamic data.

2) frame model predictor
- read data from dynamic model
- predict all object parameters one frame ahead

3) matching process
- match measured frame and predicted frame model
- allow for object merges and splits

4) correction of frame model
- adjust predicted frame model according to match results

5) motion estimation
- estimate motion parameters using dynamic model and corrected frame model

6) update dynamic world model

This scheme is of rather general nature and complies with conventional control theory [1]. Please note that according to previous definitions this scheme covers only motion detection and estimation, accumulation of motion data, and short term prediction for own use. Higher level motion (pattern) interpretation is left for other processes accessing the dynamic world model.

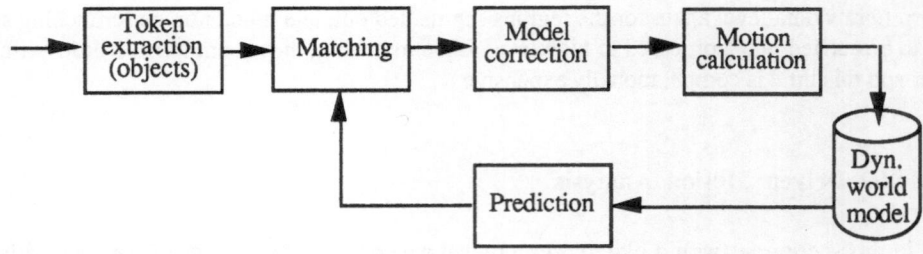

Figure 4   *Block diagram for model driven motion detection and estimation using high level tokens (objects)*

6 A Sample Implementation

A laboratory system has been implemented along the lines of the previous section [13,8,9]. Object feature vectors are input to the system, and they can be gathered from systems like the Edinburgh Fast Interval Processor [30]. This performs a very coarse image analysis, but data can be generated at video rate. The feature vectors contain position of centroid, size measures, and application dependent object features.

Apart from specifications on constraints on the scenario necessary to make the preprocessing work, the following design criteria for the motion analysis system were given:

a)   Objects (= tokens) of the observed frame are to be matched with a predicted version (model) of the same frame.

b)   An object in the observed frame may correspond to two or more currently occluding objects in the model .

c)   An object in the predicted frame model is a **single** object (and maintained as such in the dynamic model) if and only if it has previously been observed as such.

d)   The maximum number of objects per frame is limited to say 10 (for time being).

e)   Standard hardware and software should suffice for implementation, and time of execution should be less than one frame interval.

In this system the dynamic model is built from the feature vectors received from the preprocessing system, combined with the dynamic information obtained by the token matcher (velocity and acceleration). Based on this model the next frame is predicted currently using a second-order Taylor expansion, and the actual matching is performed using probabilistic relaxation [9]. The matching result is used for update of the predicted frame model and for the calculation of dynamic information updating the dynamic model.

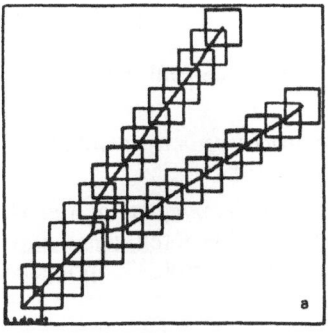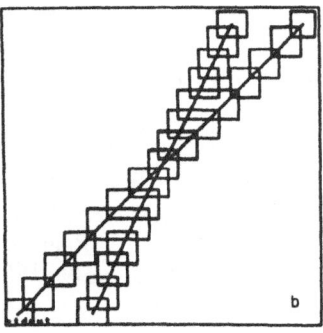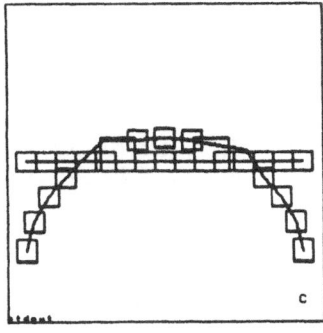

Figure 5   *These image sequences are illustrated with superimposed input images and trajectories showing system interpretation of the motion patterns. Objects in the input images are represented by their rectangles*

This approach can handle occlusions, and similar ambiguous situations. To illustrate this, a number of test sequences are reproduced in Figure 5. In Figure 5a two objects occlude initially, but as the sequence progresses the objects move apart. It is seen how the system replaces the single trajectory representing the occluding objects with the two separate ones as soon as the objects split. In Figure 5b two objects are separate initially, occlude in the middle, but separate again in the latter part of the sequence. From the traces of the model centroids we see how the system maintains the separate trajectories during the occlusion. In Figure 5c a similar situation is illustrated for one object performing straight line motion while the other moves in a semicircle. Here two occlusions occur, but the traces clearly show that the system copes. For more details see [8,9].

7 Summary

One of the prerequisites for dynamic robot vision is computationally efficient motion analysis. For the analysis for methods reported in the literature a framework which provides a break down of motion analysis techniques has been formulated.

We have discussed various alternative approaches to motion detection and analysis and found that by far the most work has been put into the motion detection type of problems. All methods rely heavily on integration into an image analysis system for pre- and/or postprocessing regarding spatial information.

Image differencing and the low and high level token matching approaches were the only methods presently considered realistic for real time operation. Aspects of higher level motion analysis were discussed, and the essence put into a scheme for a *model driven* approach to real time operation.

A high level token matching version of the model driven scheme has been implemented, and its basic performance was demonstrated on data with occlusions. But under conditions of a constrained scenario, real time motion analysis for robot vision seems feasible without the need for very sophisticated hardware.

References

[1]   K.J. Aaström, B. Wittenmark: *Computer controlled systems,* Prentice-Hall, Englewood Cliffs, NJ, 1984

[2]   J.K. Aggarwal, L.S. Davis, W.N. Martin: *Correspondance processes in dynamic scene analysis.* IEEE Proceedings, Vol. 69, No. 5, pp. 562-571, May 1981

[3]   J. Aggarwal: *On the computation of motion from a sequence on monocular or stereo images, - an overview.* In: A.K.C. Wong, A. Pugh (eds.): Machine intelligence and knowledge engineering for robotics applications NATO ASI Series F, Vol. 33, Springer-Verlag, Berlin-Heidelberg-New York, 1987

[4]   R.L. Andersson: *Real time gray scale video processing using a moment generating chip,* IEEE Robotics and Automation, Vol. RA-1, No. 2, June 1985

[5]   J. Arnspang: *Determining optic flow* Ph.D. thesis, Dept. of Computer Science, Copenhagen University, Denmark, January 1988

[6]   B.F. Buxton, H. Buxton, B.K. Stephenson: *Parallel computations of optic flow in early image processing* IEEE Proceedings, Vol. 131, No. 6, pp. 593-597, Oct. 1984

[7]   C.K. Chow, J.K. Aggarwal: *Computer analysis of planar curvelinear moving images* IEEE Trans on Comput., Vol. C-26, pp. 179-185, February 1977

[8]   H.I. Christensen: *Monitoring moving objects in real-time* M.Sc. Thesis, Inst. of Electronic Systems, Aalborg University, June 1987

[9]   H.I. Christensen, E. Granum: *On token matching in real-time motion analysis.* In: Proc. BPRA Pattern Recognition, 4th Int. Conf., Cambridge, Mar. 1988. J. Kittler (ed.), Lecture Notes in Computer Science, Vol. 301, Springer-Verlag, Berlin-Heidelberg-New York, pp. 448-457, 1988a

[10] J. Ciccotelli, M. Dufaut, R. Husson: *Control of tracking systems by image correlation.* Robotics, Vol. 5, pp. 201-206, 1987

[11] J.L. Crowley, P. Stelmaszyk, C. Discours: *Measuring image flow by tracking edgelines,* 2nd ICCV, Florida, Dec. 1988

[12] O.D. Faugeras, F. Lustman, G. Toscani: *Motion and structure from motion from point and line matches,* Proc. 1st ICCV, London, pp. 35-54, June 1987

[13] E. Granum, K.H. Munk: *Monitoring moving objects in real-time,* BAG-memo 8607, Inst. of Electronic Systems, Aalborg University, March 1986

[14] D.J. Heeger: *Optical flow from spatio-temporal filters,* Proc. 1st Int. Conf. on Computer Vision, IEEE, pp. 181-190, London, June 1987

[15] B.K.P. Horn: *Robot vision,* McGraw Hill, New York, 1986

[16] B.P.K. Horn, B.G, Schunck: *Determining optical flow.* Artificial Intelligence, Vol. 17, pp. 185-204, Aug. 1981

[17] T.S. Huang (ed.), *Image sequence analysis,* Springer Verlag, Berlin-Heidelberg-New York, 1981

[18] T.S. Huang (ed.), *Image sequence processing and dynamic scene analysis,* Springer Verlag, Berlin-Heidelberg-New York, 1983

[19] R. Jain, H.H. Nagel: *On analysis of accumulative difference pictures from image sequences of real world scenes,* IEEE Trans. on PAMI. 1, pp. 206-213, 1979

[20] B. Jähne: *Image sequence analysis of complex physical objects: nonlinear small scale water surface waves,* Proc. 1st ICCV, London, pp. 191-200, June 1987

[21] M. Jenkin: *Tracking three dimensional moving light displays*. Proc. Workshop Motion Representation Contr., Toronto, Ont., Canada, pp. 66-70, 1983

[22] M.K. Leung, Y-H. Yang: *A region based approach for human body motion analysis*, Pattern Recognition, Vol. 20, No. 3, pp. 321-339, 1987

[23] W.N. Martin, J.K. Aggarwal: *Survey, dynamic scene analysis*, Computer Graphics and Image Processing 7, pp. 356-374 (1978)

[24] H.H. Nagel: *Image sequence analysis: what can we learn from applications?:* In: T.S. Huang (ed.): Image Sequence Analysis Springer Verlag, New York, pp. 19-228, 1981.

[25] H.H. Nagel: *Overview on image sequence analysis.* In: T.S. Huang (ed.): Image Sequence Processing & Dynamic Scene Analysis Berlin-Heidelberg-New York, NATO ASI serie F. Vol. 2, pp. 2-39, 1983

[26] J. O'Rourke, N.L. Badler: *Model-based Iimage analysis of human motion using constraint propagation*, IEEE Trans. on PAMI, Vol. PAMI-2, No. 6, Nov. 1980

[27] R.F. Rashid: *Towards a system for the interpretation of moving light displays*, IEEE Trans. on PAMI, Vol. PAMI-2, No. 6, Nov. 1980

[28] J.W. Roach, J.K. Aggarwal: *Determining the movement of objects from a sequence of images*, IEEE Trans. on PAMI, Vol. 2, No. 6, November, 1980

[29] I.K. Sethi, R. Jain: *Finding trajectories of feature points in a monocular image sequence*, IEEE Trans. on PAMI, Vol. 9, No. 1, pp. 56-73, jan. 1987

[30] G.A. Shippey, R.J.H, Bailey, A.S.J. Farrow, D.R. Rutovitz, J.H. Tucker: *A fast interval processor*, Pattern Recognition, Vol. 14, Nos. 1-6, pp. 345-365, 1981

[31] J.K. Tsotsos, J. Mylopoulos, H. Dominic Covvey, S.W. Zucker: *A framework for visual motion understanding*, IEEE Trans. on PAMI, Vol. PAMI-2, No. 6, Nov. 1980

[32] S. Ullman: *The interpretation of visual motion*, MIT Press, Massachusetts, 1979

[33] A. Verri, T. Poggio: *Against quantitative optical flow*, Proc. 1st ICCV, London, pp. 171-180, June 1987

[34] A.M. Waxman, B. Kamgar-Parsi, M. Subbarao: *Closed-form solutions to image flow equations*, Proc. 1st ICCV, London, pp. 25-34, June 1987

[35] A.M. Waxman, J.J. LeMoigne, L.S. Davis, B. Srinivasan, T.R. Kushner, E. Liang, T. Siddalingaiah: *A visual navigation system for autonomous land vehicles*, IEEE Journal of Robotics and Automation, Vo. RA-3, No. 2, april 1987

[36] A.M. Waxman, K. Wohn: *Contour evolution, neighbourhood deformation and global image flow: planar surfaces in motion*, Int. Journal Robotics Research 4., No. 3, pp. 95-108, 1985

[37] J. Wiklund, G. Granlund: *Image sequence analysis for object tracking.* Proc. 5th Scand. Conf. on Image Analysis, Stockholm, June 1987

[38] S. Yalamanchili, W.N. Martin, J.K. Aggarwal: *Extraction of moving object descriptions via differencing.* Computer Graphics and Image Processing, Vol. 18, pp. 188-201, 1982

2D and 3D IMAGE SENSORS

Serge MONCHAUD
National Institute of Applied Sciences
Applications laboratory of advanced electronics techniques
I.N.S.A. - L.A.T.E.A.
35043 Rennes Cedex (France)

ABSTRACT

The growing development of robots which do more and more complex work in unstructured environments makes a compact 2D and 3D vision system necessary 3D vision is either very useful or indispensable to resolve some problems connected with autonomous movements. These systems are impeded by the absence of real 3D sensors collecting panoramic range data at medium distance (O to 10 meters) in a large volume (up to 100 m^3). We describe a certain number of solutions to this general problem, first we describe a 3D laser triangulation system implemented for a mobile robot. This system is capable of panoramic vision over a full 360˚ range around the robot. Numerous sorts of range finders (optical and acoustical) are connected to a 2D machine vision (AVISO ITMI SYSTEM) using various types of cameras (vidicon, CCD,PSD). In this case, we can use all the software package developed for this 2D machine vision (CAIMAN or CALIFE).

To add the intelligence needed for a reliable measurement system, a calibration procedure has been designed. This system could be tested in various fields of application:
- arm transfer in the neighborhood of an object to be manipulated, trajectory planning, and adaptative positioning, are typical tasks for an intelligent robot system.
- absolute location of a mobile robot in a crowded surrounding,
- audiovisual field for incrusting objects or actors in a 3D synthetic picture.

In some cases, an optoelectronic remote tracking and measuring system must be developed to track the location of a moving target (in three dimensions).

KEYWORDS : Robotic sensors - Vision and localisation.

NATO ASI Series, Vol. F 63
Traditional and Non-Traditional Robotic Sensors
Edited by T. C. Henderson
© Springer-Verlag Berlin Heidelberg 1990

As with many new techniques, artificial vision has encountered many failures. In fact, there are people who still do not believe that the technique is now operational. What is certain is that the choice of a vision system is a difficult problem. First of all, know-how in the development and the applications plays an important role. The first vision systems were installed in industry in the early 1980s. The great progress made in both electronics and computing helped these systems out of the research laboratories.

Since then, much has been written on the subject of artificial vision, but there is little description of real industrial applications. Recently in other techniques (robotics for example) the first users tried to reach the extreme limits of the possibilities of their equipment, with the idea that if a system succeeded in complex situations, it would be able to solve simple ones. This approach spread rapidly, due partly to the over-optimistic attitude of some constructors, but met with little success [1]. According to some statistics, made by the Vision Club and Robotic Axes in France, the vision market over the past 3 years has progressed by approximatly 30% per annum. The last data published gives the number of vision systems in France, at the end of 1987, there were 549 units with 178 installed in 1987 itself, with a increase of 28% in applications. In France, the first industrial sector interested by vision is still the electronics industry which has installed 19% of the systems ; the car industry follows with 13% of the applications. An important figure concerns the use of vision. It is undoubtedly inspection control which occupies the fore front with more than 83% of installed equipment. In 1987, the vision market represented about 50MF, in 1986 it was estimated at 35MF,which gives an increase of more than 40%. For 1988, the forecasts estimate 80MF for vision alne excluding teaching and specific developments. When these applications are included the market for artificial vision attains the figure of 80- 100MF for the year alone. This data shows that artificial vision is beginning to penetrate industries. It is very interesting to analyse the position of this technique lessthan 10 years after its introduction. In fact while some technical problems remain, most of the difficulties met at the beginning have now been solved, or are about to be. It is difficult to examine the rapid evolution of artificial vision without some analysis of the techniques. In the beginning, computer vision was developed to copy the human sense of sight. This eye is present throughout the cycle of the product processing to verify the aspect of the parts (quality control), but also to count the different elements (quantity control) or to locate them in space (localisation with command of the robot), etc. The advantage of the human eye is that it allows the different visualisations and other work to be carried out simultaneously. The machine, which never tires, is superior to the human. What is more, some measurement controls are impossible for the naked eye.

For correct processing, a vision system must have two parts, one for perception, the other for understanding decision.

Until the last few years, computer vision research was mainly on the 2D techniques rather than three dimensional (3D) computer vision. However in many applications it has proved impossible to successfully solve the specific problems using only 2D techniques. The determination of 3D shapes and object surface profiles is a highly important subject for a wide range of applications (thir generation robots, CAD, picture synthesis and picture analysis, pictorial data base...). Consequently, the evaluation of two dimensional pictures of the object appears inadequate in the face of real questions. Moreover the 3D shape of objects does not vary with changes of light or when the object is dirty. That is why it is fruitful to look for efficient 3D sensing systems [2].

PROBLEMS CONNECTED WITH THE DATA QUANTITY AND PERIVISION

One of the principal problems in vision is the quantity of data for a picture. By taking a picture window of 512 x 512 pixels, approximately 263,000 bytes of memory are necessary ; if the coding of pixels is made on 16 to 256 grey levels, this quantity must be multiplied by this factor. In fact, in the majority of cases, the final objective is to decide if the part is to be accepted or rejected (binary decision). This large amount of data results in very long treatment times, even with a big computer, and considerable memory space is necessary. The objective is to reduce the information, keeping only what is pertinent. Many vision systems now used depend on a reduction of information volume ; by discretisation we can make a contraction to 8192 bytes of memory. The software treatment of these contours does not raise insuperable problems. Simple but effective algorithms are able, in only one picture acquisition, to calculate many properties for each contour (gravity center, area, perimeter, number of holes, etc.) - see figure 1.

The first vision systems installed in industry corresponded to the above but the biggest difficulties had not yet been explored. The biggest one is the perivision problem.

Perivision problems are of two different kinds : on one hand the difficulties linked to optics and lighting, on the other the lack of communication. The lighting of the place, associated to optics, is a capital point for application success. The cost of good lighting, associated to correct optics, is high, it can reach approximately 20% of the total cost. Until about 1985 this aspect was often neglected since it was thought that good filtering algorithms could restore variations in lighting. However, this is not so, and in fact in more and more new applications we find systems with constant lighting using, for example, a complete cover.

Or communication problem is complex also. For a long time, the vision systems coming from the research centers were intended to be used as independent tools having no connection with the industrial environment. There were no exchanges between the vision organ and the other surrounding instruments such as programming controllers. Now the links with the surroundings have been increased, so vision begins to be a tool for industrial applications. The third aspect is the fact that researchers had developed a reliable product, but one which was difficult to program for the first industrial teams. The man machine interface were very bad. In order to market their product, the first companies have made a considerable effort so that the input of necessary data for the vision system can be done by the use of available interactive menus which may be programmed in a very simple fashion, for example using the "icons" proposed by the Apple Computers, Inc.

THE CHOICE OF THE CAMERAS

In computer vision, traditionally we speak of two types of camera : cameras with Vidicon tubes and semiconductors cameras (or, solid-state cameras). These solid-state cameras came onto the market few years ago. There is a mosaic of photoelements (pictures elements = pixels). Despite a loss of resolution (not enough pixels) and a higher cost, for the moment, they are more robust, smaller in size, and, most important there is little geometric distortion. In industrial vision, with the same sensors, we can reach an accurate measurement by techniques named "sub-pixels" [3]. Let us consider, for instance, an object measuring 50 mm, the matrix camera resolution, presently available, is limited to approximaly 500 points. The user cannot hope to make different dimension controls on these objects or position them with a precision above ± 0.1 mm. Sub-pixel technology, using the same sensor, attains over 0.01 mm ; it is consequently 10 times more precise. This use of sub-pixel techniques endows industrial vision with a large number of new potential applications. The control "go, no go" of a product could be replaced by an fine analysis of the dimensions and their evolution and stop the industrial process before the product is declared worthless. Instead of using several cameras for the inspection of a large product in order to obtain the necessary resolution only one camera would be needed, thus reducing the number of elements exposed, and the maintenance difficulties. Programming is taught. The operator adapts the position of the camera until the picture of the part is very clear. With the help of an optical pencil, he limits some zones of the part on which he hopes to make the different controls. Next he asks for some specialised tools in the library of the system and, using the optical pencil, positions them on the spots he wants to measure or inspect in the zones previously defined. The system calculates the measurements and memorizes them. The operator must now give the precision required on these measurements, to accept or reject a part in inspection during the production process. The calculations are made on a specialised computer provided by the company. This computer can pilot up to 4 cameras with 128 input/output (for programming controllers) and two RS 232 interfaces. This system MVP 1000 from ITRAM Factory systems is little affected by lightvariations and the positions of the part to be inspected could be modified with reference to the position put in the memory, within limits compatible with various industrial processes. After many years of research and development and with the appearance of new components, , FFT engines, the ITRAM engineers have managed to find a real and unsimplified solution to this complicated function.

Many manufactures now offer dedicated boards compatible with personal computers. The complete list is impossible to give ; we may mention : MATROX, DIGITAL DESIGN, I2S, EIA, etc. Each one presents its product in a hardware and software package:
- MATROX with the MVP-AT systems - DIGITAL DESIGN with the VISIONIX System loaded by a PC with the development system of the vision application - I2S with the PC Scope system - EIA with the SUPERCAM incorporating the software GIPS VISION connected with a PC (or SUN Station) used as a development tool. In all these cases the picture functions incorporate real-time ALU image operations, and flicker-free non-interlaced display output. Its performs convolutions, histograms, averaging, subtraction and many other operations at speeds which until now have only been available in much more expensive designs. An optional neighborhood processor module provides pattern matching and morphological transformations. This goes to show that considerable effort is being made to widen the market of industrial vision by means of unsophisticated personal computer support.However two fundamental questions remain unanswered :
- The world around us is a 3D world, modern machine vision should have 3D information inputs.
- The world around us has a 360° field of vision, modern machine vision should have panoramic vision capacity.
A compact 2D and 3D vision system is now an absolute necessity. 3D vision is either very useful or indispensable to resolve the problems connected with autonomous movement. These systems are impeded by the absence of real 3D sensors collecting panoramic range data at medium distance, in a large volume, and panoramic vision. Now we will describe a certain number of solutions to this general problem.

3D PANORAMIC VISION SYSTEM BY THE COUPLING OF RANGE-FINDING TECHNIQUES

Our work deals with the study of sensors, which we will consider as intermediate between high level vision and local perception. Among the numerous applications covered by this type of perception system, we have considered its implementation on a mobile robot. The specific data of this problem have led us to couple complementary devices of perception on a rotary plate.

- realization of the 3D laser rangefinder laser deflection is obtained by an electromechanical device composed of two orthogonal swivel mirrors and one mirror set at 45 degrees. From an electrical two galvanometers the driven by two DC voltages, the servo-control position is achieved by two interfaces boards. This system allows an azimuthal and zenithal deflexion of 40 degrees and its bandwidth is approximately 150 kHz. One disadvantage of this kind of deflexion is that the azimuthal trajectory is hyperbolic. Nevertheless, its wide field of vision makes panoramic vision possible.
The electronic interface is made up of two position servo-control boards belonging to the mirror and two command boards performing the digital analog conversion. The deflexion of the laser beam can be fully programmed due to the connection of these interfaces to a microprocessor board.
- cameras
By using a CCD camera and a position sensor (PIN technology). It should be noted that in robotics the lens gives an even better performance in that :
- the focal length is minimal in order to have the greatest possible depth of field
- the lens is as bright as possible.
Precision and acquisition times of these systems for various panoramas of different scenes have been given in former studies [4,5]. To our knowledge at least, four other research groups in the world are trying to solve the problem of panoramic three-dimensional range sensors.
In CANADA (CNRC OTTAWA) in the division of Electrical Engineering, in the first attempt to evaluate the real performances of the 3D camera, the range sensor was rotated around a vertical axis to obtain full panoramic (360°) range and intensity images of a room. The range varied was between 0.5 and 4 m. Figure 3 shows the actual experimental set-up used to investigate the performance of the range sensor. A pulnix model TM 540 CCD camera was utilized. The focal length of the camera lens is f = 17mm with dn number = 2. The laser-beam projector consists of a LT 024 MF laser diode (780 nm), a Sharp collimating lens and a small galvanometer to quickly scan (faster than the integration time of the CCD) the laser beam along a vertical line. Only one laser line was projected. The whole optical set-up was mounted on a motor-driven rotary positioning stage, to provide the 360° scan [6].
In the USA, five years ago, a division of the naval research laboratory artificial intelligence center illustrated a system-particular way of moving the elements to implement a very wide field scanning rangefinder [7].
In France in the LAAS of Toulouse the robotic research group have built a rotating platform supporting both a laser (time-flight principle) and a 2D CCD camera. Some results have been presented on parts of panoramic range pictures.
Now we can say that the problem of collecting a complete panoramic range picture around a robot is on the way to being solved but complex problems still remain when we try to control the navigation for autonomous vehicles and mobile robots. In fact this control requires the integration of different perception systems. All these previous systems of localisation

use preloaded information and absolute referencing systems, which is a very
limited solution because it is necessary for the vehicle to evolve in a
static, highly structured environment. That is why we now present our new
3D localisation sensor for industrial measurement. This instrument is a
remote 3D measurement system with autonomous infrared emitter for moving
robots.

A NEW 3D LOCALISATION SENSOR FOR INDUSTRIAL ENVIRONMENT

Our 3D localisation and tracking system is basically composed of two
PSD motorized cameras coupled with a pulse modulated infrared emitter. Each
camera possesses three degrees of freedom (two rotations and one
translation) in order to trace automatically the 3D trajectory of the
moving target. Object motions up to 2.5 m/s can be tracked in a volume of
about 100 m^3. This 3D measurement system has the following capabilities :
both real time localisation and tracking of the moving robot, flexibility
of use (2D or 3D with fixed or mobile cameras), possibility of using
several cameras coupled to the same autonomous infrared emitter fixed on
the robot.
Precision and acquisition times of these system for various panoramas of
different scenes have been given in former studies [8].

CALIBRATION AND REAL TIME CORRECTION OF THREE DIMENSIONAL MEASUREMENTS

Much work has been devoted to this critical problem [6,8,9,10].
Sensor calibration consists of identifying the transformation between the
image referential and the mesure referential and in determining the
reference plane equation in one of these referentials. Several methods have
been proposed to solve this problem. They fall into two categories, which
can be distinguished by the way the measurement referential is defined. In
the first category, origin and axes of the measurement referential are
derived from the rangefinder geometric components : optical axis,
photosensitive surface, height plane, etc. Computation of 3D coordinates
therefore requires relations using the rangefinder's internal geometrical
characteristics : angle between optical axis and light plane, distance
between photosensitive surface and optical center and magnification factor
on the two axes of the image referential. From a practical view point, this
leads either to building a sensor with very well defined characteristics,
or to identifying these characteristics once the sensor hase been
constructed. We have been working on this basis. In the second category,
the measurement referential is concretized in a standard reference when
calibrating regardless of how this referential is positioned and oriented
relative to the rangefinder's geometric elements. This approach enables us
to define the referential so that simple numerical values are obtained in
automatic operation. Above all, calibration requires no external optical
devices unskilled operators can manage it [9,10].

CONCLUSIONS

2D and 3D image sensors form a subject too vast to be totally
illustrated and detailed in few pages. My topic is completely different.
First of all, I hope to eliminate some mistakes, for example : that 3D
vision is more difficult than 2D vision or that panoramic vision is more
complex than vision within a reduced field. The present state of things
shows that it is impossible to be conclusive on these subjects. What is
certain is that it is necessary for researchers in vision exchange their
observations and their results as quickly as possible in order to permit

progress in these techniques and to avoid trying to solve insoluble problems.

Another very simple idea is that when some results are obtained in one place we try to use them, as soon as possible in surrounding techniques. An example will illustrate more clearly. Now we have industrial products that run well in 2D vision. Special software packays for image processing run on various computers (VAX, SUN,etc.) Some of them run on PCs. For example in France the firm CAP SOGETI - (ITMI/LIFIA in Grenoble) offers the software package CAIMAN, in the LAAS of Toulouse the software CALIFE runs on SUN MACHINE. A Japanese compagny offers a complete 2D software package named PCNFS. On the other hand, the applications of robot 3D vision have no potential limits. Now a variety of techniques have been proposed to obtain 3D informations. None of these techniques is "best in general" as each has its own drawbacks.

Finally, not enough has been done to facilitate the programming of vision tasks. This situation will get worse as the number of techniques at all levels increases. If we are able to develop efficient 3D vision techniques but are unable to simplify the programming of their applications, our efforts may fail because the cost of vision task programming may be prohibitive [2].

In our opinion, to save both time and money it is necessary to employ, whenever possible, all the developments made for 2D vision and also introduce the complementary elements of 3D vision when it is necessary. For example a depth map given by our multisensory rotating platform assembls, the contour picture given by a 2D vision machine. So we can use all the software routines which are already employed in 2D vision (scene analysis) in the case of the range map analysis.

It is for this reason that we built our workstation (2D vision system with rangefinding capabilities) around a 2D passive machine vision connected to:

  - various cameras (VIDICON and CCD types),
  - various range finders :
    . a panoramic multisensor platform,
    . a 3D vision system by range finding,
    . a 3D automatic target tracking system.

This complete experimental system, the only one of its kind in France, allows a real exploration of artificial vision by computer. At least, we hope so!

ACKNOWLEDGMENT

We wish to thank Mme Ann Cochennec for her collaboration on the english version of this paper.

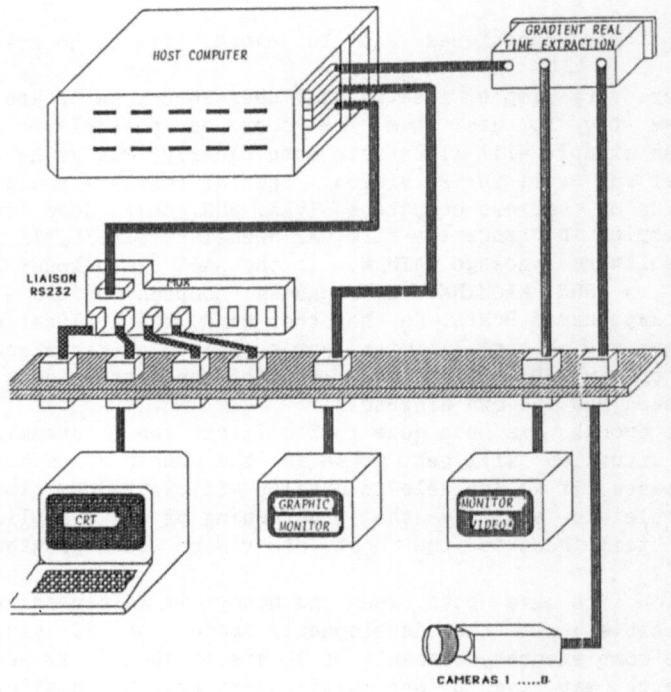

*Figure 1 : Aviso System (ITMI) Gradient Real Time extraction
(technical documentation)*

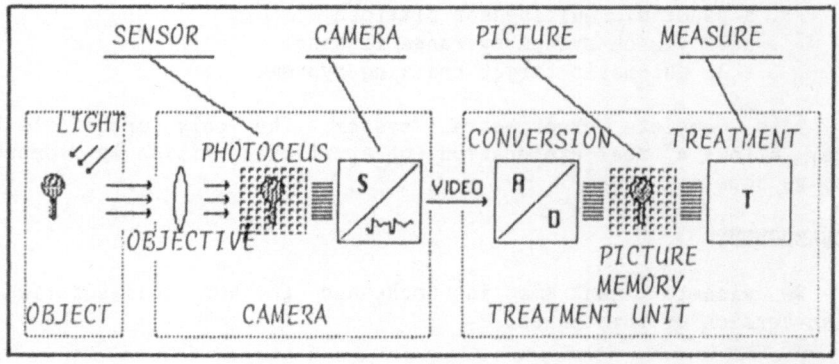

Figure 2 : Resolution in industrial vision [3]

Figure 3 : A wide-field scanning triangulation rangefinder for machine vision [7]

Figure 4 : A new 3D localisation sensor for industrial measurement [8]

REFERENCES

1. Segaf H. : Vision industrielle choisissez d'abord le savoir faire. Mesures numéro 602 27/02/89 - ISSN 0755-219X, pp. 80-84

2. Nitzan D. : Three dimensional vision structure for robot applications. IEEE transactions on pattern analysis and machine intelligence vol 10 n° 3 may 1988, pp. 291 - 309

3. Lelièvre D. : Vision industrielle, à capteur égal mesure plus précise, Mesures n° 13, 19.10.87, ISSN 075J-219X, pp. 27-35

4. Monchaud S. : Contribution to range finding techniques for third generation robots. Intelligent Autonomous Systems Conference Amsterdam 8.11, December 1986, pp. 459-469

5. Noureddine F. : Système de vision panoramique 3D par télémétrie. Doctorat INSA Rennes (France)

6. Blais F., Rioux M., Domey J. : Compact three dimensional camera for robot and vehicle guidance (paper submitted to Optics ad Lasers in Engineering Review).227-229.

7. Pipitone J. F Marchall G. T. : A wide field scanning triangulation rangefinder for machine vision. International Journal of Robotics Research, vol 2, n° 1, 1983, pp. 39-49

8. Shi J., Monchaud S., Prat R. Domey J. : 3D object localisation and automatic tracking, Rovisec 7. Advanced Sensor Technology 2-4 february 1988, Zürich Switzerland, pp. 197-207

9. Bumbaca F., Blais F., Rioux M. : Real time correction of three-dimensional non linearities for a laser range finder. Optical Engineering, April 1986/vol 25, n° 4, pp. 561-565

10. Faure J.M. Yang Y.: Calibrating a 3D visual sensor : a factory-oriented approach. Rovisec 7. Advanced sensor technology 2-4 February 1988, Zürich, Switzerland, pp. 21-25

Tactile Sensing Using an Optical Transduction Method

Howard R. Nicholls,
AI & Robotics Research Group,
Department of Computer Science,
University College of Wales Aberystwyth,
Penglais,
Aberystwyth,
Dyfed SY23 3BZ,
United Kingdom

Abstract

This paper describes the characteristics of a very high resolution tactile sensor for robotics. The sensor uses a pressure intensity to light intensity transduction method that produces very detailed images of contact. A robot finger-mounted sensor incorporating this method has been built which produces a video output suitable for feeding into a conventional image processing system. The principle and design of the tactile sensor is explained, the main parameters affecting its performance are identified, and experiments using this sensor with an image processing system are described. An example application, using the sensor for grasp error compensation and for inspection, is reported on. Finally, some attributes of tactile sensing are discussed.

1 Introduction

Tactile sensors determine various aspects of the environment through direct contact with objects within that environment. Such sensors can be used to ascertain the distribution of forces over an area, and can also be used to sense shape parameters, location, and orientation of objects touching the sensor. The simplest form of tactile sensor consists of a single touch probe, but effort has also been put into developing sensors that consist of an orthogonal grid, or array, of sensing elements. Each element is known as a **tactel**, a term derived from 'tactile element'. Such sensors produce as output a digital image that can be input to a conventional image processing system, as might be used for analysing images from a vision camera.

Until recently, most sensors have had fairly low spatial resolution (e.g. 16 x 16 elements), compared to that of vision sensors. As is shown in a recent survey,[1] interest in tactile sensing is increasing, although the technology is still in an immature state. Few examples of tactile sensing

NATO ASI Series, Vol. F 63
Traditional and Non-Traditional Robotic Sensors
Edited by T. C. Henderson
© Springer-Verlag Berlin Heidelberg 1990

are to be found on the factory floor, partly because of the expense and low resolution of many sensors.

This paper describes the operation of a very high resolution tactile sensor array; it is able to detect detailed surface variations which are beyond the capabilities of most other tactile sensors. The operational characteristics of the sensor are identified, and an example application is described to show the practical use of this device.

2 The Optical Transduction Method

The transduction method used in the sensor developed at the University College of Wales, Aberystwyth uses an optical phenomenon to form and direct the tactile image onto a camera imaging system.

The tactile sensor uses the properties of optical reflection between media of different refractive index. Consider light travelling from a medium A of refractive index n_A to a medium B of refractive index n_B. Total internal reflection occurs if $n_A > n_B$. Conversely, diffuse relection occurs if $n_A < n_B$. The sensor is designed to exploit this phenomenon.

The transducing structure is composed of a clear plate, a light source, and a compliant membrane stretched above, but not in close contact with, the plate (see figure 1).

Let n_0, n_1, and n_2 be the refactive indexes of air, plate, and the membrane respectively, where $n_0 < n_1 < n_2$. Light is directed along an edge of the plate, so that it shines into the plate body. In the unloaded state, the plate is surrounded by air. Since $n_1 > n_0$ the light is totally internally reflected inside the plate. Thus, very little of the light entering along the edge of the plate actually leaves it, and so the lower surface of the clear plate, which is the imaging area, appears dark.

However, when the membrane is brought into contact with the plate, as would be the case when an object is placed on the sensor surface, diffuse reflection occurs on the plate-membrane boundary. This is because $n_1 < n_2$, the reverse of the original situation. Light thus passes out through the lower surface of the plate at the point where the membrane has contact. This can be directed onto an imaging device such as a CCD array. The image will have a bright patch corresponding in position and size to the part of the object in contact with the sensor surface.

The intensity of the bright patch is proportional to the magnitude of the pressure between object and plate. Consider the area of plate viewed by a single pixel of the imaging camera: the amount of light reaching the pixel depends on the number of elementary contact points between the clear plate and membrane microsurface. As relative pressure between plate and object increases, more points on the membrane microsurface come into contact with the plate and so single pixel intensity increases with pressure. This principle applies to the surface as a whole. Examples of images obtained in this way are shown in figure 2, and also in Mott, Lee, and Nicholls.[2]

This explanation is corroborated by evidence obtained experimentally, and also by the theory of plastic contact of two surfaces. Betts et al. give a good experimental analysis of a similar design used in medical applications, and show that the pressure-intensity relationship is linear for the materials which they use.[3] They do not give the name of the material which they use, however. The microtexture explanation of the increase in light intensity for applied pressure relates well with earlier work carried out in studies of contact between two rough surfaces of hard materials. See papers by Pullen and Williamson,[4] and Childs[5] for further details on this subject.

The explanation of the sensor's behaviour above requires a clear medium for transmitting the internally reflected light. In practice, acrylic, glass, and perspex have been found to work successfully.

3 Implementation of the Optical Transduction Method

3.1 Introduction

A gripper-mounted tactile sensor using this optical transduction method was designed and constructed, based on a modified commercial CCD camera system. The sensor was built inside a finger that bolts on to a conventional parallel jaw robot gripper. The other gripper finger is a passive unit. The sensing area forms part of the grasp region of the gripper, so that anything held by the gripper will be in contact with the tactile sensor.

The sensing area is formed by an acrylic window in the finger body, covered by a compliant membrane, and edge illuminated by three red light emitting diodes. The tactile image from the sensing area is directed via a mirror and prism onto a CCD array inside the finger (figure 3), to provide a very high resolution image. This arrangement makes the sensing finger very compact, and fairly light (240g) since the finger body is made from machined aluminium.

A single cable connects the finger to an electronics module, providing d.c. power and drive signals. The electronics module is the camera body of an off-the-shelf CCD camera system. The lens and CCD array were removed from the camera body, and replaced by a 15-way D-connector into which plugs the lead from the tactile sensing finger. This module drives the CCD array and converts the output from it into an analogue video output. The CCD array was remounted inside the finger body, such that it collects the image from the underside of the acrylic plate. A photograph of the completed sensor mounted on a parallel action pneumatic gripper is provided in figure 4.

The number of elements on the CCD array is 384 horizontal by 491 vertical elements, matched to a sensing window size of 17.6 x 13.2mm. The area represented by a single tactile element (tactel) is $0.0012mm^2$.

Some experiments to characterise the performance of this sensor are described below.

4 Experiments Using the Tactile Sensor

4.1 Introduction

This section describes a set of experiments carried out using a single tactile finger. The video output of the sensor was connected to an 8 bit, 512x512 pixel framestore that is accessed by a minicomputer. An image processing package was used to extract relevant information from the tactile image.

The experiments carried out were designed to investigate various properties of the sensor, including the nature of the image background, the response to single and multiple impulses, and the the sensor's response to increasing applied forces. The most critical factor which affects the sensor's performance is the compliant membrane; consequently, each experiment was repeated using various samples of rubber for the membrane.

The samples of rubber used were:-

- latex red, 0.18-0.25mm (7-10 thou) thick
- latex red, 0.31-0.38mm (12-15 thou) thick
- latex red, 0.43-0.51mm (17-20 thou) thick
- latex jade green, 0.18-0.25mm (7-10 thou) thick
- latex vibrant-red, 0.31-0.38mm (12-15 thou) thick
- latex black, 0.36-0.41mm (14-16 thou) thick
- white balloon rubber
- RAPRA grey, 1.02mm (40 thou) thick

All the latex rubbers are available in sheet form. The balloon rubber is also latex, and was a sample that gave promising performance after trying many different types of rubber early on in the project, and so was included for comparison. The RAPRA grey sample was supplied by the Rubber and Plastics Research Association, U.K., as a test piece.

The experiments are described below along with some relevant graphical results.

4.2 The Tactile Image Background

Experiments to examine the background intensities of a tactile image with no applied impulse were carried out in ambient office lighting with the sensing pad of the sensor facing towards the floor. A tactile image for each rubber sample was digitised, and the average intensity calculated. Figure 5 shows the average background intensity for each of the rubber samples, and figure 6 compares the standard deviation from the mean for each sample.

General experimentation showed that latex red, vibrant-red, and green are translucent, thus it is possible to get a dark 'shadow' on the image before contact has been made. This effect actually produces the reverse effect to that of proper contact between sensor and object. The 'shadow' is a dark region in the background, whilst contact areas show as bright regions, relative to the normal background intensity range. It is therefore easy to distinguish between the two effects, although it is desirable that the rubber samples are completely opaque. For each experiment, care was taken to ensure that ambient lighting remained constant for each of the samples to reduce spurious results arising from this phenomenon.

For the same colour, red, it was found that the background intensity of the image decreased with increasing rubber thickness. This is due to a reduction in translucency because of greater thickness. Green, white, and grey rubbers all had considerably lower background intensities than the red ones.

4.3 Response to a Single Impulse

To investigate the effect a single impulse has upon a tactile image, the tip of a needle was applied to the sensor with a force of 5g. Since the tip of the needle is larger than one tactel, a region of the tactile image was affected; the region was centred in a 30x30 window and the maximum intensity in this window found and recorded.

Vibrant-red and black rubbers did not respond at all to the impulse, nor indeed to any impulse, and so were left out of further experiments. The vibrant-red rubber is much stiffer than the other samples, and its failure to perform adequately is most likely due to the material's inability to extrude into close contact with the acrylic microsurface. The failure of the black rubber is probably due to it absorbing any diffusely reflected light from the acrylic, rather than reflecting it out of the lower plate surface, since it has a very dark, matt surface. A plot of the various responses is given in figure 7.

4.4 Dual Impulse Separation

The response of the tactile sensor to two closely separated impulses was examined. A pair of caliper jaws were applied evenly to the sensor surface and the tactile image digitised. This produced two regions within the tactile image until the caliper jaw separation could not be resolved by the sensor, at which point they merged into a single region.

As one might expect, it was found that the thinner rubbers were the most sensitive. Both 0.18mm (7thou) thick rubbers, red and green, could detect jaw separation down to 0.3mm. For distances less than this, the two regions merged together such as to be indistinguishable by the image processing algorithm.

At the other end of the scale, the 1.02mm (40thou) thick RAPRA rubber could only detect separations of greater than 1.1mm. The minimum separations detected for the different rubbers is shown in figure 8.

4.5 Response for Increasing Applied Force

To obtain an indication of the characteristics of the sensor, the image of a region produced by a single impulse applied under force was digitised. The maximum intensity within this region was calculated and recorded. This was done for a sequence of impulses, increasing in force. A graph was plotted to show the response of the sensor to applied forces. Each application of force to the sensor was independent from all the others, so hysteresis effects, creep, etc. did not affect the readings. The experiments were carried out for each of the usable rubber types, i.e. vibrant-red and black were excluded. The resultant graphs are shown in figure 9.

The graphs showing the rise in intensities for increasing applied force all have similar shape. They consist of a sharp rise for small force increments, then a tailing off to a plateau. Essentially, when considering the maximum response intensity, the sensor response is monotonic.

As previous experiments have shown, the thinner rubbers are the more sensitive. The green rubber could produce a response for as low a force as 1.19g. However, the largest force that this rubber could sense before cutoff was about 5g. The RAPRA grey sample produced the highest sensed force before cutoff, at about 25g. Saturation of an image occurs for fairly low applied forces, certainly less than that produced by a simple robot gripper.

5 Discussion

The implementation of the optical transduction method described above demonstrate its feasibility as a technique for robot tactile sensing. The critical component of a sensor using this transduction method is the rubber membrane. The behaviour of the membrane dictates the resolution and sensitivity of the sensor. Whilst other parameters, such as illumination level, shielding from stray light, etc., may affect the image, incorrect membrane selection can cause complete non-functioning of the sensor. The black latex rubber is an example of this.

The experiments with the sensor show that colour is a significant parameter in membrane functioning. Similarly made rubbers of the same thickness but different colours behave differently, as shown by the performance of latex red, vibrant-red, and black rubbers. There are several reasons for this. Firstly, the transduction method uses optical reflection, and so the reflective properties of the rubber are significant. One would expect a dull black rubber to behave poorly, since it has low surface reflectance, and the experiments showed that such a rubber was unsuitable for this type of tactile sensor.

In addition, rubbers of different colours have different molecular structures, although they may be of the same general type, such as latex. The molecular structure of a rubber influences its mechanical behaviour, determining factors such as the hardness, susceptibility to creep, elasticity, and so on. Stiffer rubbers are less able to achieve close contact with the acrylic microsurface than more flexible ones. They do not readily extrude into microcavities, and this affects the intensity of the tactile image. One would expect rubbers of different colours to behave slightly differently, since they have slightly different molecular structures, and this is supported by the empirical evidence. Red, green, black, and vibrant-red all had differing background intensities, and behaved in different ways.

Related to the above point is the fact that the amount of plasticiser in a rubber affects the viscoelastic properties of the material. Informal experimentation with a wide variety of samples showed that rubbers with a larger proportion of plasticiser were able to extrude onto the glass plate easily. This causes the saturation level for the sensor to be low, typically around 2g; i.e., forces above this do not cause an increase in image intensity.

General experimentation demonstrated that there is an inverse relationship between rubber thickness and sensitivity. The very thin rubbers are able to pick out very fine textures on sample objects, whilst this detail was lost on the thicker samples. Impulses as small as $0.124mm^2$ were detected using a 0.18mm thick (7 thou) rubber. The rubbers also have a small dynamic range, with the minimum distinguishable force being 1.19g, rising to a maximum of 25g between any of the samples. The response curve is monotonic, though not linear, reaching a plateau at the saturation force for the rubber.

Another parameter that is of significance, though not formally tested, is the illumination of the clear plate. Varying the illumination level alters the image intensity: decrease the illumination level and the image intensity also decreases, and vice versa. The wavelength of the illumination is also probably a determining factor of the sensor's output, particularly for sensor designs using CCD arrays. The tactile sensor uses three red l.e.d.s, chosen primarily because they have the highest illumination level for the requisite physical size. Red rubber also works well on this sensor. An examination of the spectral response curve for the CCD imaging array shows that its peak is in the infra-red part of the spectrum. The wavelength of the light relected off the red rubber membrane is near the most sensitive part of the CCD's response, since it is at the red end of the spectrum. It is reasonable to suppose that part of the explanation for the high intensity of the tactile images is due to compatibility between the illumination wavelengths, the reflectivity of the rubber, and the CCD spectral response. Formal experimentation will show if this is the case, but casual use of the sensor suggests it to be so.

The membrane texture also affects the sensor's behaviour. The latex rubbers are essentially smooth, without any intrinsic pattern or texture to them. They behave well but with small dynamic range. Some textured rubbers were tried, but the texture strongly influenced the output pattern and were considered unsuitable. An extremely fine textured rubber might give a wider dynamic range, and be less likely to stick to the clear plate, but there will always be the imposition of the rubber's texture onto the image. Some other investigators in this field, (e.g. Begej[6], Tanie et al.[7], and King et al.[8]) have tried textured rubbers, with varying degrees of success.

One reason for experimenting with textured rubber is to prevent it from sticking to the clear plate after the impulse has been removed. The author has rejected a number of rubber samples because of their propensity for sticking, but didn't find this to be a major problem with the latex rubbers providing they were cleaned first and the acrylic plate was free of any grease.

The practical experience gained from experimenting with this sensor and various rubber types lead to the production of the following requirements list for the properties of a suitable membrane material. The latex rubbers met some of these criteria, but are not the perfect solution. The requirements are in no particular order of priority, since they are all significant.

- The membrane must respond quickly to surface impulses (< 10ms).

- The membrane must recover quickly after the release of pressure on it (< 10ms).

- No marks or deposits must be left on the clear plate by the membrane.

- A monotonic, preferably linear, increase in elementary contact points between membrane and clear plate as pressure increases on the upper membrane surface is required.

- Small amounts of plasticiser in the membrane produce better results, since with large amounts the material extrudes onto the clear plate easily, producing saturation for low pressure.

- Normal ambient temperature ranges must not significantly affect the behaviour of the membrane.

- The membrane must be opaque, with the surface in contact with the clear plate being reflective.

- There should be minimal 'creep' effects from the viscoelastic properties of the membrane for contact times of several minutes.

- The membrane must be robust enough to remain undamaged under significant overload.

- The membrane must to be durable and not wear out after a few hours of operation, or a few hundred cycles.

Sensors based on the optical transduction method have promise, due to their high resolution and simplicity. They are best suited to detecting fine surface detail, rather than smooth planar surfaces, and hence suggest themselves for inspection and verification tasks, recognition, location evaluation, and so on. The advantages of this transduction method include its very high resolution, compatibility with vision sensing technology, freedom from electrical interference problems, low cabling requirements, and the facility for having the processing electronics remote from the sensor. However, the disadvantages are that it depends upon an elastomer, which may not be robust and can cause hysteresis, creep, etc.

6 An Example Application

To demonstrate the utility of mounting such a sensor on a robot gripper, an example robot task was designed and implemented. A brief description of the task is given below.

The task chosen was a simple pick and place operation. The component selected was the rear axle of a bicycle, which has a thread on the shaft. This component is one of several available with appropriate jigs in a robot workcell in the laboratory at Aberystwyth. The robot used was a Unimate PUMA, which had as its gripping system the tactile sensor and a pneumatic actuator.

The axle was positioned in a jig in a vertical plane. However, the centreline of the axle was deliberately inclined at an angle to the vertical within this plane. This angle is referred to as the angle of inclination. It was changed between cycles of the task, thus the axle was not presented to the robot in a known orientation.

The robot was programmed to move from a start location to a location above the axle in its jig. The robot then moved down to a location and orientation whereby the tactile sensor surface plane was parallel to the plane in which the axle was presented and such that the axle would be grasped by the gripper jaws when they were closed. The axle was then grasped and the resultant tactile image digitised. The centreline of the axle was therefore at an angle to the centreline of the tactile sensor, and also offset from the centre of the sensing region. The robot's tool centre point was defined to be the centre of the sensing region surface when the jaw separation was zero.

Various parameters were then extracted from the tactile image of the grasped axle:-

- The angle between the axle's centreline in the tactile image and the centreline of the image itself. If these lines are colinear, the resultant angle is zero. This angle corresponds to the angle of inclination.

- The offset from the centre of the image to the centreline of the axle. This is measured along the x-axis of the tactile image.

- The pitch of the thread. This was taken as the average distance, in tactels, between the centres of gravity of successive regions produced by successive thread crests.

- The detection of any missing parts of the thread. The successive distances were checked to see if any were outside a specified range, if so then the thread was assumed to be damaged.

The angle and offset parameters were then passed on to the robot controller which incorporated them in a tool transform. This enabled the orientation and position of the axle to be adjusted such that it was in a suitable state for the task to proceed. The robot then moved to another location, vertically above a hole in another jig. It then successfully inserted the axle into the hole provided in the jig. This cycle was repeated successfully for various axle presentations.

This simple task illustrates that a number of different functions can be performed based upon a single tactile image. Grasp monitoring and correction were achieved as well as inspection and measurement. The same image processing system has been used for analysing images from vision sensors, thus the tactile sensor integrates as a multi-functional item into a multi-sensor system.

7 Some Aspects of Tactile Sensing

The previous sections explained the operation of a specific sensor, and its use in an example application. This section discusses some aspects of tactile sensing which should be considered when choosing a particular sensing modality. The following attributes, originally identified by Overton,[9] and expanded upon here, characterise the main aspects of the low-level tactile domain. The entity which produces a non-background reading on a tactile sensor is referred to as a **stimulus**.

Explicit Three-Dimensional Information
The data from a tactile sensor can represent explicit three dimensional information about a contacting stimulus. The readings from a sensor can be interpreted in terms of force, but also in terms of a displacement normal to the sensor surface plane. For example, a linear potentiometer can serve as a force sensor if suitably calibrated, or alternatively can be used to measure linear displacement. An array of such sensors would enable a three-dimensional surface map to be built up. Single-camera vision systems do not produce direct three-dimensional data, the readings from them represent the projection of the solid object onto a two-dimensional plane. Any depth data must be inferred from image intensity variations, which may actually be caused by illumination variations, surface reflectance properties, etc. Visual depth data is not always reliable, and binocular stereo camera systems, for instance, may be required to determine accurate depth information.

No Specularity
Specularity is the occurrence of bright patches of light in the image because of direct reflection from the illuminating source. This can be a problem in vision systems, particularly if there are physical constraints on the positioning of the illumination sources. Tactile sensors do not have this problem, since they are not based on direct optical reflection from the object. The tactile analogue of visual specularity is perhaps electromagnetic interference from the surroundings, but this is only relevant to certain types of sensor hardware, and is not a fundamental tactile attribute.

No Shadows
Shadows in vision sensing can be a problem, since they will show as dark areas, maybe obscuring features of interest in the scene. Tactile sensors by their nature do not suffer from this problem, the intensity readings obtained being derived solely through contact between sensor and stimulus.

Direct Measurements of Stimulus
There is a direct mapping between stimulus shape and image, since non-background intensities are only present in the tactile image when there is actual contact between sensor and stimulus. This

applies wherever the tactile sensor may be in space. The visual sensing case is again different, because of the mapping from three dimensions to two dimensions. Although size parameters can be calculated from a visual image, the sensor's distance from the viewing surface must be known to enable the calculation of the physical area one pixel occupies. If the camera is moved to a new location, the system must be re-calibrated. Tactile measurements are sensor position independent, visual measurements are not.

No Perspective Distortions
Tactile measurements are direct, and do not introduce any distortion into the image. Vision sensing can have difficulties arising from distortion of the objects in the image due to perspective.

Ease of Separation of Stimuli from Background
The segmentation of features in a tactile image from the image background is a straightforward task. If there is no contact between sensor and any stimuli, then the background image intensity is at some uniform value. Any tactels having values different from this background level must therefore represent contact points. The task of segmenting the image into regions corresponding to areas of contact and regions representing no contact can be based upon a simple thresholding technique. The vision sensing case can be more complex, because of the specularity and shadow problems mentioned above, as well as illumination level variations, surface reflectance changes, and so on.

Tactile Occlusion is Different from Visual Occlusion
Occlusion in tactile sensing is a situation in which two or more stimuli are in contact with the sensor, and one stimulus is preventing part of at least one other stimulus from touching the sensor surface. The resultant tactile image will be composed of regions corresponding to the surfaces of several stimuli, but the image of one of the stimuli will be overlaid on the region of another stimulus. For example, if a very thin object, say a ring washer, is placed on a tactile sensor and then a large rectangular block is laid on the washer, pushing the washer into the sensor, a composite tactile image will result. This image will contain a rectangular feature with a circular feature within it. The washer will occlude the rectangular block's image. Occlusion occurs in visual images also, but may be ambiguous. Visual occlusion occurs when one object blocks the line of sight between the camera and another object. This could be because one object is touching the other and so a region is obscured from sight. The same effect happens when they are quite separate from one another but are aligned with the camera. It is difficult to differentiate between these two situations.

Tactile Sensors Provide Local Views
The view of the environment provided by tactile sensors is very much localised, because such sensors only give responses to stimuli which are actually touching the sensor surface. Events happening close to, but not in contact with, a tactile sensor will not be registered at all by that sensor. The tactile sensor will often be smaller than the object being sensed, and will only detect one or a few faces of the object. Robot gripper-mounted sensors in particular are likely to be smaller than the object being grasped. The information provided by a single tactile image is

therefore highly localised within the robot's environment, as opposed to vision sensing which can supply a global view of a workcell, for example. The local view can be supplemented by using multiple tactile images obtained through an active sensing strategy. This requires the integration of tactile images and manipulator trajectories. The local view can be used to supplement global information supplied from vision sensors, providing a detailed image which is perhaps impossible to gain through global visual observation because of occlusion.

Tactile Sensing May Produce Physical Distortion

By its very nature, tactile sensing will involve physical interaction with a sensed object. This interaction can result in the physical deformation of the object under examination, or perhaps its movement in space. Experiments that sense hardness, for example, involve advancing the tactile sensor at small increments into the sampled object and recording tactile images at each stage. The changes in the tactile images indicate the hardness of the sample. The act of poking the object of interest may deform it, in some cases irreversibly. Alternatively, the tactile sensing act could physically move the object under inspection. Hillis, for example, investigates the stability of some objects under investigation by pushing them with a tactile finger.[10] This could cause complications in a robot workcell, as inspected components could be moved off a reference location. This interaction between tactile sensor and stimulus is in contrast to the passive nature of vision sensing.

Vision sensing has many applications and is in practical use on the factory floor. The above comparison shows that some of the problems present in vision are not present in tactile sensing. Tactile sensing has other limitations, however, such as only providing local views, and having the potential to distort a sensed object. The two sensing modalities complement one another rather than compete; for example, vision provides global views whilst tactile provides local views; sensing through touch can provide information such as force, hardness, and thermal conductivity which vision cannot.

8   Conclusion

This paper has discussed a high resolution tactile sensor that uses an optical transduction technique. The sensor uses an elastomer which essentially determines the behaviour of the sensor; certain latex rubbers were found to provide very detailed tactile images and so this sensor has potential as an inspection device. It was used in an example application to show that location, orientation, recognition, and inspection tasks can be performed using a single tactile image. Thus, tactile sensing should be considered as a viable sensing modality. Vision sensing and tactile sensing were also compared, and it is concluded that these modalities complement rather than compete with one another.

Acknowledgements

Part of this work was carried out with the aid of funding from the Science and Engineering Research Council under CASE Award KB001, and also with support from the former British

Robotic Systems Ltd. Thanks is also due to Mr. Wynford Davies of the Computer Science Department for implementing the example application. The photographs in figure 2 also appear in Sensor Review, vol. 7, no.4.

References

1. Nicholls, H.R., and Lee, M.H.: A survey of robot tactile sensing technology. Int. Journal of Robotics Research, vol.8, no.3, pp. 3-30, 1989.

2. Mott, D.H., Lee, M.H., and Nicholls, H.R.: An experimental very high resolution tactile sensor array. Proc. 4th Int. Conf. on Robot Vision and Sensory Control 1984, London, pp. 241-250. Kempston, Bedford UK: IFS(Pubs) 1984.

3. Betts, R.P., Duckworth, T., and Austin, I.G.: Critical light reflection at a plastic/glass interface and its application to foot pressure measurements. Journal of Medical Engineering and Technology, vol.4, no.3, pp. 136-142, 1980.

4. Pullen, J., Williamson, J.B.P.: On the plastic contact of rough surfaces. Proc. Royal Society, London, part A, no. 327, pp. 159-173, 1972.

5. Childs, T.H.C.: The persistence of roughness between surfaces in static contact. Proc. of the Royal Society, London, part A, no. 353, pp. 35-53, 1977.

6. Begej, S.: An optical tactile array sensor. Univ. Massachusetts at Amherst, Dept. Computer and Info. Science report COINS 84-26, 1984.

7. Tanie, K., Komoriya, K., Kaneko, M., Tachie, S., and Fujikawa, A.: A high resolution tactile sensor. Proc. 4th Int. Conf. on Robot Vision and Sensory Control 1984, London, pp. 251-260, Bedford UK: IFS(Pubs) 1984.

8. King, A.A. and White, R.M.: Tactile sensing array based on forming and detecting an optical image. Sensors and Actuators, vol.8, no.4, pp. 49-63, 1985.

9. Overton, K.J.: The acquisition, processing, and use of tactile sensor data in robot control. Univ. Massachusetts at Amherst, Dept. Computer and Info. Science report COINS 84-08, 1984.

10. Hillis, W.D.: A high resolution imaging touch sensor. Int. Journal Robotics Research, vol.1, no.2, pp. 33-44, 1982.

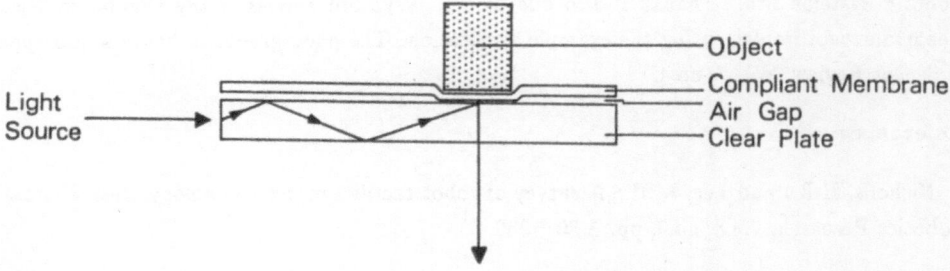

Figure 1. principle of operation

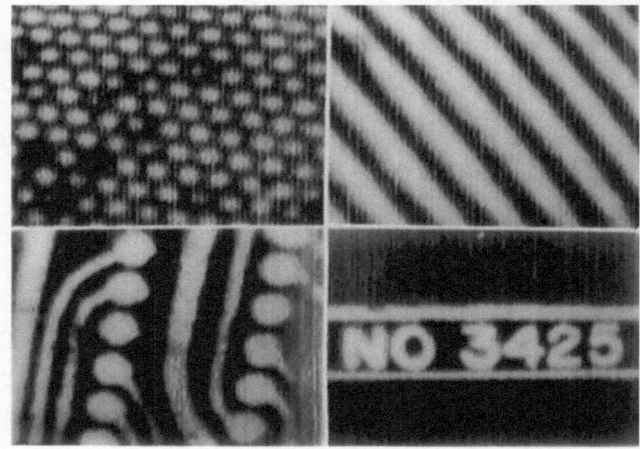

Figure 2. sample images – surface of metal file; ribbon cable;
pcb track; raised lettering

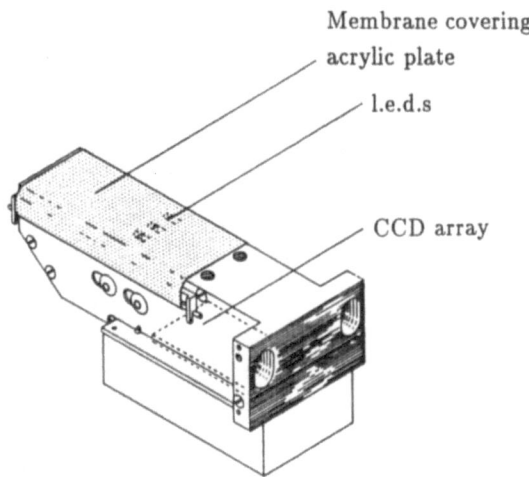

Membrane covering
acrylic plate

l.e.d.s

CCD array

Figure 3. diagram of sensor

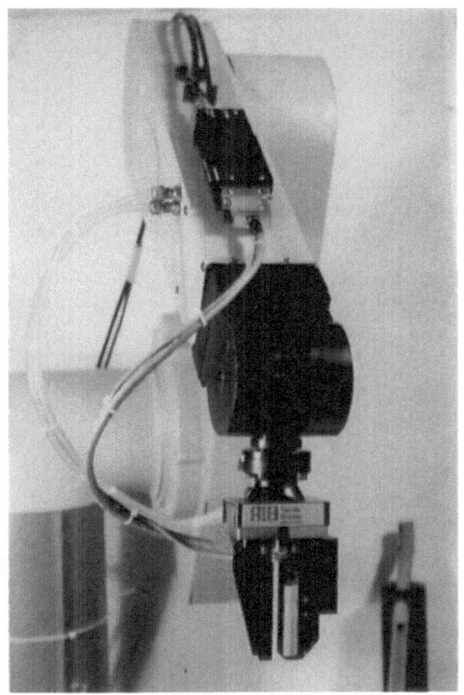

Figure 4. sensor on robot

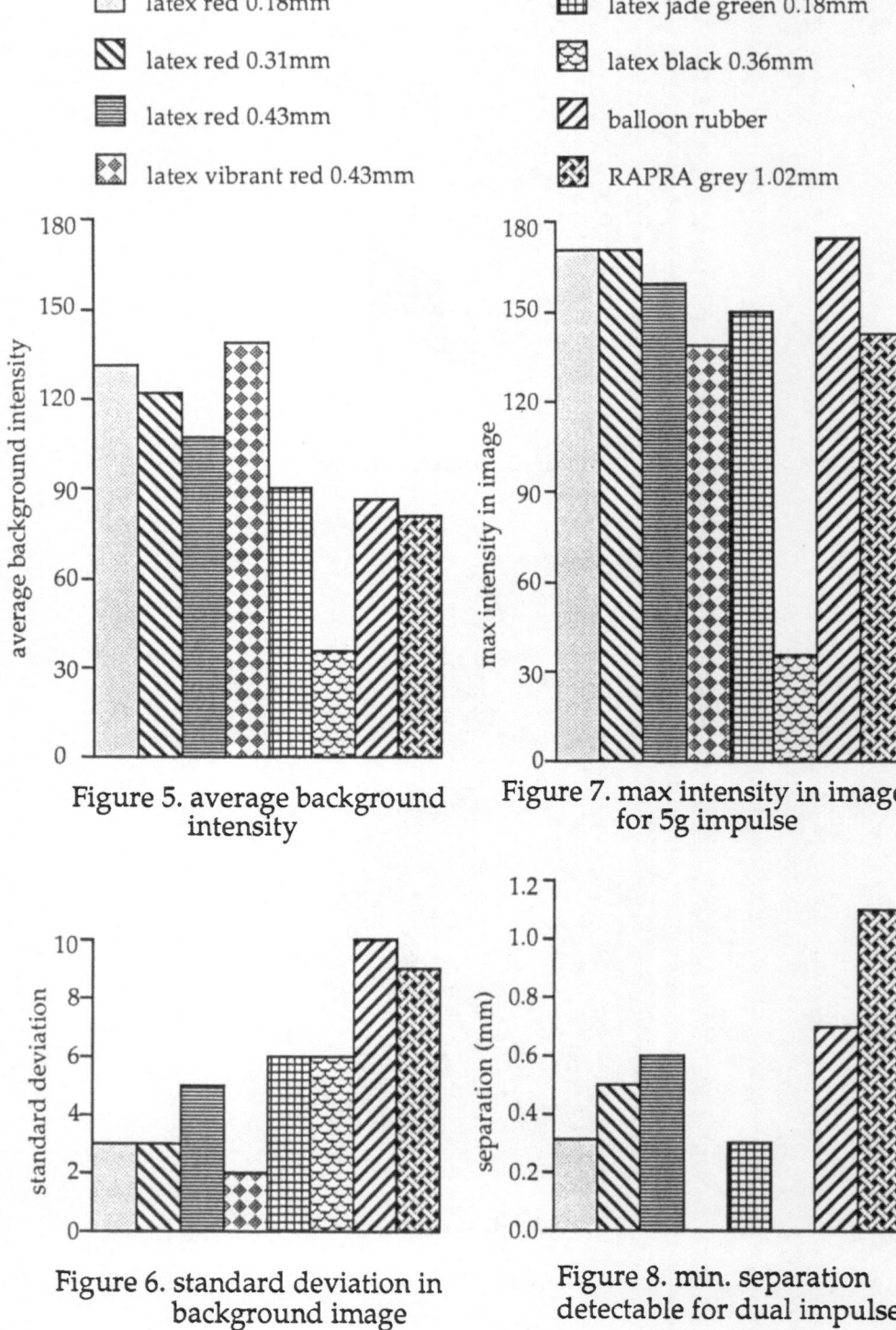

latex red 0.18mm
latex red 0.31mm
latex red 0.43mm
latex vibrant red 0.43mm
latex jade green 0.18mm
latex black 0.36mm
balloon rubber
RAPRA grey 1.02mm

Figure 5. average background intensity

Figure 7. max intensity in image for 5g impulse

Figure 6. standard deviation in background image

Figure 8. min. separation detectable for dual impulse

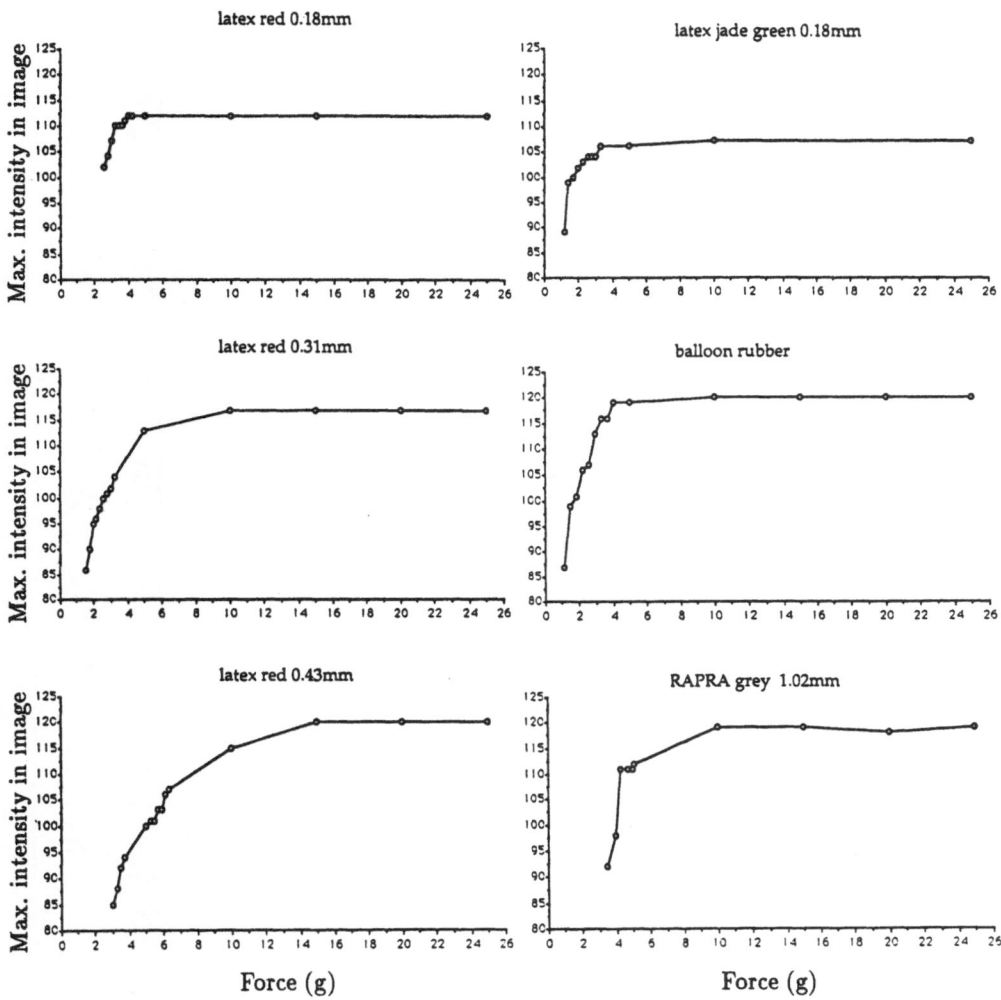

Figure 9. max. intensity in image for increasing force

ABSOLUTE POSITION MEASUREMENT
USING PSEUDO-RANDOM ENCODING

Emil M. Petriu
University of Ottawa
Department of Electrical Engineering
Ottawa, Ont., K1N 6N5
Canada

1. Introduction

Many domains in robotics would definitely benefit from an efficient digital absolute position measurement capability. Absolute shaft encoders are attractive for joint control applications as their position is recovered immediately when power is restored after an outage, and they do not accumulate errors as incremental transducers often do [1]. The ability to measure absolute position would be a notable asset for *automated guided vehicle* (AGV) navigation. This is especially significant in situations where they have to avoid obstacles that may appear on their paths.

Absolute position measurement usually requires that a distinct n-bit code is marked on each quantization interval of a scale (Fig. 1). The position "p" of the pointer relative to the origin of this scale is then estimated by reading the specific code written on the quantization interval currently facing the pointer:

$$p = \sum_{k=1}^{n} x(k) \cdot 2^{k-1} \cdot q.$$

Of course, the resulting number of code tracks on the scale increases proportionally with the desired measuring resolution, prohibiting the use of this "natural" absolute encoding method for the very high resolution applications required by the contemporary industrial practice.

NATO ASI Series, Vol. F 63
Traditional and Non-Traditional Robotic Sensors
Edited by T. C. Henderson
© Springer-Verlag Berlin Heidelberg 1990

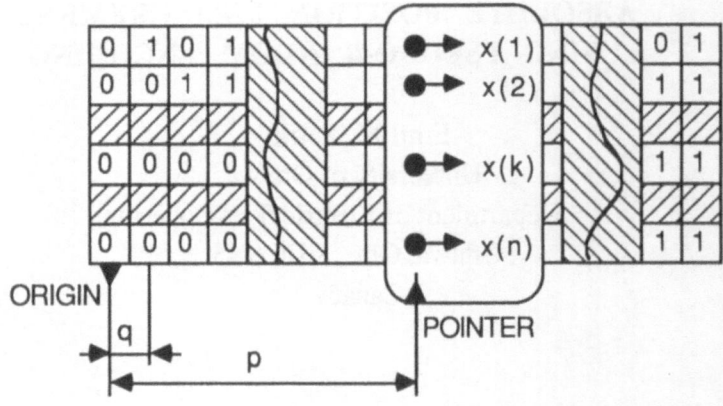

Figure 1. The natural approach to absolute position encoding.

As it will be shown further , the **pseudo-random encoding** of the scale provides a more efficient alternative for the absolute position measurement. The new encoding method is based on the properties of *pseudo-random binary sequences* (p.r.b.s.) and has the notable advantage of requiring only one bit of code per quantization interval. This advantage makes absolute position measurement more accessible for very high resolution applications.

This paper represents a synthesis of the pseudo-random encoding techniques recently developed by the author for absolute position measurement applications.

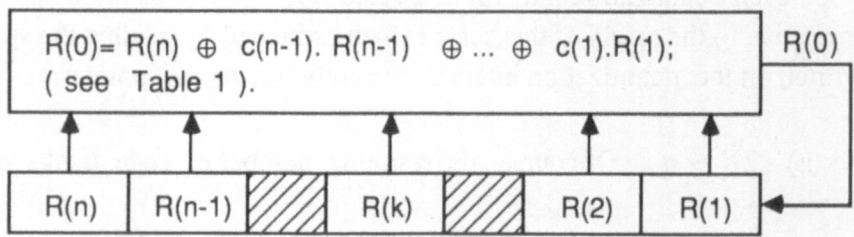

Figure 2. Direct-feedback shift register.

2. Pseudo-Random Encoding

In this case, the scale is encoded (one bit per quantization interval) with the terms of a p.r.b.s. $\{ S(p) \mid p = 0,1, ..., 2^n - 2 \}$ generated by a modulo-two direct-feedback n-bit shift register (Fig. 2 and Table I).

Table I. Direct- and reverse-feedback equations for some pseudo-random shift registers.

Shift register length n	Feedback for direct p.r.b.s. $R(0) = R(n) \oplus c(n-1) \cdot R(n-1)$ $\oplus ... \oplus c(1) \cdot R(1)$	Feedback for reverse p.r.b.s. $R(n+1) = R(1) \oplus b(2) \cdot R(2)$ $\oplus ... \oplus b(n) \cdot R(n)$
4	$R(0) = R(4) \oplus R(1)$	$R(5) = R(1) \oplus R(2)$
5	$R(0) = R(5) \oplus R(2)$	$R(6) = R(1) \oplus R(3)$
6	$R(0) = R(6) \oplus R(1)$	$R(7) = R(1) \oplus R(2)$
7	$R(0) = R(7) \oplus R(3)$	$R(8) = R(1) \oplus R(4)$
8	$R(0) = R(8) \oplus R(4)$ $\oplus R(3) \oplus R(2)$	$R(9) = R(1) \oplus R(3)$ $\oplus R(4) \oplus R(5)$
9	$R(0) = R(9) \oplus R(4)$	$R(10) = R(1) \oplus R(5)$
10	$R(0) = R(10) \oplus R(3)$	$R(11) = R(1) \oplus R(4)$

The absolute position identification is based on the well known p.r.b.s. **window property** [2]. According to this any n-tuple $\{S(p+n-k) \mid k=n,...,1\}$ scanned by a window $\{x(k) \mid k=n,...,1\}$ is unique and may fully identify the current position "p" of the window (Fig. 3), [3]-[5]. Two major problems have to be solved in order to qualify this encoding method for practical applications :

  i. the correct scanning of the pseudo-random code track;
  ii. the conversion of the pseudo-random n-tuple $\{X(k) \mid k=n,...,1\}$ into a more convenient natural binary code $\{P(k) \mid k=n,...,1\}$;

Some original solutions to these problems will be discussed in the next sections.

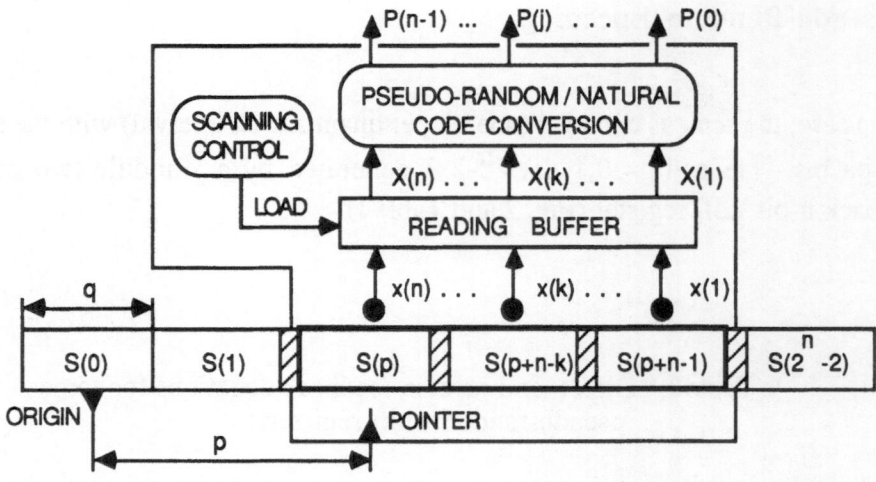

Figure 3. Pseudo-random encoding: the absolute position "p" is
fully identified by the n-bit window.

3. Scanning the Pseudo-Random Code Track

A problem facing all absolute position measuring techniques is that of the
reading ambiguities which may occur when a transition between two consecutive
positions takes place. A straightforward solution to this would be to use an
external synchronization track laid along the position code track(s).

An **immediate implementation** of the synchronization track is shown in
Fig. 4. Sampling strips of value 1 are arranged in such a way as to fall in the
middle of each quantization interval of the p.r.b.s. track. Whenever such a strip
arrives in the front of the supplementary synchronization head AUT (facing the
pointer) the information delivered by the n code reading heads is considered to
be stable and ready to be stored in the code reading buffer. Although it solves the
reading ambiguities problem, this solution leads to a systematic **hysteresis
error** $e = q_0 / 2$, where "q_0" is the width of the space "0" between two

sampling strips. As shown in Fig. 4, the way in which this error affects the measured position, (represented by the current contents of the reading buffer), is a function of the direction of the relative displacement between the pointer and the scale.

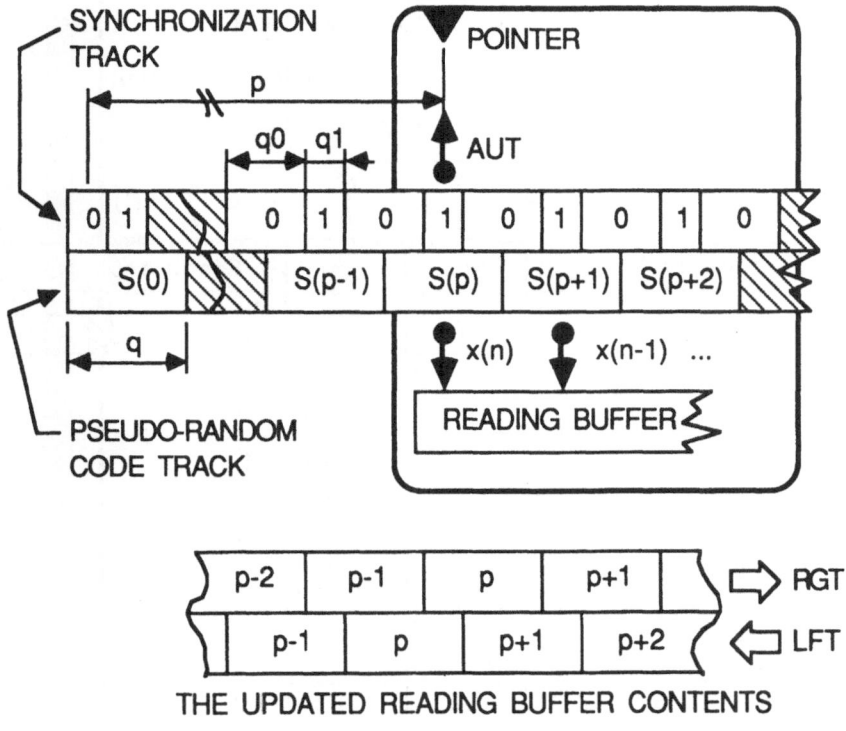

Figure 4. Sampling-type synchronization track.

A **more effective solution** [6] is shown in Fig. 5. It employs a simpler layout of the synchronization track consisting of a sequence of alternate "0" and "1" intervals each having the same width "q" as the quantization intervals of the p.r.b.s. code track. Two reading heads, AUT and VER, are employed to investigate the synchronization track. The AUT head is used to detect the transitions between two consecutive quantization intervals. The second head VER, placed at a distance q/2 from AUT, is used together with AUT to determine the direction of the relative displacement between pointer and scale.

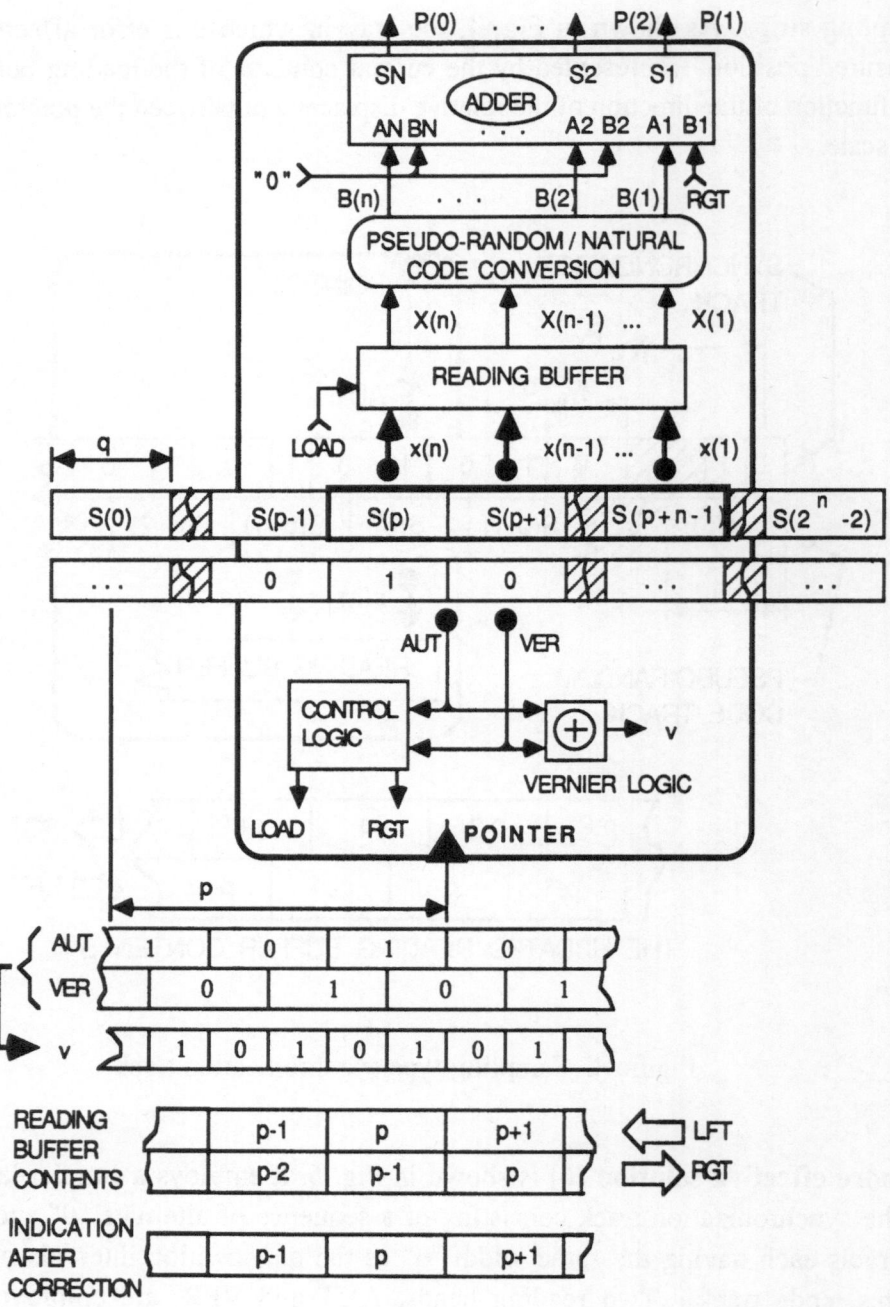

Figure 5. New synchronization track design.

The distance of q/2 between the two synchronization heads allows them to work also as **vernier** increasing the overall measuring resolution by a factor of two. The resulting value of the measured absolute position is then: $p = (2p + v) \cdot q/2$.

The n code reading heads $\{x(k) \mid k=n,...,1\}$ are shifted with q/2 relative to the AUT head ensuring that their outputs are stable when AUT detects a transition between two consecutive positions. Due to this arrangement, a systematic error of "-1" will always happen when the pointer moves to the right. Let us consider the situation shown in Fig. 5 representing the pointer at the transition point between the positions "p" and "p+1". Regardless of the direction of motion, the reading buffer will be loaded with $\{x(k)=S(p+n-k) \mid k+n,...,1\}$ representing the pseudo-random n-tuple for the position "p". If the transition occurs when the pointer moves to the left, the new position will be "p" which will be correctly identified by the updated buffer contents. If the transition is produced by the pointer's motion to the right, the new position will be "p+1" while the updated buffer contents will still be "p". A correction of "+1" is then required in this case. It results that this code scanning technique needs to know the direction in which the pointer moves relative to the scale. A special **control logic** is used to determine this direction, and based on it further generate the buffer synchronization pulse LOAD as well as the final correction signal RGT (Fig.5).

It can be easily seen that this scanning technique requires the same number of reading heads (n+2) as the straightforward technique for the same overall measuring resolution (n+1). The new scanning method is however superior because it requires a simpler synchronization track and eliminates the hysteresis error.

4. Pseudo-Random / Natural Code Conversion

A translation from the pseudo-random binary code into the more convenient natural binary representation is always necessary for practical applications. The duration of this code conversion is critical for the absolute position measurement cycle. The conversion methods vary in the extent of their degree of parallelism and corresponding time/equipment cost. A strictly **parallel** solution would use a code conversion table stored in ROM. This is equipment expensive for applications requiring high resolution position measurement. At the other extreme, a strictly **serial** translation exploits the reversibility of the p.r.b.s.

generating algorithm [4]. This method is based on the idea that it is possible to find the natural value associated with any pseudo-random n-tuple by simply counting the number of reverse feedback shifts (Table I) that it takes for the given n-tuple to arrive back into the "zero position" pseudo-random pattern. In this case the solution is less equipment expensive but more time expensive for the high resolution measurements.

Fig. 6 shows the conversion time cost variation for the different implementations of the code conversion function.

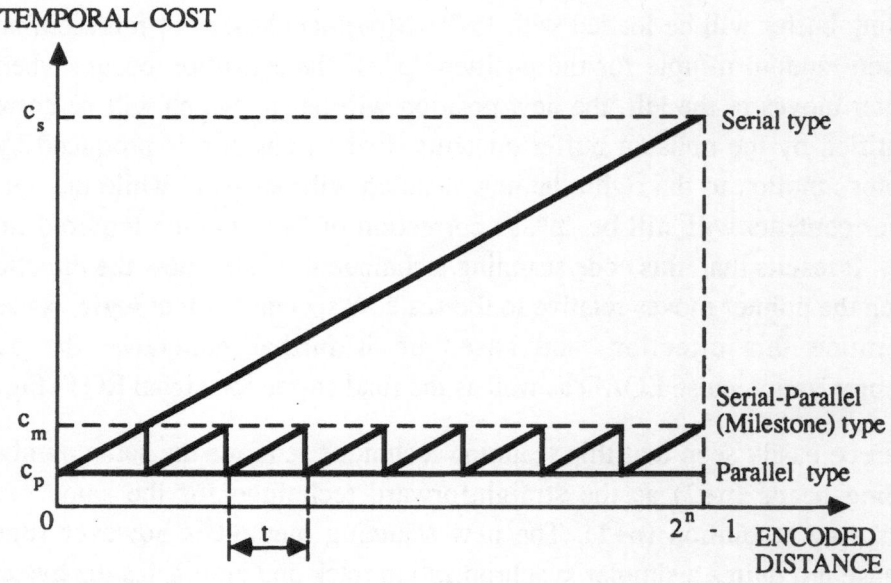

Figure 6. Temporal cost for different code conversion methods.

A compromise (**serial-parallel**) solution [7] is exemplified in Fig. 7. Consider a pseudo-random encoded track where certain positions (uniformly distributed with a period of "t") are employed as **milestones.** The code conversion for any position $p = m \cdot t + r$, where $m \cdot t$ is the position of the nearest "down the track milestone" and $Q(m)$ and "r" represents the position relative to this milestone, will be discussed. The natural code for "r" is found by counting the steps required to arrive by successive back-shifts from the initial code to the nearest milestone $Q(m)$. All intermediate states of this serial shift-back operation are checked in parallel against all possible milestone pseudo-random

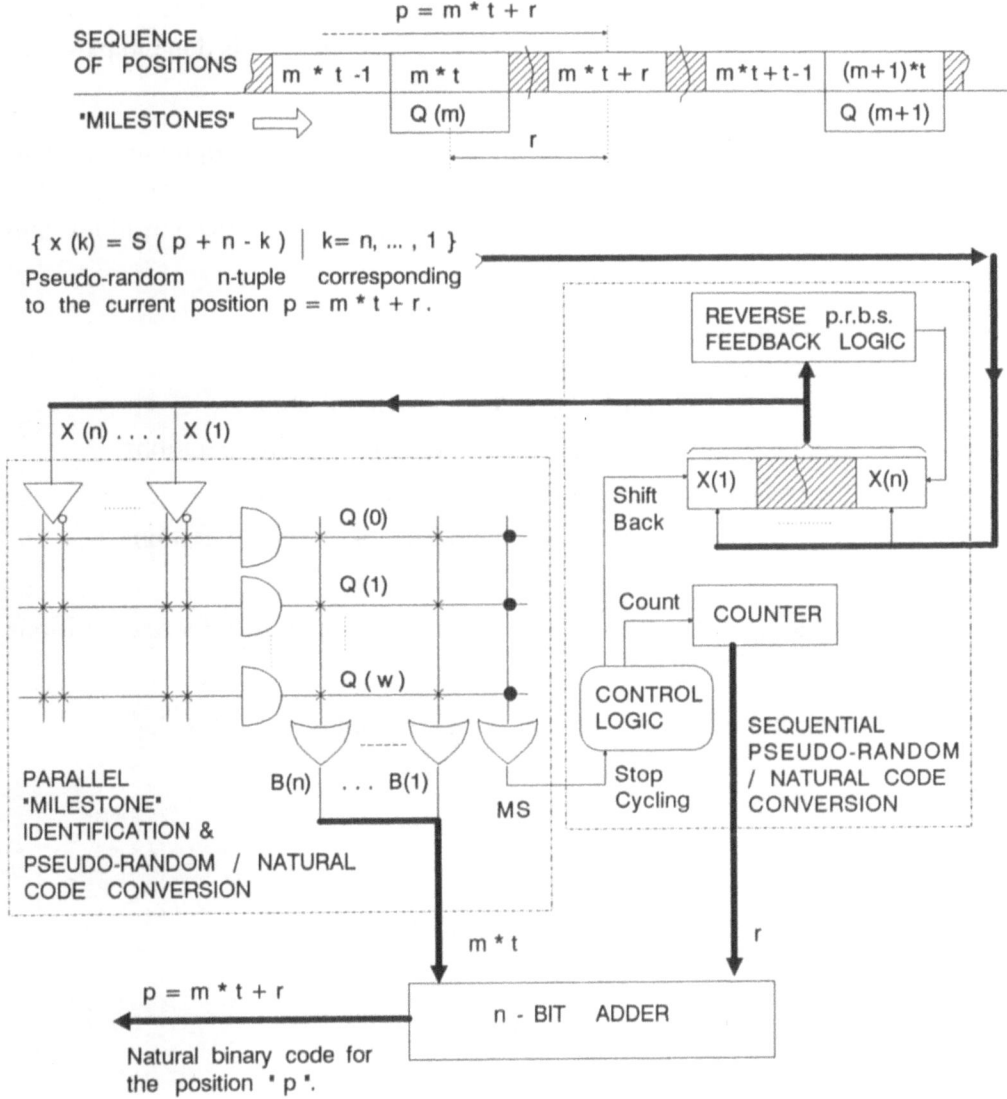

Figure 7. Serial-Parallel Code Conversion.

patterns. Thus with this method the code conversion of the relative position "r" distance is found serially while the milestone code conversion is done in parallel.

The equipment and the temporal costs are used to decide on the optimal degree of parallelism to be used for a specific implementation of the serial-parallel conversion method. Given below are the definitions of the variables and constants required for this cost analysis:

$2^n - 1$ = the total length of the p.r.b.s.

t	=	the distance between successive milestones
k_1	=	the equipment cost associated with each milestone
k_2	=	the basal equipment cost for the serial back-shift operations
k_3	=	the basal temporal cost for a fully parallel solution
k_4	=	the temporal cost associated with each back-shift operation

Equipment and temporal cost equations may be developed with these definitions as a function of "t" as follows:

equipment cost = $k_1 \cdot$ milestones number + $k_2 = k_1 \cdot$ int $(2^n - 1 / t) + k_2$

temporal cost = $k_3 + k_4 \cdot t$

The total cost of the serial-parallel conversion will be then:

total cost = equipment cost + temporal cost

total cost = $k_1 \cdot$ int $(2^n - 1 / t) + k_2 + k_3 + k_4 \cdot t$

Fig. 8 shows a graph of the equipment cost, temporal cost, and total cost as a function of "t". The temporal cost associated with the serial back-shift matching operation linearly increases with the independent variable "t" while the equipment cost is inversely proportional with this variable. The combined total cost graph does have a minimum between 1 and $2^n - 1$. The value of "t" which minimizes the total cost function is the following:

$t_{opt} = (k_1 / k_4 \cdot 2^n - 1)^{1/2}$

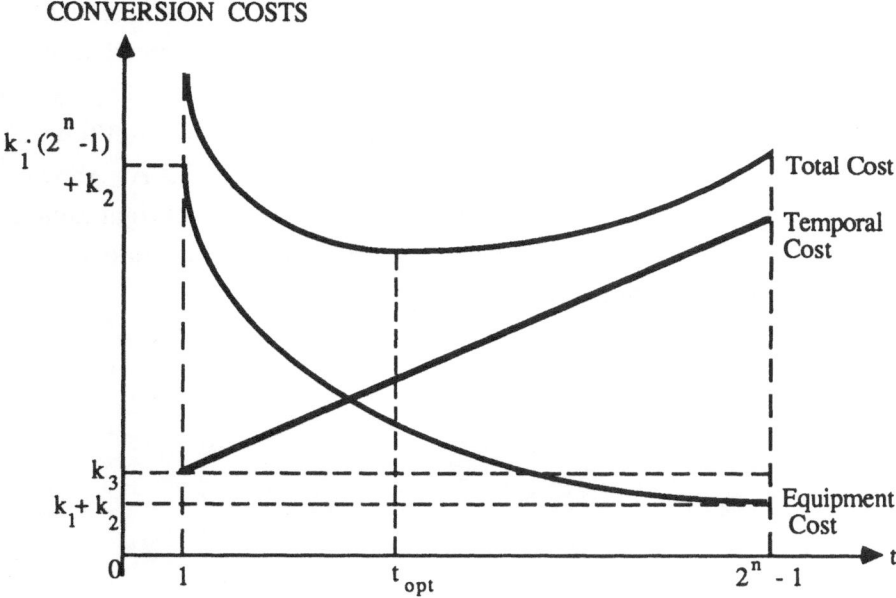

Figure 8. Serial- parallel code conversion costs as a function of the distance "t" between milestones.

Depending on the relative ratios of the constants k_1 and k_4 (for a given "n") the optimal distance between milestones will vary. This cost analysis provides a guide for choosing the number of milestones (i.e. the degree of parallelism) for a specific application. A higher degree of parallelism is requested when the temporal cost is relatively higher than the equipment cost. A higher measuring resolution will always lead to an increase in the number of milestones in order to maintain the same code conversion speed.

5. On Board Synchronization for AGV Absolute Position Recovery

The proper scanning of the p.r.b.s. code track always necessitates a means of synchronization . As discussed in Section 3 it is usually implemented as a supplementary track laid along the p.r.b.s. track.

When used for AGV absolute position recovery, the cost of this off board synchronization track may represent a significant portion of the total equipment cost of the position measurement feature. This section presents an original **on board synchronization** method that eliminates the need for the physical synchronization track. As shown in Fig. 9, this results in a minimal complexity of the track configuration laid on the floor which is reduced to only one guide-path (mandatory for all AGVs) and the one-bit wide p.r.b.s. code track.

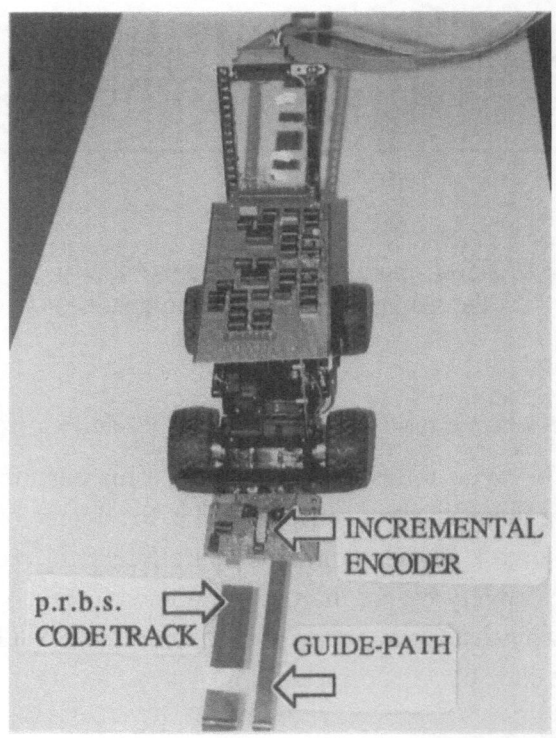

Figure 9.   Experimental AGV.

The new synchronization technique, illustrated in Fig. 10, employs an incremental encoder for the on board generation of the code reading synchronization signals. This encoder has a resolution "y" appropriately selected as a submultiple of the encoding resolution "q" of the pseudo-random code track: y=q/c. As the AGV moves along its guide-path the incremental encoder will generate an output signal SYNC. This may for convenience be conceived as a **virtual synchronization track** alongside the other two physical tracks laid on the floor: the guide-path and the p.r.b.s. code track.

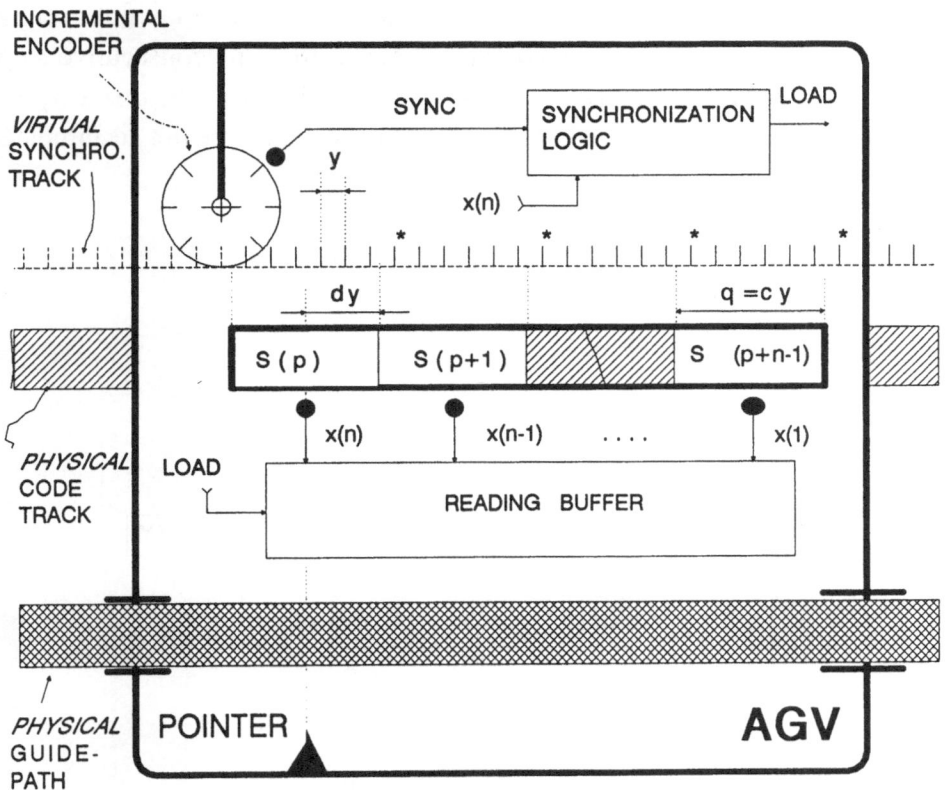

Figure 10. The "on-board" synchronization principle.

A synchronization logic circuit will select the specific SYNC pulse that occurs when the AGV code reading heads arrive in the middle of the quantization step. This specific pulse (star marked in Fig. 11) propagates as the LOAD signal used to update the reading buffer containing the pseudo-random positional information. To avoid the cumulative errors inherent in the incremental encoding , an external initialization mechanism was introduced. The **synchronization logic** detailed in Fig. 11 is implemented as a "modulo c" counter that generates a LOAD pulse after every "c" SYNC pulses. Any time a quantization step transition is detected by the x(n) reading head, this counter is preset with the "c-d" value in such a way that the LOAD pulse will occur "d" SYNC pulses after the beginning of each quantization step. Adopting a proper d=c/2 value will ensure that the information provided by the code reading heads is always stable when the reading buffer is updated. This resynchronization of the modulo "c" counter controlled by the physical code track may not occur at every quantization step, but it will definitely occur after "n" consecutive steps (where "n" is the AGV position measurement resolution). In this way the slippage errors that may occur with the incremental encoder are not accumulated and their effect is limited.

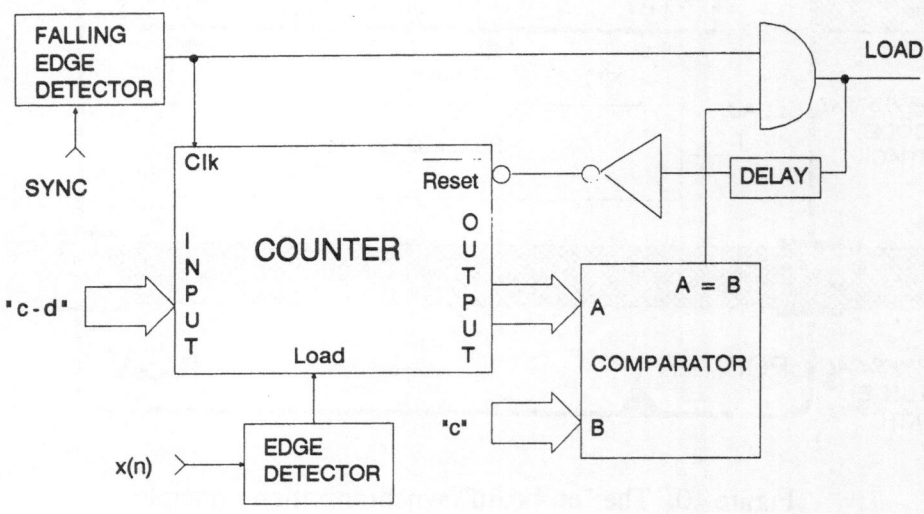

Figure 11. Synchronization logic.

Being directly driven by the AGV positional parameters, this synchronization method is time invariant; hence, it is also speed and acceleration insensitive. As a drawback, the on board synchronization may introduce some limited slippage error. The actual value of this error may be reduced by increasing "c", the resolution ratio between the p.r.b.s. track and the virtual synchronization track.

6. Conclusion

The code scanning and the code conversion techniques discussed in this paper contribute to the promotion of pseudo-random encoding as a competitive alternative for the implementation of high resolution absolute encoders. This encoding technique also offers a promising solution to the measurement of AGV absolute position anywhere on the guide-path. It can be implemented as a stand-alone function to cost-effectively upgrade existent optically-guided industrial AGVs.

References

1. Doebelin, E.O.: **Measurement systems: Application and design.** 3rd ed. New York: McGraw-Hill 1983

2. MacWilliams, F.J., and Sloane, N.J.A.: "Pseudo-Random Sequences and Arrays," **Proc. IEEE**, Vol. 64, No. 12, pp. 1715-1729, Dec. 1976

3. Arazi, B.: "Position Recovery Using Binary Sequences," **Electronics Letters** , Vol. 20, pp.61-62, 1984

4. Petriu, E.: "Absolute-Type Pseudo-Random Shaft Encoder with Any Desired Resolution," **Electronics Letters** , Vol. 21, No. 5, pp. 215-216, 1985

5. Petriu, E.: "Absolute Type Position Transducers Using a Pseudorandom Encoding," **IEEE Trans. Instrum. Meas.,** vol. IM-36, No. 4, pp. 950-955, Dec.1987

6. Petriu, E., Basran, J.S.: "On the Position Measurement of Automated Guided Vehicles Using Pseudo-Random Encoding," **IEEE Trans. Instrum. Meas.,** (to appear), vol. IM-38, No. 3, June 1989

7. Petriu, E.: "New Pseudorandom / Natural Code Conversion Method," **Electronics Letters** , Vol. 24, No. 22, pp 1358-1359, 1988

Silicon Tactile Image Sensors

P.P.L.Regtien
Delft University of Technology
Department of Electrical Engineering
Mekelweg 4, 2628 CD Delft
The Netherlands

Introduction

Tactile sensing is one of the three possible means through which a human being can perceive image information. Unlike the two others, visual and acoustic perception, tactile sensing requires physical contact with the objects to be perceived or recognized, the information being carried by dynamometric parameters such as force and torque.

Tactile perception is a composition of several sensations received through the sense of touch. The most primitive of these sensations is the detection of an object's presence. Localisation of the object follows from additional proprioceptive information at the moment of touch. Many other characteristics can be detected, even with a single tactile element (taxel): dynamic scanning perpendicular to the object's surface reveals information about the elastic properties; tangential scanning about its shape and texture.

Image information can also be achieved in a static way using a matrix of taxels. The human hand contains a multi-dimensional tactile system, enabling efficient acquisition of both static and dynamic image information. This tactile system consists of a high-density matrix with tacile elements responding to both force and torque, embedded in a compliant tissue and integrated with a multiple degree-of-freedom actuator. With a little training, a human being can perform extremely complex manipulations based only on tactile perception, for instance fastening shoelaces.

In robotics, the use of image information is rapidly growing. Camera-based vision systems perform such tasks as localization, recognition and quality inspection of product parts. Optical and acoustic range-finding systems are examples of scanning imagers. Early studies [1] have indicated the need for tactile information in robotic applications, in particular fine motion control in product handling and automatic assembly processes.

NATO ASI Series, Vol. F 63
Traditional and Non-Traditional Robotic Sensors
Edited by T. C. Henderson
© Springer-Verlag Berlin Heidelberg 1990

There is, up to now, a wide discrepancy between the tactile abilities of biological systems and of robots. The few tactile sensors that are commercially available at present [2] have poor tactile capabilities compared to the human finger. The main difficulty encountered in the design of a tactile sensor is the construction of a high-density matrix of force sensitive elements with low hysteresis and skin-like properties (thin, compliant, wear resistant, and easy to mount on the gripping surfaces of an actuator.

Requirements

Any force sensor is basically a displacement sensor: when a force is applied to a sensor material, this material undergoes some deformation according to its elastic properties. From the deformation, which is either macroscopic (for instance a spring) or microscopic (as in piezoelectric materials), another measurable effect is derived to determine the displacement. The first one of this series of conversions (transduction from force to displacement) determines to a large extent the characteristics of the entire tactile sensor: range, sensitivity, hysteresis and compliancy are determined mainly by the choice of the elastic sensing material, by which the force is converted into a displacement.

General requirements on tactile sensors are low hysteresis and a high spatial resolution. Other specifications are strongly application dependent. The strong hysteresis of many tactile materials forces thresholding the sensor signals, resulting in a binary image of the touched object. This is no severe limitation in such applications as the localisation and recognition of objects with a known shape. Delicate assembly tasks with compliant tactile sensors require a higher force resolution, although a lot of information can be derived from changes in a binary tactile image. High force resolution is necessary when using compliant sensors in high precision fine motion control. In such an application, non-compliant tactile sensors are preferred, because the position of the object when fixing it into the gripper will be preserved. However, compliant sensor surfaces facilitate control and are less susceptible to damage when handling rigid objects.

The next sections give an overview of the major principles of tactile sensors, with a discussion of the possibility of applying silicon technology or compatible technologies.

Displacement-based tactile sensors

Probably the oldest type of tactile sensor is the pen matrix [3]. The displacement of a tactile pen can be relatively large. Consequently, this type is very useful for 3D recognition or testing of industrial parts [4-6]. The pen position is measured according to well-known principles for displacement sensors.
A variation of this type is the matrix of steel balls [7]. This construction can be used for testing the roughness of an object.
In 2D applications, a pen matrix may be combined with IC technology. The pens transfer the force to an integrated circuit containing a matrix of pressure-sensitive elements, for instance piezojunction transistors, piezoresistors or piezoelectric elements.
Disadvantages of pen matrix sensors are the complex mechanical construction and the relatively large dimensions of the device. The spatial resolution is restricted by the mechanical dimensions of each tactile pen.

A number of optically based tactile sensors have been proposed, most of which are based on the modulation of a light path by a displacement of an elastic part of the sensor [5,8-10]. The light can be supplied individually to each taxel (by a glass fiber bundle) or to them collectively, using a diffuser. The output signal is picked up by individual glas fibres, one for each taxel. Another technique is the conversion of a variable contact area to an optical image using a camera.
The spatial resolution of these tactile sensors is limited by the dimensions of the optical components, and is highest when glass fibers are used. The total dimensions of such a sensing system remain rather large, however, because many optical components (laser, optical scanners) can not be integrated with the electronics.

All rubber-based tactile sensors suffer from a substantial hysteresis. A material that behaves in an almost perfectly elastic way happens to be

silicon. Silicon micro-machining has become a common technique [11].
Membranes etched out of silicon are used for a variety of pressure sensors
[12-15]. Lithography allows the construction of a matrix of pressure
sensitive elements on a single wafer. The deformation of a membrane can be
measured either capacitively, with integrated strain gauges, or optically.
The capacitive type consists of two parts, a glass substrate and a processed
silicon wafer attached face-to-face by anodic bonding. On the glass
substrate a matrix of square or circular electrodes is deposited. The
silicon wafer contains a matrix of membranes, which are in line with the
electrodes on the glass substrate. An electrode on the membrane completes
the flat capacitor, whose air gap varies with an applied force. The
advantages of this structure are: no hysteresis; high sensitivity (both the
air gap and the membrane can be very thin); overload protection (an
isolating layer on top of the electrode on the glass substrate limits the
deflection of the membrane and prevents the shortcircuiting of the
electrodes) and high spatial resolution. With a mesa (Fig.1), the force can
be transferred to the membrane [16].

Deformation of elastic sheets

The sensors of this group are characterized by a change in the thickness of
a sheet of elastic material, such as rubber. Spatial resolution is achieved
by 2D-scanning of the thickness, using a grid of point electrodes or two
layers of line electrodes. The strip electrodes offer row-wise and column-
wise scanning of the sensor matrix, reducing the number of sensor wires in
an mxn matrix to m+n+1. Care should be taken to prevent phantom images. Such
difficulties can be circumvented by the individual processing of taxel
data, a technique that requires IC (-compatible) technology. On-chip
conversion of the image data to a serial format further reduces the number
of sensor pins.

The thickness of the elastomeric layer can be measured in various ways, for
instance acoustically [17], capacitively [18-20] or optically.
Compatibility with IC technology requires both the source and the detector
of the measurement signals to be at one side of the sheet, which is possible
in all of these examples.

Use of the acoustical principle of thickness measurement requires objects
with a high acoustic impedance, in order to have sufficient sound reflection
at the top layer. Source and detector may consist of piezoelectric elements
connected between the wafer and the elastic layer. Piezoelectric polymers
are preferred over ceramic materials because they enable a better impedance
matching to the elastomer and because of their availability in thin sheets.
As in most ultrasonic transducers, special attention must be given to the
backing of the transmitter and the receiver, in order to get a proper
sensitivity. The displacement resolution is determined by the elasticity and
the acoustic velocity of the elastomer. Assuming a 1 mm thick elastic layer
with an acoustic velocity of 1000 m/s and a displacement resolution of 1%,
the measurement of the time-of-flight of an acoustic pulse should have a
resolution of about 200 ns, requiring an even shorter pulse duration.

Making the same assumption with respect to layer thickness and displacement
resolution, the capacitive type requires a measurement resolution in the
order of 10 fF, depending on the size of the taxels. The small signal
changes involved in this type of sensor make the use of IC-technology
mandatory. A typical difficulty encountered in most silicon-based tactile
sensors is a reliable connection of the sheet to the wafer, without
additional signal loss or hysteresis.

A common advantage of this type of sensor is the separation of the
transduction from force to displacement and the displacement measurement
itself, which facilitates optimal choice of sensing materials. This
advantage is lacking in the next two groups of tactile sensors.

Piezoresistive sensors

In these types of sensors an applied force causes a geometrical deformation
of an elastic material upon which this material changes its conductivity.
Most semiconductors exhibit this property, and is employed in semiconductor
strain gauges. Such strain gauges are the basis of a number of commercially
available load cells. Basically, a tactile imager can be constructed as a
matrix of miniature load cells. The spatial resolution of such a tactile
imager is limited by the rather large size of such individual elements.

A much better resolution can be achieved by piezoresistive elastomers and polymers. Intrinsically non-conductive, elastomers can be made conductive by impregnating them with metal or carbon particles. Several manufacturers produce such materials, in a variety of sizes and shapes, like sheets and rods. A carbon fibre also shows piezoresistivity and therefore is, in principle, also a candidate for tactile sensing material [21]. Several researchers have proposed tactile sensors based on piezoresistive elastomers [22-24].

The sensitivity of piezoelastomers is rather large: the conductivity may change by several orders of magnitude even at moderate pressures. The main drawback of such piezoresistive materials is the large hysteresis, limiting the use of tactile sensors to binary images only.

The principle can be realized using IC technology. Fig.2 shows the details of an experimental construction [25], that has been integrated with individual signal processors for two-dimensional convolution of the tactile data.

Piezoelectric sensors
<u>Piezoelectric sensors</u>

A piezoelectric material responds to a force by the generation of a surface charge (or a voltage difference), at negligible deformation. Polarized ceramics, although cheap and very useful in a large number of sensor applications, are not well suited for use in tactile image sensors: they show a too large leakage, preventing static measurements of pressure, and are not easy to machine. Polarized polymers have excellent piezoelectric and acoustic characteristics, are pliable, and are available in a variety of thicknesses [26]. Therefore, such a polymer is an excellent sensor material for tactile imagers [27,28].

A major drawback of piezoelectric materials is the inherent pyroelectricity, making the sensor respond to temperature changes and gradients as well. A good thermal conductivity of the backing may reduce largely thermal interference with the force signals.

Charge is generated everywhere on the film where a force is applied. The self-generating nature of the transduction process complicates readout by scanning rows and columns: each taxel requires individual charge readout.

Here, again, IC-technology offers a solution to this problem. A matrix of charge-sensitive transistors (i.e. MOSFETs) can easily be integrated onto a silicon wafer. The piezoelectric material is connected either on top of the silicon or next to it on a common substrate [29].

Conclusions

Silicon IC-technology offers a good solution to some severe problems encountered in the construction of high-resolution tactile image sensors. The spacing between the tactile elements (taxels) can be reduced significantly compared to discrete elements. A substantial reduction in the wiring is achieved by on-chip scanning, multiplexing or individual signal processing of the tactile signals. A number of sensing principles can be used, such as elastic deformation, piezoresistive and piezoelectric effects. In particular sheets of piezoresistive or piezoelectric polymers offer good prospects for their use in tactile sensors. Silicon has some favourable properties that makes it a very attractive material for use in tactile sensors: the wafers are thin, can contain local signal processing (for each taxel), and it allows a high taxel density. Silicon can be either the sensor material itself, or act as a substrate for tactile sensor sheets. There are also some drawbacks to this material: it is brittle, is only flat in shape and has a limited temperature range of about 150 C.

References

[1] L.D.Harmon: "Automated Tactile Sensing", Int.J.Robotics Research, 1 (1982) 2, pp.3-32.
[2] P.Dario, D. de Rossi: "Tactile sensors and the gripping challenge", IEEE Spectrum, Aug.1985, pp.46-52.
[3] S.Takeda: "Study of artificial tactile sensors for shape recognition", Proc. 4th Int. Symp. on Industrial Robots, Nov. 1974, Tokyo, pp.199-208.
[4] C.J.Page, A.Pugh, W.B.Heginbotham: "Novel techniques for tactile sensing in a three-dimensional environment", Proc. 6th Int. Symp. on Industrial Robots, March 1976, Nottingham, U.K., pp.C4-33 to C4-46.
[5] B.V.Jayawant, J.D.McWatson: "Array sensor for tactile sensing in robotic applications", IEE Colloq. on Solid State and Smart Sensors, March 1988, London, IEE Digest 1988/39, pp.8/1-8/4

[6] G.Stute, H.Erne: "The control design of an industrial robot with advanced tactile sensitivity", Proc. 9th Int. Symp. on Industrial Robots, 1979, Washington DC, pp. 519-527.

[7] H.J.Warnecke, M.Schweizer, E.Abele: "Cleaning and casting with sensor-controlled industrial robots", Proc. 9th Int. Symp. on Industrial Robots, 1979, Washington DC (USA), pp.535-544.

[8] J.Rebman, K.A.Morris: "A tactile sensor with electrooptical transduction", Proc. 3rd Int. Conf. on Robot Vision and Sensory Controls, 1983, pp.210-216.

[9] R.M.White, A.A.King: "Tactile array for robotics employing a rubbery skin and a solid-state optical sensor", Proc. 3rd Int. Conf. on Solid-State Sensors and Actuators, Philadelphia, Penn., (USA), June 1985, pp.18-21.

[10] S.Begej, "Planar and finger-shaped optical tactile sensors for robot applications", IEEE J. of Robotics and Automation, $\underline{4}$,(1988), pp.472-484.

[11] G.Delapierre, "Micro-machining: a survey of the most commonly used processes", Sensors & Actuators, $\underline{17}$, (1989), pp.123-138.

[12] K.J.Chun, K.D.Wise: "A capacitive silicon tactile imaging array", Proc. 3rd Int. Conf. on Solid-State Sensors and Actuators, Philadelphia, Penn., (USA), June 1985, pp.22-25.

[13] Y.L.Lee, K.D.Wise: "A batch-fabricated silicon capacitive pressure transducer with low temperature sensitivity", IEEE Trans. on Electron Devices, $\underline{ED-29}$, 1982, pp.42-56.

[14] H.Tanigawa, T.Ishihara, M.Hirata, K.Suzuki: "MOS integrated silicon pressure sensor", IEEE Trans. on Electron Devices, $\underline{ED-32}$, 1985, pp.1191-1195.

[15] A.Hanneborg, T.-E.Hansen, P.A.Ohlckers, E.Carlson, B.Dahl, O.Holwech: "An integrated capacitive pressure sensor with frequency-modulated output", Sensors & Actuators $\underline{9}$, 1986, pp.345-351.

[16] R.A.Buser, N.F.de Rooij: "Monolitisches Kraftsensorfeld", VDI-Berichte Nr.677, 1988, pp.115-118.

[17] A.R.Grahn, L.Astle: "Robotic ultrasonic force sensor arrays", Robots 8, Conf. Proc. Soc. of Manufacturing Engineers, Dearborn, 1984, pp.22-1 to 22-18.

[18] R.A.Boie: "Capacitive impedance read-out tactile image sensor", Proc. Int. Conf. Robotics, March 1984, pp.370-378.

[19] P.P.L.Regtien, R.F.Wolffenbuttel: "Tactile imaging sensor", Proc. Sensor'85, May 1985, Karlsruhe (FRG), pp.1.51-1.58.

[20] R.F.Wolffenbuttel, P.P.L.Regtien: "Integrated capacitive tactile imaging sensor", Proc.16th Int.Symp. on Industrial Robots, Brussels, 1986, pp.633-641.

[21] M.H.E.Larcombe: "Why carbon fibres can give robots a strong sense of grip", Sensor Review, $\underline{1}$ (1981), pp.58-59.

[22] W.E.Snyder, J.St.Clair: "Conductive Elastomers as sensor for industrial parts handling equipment", IEEE Trans. Instr. & Meas., $\underline{IM-27}$, (1978) 1, pp.94-99.

[23] B.E.Robertson, A.J.Walkden: "Tactile sensor system for robotics", Proc. 3rd Int. Conf. on Robot Vision and Sensory Controls, Nov. 1983, pp.572-577.

[24] N.Ghani, Z.G.Rzepczynski: "A tactile sensing system for robotics", in: L.O.Hertzberger, F.C.A.Groen (eds.), "Intelligent autonomous systems", North Holland, Amsterdam 1987, pp.441-445.
[25] M.H.Raibert, J.E.Tanner: "Design and implementation of a VLSI tactile sensing computer", Int.J.Robotics Research, 1 (1982),3, pp.3-18.
[26] G.Gerliczy, R.Betz: "SOLEF PVDF biaxially oriented piezo and pyroelectric films for transducers", Sensors & Actuators, 12, 1987, pp.207-224.
[27] R.G.Schwartz, J.D.Plummer: "Integrated silicon-PVF2 acoustic transducer arrays", IEEE Trans Electron Devices, EP-26, (1979) 12, pp.1921-1931.
[28] P.Dario, C.Domenici, R.Bardelli, D. de Rossi, P.C.Pinotti: "Piezoelectric polymers: new sensor materials for robotic applications", 13th Int.Symp. Industrial Robots, April 1983, Chicago, Ill., (USA), pp.14-33 to 14-49.
[29] D.L.Polla, W.T.Chang, R.S.Muller, R.M.White: "Integrated zinc oxide-on-silicon tactile sensor array", Proc. IEDM, 1985, pp.133-136.

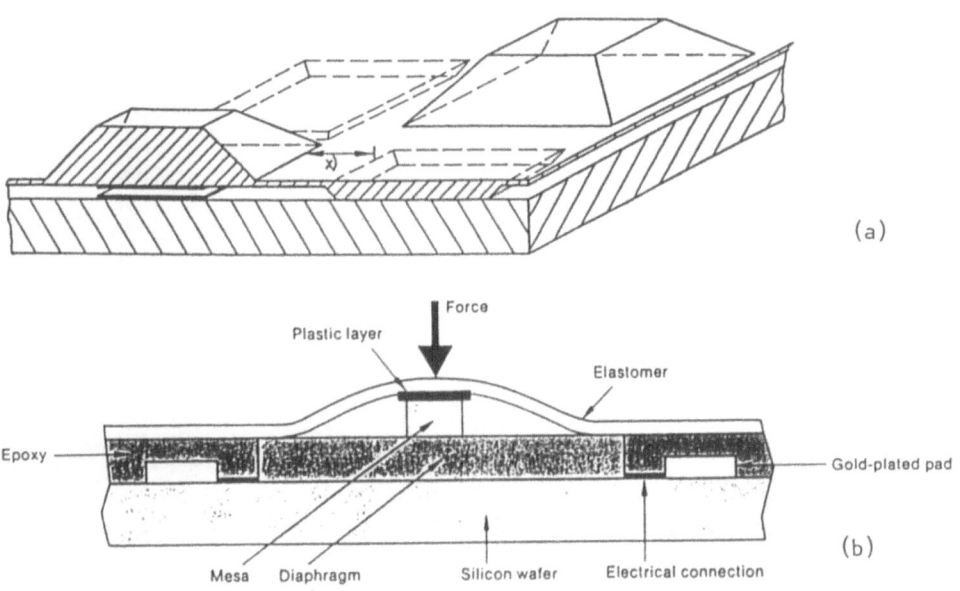

Fig.1. from (a) [16], (b) [2]

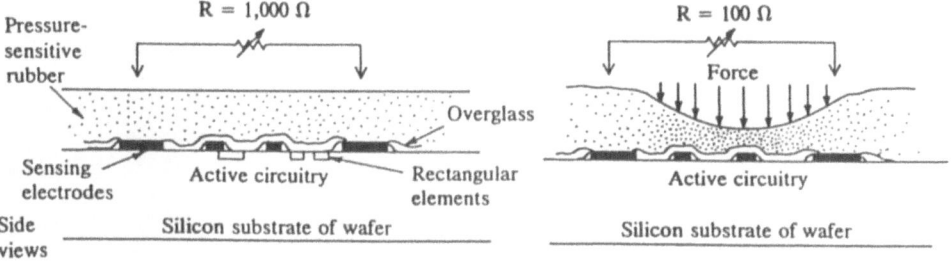

Fig.2. from [25]

Multiresolutional Laser Radar

Rudolf Schwarte
Institut für Nachrichtenverarbeitung (INV)
Zentrum für Sensorsysteme (ZESS)
University of Siegen
5900 Siegen
W–Germany

Abstract

The INV–laser radar working group has spent considerable effort in the design and implementation of an advanced 3D–sensor, envisaging medium and large scale applications in robotics and automation, especially real–time measurements of shape, position, orientation and movement of three–dimensional objects as well as environment perception, path finding and docking of robotic vehicles. Using laser pulses instead of microwave pulses the operating principle is quite similar to that of a microwave radar. Realization of the projected laser radar, however, imposes a number of very crucial points.

This paper describes the main difficulties being encountered and presents elaborated solutions like an improved sensor head with so called mirror optics, a new fiber optic concept for ultimate precision, time windowing facilities for target selection, amplitude control, closed–loop real–time data processing for contour tracking and, finally, fast contour estimation procedures with special Kalman–jump–filter algorithms. The INV–laser radar is able to measure three coordinates, reflectivity and range rate of passive 3D–surface points at a primary measuring rate of some 10.000 Hz. After data processing the output data rate is reduced to some 100 Hz performing mm–resolution.

Experimental results of the INV–laser radar demonstrate an extraordinary precision, the ability to find and identify simple 3D–objects by measuring cross–sections and automatic 3D–edge resp. contour tracking as well as the process of taking a 3D–picture of a small workpiece by raster scanning. Since the latter mode consumes too much time it is more efficient first to analyze the 3D–scene by means of a 2D–vision system. Thereafter the laser radar is directed only to the interesting edges and structures in order to add the corresponding depth information.

NATO ASI Series, Vol. F 63
Traditional and Non-Traditional Robotic Sensors
Edited by T. C. Henderson
© Springer-Verlag Berlin Heidelberg 1990

1 Introduction

Since most of the tasks in production, inspection and transportation are of three–dimensional geometric nature adequate 3D–sensors are inevitable. Forced by the international industrial competition in flexible production automation there is an urgent need for better 3D–sensors, i.e. more versatile, intelligent, precise and faster.

Far away from the high performance of biological vision systems there are a number of competing 3D–machine vision systems, practically all of them candidates to intensive investigation and further development.

Most important optical 3D–measurement procedures are

- **Laser triangulation**, a fast, precise and powerful tool for short ranges [1] Triangulation sensors for 2D– and 3D–applications are already commercially available. [2] describes a triangulation laser radar mounted on an autonomous mobile robot.
- **Structured illumination**. Use of multiple triangulation, e.g. by means of light stripes and CCD–receivers
- **Laser interferometry and holographics**, no absolute range measurements but extreme resolution of sub–micrometers using coherent light.
- **Defocussing procedure**, a sort of triangulation, spending more time for higher resolution
- **Stereo–vision**. Apart from special applications 3D–stereo vision still lacks from real–time–capability and reliability. In [3] and [4] a new approach is presented providing depth information of vertical edges at video frame rates.
- **CW–Laser radar**. Time–of–Flight (TOF) resp. distance is measured by means of RF–phase shift of the target reflection of a CW–modulated laser beam. [5] may stand for an advanced system developed for robotics applications.
- **Pulsed–laser radar**. Direct TOF–measurement of the reflected laser pulses for 3D–applications has been investigated in [6 to 10].

The latter procedure offers some advantageous aspects, e.g. direct and absolute range measurement, no shadowing effects, extremely large measuring range from cm to several 100 m on non–cooperating targets and remote sensing as well as EMI–safety using a fiber driven sensor head. Thus the INV–working group decided to design and to implement a new versatile, precise and flexible 3D–sensor for robotics, automation and in–door navigation on the base of a pulsed laser radar. Experimental investigations revealed severe problems. Finally appropriate solutions have been found, leading to new concepts and facilities.

2 Laser–Radar Concept and Realization

2.1 Overall system design

Fig. 1 shows a block diagram of the implemented laser–radar system. The principle of operation is relatively simple. The time control unit generates a pulse to trigger the laser transmitter and to start the time–of–flight evaluation. The laser diode fires a short pulse (about 15 watts, 5 ns–duration, 500 ps–rise time and 905 nm wave–length) through a glass fiber to the optical sensor head. If a target at distance R is hit a small fraction of the pulse energy may be reflected and received by the sensor head and fed through a glass fiber to the avalanche–photo–diode receiver. The output voltage $U_E(t)$ is given to a constant fraction discriminator (CFD) in order to derive a stop–signal which stops the TOF–evaluation counter. Besides the pure TOF = $2R/Co$ from the sensor head to the target and back there are additionally delays of the transmitter, receiver, CFD and fiber optics, which have to be eliminated from the total Start–to–Stop time. This is supplied by reference measurements using the reference fiber indicated in Fig. 1. The selection of either target or reference–pulse measurements is managed by means of time windowing the TOF–evaluation unit.

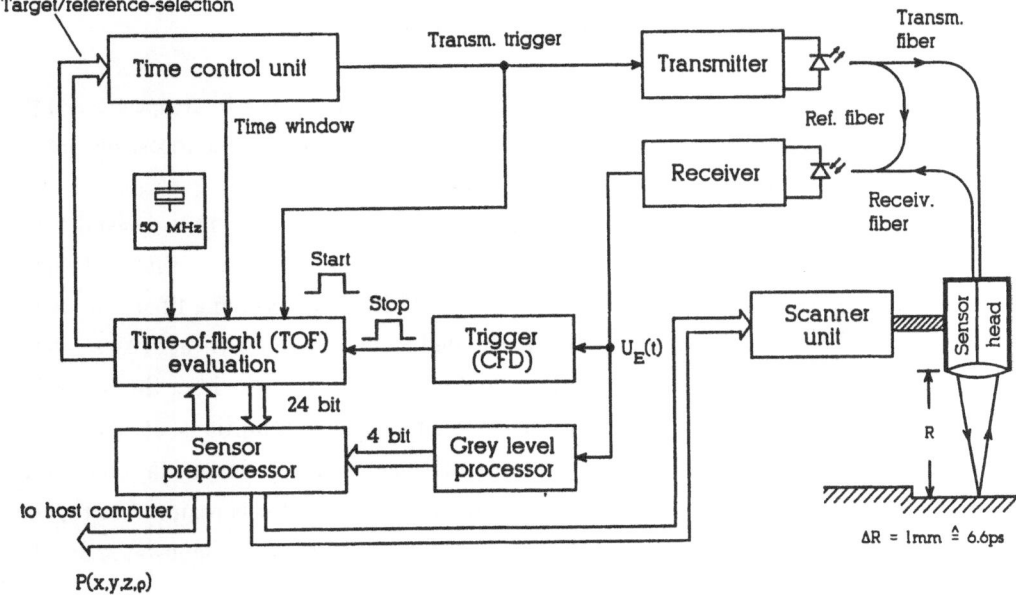

Fig. 1   Block diagram of the implemented laser–radar system

The sensor preprocessor has to evaluate the target–point coordinates from the 24 bit–range values and the scanner position at a measuring rate of up to 20 kHz. Simultaneously the reflectivity resp. the grey level of the 3D–points is taken into account by 4 bit values. Thus the amplitude of the target reflection can be controlled in order to eliminate too small and too large levels causing erroneous range data.

It is obvious that the sensor preprocessor requires a high performance parallel architecture for real–time processing all the tasks of target/reference–data extraction, connection of corresponding range and scanner data, amplitude control, test of reasonableness, time windowing, edge detection, fast contour estimation and contour tracking. [11] describes the parallel data processing system, based on 68XXX– and TMS 320/20/C25 processors, including aspects of fault tolerance. Besides fast signal–processing demands there are further crucial points to be solved.

Especially a desired range resolution of 1 mm which is equivalent to a turn–around– TOF of $6.6 \cdot 10^{-12}$s $= 6.6$ picoseconds causes extraordinary requirements to timing circuitry. The application of an optical reference is obligatory. But there still remain several time errors to be compensated or eliminated.

2.2 Fiber optic concept and mirror optical sensor head

Most of the problems mentioned above can be solved by means of a fiber optic arrangement (Fig. 2) in connection with a special design of the sending and receiving optics forming a "mirror–optical"–sensor head in Fig. 3 [13, 14].

Fig. 2 demonstrates how the small optical reference pulse (1% of power) is derived from the transmitting fiber (400 μm core diameter) over directional coupler 1 into the reference fiber (200 μm) and then over directional coupler 2 into the receiving fiber (600 μm). All fibers are of the step–index PCS (plastic cladded silica) – type with numerical apertures of about 0.3. Mode mixers have to be introduced in order to get rid of the temperature and operating–point sensitive mode–dependent coupling conditions at the laserdiode/transmitting fiber interface and to approximate an homogeneous modal phase and power distribution within the fibers. A controlled optical attenuator is required to reduce the wide range of reflected pulse power extending from milliwatts to nanowatts due to target distance and reflectivity.

Construction and function of the sensor head is outlined in Fig. 3. There is only one lens. The space between the fiberends and the lens is symmetrically halved by means of a thin two–sided mirror.

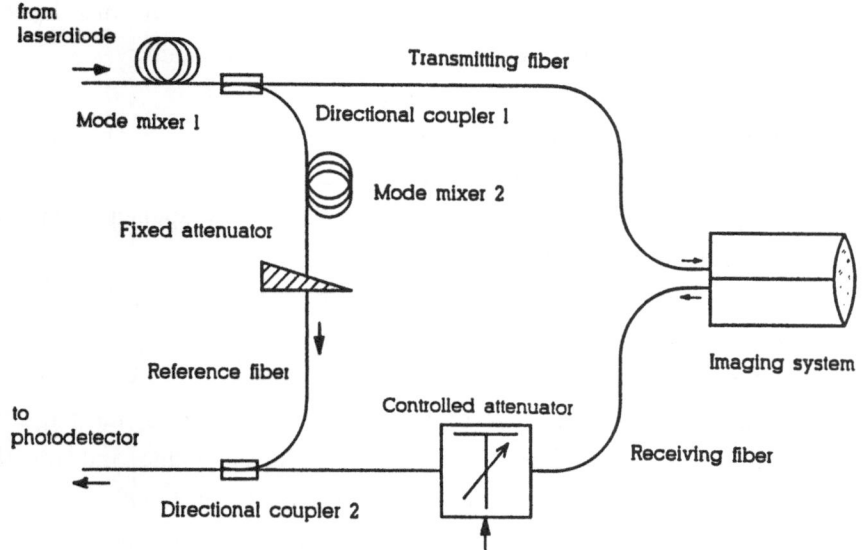

Fig. 2   A fibre–optic concept for the pulsed laser radar

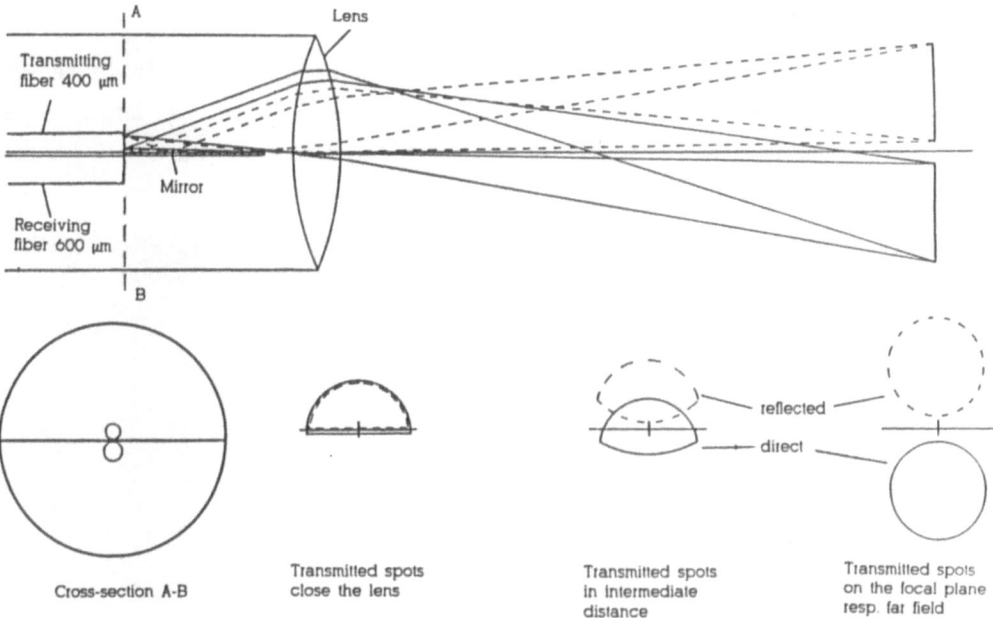

Fig. 3   Design and functioning of the mirror–optical sensor head

The transmitting fiber lies on one side close to the mirror and the receiver fiber lies opposite both ending at the focus plane as close as possible to the focus point.

Thus the transmitting area i.e. fiber end is located in the mirror image of the receiving area and vice versa, automatically providing an optimal focus situation. A thorough description and analysis is found in [15]. Compared to the usually applied parallel or coaxial optics the advantages are manifold:

- There is no blind angle or blind range. The measuring range extends almost from the lens to infinity.
- Overlapping of the transmitting and receiving beams is representative for precise TOF–measurement. The cross–sections of the receiving beam is symmetrical to that of the transmitting beam being illustrated in Fig. 3. Solid lines indicate the direct and dashed lines the reflected imaging of the transmitting fiber end.
- The amplitude dynamics as a function of distance is reduced considerably within the m–range.

The reason for using a greater receiving fiber diameter (600 μm instead of 400 μm) are aberrations enlarging the transmitted illumination spot in practice. An improved version of this construction is presented later with respect to Fig. 11.

In Fig. 4 some mirror–optical sensor heads for different range applications are shown. Since the minimum received power is about 10 nW with an APD–receiver larger ranges require larger lens diameters as simply seen from the ideal power formula.

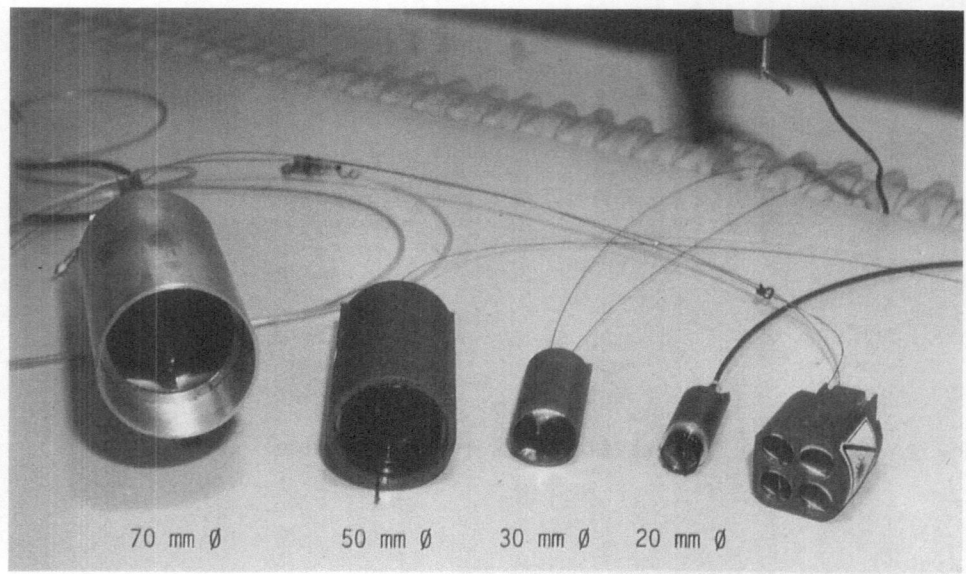

Fig. 4   Some mirror–optical sensor heads

For a lambertian reflector with reflectivity Rho, a receiving area A_{rec} and distance R the relation of the transmitted and received power P_{rec} resp. P_{trans} is approximately given by

$$P_{rec} = \frac{A_{rec} \cdot Rho}{\pi \cdot R^2} \cdot P_{trans}$$

At the right hand Fig. 4 shows a multiple sensor head, investigated for contour tracking of planes. This facility of supplying several sensor heads with one laser–radar system is enabled through windowing techniques and multi–target selection to be outlined in the following section.

2.3. Time–of–flight evaluation and signal processing

In order to determine the TOF with ps–precision despite fast amplitude variation of the target reflections, e.g. during scanning, the conventional constant fraction discriminator CFD could be improved by adding an offset–comparator as shown in Fig. 5a. By that way the zero–crossing–detector (here Plessey SP 9685) gets a positive offset–voltage at the moment of pulse arrival. The experimental result in Fig. 5b demonstrates, that the amplitude dependent time–walk could be diminished considerably.

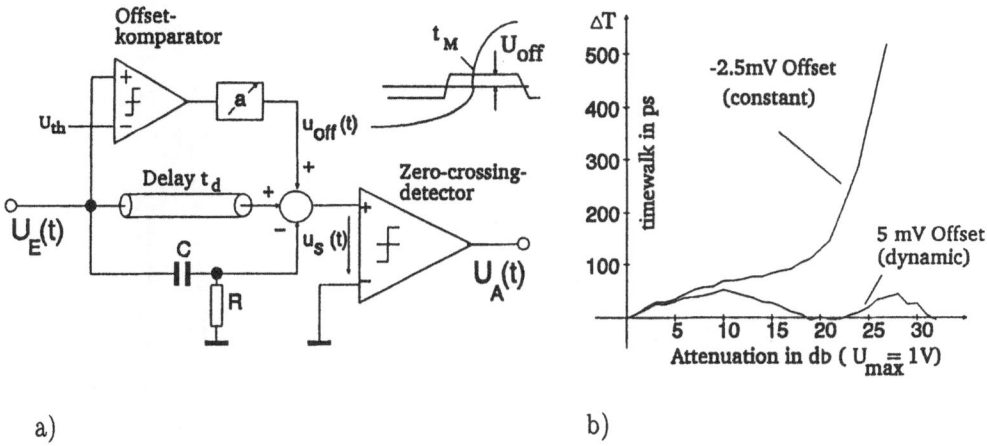

a)   b)

Fig. 5   Improved Constant–Fraction–Diskriminator (CFD) for Time–of–flight (TOF) evaluation. a) Circuit design b) Amplitude dependant time–error with and without offset–switching

Assuming of a tolerated time error of e.g. $\pm$ 25 ps the amplitude range was increased by about 26 dB. [15] comprises a detailed description of this matter as well as of the further TOF–evaluation circuitry like

- partial TOF–stretching by a factor of about 1000 using a dual–slope time–to–digital converter thus
- enabling a TOF–quantisation to be performed at ps–resolution using conventional 50 MHz–counting;
- a pulse–amplitude peak detector required for the optical attenuator control and a digitizer for amplitude control resp. grey–level processing;
- time–windowing for reference resp. target selection enabling system self–test and control.

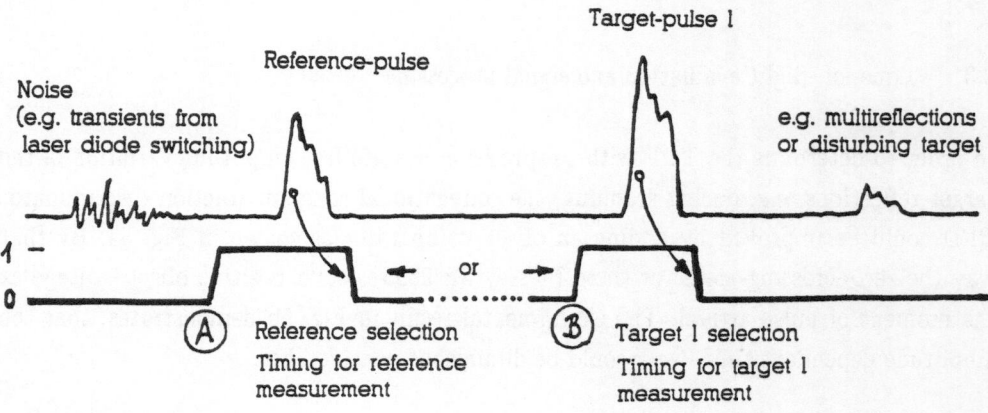

a) Time windowing

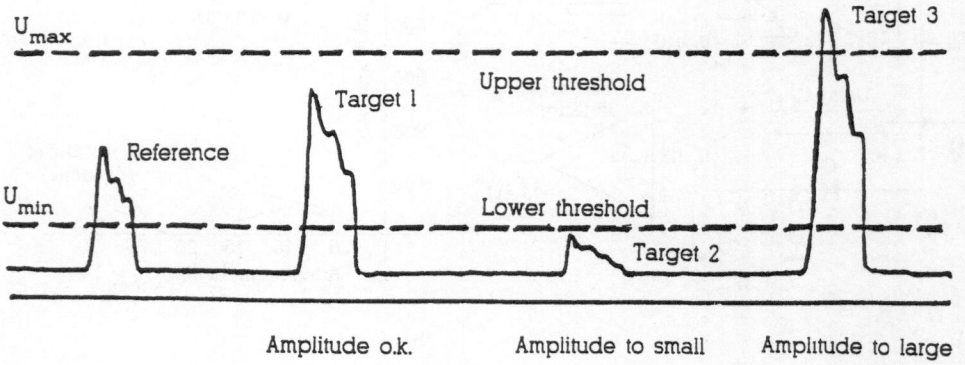

b) Amplitude windowing

Fig. 6 Signal processing for target selection and amplitude control

Fig. 6a illustrates how to select the reference pulse or the desired target. The leading edge of the gating pulse opens the time window, e.g. at "A" for reference measurement resp. at "B" for target 1 measurement. The first pulse of sufficient amplitude shuts the time window causing the trailing edge.

In practice this facility allows range measurements through non solid targets like a curtain or through a glass window or calibration by means of a second reference fiber.

Fig. 6b demonstrates the amplitude selection. Pulse reflections with amplitudes smaller than the lower or larger than the upper threshold are not accepted for further processing. Digitizing the admitted amplitude range between provides to pick up grey–level 3D–images.

3 Experimental Results

3.1 Range resolution and accuracy

Due to noise and jitter of the whole chain of electronic components the single shot TOF–measurement suffers from a standard deviation of about 60 ps resp. 9 mm at a measuring rate of 10 kHz – even providing low background illumination or the use of an interference filter of 905 nm. For higher accuracy more time has to be spent, e.g. an averaging process over 100 range values improves the standard deviation by the factor ten at a data rate of 100 Hz. Fig. 7 demonstrates the extraordinary accuracy achieved with a pulsed laser radar. Almost independent from the absolute distance in the meter— range a non cooperating target being shifted within mm–steps can be measured absolutely at submillimeter resolution.

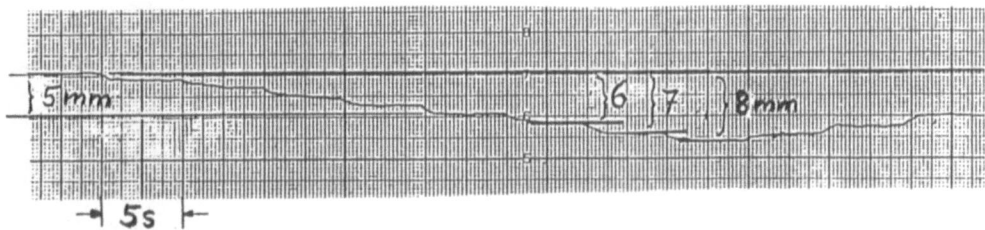

Fig. 7   Distance measurements to a passive target slowly being shifted within mm— steps – demonstrating the absolute accuaracy of a pulsed laser radar

3.2 3D–Edge resp. contour tracking

Software for edge–finding and tracking has been implemented in order to get immediate
information about object type and position in robotics, faster than raster scanning is able
to perform. For example in a simplified task an unknown type of cylinder ring has to be
determined as shown in Fig. 8. At first the sensor head scans across the scene to find the
object. After meeting an edge the system starts an analysis–task concerning the contour
recognition in the x/y–plane by tracking clock–wise one round. Thus the outline circle,
the diameter and the center point are provided. Thereupon the cross–section is measured
by only one scanning motion – here without edge restoration by deconvolution [15]. Due
to the extraordinary depth of focus of this laser radar system such 3D–object information can be acquired at distances from centimeters to several meters.

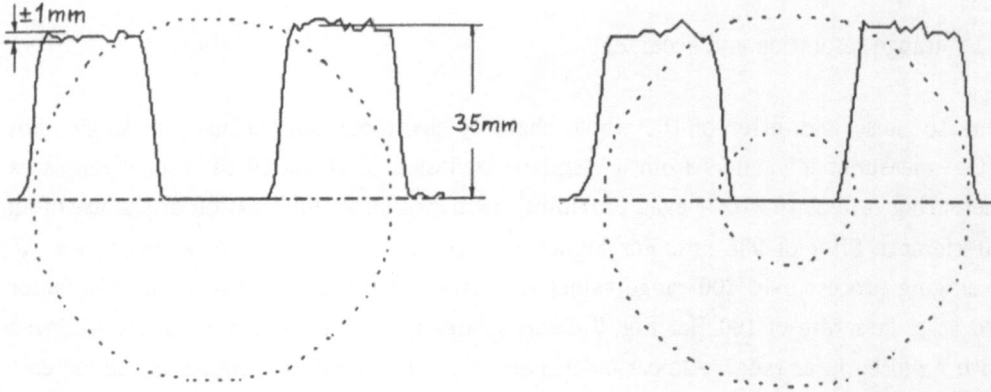

Fig. 8   Automatic objekt finding, 3D–edge tracking and cross–section evaluation of a
          cylindre ring (here without edge restoration by aperture deconvolution)

3.3 Example of 3D–imaging by raster scanning with and without special data proces-
       sing

For 3D–object–recognition of arbitrary objects a raster scan mode may be applied. Fig.
9a shows the workpiece taken for example. First the analysis–tasks mentioned above
provides a coarse dimensioning of the 3D–object. A special raster–scan–task uses this
information to determine the field of view for object acquisition. The quantity of scan–
lines depends on the coarse dimensions of the object being found.

For measurement–data processing we distinguish three modes.

1. Processing of a cluster of range values related to only one position measurement in the x/y–plane providing a time–averaged range value.

2. A number of range–values along the scan–line are taken into account for data processing in time and one dimension in space.

3. Several adjacent scan–lines are incorporated in contour estimation over time and two dimensions in space.

Fig. 9 illustrates two modes of object acquisition. In Fig. 9b a recursive averaging algorithm with respect mode 1 and in Fig. 9c Kalman–filter algorithms enhanced by means of an edge detector have been applied.

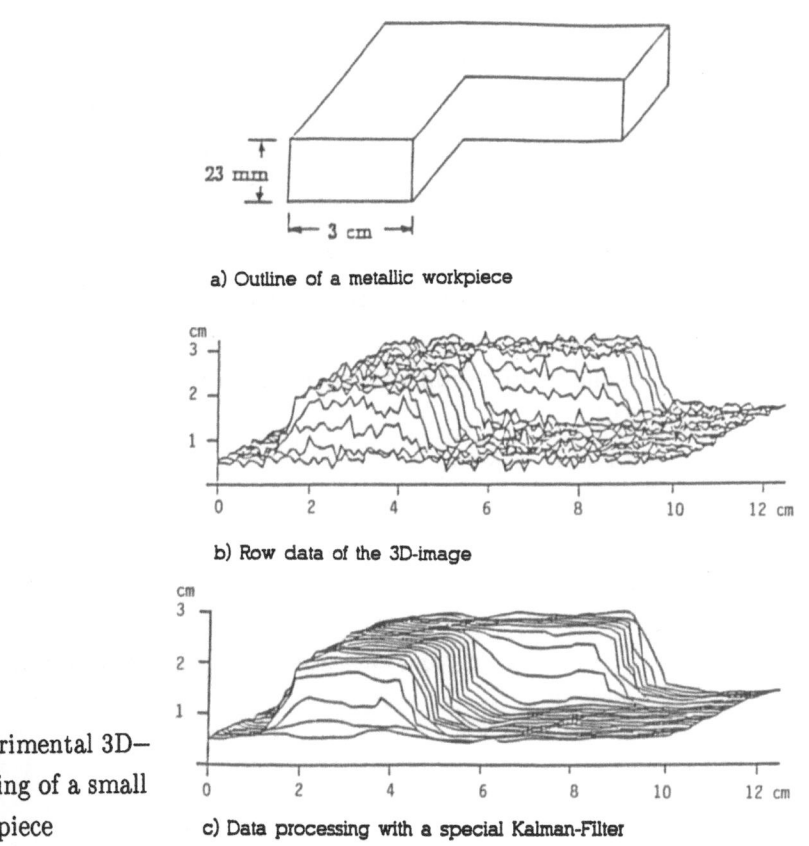

a) Outline of a metallic workpiece

b) Row data of the 3D-image

Fig. 9 Experimental 3D–imaging of a small workpiece

c) Data processing with a special Kalman-Filter

Based on preliminary studies of Loffeld [12, 18, 19] new results in on–line contour estimation have been achieved recently. Using a sort of switched Kalman–filter algorithms – a so–called Kalman–jump filter [20] – proves new facilities in on–line restoration of arbitrary 3D–contours with unknown jumps, sharps bends and steep slopes as demonstrated in Fig. 10. The test contour of Fig. 10a is corrupted by Gaussian noise, simulating the measurement noise. Without any jump standard–Kalman filtering is applied. Special edge detection algorithms estimate time and amplitude of jumps within only 10 to 20 samples, reducing the total noise energy of this contour to less than 10 percent in Fig. 10b.

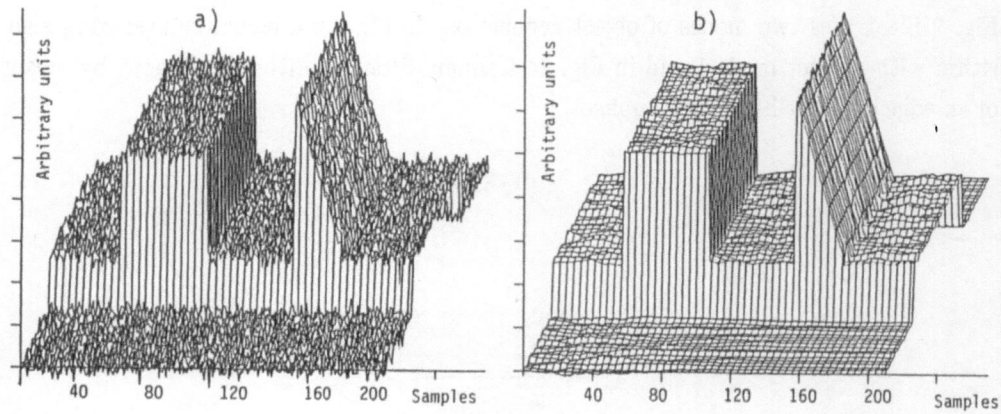

Fig. 10   Contour estimation with a so–called Kalman–jump filter [20]
 a) Simulation of the measurement noise superposing a technical contour
 b) On–line contour estimation by means of the new Kalman–jump filter

4. Conclusions and Future Aspects

The present work on the described laser–radar systems will be continued, especially by a new firm, founded in order to manufacture laser–radar instruments and systems for industrial applications.

The preliminary state–of–the–art and envisaged improvements of the INV–laser–radar may be summarized.

4.1. Summarized results

The new laser–radar concept has been implemented with encouraging results, providing some improved facilities to non–tactile medium and large scale 3D–object measurement, in particular:

- Reduced electronics to a small–sized 2–3 liter box
- Range resolution of about 1 mm at a processed data rate of 100 Hz independent of the absolute range
- Range rate resolution in the mm/s–region
- Measuring rate up to 20 kHz
- Self–test routines using the internal reference fiber
- Measuring range from cm to some 100 m on passive targets or to some 10 km upon retroreflectors
- A small, flexible sensor head provides EMI–and explosion–safety applications and simplified scanning facilities
- Time windowing performs versatile target selection and multi–sensor–head applications, e.g. multi–target or multi–direction tracking
- Amplitude windowing improves accuracy and peak detection delivers grey–level information in addition to 3D–contours.
- New special Kalman–filter algorithms enable considerable noise reduction in on–line 3D–contour acquisition even at the presence of unknown steps.

4.2   Envisaged improvements

1)   The efficiency of the sensor head can be increased considerably by means of a $90°$– two–sided mirror instead of the $180°$–mirror of Fig. 3. In Fig. 11 this two principles are compared, illustrating the cross–section of the focus plane of the sensor far–field illuminations. Arranged in the $90°$–corner the transmitting fiber produces a quartet of spots due to reflections. Thus the receiving area is increased and the scattering losses are reduced by an optimally arrangement of three receiving fibers which have to be connected to the same photodiode.

2)   The reference pulse in Fig. 2 has not to be derived from the measuring pulse since laserdiodes internally produce two synchronized pulses at both ends of the resona-tor crystal. We open the laserdiode case in order to attach the measuring–puls fiber to the front end and the reference fiber to the rear end of the resonator, thus elimi-nating coupler 1 in Fig. 2 and its attenuation and mode distortion effects.

3)   A problem of eye–safety occurs due to increased laser pulse power required for range measurements at more than about 10 meters.

It could be solved using laser diodes at about 1.55 μm instead of 0.9 μm. As far as no high–power pulsed types are available a wayout is to apply CW–types and pulse compression techniques.

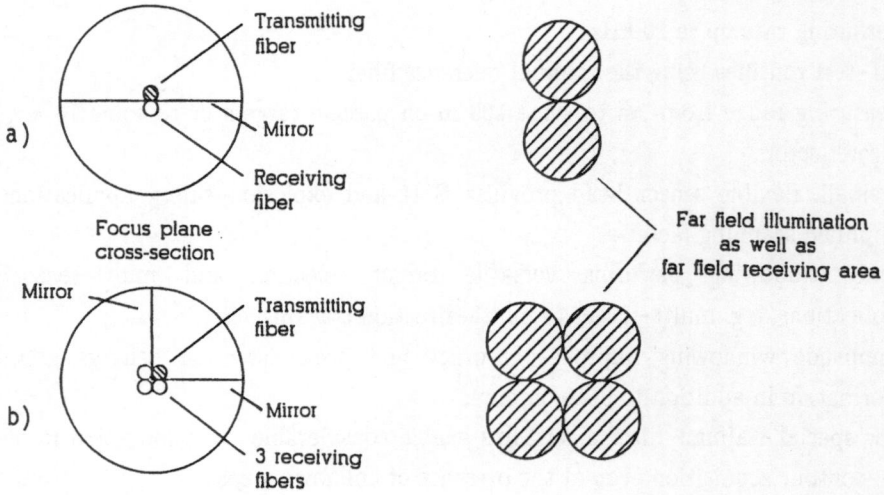

Fig. 11   Improvement of the mirror–optical sensor head

a) Realized 180°–mirror optics   b) High efficiency 90°–mirror optics

4.3 Integration of the INV–laser–radar into an hybrid system

The described laser–radar may serve as a stand–alone system in many automation tasks like

- medium and large scale 3D–object measurement, handling and recognition
- in–door and out–door navigation
- robotic and autonomous mobile vehicle guidance
- 3D–environment perception
- obstacle detection and avoidance
- path finding

First investigations prove that a significant increase of performance can be achieved by adding a commercially available 2D–CCD–camera and image–processing system. After scene analysis by a 2D–supervisor vision system – as shown in Fig. 12 – the TOF–radar system is pointed directly to acquire the task–oriented range information. Using appropriate software routines this combination offers a powerful tool with respect to time critical and real–time 3D–applications and closed–loop motion.

For fast synchronizing of the 2D– and the 3D–system we found the laser spot can be applied.

Since the accuracy of the TOF–laser–radar system may not be sufficient for most short range 3D–tasks there is a low–cost way to additionally apply laser triangulation.

Using the existing scanner system to deflect a thin CW–laser beam in cooperation with the additionally integrated triangulation–CCD–system as shown in Fig. 12 this hybrid system performs short range 3D–imaging at about 10 to 100 μm resolution too. Within the overlapping range from e.g. 20 to 100 cm due to optical base and design there is an increased reliability by means of system redundancy. Enhanced by that way the overall system additionally will provide

- small object inspection and recognition
- robotic vehicle docking
- surface inspection
- meticulous operations
- assembling tasks

Time saving software routines mentioned above for rational range finding of interesting structures within the 2D–camera's field–of–view may be applied to the TOF– as well as to the triangulation–laser–radar system.

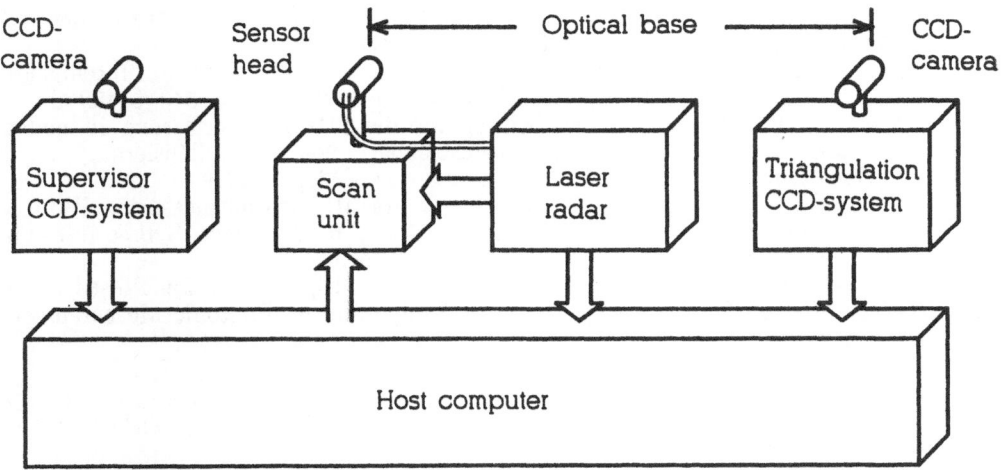

Fig. 12   Enhanced 3D–sensor system with a supervisior–2D–CCD system supported by a TOF–laser–radar and by a triangulation laser radar using the same scan unit

5. Acknowledgments

I am very endebted to my coworkers at INV, V. Baumgarten, B. Bundschuh, K. Hartmann, O. Loffeld, L. Tran Duc, D. Ley, R. Klein and to the development crew of the firm SenTec GmbH, Siegen, I. Aller, W. Graf, J. Klicker, J. Schaumann, H. Weber, for their engaged work contributing to this paper.

References

[1]   G. Hirzinger: Eine neue Generation von Robotersensoren und ihre Integration in multisensorielle Greifersysteme, Kap. 2.3 in "Intelligente Sensorsysteme", Editor J. Rogos, Springer Verlag 1989, ISBN 3–540–51488

[2]   R. Hinkel, T. Knieriemen: Environment Perception with a Laser Radar in a Fast Moving Robot, IFAC Symposium on Robot Control, Karlsruhe 1988

[3]   Löcherbach, E.: Ein Beitrag zur On–line Entfernungsmessung mittels Stereo–Bildauswertung, Doctorial thesis at the University of Siegen, 1986

[4]   Lotz, R.: 3D–Vision mittels Stereobildauswertung bei Videobildraten, Doctorial thesis at the University of Siegen to be published.

[5]   Grabowski, R., Wurll, P.: Dreidimensionale Abtastung industrieller Szenen mit optischer Lotung, 7. DAGM–Symprosium, Erlangen 1985

[6]   Määttä, K., Kostamovaara, J., Myllylä, R.: A laser rangefinder for hot surface profiling measurements, Proc. of Laser Technologies in Industry, SPIE Conference, Vol 952 A&B, 8 pp., (1988)

[7]   Kompa, G.: High Resolution Pulsed Laser Radar For Contour Mapping, Conf. Proceedings, 28th. Midw. Symp. on Circuits and Systems, 1985, P. 527–530

[8]   Schwarte, R.: Performance Capabilities of Laser Ranging Sensors', Proc. of ESA Workshop SPLAT, Les Diablerets, March 1984

[9]   Schwarte, R.: A New Concept for a Precise and Versatile Laser Range Finder and Optical Radar, Conference Proc. Laser 85, München 1985

[10] Schwarte, R. et al.: Laserradar mit Impulszeitmessung, Kap. 2.1 in "Intelligente Sensorsysteme", Editor J. Rogos, Springer Verlag 1989, ISBN 3–540–51488–0

[11] Hartmann, K.: Sensordatenverarbeitung für die 3D–Objekterfassung in einem Laserradar–System unter Berücksichtigung von Parallelisierung und Fehlertoleranz, Doctorial thesis at the University of Siegen, 1989

[12] Loffeld, O.: A New Switched Kalman Filter For 3D–Contourmeasuring With A Laser Diode Range Finder, Proc. ASST 1987, 6. Aachener Symposium für Signaltheorie, Informatik Fachberichte Nr. 153, Springer 1987

[13] Graf, W.: Fiber Optics and Receiver Design for A High Resolution Pulsed Laser Radar, Proc. IMEKO 13th International Symposium on Photonic Measurement, Hamburg, 1987

[14] Schwarte, R., German Patent Nr. 3419320 and US–Patent Nr. 4,737,624

[15] Bundschuh, B.: Modellierung und systemtheoretische Beschreibung eines laseroptischen 3D–Kontur–erfassungssystems, Doctorial thesis to be published at the University of Siegen

[16] Baumgarten, V.: Doctorial thesis to be published, Elektronische Zeitmessung im Pikosekunden–Bereich für ein Laserpulsradar.

[17] Schwarte, R., German Patent Nr. 3419117 and US–Patent Nr. 4,734,587

[18] Loffeld, O.: Ein neuartiges Switched Kalman–Filter mit geringer Wortbreite für die hochauflösende Entfernungsmessung nach dem Laserpuls–Laufzeitverfahren, Doctorial thesis at the University of Siegen, 1986

[19] Loffeld, O.: Thesis for Habilitation, University of Siegen, 1989 to be published

[20] Tran Duc, L.: Doctorial thesis to be published, Ein Kalman–Jump–Filter zur berührungslosen 3D–Konturerfassung

Olfaction
Metal Oxide Semiconductor Gas Sensors
and Neural Networks

M. W. Siegel
Carnegie Mellon University
The Robotics Institute
Pittsburgh, PA 15213
USA

Introduction

Primitive mobile life forms, and the mobile cells of higher animals, derive their motivation and direct their navigation by *chemical* senses. In the simplest cases these creatures are hard-wired to swim toward nutrient concentration gradients, and to swim against irritant concentration gradients. The vestiges in humans of the sometimes extremely sensitive, selective, and differential chemical senses of primitive forms are taste, our ability to detect and identify four classes of chemicals in water solution on our tongues, and smell or olfaction, our ability to detect and identify many gases, vapors, and complex mixtures in the air passing through our noses.

Primitive "autonomous" mobile robots derive their motivation from human directive, and direct their navigation largely via artificial visual and acoustic senses. Their autonomy is thus in the discretion they are allowed in planning and executing getting from here to there; the heres and the theres generally are dictated more-or-less explicitly by humans. An evolutionary path for sophistication of the navigational abilities of mobile robots looks self-evident, and in fact such evolution is well underway. On the other hand an evolutionary path along which mobile robots might even begin to progress toward replacing motivation by human fiat with self directed motivation is difficult to conceive, at least if the senses that drive motivation are to remain as complex as vision and hearing.

But self directed motivation of mobile robots based on one or more chemical senses is easy to imagine, and indeed is probably within the capability of existing technology. For example, we could build right now a mobile robot that would meander around a chemical plant, sniffing as it goes for gas leaks (or the vapors of liquid leaks), navigating toward them while avoiding hazards visually, yet always motivated as to overall direction (an undetermined endpoint notwithstanding) by the chemical concentration gradient, with end-point navigation perhaps directed to the offending pipe fissure or open valve by acoustic homing toward the source of the hiss, and effecting simple repairs or summoning human assistance (with appropriate hazard warning) based on the fusion of chemical, visual, and acoustic sensory information in context. Similar scenarios can easily be imagined for firefighting robots, prospecting robots, rescue robots, contraband interception robots, and others.

Toward this end, my colleagues and I have been studying chemical sensing, not with the traditional goal of inventing alternative or improved instruments for doing precise chemical analysis, but rather from the perspective of exploring alternative approaches to an artificial sense of smell to motivate autonomous activity.

First, I describe the characteristics of metal oxide semiconductor (MOS) resistors used as gas sensors, particularly for combustibles. Response, as a change in resistance, depends on combustible gas identity and concentration, sensor temperature, and also on the concentrations of other gases and vapors present, *i.e.,* response to a mixture is *not* a linear superposition of responses to the individual components. These sensors inherently respond to broad classes of compounds, but differential responses conducive to signature identification can be induced by variation of parameters such as temperature and catalyst nature and concentration. I then survey the olfactory system, in contrast with other biological sensory and sensor interpretation systems, in analogy with MOS gas sensors, and in

NATO ASI Series, Vol. F 63
Traditional and Non-Traditional Robotic Sensors
Edited by T. C. Henderson
© Springer-Verlag Berlin Heidelberg 1990

analogy between its neurological organization and the organization of simple artificial neural networks. I then describe the class of large area thick film MOS resistors that we have constructed with a spatial gradient in relative sensitivity induced by differential heating. Multiple resistance measurements along the temperature gradient yield a multi-feature signature that we analyze and interpret using artificial neural networks. In particular, I contrast the approach that emphasizes binary or nearly binary unit activities, and thus forces every mixture into a class distinct from the classes of its single species components, with an approach that is less rigidly binary, encouraging creation of only enough classes to account for the individual species present, and putting mixtures into multiple classes in proportion to the concentration of the mixture components in each class.

Metal Oxide Semiconductor (MOS) Gas Sensors

The resistance of MOS gas sensors [1] is known empirically to have a power law dependence on oxygen concentration directly and on combustible (reducing) gas concentration inversely. For first order reactions, interpretable as one molecule of combustible species X reacting with one O_2 molecule

$$X + O_2 \longleftrightarrow XO_2 \tag{1}$$

the behavior is well described by

$$R = R_o \left(\frac{[O_2]}{1 + K_1^X [X]} \right)^\beta \tag{2}$$

Clifford and Tuma [2, 3] have observed this power law response for a variety of commercial MOS sensors, and they have found the constant β for Figaro Corporation Taguchi Gas Sensor (TGS) #812 to be between 0.25 and 0.55, depending unpredictably on the particular sensor in hand.

Some insight into this behavior is provided by a seat-of-the-pants model of the interplay between conduction electrons, oxygen, and reducing gases in and on metal oxides such as SnO_2. The oxides are manufactured with the normal metal-oxygen stoichiometry disturbed in the direction that leaves them, in the absence of atmospheric oxygen, slightly reduced, i.e., $SnO_{2-\varepsilon}$ rather than SnO_2. The small excess free metal content contributes conduction electrons that make for much lower electrical resistance than is the norm for fully oxidized metals. Atmospheric oxygen captures free electrons or depletes free metal, increasing the electrical resistance. Reducing gases deplete oxygen or restore the free metal, reducing electrical resistance. Thus resistance increases with oxygen partial pressure and decreases with reducing gas partial pressure.

A simple generic equilibrium picture [4] is adequate to transform $[O_2]_{atm}$ and $[X]_o$ concentrations outside the sensor and $[e^-]_{vac}$ electron density inside the sensor under vacuum conditions into $[O_2]$, $[X]$, and $[e^-]$ inside the sensor with the atmosphere admitted. The three reactions

$$O_2 + e^- \xleftrightarrow{k_1} O_2^- \tag{3}$$

$$X + e^- \xleftrightarrow{k_2} X^- \tag{4}$$

$$X + O_2 \xleftrightarrow{k_3} XO_2 \tag{5}$$

yield, by eliminating $[O_2^-]$, $[X^-]$, and $[XO_2]$ from the three equilibrium equations and three conservation equations implied by reactions 3, 4, and 5:

$$[e^-] = \frac{[e^-]_{vac}}{1 + k_1 [O_2] + k_2 [X]} \tag{6}$$

$$[O_2] = \frac{[O_2]_{atm}}{1 + k_1 [e^-] + k_3 [X]} \tag{7}$$

$$[X] = \frac{[X]_o}{1 + k_2 [e^-] + k_3 [O_2]} \tag{8}$$

There is, of course, no general closed form simultaneous solution to equations 6, 7, and 8; however with the reasonable conjecture

$$\frac{R}{R_o} = (\frac{[e^-]_{atm}}{[e^-]})^\beta \tag{9}$$

equation 2 can be convincingly argued to be a plausible solution in physically realistic limits.

In many important cases the response is not first order. For example H_2 is observed to obey

$$R = R_o(\frac{[O_2]}{1 + K_2^{H_2}[H_2]^2})^\beta \tag{10}$$

and is thus phenomenologically equivalent (as expected) to

$$2H_2 + O_2 \longleftrightarrow 2H_2O \tag{11}$$

The response to CO depends on the concentration of water vapor, and thus vice versa the response to water vapor depends in part on the concentration of CO:

$$R = R_o(\frac{[O_2]}{1 + K_1^{CO H_2O}[H_2O][CO]})^\beta \tag{12}$$

The notations $K_2^{H_2}$ and $K_1^{CO H_2O}$ re-iterate that reactions leading to equations 10 and 12 are respectively second order in $[H_2]$ and first order in each of $[CO]$ and $[H_2O]$.

In general, terms of many orders, describing multiple reaction and response modalities, are simultaneously effective. All the parallel channels are summarized by

$$R = R_o(\frac{[O_2]}{\sum_{i=0}^{\infty}\sum_{j=0}^{\infty}\sum_{k=0}^{\infty} \cdots K_{ijk\ldots}^{XYZ\ldots} [X]^i [Y]^j [Z]^k \cdots})^\beta \tag{13}$$

with the understanding that $K_{000\ldots}^{XYZ\ldots} \equiv 1$ for all mixtures X, Y, Z, $\cdots$. Since reaction orders that are ratios of small integers are common, the indices $i, j, k, \ldots$ symbolize the rational numbers, not just the integers. However, in practice, few processes are observed at orders greater than second.

The rate constants $K_{ijk\ldots}^{XYZ\ldots}$ depend on the rate of collisions, and on the probability that on any given collision the reaction will actually occur. The rate of collisions depends on the translational speed which is proportional to the square root of the temperature. The reaction rates then depend on the products of the rate constants and powers of the number densities of the reactants appropriate to the orders of the reactions. The number densities are proportional to the concentrations $[X], [Y], [Z], \cdots$, i.e., the $K_{ijk\ldots}^{XYZ\ldots}$ essentially measure the reaction rates per unit concentrations. The probability that a collision will result in a reaction depends primarily on an activation energy E_a which is supplied by thermal motion. The reaction probability is then given by the Boltzmann factor $e^{-E_a/kT}$. The temperature range of interest is usually rather small, rarely more than a factor of two or three between room temperature ($\approx 300\ K$) and thermal destruction of the sensor. Activation energies are typically $\approx$ 1.5 eV, some sixty times the energy of thermal motion at room temperature. Thus between room temperature and thermal destruction the collision frequency changes by less than a factor of two, whereas the Boltzmann factor may change by several orders-of-magnitude. The dependence of the rate constant on temperature is thus well approximated by

$$K_{ijk\ldots}^{XYZ\ldots}(T) \approx K_{ijk\ldots}^{XYZ\ldots}(0) e^{-E_a/kT} \tag{14}$$

I have given no indication, so far, of the magnitude of the constants in the model as it might be applied to a real material such as SnO_2, i.e., of the sensitivity of MOS sensors to gases of potential interest. This is the good news: high sensitivities are easy to obtain, e.g., the resistances of the TGS #812 sensors discussed by Clifford and Tuma typically drop from $10^4\ \Omega$ to $10^3\ \Omega$ in response to a change in $[H_2]$ from 100 ppm to 1000 ppm. The model suggests no reason to expect any intrinsic specificity, other than the low level specificity of response with decreasing resistance to reducing gases, increasing resistance to oxidizing gases, and not at all to more-or-less inert gases. This then is the bad news: MOS sensors are not particularly selective, so while they are good for indicating the presence of *some* combustible gas, they are, one at a time, unsuitable for reporting *which* combustible gas they have detected. On the other hand arrays of sensors with even small differential sensitivities from element to element offer the opportunity both to identify and to quantitate gases and gas mixtures by a variety of signature separation methods.

Olfaction and Other Senses

In contrast to the visual and auditory senses, the human chemical senses of taste and smell are poorly understood. This has more to do, I think, with the essentially parametric nature of sight and hearing *vs* the essentially non-parametric nature of taste and smell than it has to do with any particular complexity of biological implementation at the cellular transducer level. In fact the chemical sense organs are structurally simpler, more primitive and apparently less differentiated than are their optical and acoustic counterparts, so the relative levels of our understanding would be reversed were understanding dominated by physiology.

The three types of color vision sensory cells are, via their incorporation of red, green and blue color pigments, sensitive to three overlapping but distinct bands of the optical continuum. Auditory sensory cells cover more but narrower bands, each receptor mechanically tuned to a specific vibrational frequency. Thus color vision and sound are *parametric* sensory modalities: each cellular transducer can be characterized by a single parameter, center frequency of its optical or acoustic response respectively. Taste, with salt, sweet, sour, and bitter constituting four primaries, is often compared with color vision and its red, green, and blue primaries, but the analogy is superficial. Whereas color is a *measurement* along a continuum spanning the optical frequency range from 4×10^{14} Hz (red) to 1×10^{15} Hz (blue), and pitch is a *measurement* spanning the acoustic frequency range from somewhere above 16 Hz (C_o) to somewhere below 32 kHz (C_{11}), there is no such continuum along which salt, sweet, sour, and bitter fall.[1] The Doppler shift can transform red into blue, or buzz into beep, but no relativity effect can turn salty (presence of sodium-like metal ions) into sweet (presence of certain carbohydrates with OH appendages). Thus taste is a more like a *classification* than a *measurement*.

Whether smell is more like a *classification* or more like a *measurement* turns out to be unresolved. Amoore [5] correlated elaborate odor classification trials by human subjects with stereochemical data. His results support a geometrical lock-and-key classification model reminiscent of Lucretius: a few tens of primaries (camphor, musky, floral, peppermint, ether, ...) each corresponding to a molecular shape (sphere, disk, disk-plus-tail, wedge, rod, ...) from which a set of shape-sensitive receptors can be inferred. Small molecules might mate with different receptors by presenting themselves in different orientations, and huge molecules might fit different side group protuberances on their complex surfaces into a variety of receptors. Thus a single type of intricately shaped molecule might evoke an olfactory symphony, whereas a mixture of chemically different but more-or-less identically shaped molecules might stimulate but a single note.

On the other hand Gesteland [6], with the Lettvin and Pitts of frog's-eye fame [7], studied the frog's olfactory epithelium and concluded that every cell was sensitive (with either an excitatory or inhibitory effect), in varying proportion from cell to cell, to a large subset if not all the chemicals that the frog could smell:

> "... if we choose pairs of odours which smell similar to each other or which are sterically similar by Amoore's categories, we can often find a single fibre clearly discriminating between them without too much searching ... it would be possible then, by having very many fibres, to distinguish very many compounds, for we would treat the characteristic of every fibre as a separate dimension."

---

[1]Not quite: sour being the taste sensation of acids and bitter being the taste sensation of bases, sour and bitter do fall on the *pH* continuum. But the deadband encompassing the neutral region around *pH 7* (water) is so large that it cannot really be said that the sour-bitter components of taste measure *pH*.

Although Gesteland *et al* are careful to remind us that frogs are frogs and people are people, the contrast between their conclusions drawn from direct measurements on cells stimulated by odors and Amoore's conclusions drawn from human perception experiments is so stark as apparently to leave the door open for all manner of speculation, which Lettvin and Gesteland [8] do unabashedly in a companion paper to Gesteland, Lettvin, and Pitts [6].

Whatever lack of resolution there is about the olfactory transducers' discriminatory abilities by chemical selectivity (in contrast to computational discrimination), the structure of the wiring diagram of nerve fibers connecting the olfactory sensor layer with the olfactory cognition region of the brain is noncontroversial. With unusual simplicity, presumably due to the antiquity of the machinery, a single layer of interconnections separates the sensory apparatus in the nose from the cognitive apparatus in the brain. The roof of the human nasal cavity contains sensory patches, each about *5 cm^3* in area, comprised of about *50 x 10^6* ciliated olfactory cells. Each cell sends a single nerve fiber directly up, through the cribriform plate that separates nasal cavity from brain, into the adjacent olfactory bulb on the brain side of the plate. About *1000* primary receptor neurons converge on each secondary mitral cell. Axons from the mitral cells connect the olfactory bulb to the cognitive areas of the brain. In short, the olfactory apparatus looks a lot like a three-layered perceptron [9, 10] with a layer of sensory units corresponding to the cells of the olfactory patch at the top of the nasal cavity, a layer of associative units corresponding to the mitral cells of the olfactory bulb on the other side of the cribriform plate, and a layer of response units[2] in a cognitive area of the brain found in a fold of the cortex on the under surface, near the temporal lobe [11].

The correspondence between the physiology of the olfactory machinery and the rubric of artificial neural network models of sensory perception is close, but the match to *simple* models, *e.g.,* layer-to-adjacent-layer connections with a few to a few tens of connections per unit, is arguable. It is certainly obvious that the thousand-fold convergence from sensory to associative layer goes far beyond anything that is typically simulated. Whether the sensory layer is equipped with one class of sensors each of which has a different set of sensitivities to essentially every odorous chemical, or alternatively, equipped with several classes of sensors that pre-classify according to some chemical criteria, *e.g.,* molecular shape, but within each class may have a broad range of sensitivities to a variety of species, is also unclear.[3] The MOS gas sensor that we have built and are about to describe has multiple sensory elements that differ from each other quantitatively rather than qualitatively, and are thus analogous to either the former one class olfactory model, or to one class or channel of the latter multiple class olfactory model.

Sensor, Apparatus, and Measurements

Our sensors are fabricated on alumina ceramic substrates intended for mounting hybrid circuits. The size and shape of the substrate is compatible with a 20-pin dual-in-line integrated circuit. A gold contact pad and electrode pattern is printed, then fired at high temperature onto the substrate, as shown in Figure 1. Next a layer of a modified commercial SnO_2 semiconductor ink is screen printed over the electrodes. The modified SnO_2 ink and the screen printing technique provide a cost-effective method to fabricate a high quality sensor array. Catalysts such as fine Pt metallic powder are blended into the commercial inks to enhance sensitivity. The organic binders in the ink are burned away by re-firing. Detailed descriptions of the fabrication procedure, which is actually carried out by our collaborators at Oak Ridge National Laboratories, is given in [12, 13, 14].

---

[2]I prefer the graphic *sensory-*, *associative-*, and *response-layer* terminology employed in the early perceptron models of Rosenblatt [9] and Block [10] to the *input-*, *hidden-*, and *output-layer* terminology used in most of the current literature on simulated artificial neural networks.

[3]It is clear that the specialized olfactory apparatus of certain invertebrates, *e.g.,* moths that respond to a single molecule of their species sex attractant, illustrates an entirely different approach to chemical sensing, an approach in which astounding sensitivity has been purchased by dedicating enormous resources to apparatus with but a single narrow function.

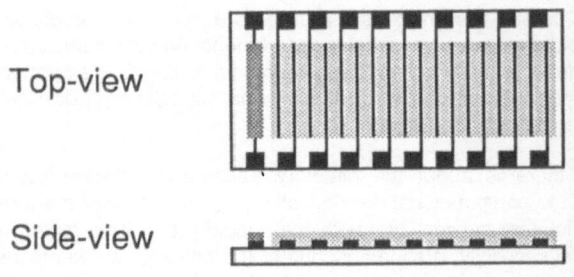

Figure 1. INTEGRATED GAS SENSOR ARCHITECTURE

The gold metallization is shown in black, the SnO_2 sensing layer shown crosshatched. The disconnected element on the left is an on-chip heater.

For controlled testing the sensor is mounted in a holder that incorporates pressure contact probes, electrical heaters, and temperature sensing devices. A temperature gradient is maintained along the length of the sensor by supplementary heat applied to one end of the substrate with a small, high power ceramic heater. The temperature difference between the hot end and cold end of the sensor surface is typically about *100 C*. The assembly is installed in an air-tight metal test chamber supplied with a constant flow of either pure synthetic air or dilute sample vapor in synthetic air. Vapor samples are generated by saturating a secondary synthetic air flow in a bubbler, then diluting it in a precision flow controlled mixing system. Binary mixtures are made by turbulent mixing of the outputs of two bubblers. Vapor concentrations are computed from the relative volumetric flow rates, temperature measurements, and vapor pressure tables. A total hydrocarbon analyzer down-stream of the test chamber is used for confirmation.

Before taking any measurements, the test chamber is flushed with synthetic air at *1000 ml/min* flow rate for at least *5 hr*. Then in a typical experiment the sample gas is turbulently mixed with the main air flow and is introduced into the test chamber for *50 min*. The response patterns to various gases, vapors, and mixtures are recorded by measuring the resistances cyclically. Figure 3 shows typical data.

A simplified block diagram of the measurement and support system is shown in Figure 2. The neural network simulator is generally run off-line on the VAX, although it is PC compatible and we sometimes run it on-line.

The responses plotted in Figure 3 are $(R_o - R)/R_o$, the usual feature presented in the solid state gas sensing literature. Each sensitivity to each of the test compounds *ethanol*, *methanol*, and *heptane* increases with operating temperature in the temperature range of these experiments. At the hot end, each absolute sensitivity is three to four times higher than at the cold end, and of a similar magnitude to the TGS #812 sensor. To minimize irreversible changes due to abrupt environmental shock to the sensor, several measurement cycles are made with each sample vapor before changing to another vapor. This method, as opposed to more frequent cycling of sample type, increases vulnerability to slow drift. The drift is compensated by normalization of the data before presentation to the sensory layer of the neural network simulator.

Classification by Neural Network

A neural network is a multi-input, multi-output machine, usually simulated but sometimes realized physically, with at least two layers of neuron-like nodes designated the "input" or *sensory* layer and the "output" or *response* layer [15, 16]. Neural networks with one or more additional "hidden" or

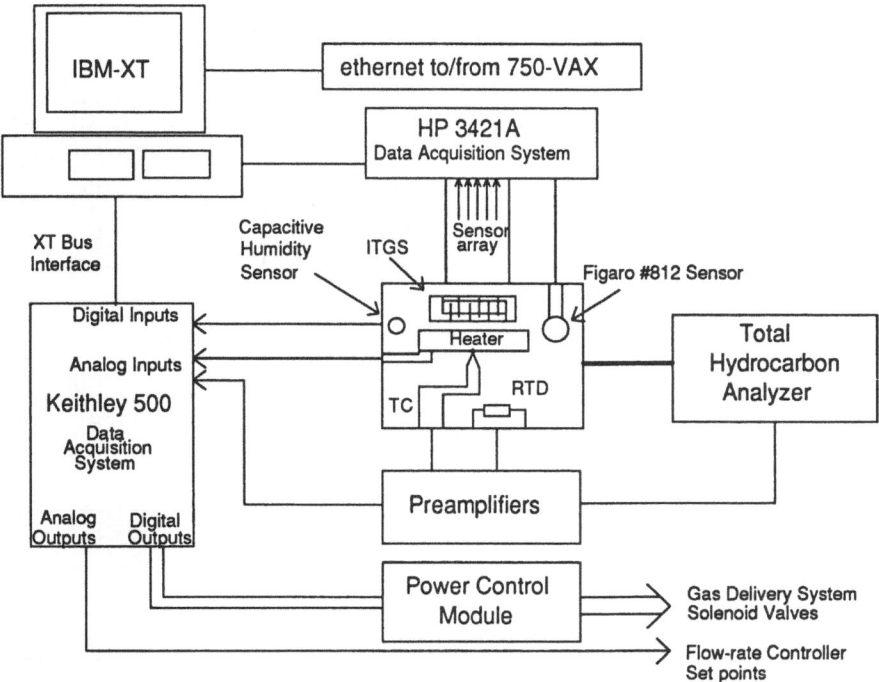

Figure 2. DATA ACQUISITION AND CONTROL SYSTEM

High precision measurement of sensor resistances is by an HP 3421A data acquisition system on the IEEE-4888 bus. Moderate and low precision monitoring of housekeeping and support sensors, and control of valves and heaters, is by a Keithley 500 data acquisition system memory mapped on the IBM-XT bus. Preamplifiers for thermocouples and RTDs are Analog Devices modules. The total hydrocarbon analyzer is a Mine Safety Appliances flame ionization detector. The FID and the Figaro TGS #812 sensor are used for checking and comparison. The capacitive humidity sensor is read by the Keithley unit at the capacitance-to-frequency output of a capacitance meter.

associative layers are of particular interest because the associative layers are able to hold internal representations that make for a higher level of abstraction than is possible without them.[4] Each layer is assigned a predetermined number of nodes or *units*. Each unit is characterized by a unique threshold (inhibition of stimulation) or bias (propensity toward stimulation), and a set of weights that characterize the strength of connections to higher and lower layers. Each unit has an output or *activity* that is a function of the sum of its inputs, *i.e.,* the activities of the previous layer weighted by the strength of the connections, offset by its threshold or bias. Rather than connecting to lower neural layers, the sensory units couple the network to some information about the external world. Similarly, rather than connecting to higher neural layers, the response units actuate some behavior in the external world. There are many variations on all the themes, including cross-connections within layers, back-connections from higher to lower layers, and connections that jump layers.

The classifying power of a multi-layer neural network depends, fundamentally, on what kinds of input-output functions it is able to represent by virtue of its topology and the definition of the input-output functionality of the individual units. The gap between the discriminant functions that a given network

[4]The classic example is the XOR problem, with two inputs and one output, and the abstraction that what associates inputs into the same class is their *dis*similarity; an XOR unit cannot be simulated by any two layer network, but it can be simulated by a three layer network with just a single associative unit.

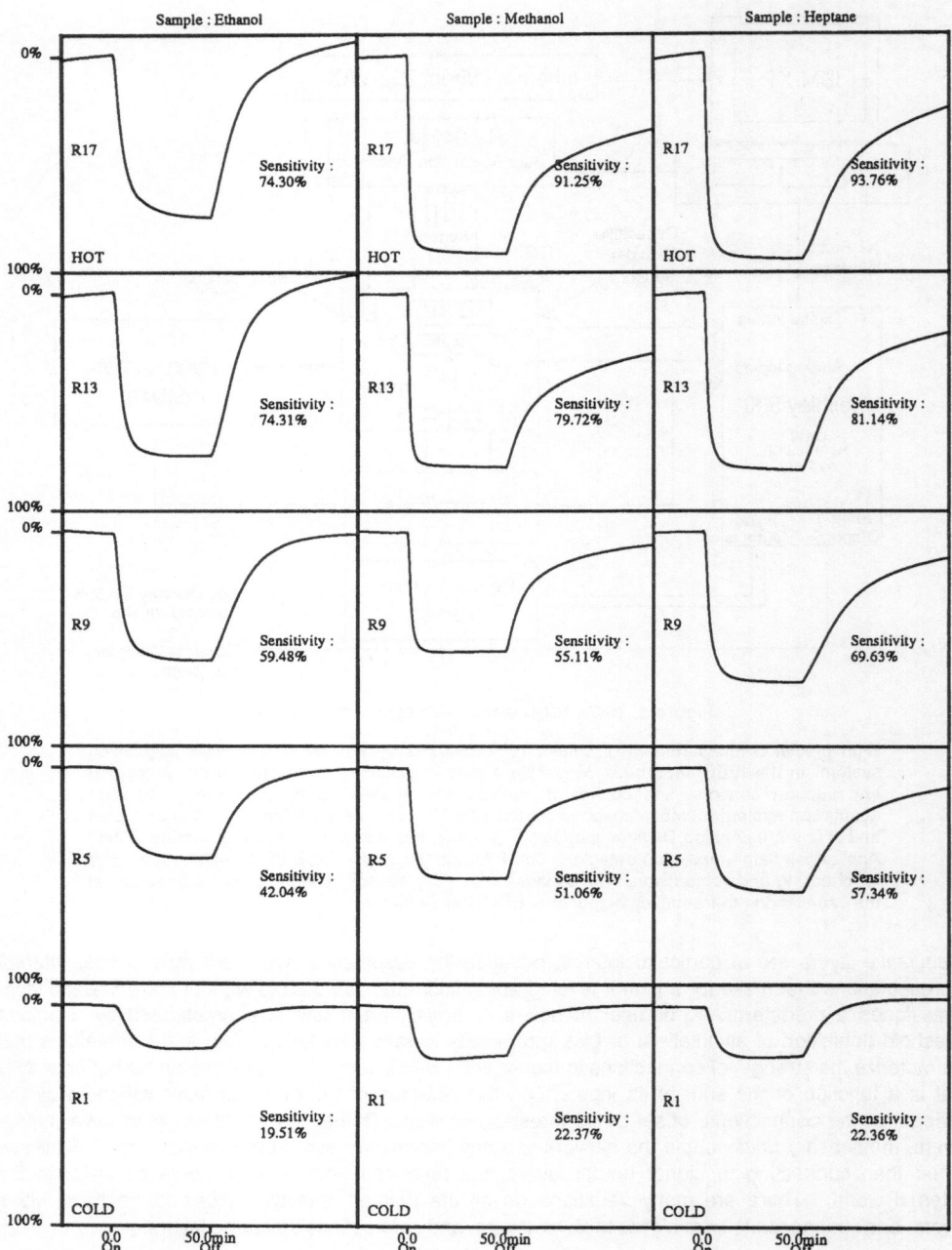

Figure 3. SENSITIVITY TO *ETHANOL, METHANOL* AND *HEPTANE* VAPORS

Sample vapor from a bubbler is mixed into an artificial air flow and introduced to the test chamber for *50 min.* By convention the sensitivity is given as $(R_o - R)/R_o$. The time constant of the rapid resistance drop at the hot end of the sensor upon *methanol* and *heptane* introduction is close to the limit set by the ratio of test chamber volume to volumetric flow rate. The longer fall times for *ethanol*, and for *methanol* and *heptane* on the cooler parts of the sensor, and the longer rise times for all samples at all temperatures upon termination of sample introduction, are characteristic of the sensor.

can represent and a particular discriminant function instantiation that successfully segments the members of a large data set into a small number of relevant constituent classes is the existence of a teaching and learning procedure that specifies the desired response to a few pieces of archtypal data, and in a reasonable number of steps each taking a reasonable time incrementally and convergently modifies the weights and thresholds until some acceptable level of performance is demonstrated. The concept long predates the discovery of an efficient teaching and learning procedure. The recently invented *back propagation algorithm* has bridged the gap by providing a practical and effective teach-learn paradigm. The principle is that the output errors are propagated back through the network from output to input, and the resulting "activities" determine the sign and magnitude of the modifications made to each threshold and weight.

My colleagues Lanwai Wong and Toshikazu Takemori have assembled, from bits-and-pieces of approaches and alternatives in the recent literature [17], a three layer neural network simulator with back propagation, with sixteen to thirty-three input units, generally five to ten hidden units, and six output units for classifying the three combustible vapors, *ethanol*, *methanol*, and *heptane*, and their more-or-less equal concentration binary mixtures viewed for this purpose as separate classes [18]. Their implementation[5] of the neural network simulator and back propagation training algorithm is illustrated in Figure 4.

During training the simulator is cyclically presented with pairs of sensor data vectors **s_data** and the corresponding desired response vectors **r_class**, and adjustments to weights and thresholds are made according to the back propagation algorithm. When the network is used as a classifier **s_data** are presented sequentially and for each an approximation to **r_class** is returned.

The **s_data**-components are the sensor responses derived from the seventeen resistance measurements made across segments of a sensor like the one depicted in Figure 1. A linear preprocessing algorithm scales and normalizes **s_data** into the sensory layer input vector **s_input**. We regard this as an elementary sort of feature extraction. Our favorite preprocessing algorithms are *(1)* normalization to the range *0* to *1* by subtraction of the minimum and dividing by the range in each data vector, which yields a seventeen element **s_input**, *(2)* differencing adjacent element resistances, or "spatial differentiation," which yields a sixteen element **s_input**, and *(3)* the combination of *(1)* and *(2)*, which yields a thirty-three element **s_input**. The approach that we took in my work with Wong and Takemori treats each individual gas and binary mixture if it were a separate class,[6] so **r_class** is a 6-dimensional vector. Each **r_class** used in training has five zero components and one unity component that corresponds to the class specified.

Each sensor unit's **s_input**-component $s\_input_{s\_unit}$ is passed through a sigmoidal thresholding and activation function of the form

$$s\_activity_{s\_unit} = \frac{1}{1 + exp\left(-(s\_input_{s\_unit} - s\_thr_{s\_unit})\right)} \tag{15}$$

Each associative unit's **a_input**-component is computed by summing the weighted inputs to that unit from all the sensory layer units:

$$a\_input_{a\_unit} = \sum_{s\_units} w_{s\_unit,a\_unit} \, s\_activity_{s\_unit} \tag{16}$$

The associative unit activities **a_activity** are computed in a manner analogous to equation 15, and the response unit activities **r_activity** are computed by similar forward propagation as illustrated in Figure 4. **r_activity**, interpreted as an indicator of the class of the corresponding **s_data**, is the answer. Operated as a classifier the algorithm stops here. During training the answer is compared with the

---

[5]They did not actually employ thresholding and an activation function at the sensory level, but for generality I have indicated the possibility.

[6]In the following section I will discuss the alternative (and I think preferable) viewpoint that this is a three class problem that is best described by structuring the network to report partial membership in multiple classes, weighted by relative concentration.

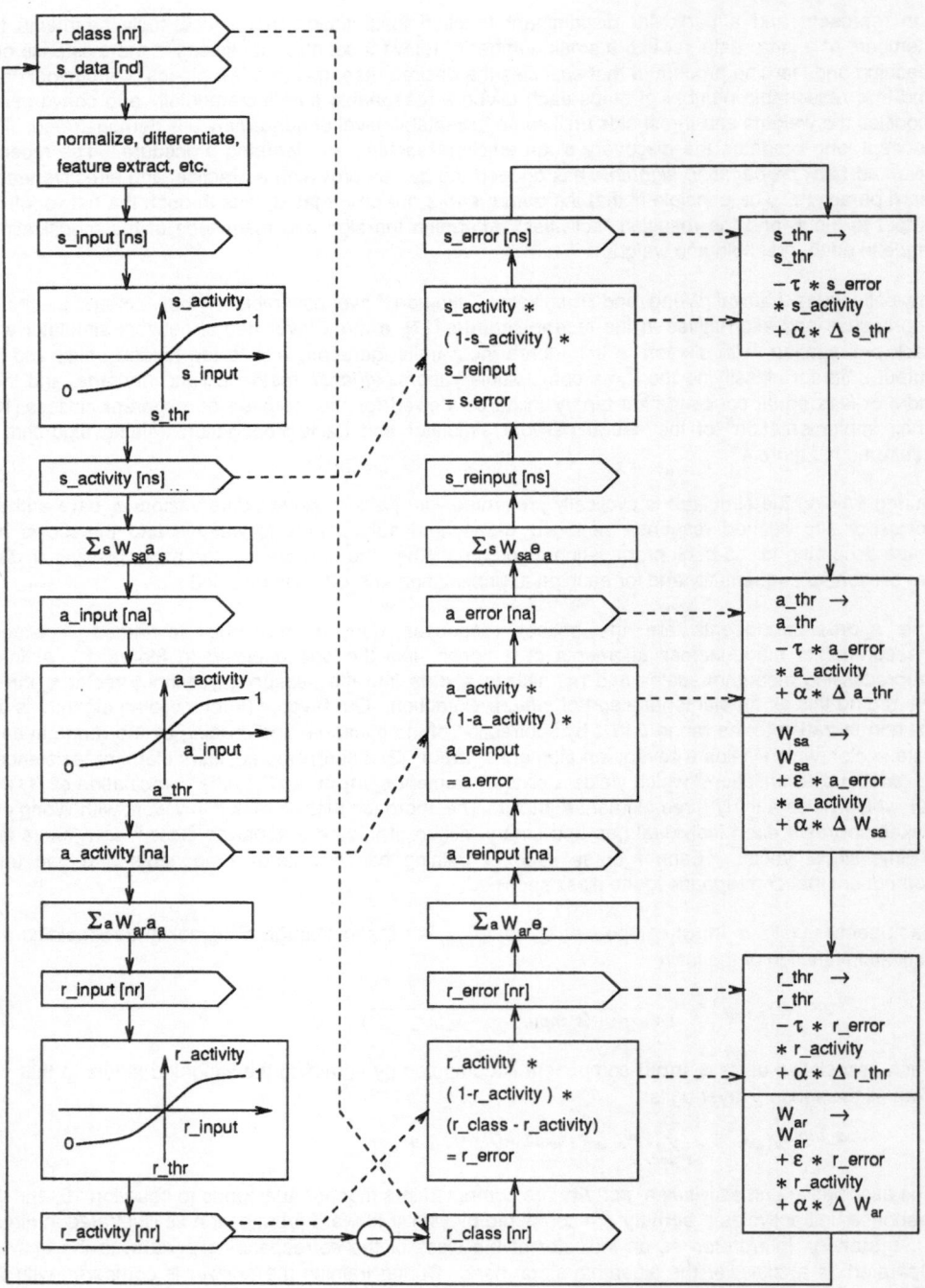

Figure 4. THREE LAYERED NEURAL NETWORK WITH BACK PROPAGATION.

s_data, an *nd*-dimensional vector, is the input. During training **r_class**, an *nr*-dimensional vector, is also input for each **s_data** vector. Output is the *nr*-dimensional vector **r_activity**. During training the difference between **r_class** and **r_activity** is back propagated through the network, and the errors used to "adapt" the weights and thresholds.

known correct answer, and the back propagated error generates the modifications to the weights and thresholds as depicted.

Back propagation differs from forward propagation only in the direction that the network is traversed, and in the replacement of the sigmoidal activation and thresholding functions that transform **u_inputs** into **u_activitys** (where **u** denotes **s**, **a**, or **r**) with an error propagation function defined by

$$u\_error_{u\_unit} = u\_activity_{u\_unit} \bullet (1.0 - u\_activity_{u\_unit}) \bullet u\_re\text{-}input_{u\_unit} \tag{17}$$

These parabolic functions of the activities are zero valued at zero and unity activities, and they peak in the middle. Like positive feedback on an amplifier, they force each unit's activity toward an extremum.

The corrections applied to the weights and thresholds include both a part proportional to the product of the unit's activity and back propagated error, and a "momentum" part that damps oscillation by incorporating into each modification a term that is a fraction of the previous modification to that weight or threshold. Training until each **r_activity**-component differs by no more than a few percent from the corresponding **r_class**-component takes a few hundred cycles through a training set of two or three **s_data** vectors from the set of each of the six sample types.

Some formal mechanism is needed to decide what class a given **r_activity** is reporting, since it is never the case that an **r_activity** has exactly one unit component and five zero components. The filter that we use to automate the decision is

$$P_{r\_unit} = r\_activity_{r\_unit} \prod_{all \, r\_units \, \neq \, r\_unit} (1 - r\_activity_{r\_unit}) \tag{18}$$

so that **P** is essentially the vector of joint probabilities P_{r_unit} that the sample is in class r_unit and not in any other class.

Using a data set of 1900 **s_data** vectors approximately equally divided among the six classes, where the data were collected in nineteen separate cycles of one hundred measurements each on seven separate days during a two week period, 1608 or 85% were correct when the seventeen normalized values were used as input. When the sixteen normalized differences between adjacent sensors elements were used as the input, 1814 or 96% were correct. When the thirty-three values corresponding to the union of the normalized and normalized difference values were used, 1747 or 92% were correct, essentially the average of the two separate values. Understanding this somewhat surprising effect awaits further work. Adding additional training data, particularly data selected from cycles on which performance is poor, improves performance statistics substantially. This is of course in keeping with the well known result that perceptron-like neural networks are guaranteed to learn their training data perfectly. However we observe that it is not only the learning of the new training data that contributes to improved overall performance: there is general improvement in performance on all data in the cycles from which the new training data are taken. This illustrates the ability of machines of this type in some situations to generalize, and to aggregate classes from disjoint sets. Experimental and performance details are elaborated in Wong, Takemori, and Siegel [18], from which parts of this and the previous section are abstracted.

Quantitation by Neural Network

When a sour tasting *HCl* solution (hydrochloric acid) is mixed with a bitter tasting *NaOH* solution (sodium hydroxide, an alkali or "base"), the *HCl* and *NaOH* react chemically, yielding a solution of the salty tasting compound *NaCl* (sodium chloride, the archetype of the "salt" class). We would thus be delighted were a taste-mimicking four response unit (*sour, bitter, salty,* and *sweet*) neural network to classify uniquely *HCl* solution in one class, *NaOH* solution in another class, and *NaCl* solution in a third class. In contrast, when the pleasantly smelling C_2H_5OH vapor (*ethanol*, the active ingredient in alcoholic beverages) is mixed with the unpleasantly smelling C_7H_{14} vapor (*heptane*, a component of gasoline), under ordinary conditions of temperature and pressure they do not react chemically, they just mix. Thus between the alternatives of a three response unit neural network with *binary* output units for the three classes *ethanol, heptane,* and *ethanol-heptane mixture* and a two response unit neural network with *continuous* output units for the two classes *ethanol* and *heptane* the latter would seem preferable, particularly if it were quantitatively accurate to interpret an output of x_e by the *ethanol* response unit and an output of x_h by the *heptane* response unit as meaning that the mixture is in the proportion $x_e : x_h$ of *ethanol* vapor to *heptane* vapor. In this section I will contrast the *neural network as classifier* approach that we took to the system of *ethanol, methanol,* and *heptane* and their binary mixtures as a six class, binary output problem, with an alternative (and I think more realistic and more useful) *neural network as quantitator* approach wherein the aim is to use the neural network method to compute the concentrations of the components of a mixture.

First note, however, that response units that are constrained to binary output values are by no means useless for quantitation. For example, we might imagine that multiple repetitive classifications could be made by a network having a probabilistic output behavior, such that the responses of its *ethanol*-response unit and its *heptane*-response unit to a mixture of relative concentration x_e of *ethanol* and x_h of *heptane* would be *1* in fraction x_e of the trials and *0* in fraction $1 - x_e$ of the trials, and (independently) *1* in fraction x_h of the trials and *0* in fraction $1 - x_h$ of the trials respectively. Such behavior would have analogies to measurements on quantum mechanical systems, where the result of each trial is constrained to be drawn from the set of system eigenvalues [19], other analogies to "flip-flop" sensors [20],[7] and further analogies in to biological neural systems with characteristic amplitude-to-frequency response. But it is interesting also to consider the alternative of neural networks with multiple continuum outputs.

In the early days of neural networks the nodes or units were usually regarded as strictly binary, *i.e.,*

$$u\_activity_{u\_unit} = \begin{cases} 0 \ if \ u\_input_{u\_unit} - u\_thr_{u\_unit} <= 0 \\ 1 \ if \ u\_input_{u\_unit} - u\_thr_{u\_unit} > 0 \end{cases} \tag{19}$$

in contrast to the continuous sigmoidal functionality (equation 15) that is currently preferred. The change was driven primarily by the desire to approximate the discontinuous Heaviside Function (equation 19) with a continuous function like the sigmoid to facilitate analyzing certain training and behavior models of neural networks, *e.g.,* models that employ analogies with collections of interacting atoms undergoing thermal relaxation from strained to equilibrium states. Their predilection toward pseudo-binary output is then driven by the slope-at-threshold of the sigmoid, scaled by a "sharpness" multiplier that can be inserted in the exponential, and also by the error weighting, in back propagation, by $u\_activity * (1 - u\_activity)$. I have recently begun to investigate the effectiveness with which the current generation of neural networks can quantitate the response of our integrated gas sensor to mixtures, with an eye towards improving that effectiveness by the adjustment of the following options: pre-processing the input data, including an input thresholding and activation function, sharpening the sigmoidal threshold function, and changing the error weighting scheme used during back propagation. This is obviously a program with many dimensions; there is only time and space here to report the approach and preliminary findings.

A measurement of the concentrations of *ethanol, methanol* and *heptane* in an arbitrary mixture can be represented as a vector in a three dimensional space: $c = x_e e + x_m m + x_h h$ where **e, m, h** are unit vectors in the *ethanol, methanol,* and *heptane* directions. The data on which the work described in the previous section is based represent measurements taken in severely constrained portions of this space: single species measurements are taken in the unit vector directions **e, m,** and **h**, and binary mixture measurements are taken along the lines $0.5e + 0.5m$, $0.5e + 0.5h$, and $0.5m + 0.5h$. For the numerical experiments I have also generated a simulated data set[8] of random binary mixtures $x_e e + (1 - x_e)m$ for a simulated *ethanol-methanol* mixture, etc. The simulation populates the **em-, eh-,** and **mh**-planes with "data."

---

[7] A "flip-flop" sensor is one in which the transducer output creates a bias that upsets the otherwise *a priori* equal probabilities that a bistable circuit will power-up in the *0*-state or *1*-state.

[8] The combining rule I used to create simulated resistances was to add the randomly weighted *reciprocals* of the measured resistances and take the reciprocal of the result. This approach models (naively, but it is a beginning) linear superposition of response to multiple gases the by addition of *conductance* changes.

Having thus assembled a data set composed of measurements on single vapors, measurements on binary mixtures of vapors in fixed (nominally equal) proportions, and simulated binary mixtures of vapors in random proportions, I then train neural networks on subsets of the three data sets and observe the performance of these networks in classifying and quantifying the data sets from which their training data was drawn, and the other data sets too. I represent the results graphically as plots of $r\_activity_{r\_unit}$ vs $r\_class_{r\_unit}$ points, without distinguishing (during the present discussion) as to which (*ethanol, methanol, heptane*) unit is responsible for which point. I then investigate classification and quantitation behavior over a *3* data-set × *3* neural-network matrix.

I will first discuss the *first* matrix element, behavior of the network trained on a subset of the single vapor samples and exercised on the full data set of single vapor samples, and (at the risk of telegraphing my punchline), the *last* matrix element, behavior of the network trained on a subset of the simulated data and exercised on the full set of simulated data. The former is shown on the left side of Figure 5 and the latter on the right side of Figure 5. These illustrate very well the meanings of *classification* and *quantitation* by a pair of neural networks of exactly the same topology, started in exactly the same initial state, and trained by exactly the same back propagation algorithm, the difference being that in the former case the **r_class** vectors were limited to the unit vectors, and in the latter case the **r_class** vectors were random two unit vector admixtures. The interesting thing about the former case is how *uninteresting* it is: *450* data points are so tightly clustered in the lower left and upper right corners that the plot looks almost empty: this net is a effective classifier. The latter case illustrates that the same paradigm can also produce an effective quantifier: *150* data points lie for the most part along the line of unit slope and zero intercept, which is the performance desired of a quantifier. The short vertical cluster in the lower left corner shows that the zero concentration components, one in each data vector, are being correctly reported as near zero. But there is still a hint of predilection toward binary behavior: when *r_class* is less than about *0.25* or more than about *0.75* there is a tendency for *r_activity* to be near *0* and *1* respectively.

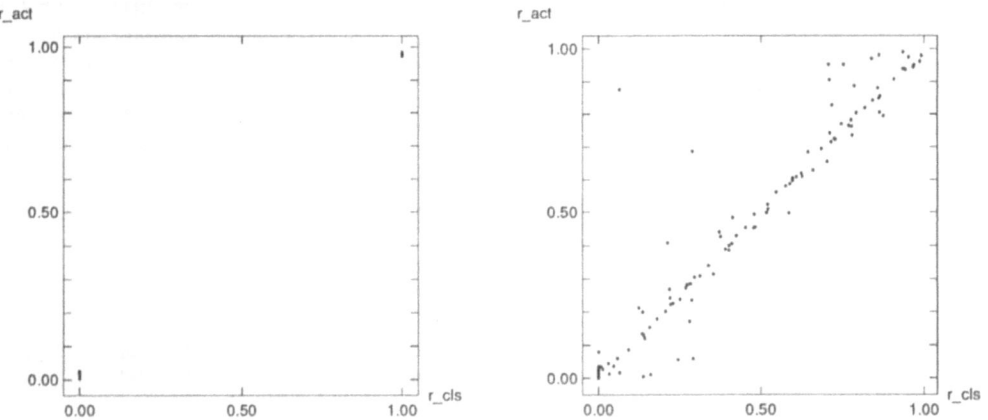

Figure 5. ACTIVITY *vs* CLASS VECTOR FOR VAPORS AND VAPOR MIXTURES

Left: trained on single components, exercised on single components.
Right: trained on random binary mixtures, exercised on random binary mixtures.

The left half of Figure 6 illustrates the behavior of the network trained on a subset of the unary and binary mixture data and exercised on a larger set of unary and binary mixture data, and the right half of Figure 6 illustrates the behavior of the same network when exercised on the simulated data set of binary mixtures of random composition. The former case shows that the network is able to classify both the unary and binary mixtures, although the distribution is much tighter for the unary than for the binary cases; nevertheless, the error rate for this network as a classifier on this data set is close to zero. The latter case is actually a very difficult test addressing the question of the extent to which reference data taken on lines in a three dimensional space can calibrate measurements made on planes containing those lines: not yet the general question of the extent to which reference data taken on lines can calibrate the volume of the space, but moving in that direction. Although the three types of points corresponding to the three binary mixture types are not distinguished in this plot, it seems clear that there are three sigmoidal curves here, each representing a gentle tendency to binarize one of the simulated mixtures according to whether its proportions are greater or smaller than the actual proportions in the real binary mixture.

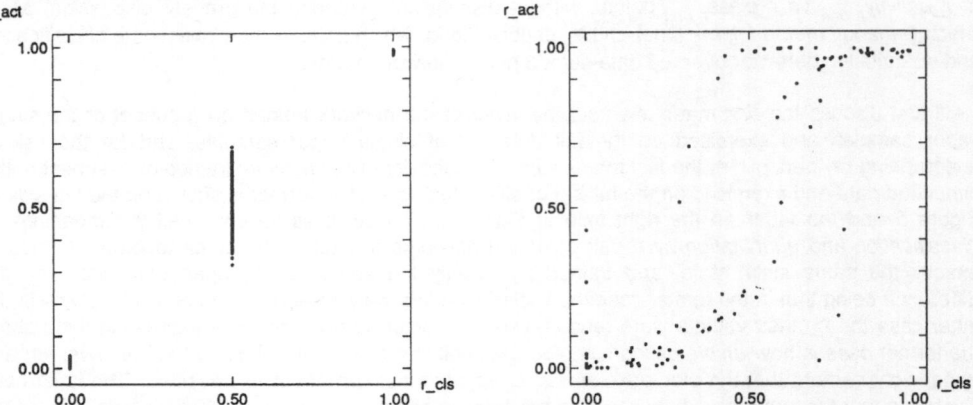

Figure 6. ACTIVITY *vs* CLASS VECTOR FOR VAPORS AND VAPOR MIXTURES

Left: trained on unary and binary mixtures, exercised on unary and binary mixtures.
Right: trained on simulated binary mixtures, exercised on real unary and binary mixtures.

How to minimize the tendency toward binary response when continuum response is what is appropriate is a goal of future work. I expect it can be done by selecting network and training algorithm parameters properly. The results to date, with entirely conventional approaches to network and training algorithm parameters, are encouraging. The remaining question is the generality with which the resulting networks can represent physically appropriate nonlinear response functions, and the extent to which they can abstract those response functions with only the shadows of physical models of sensors available and only the calibration ("training") data sets which are restricted by combinatorial enormity to data located on limited subspaces of lower dimensionality.

Symbols Used

$[X]$	Density, concentration, or partial pressure of species X. It is often useful to interpret the units as being relative to $[O_2]_{air}$, the density, concentration, or partial pressure of O_2 in nominal air.
R	Sensor's electrical resistance. [ohms]
R_o	Sensor's electrical resistance in pure air. [ohms]
K_n^X	Rate constant for n-th order reaction, *i.e.*, one that depletes $[O_2]$ in proportion to $[X]^n$. [concentration-units^{-n} x sec^{-1}]
$K_{ijk\ldots}^{XYZ\ldots}$	Rate constant for multi-order reaction, *i.e.*, one that depletes $[O_2]$ in proportion to $[X]^i [Y]^j [Z]^k \cdots$. [concentration-units$^{-(i+j+k+\ldots)}$]
β	Power law exponent describing dependence of $\frac{R}{R_o}$ on gas concentrations.
k_i	Equilibrium constant.
kT	Product of Boltzmann's constant and the absolute temperature.
E_a	Activation energy.

Acknowledgements

Paul Clifford, under the mentorship of David Tuma and Angel Jordan, brought a wealth of experience and perspective on MOS sensors to my lab during its formative years. The apparatus and code were implemented in stages by Mark Licata, David Wagner, Tamal Mukhergee, Lanwai Wong, and Toshikazu Takemori. Alan Guisewite provided us all with craftsmanship and continuity. The sensors were fabricated at Oak Ridge National Laboratories by Robert Lauf and Barbara Hoffheins. Support for developing the concept, building the apparatus, and initiating the measurements at CMU, and for the fabrication program at Oak Ridge, was provided by Cabot Corporation (Boston), championed by Law McCabe. Toshikazu Takemori's participation in our research was supported by Osaka Gas Company (Japan). An unrestricted grant provided by Mettler Instrumente AG (Switzerland) on the initiative of Rudolf E. Kubli enabled me to pull it all together for the NATO workshop.

References

1. D. Schuetzle and R. Hammerle. *Fundamentals and Applications of Chemical Sensors*. American Chemical Society, Washington, DC, 1986.
2. P. K. Clifford and D. T. Tuma. "Characteristic of Semiconductor Gas Sensors Part I: Study State Gas Response". *Sensors and Actuators 3* (1982-1983), 233-254.
3. P. K. Clifford and D. T. Tuma. "Characteristic of Semiconductor Gas Sensors Part II: Transient Response to Temperature Change". *Sensors and Actuators 3* (1982-1983), 255-281.
4. A. W. Laubengayer. Reaction Rates and Chemical Equilibrium. In *General Chemistry*, Rinehart & Company, New York, 1957, Chap. 15, pp. 239-41.
5. John E. Amoore. *American Lecture Series*. Volume 773:*Molecular Basis of Odor*. Charles C. Thomas, Springfield, IL, 1970.
6. R. C. Gesteland, J. Y. Lettvin, W. H. Pitts. "Chemical Transmission in the Nose of the Frog". *J. Physiol. 181* (1965), 525-59.
7. J. Y. Lettvin, H. R. Maturana, W. S. McCulloch, and W. H. Pitts. "What the Frog's Eye Tells the Frog's Brain". *Proceedings of the IRE 47* (November 1959), 1940-51".
8. Jerome Y. Lettvin and Robert C. Gesteland. "Speculations on Smell". *Cold Spring Harbor Symposia on Quantitative Biology 30* (1965), 217-25.
9. Frank Rosenblatt. *Principles of Neurodynamics.* Spartan Books, Washington DC, 1962.
10. H. D. Block. "The Perceptron: A Model for Brain Functioning". *Reviews of Modern Physics 34*, 1 (January 1962), 123-35.
11. Anton J. Carlson and Victor Johnson. The Spinal Cord and Brain. In *The Machinery of the Body*, University of Chicago Press, Chicago, Illinois, 1941, Chap. 10.
12. B. S. Hoffheins, R. J. Lauf, M. W. Siegel. "Intelligent Thick Film Gas Sensor". *Hybrid Circuits 14* (September 1987), 8-12.
13. M. W. Siegel, R. J. Lauf, and B. S. Hoffheins. Dual Gradient Thick-Film Metal Oxide Gas Sensors. Proceedings of the Tokyo meeting, Transducers '87, June, 1987, pp. 599-604.
14. B. S. Hoffheins, R. J. Lauf, M. W. Siegel. Intelligent Thick Film Gas Sensor. Proceedings of the Atlanta Meeting, International Society for Hybrid Microelectronics, October, 1986.
15. Marvin Minsky and Seymour Pappert. *Perceptrons*. The MIT Press, Cambridge, MA, 1969.
16. David E. Rumelhard, James L. McClelland, et al. *Parallel Distributed Processing*. The MIT Press, Cambridge MA, 1988.
17. David C. Plaut with Geoffrey Hinton and Steven Nowlan. CMU Back-propagation Simulator User's Guide. Carnegie Mellon University, October, 1987.
18. Toshikazu Takemori, Lanwai Wong, and M. W. Siegel. Gas Identification System using Graded Temperature Sensor and Neural Net Interpretation. Technical report CMU-RI-TR-20-89, The Robotics Institute, Carnegie Mellon University, July, 1989.
19. David Bohm. *Quantum Theory*. Prentice-Hall, Inc., Englewood Cliffs, NJ, 1951.
20. S. Middelhoek, P. J. French, J. H. Huijsing, and W. J. Lian. Sensors with Digital or Frequency Output. Proceedings of the Tokyo meeting, Transducers '87, June, 1987, pp. 17-24.

Development Environment for Tactile Systems

Adolfo Steiger Garção, Fernando Moura-Pires
Grupo de Robótica Inteligente
Faculdade de Ciências e Tecnologia
Universidade Nova de Lisboa
Quinta da Torre
2825 Monte da Caparica, Portugal

Abstract

A development environment for Robotic Sensor Systems is described. A new system, under development, based on a user perspective will be described, including: A graphical language; macro commands with development facilities; tools for CAD supported sensor simulation; debugging facilities, for output (display) and also using "browsers" for inspection of internal representation structures. Some examples using this environment will be provided, as well as a perspective for future work

Keywords: Perception, Sensorial Integration, Tactile Sensor, Graphical Language.

1 Introduction

An integrated robotic sensorial system has to consider tactile sensing. Touch being an active sense, movement, generalized forces and information from matrix sensors have to be taken in consideration.

We strongly believe that a versatile and open development system, allowing the integration of different representation paradigms and functionalities, is a key point to achieve successful simulation, testing and implementation of perceptional oriented architectures. As in many situations some exploratory work has previously been done. A Pascal based programming environment was first test bed. A better understanding of hidden problems as well as the tools' limitations emerged, leading to new concepts. Object oriented and reactive programming will play an important role stressing the importance of flexible representation paradigms, mainly frame oriented, as well as the need for a specific autonomous sensor information processor[1].

NATO ASI Series, Vol. F 63
Traditional and Non-Traditional Robotic Sensors
Edited by T. C. Henderson
© Springer-Verlag Berlin Heidelberg 1990

One of the major consequences of the previous experience was the proposition and subsequent development of an interactive graphic oriented language, allowing the fast definition of filters, planned as chained transformations, aimed to process sensor originated information. This major step represents only an initial move towards the specification and implementation of an adequate and integrated set of tools. Concepts like the system's internal state and its memory capabilities were established and became an instrument for study, modelling and understanding of needed dynamic analysis on sensor originated information (in active sensors like tactile ones, static information is in general a useful starting point to determine the sequence of interactions the identifying process will be based on). We established a clear identification of the role the user could interactively play, defining the state scheduling and the intermediate finite number of transformations the system was supposed to memorize.

Another natural evolution was the use of more sophisticated graphical workstations (Evans & Sutherland PS-390), [2]. Though their utilization revealed a sound coherence between the transformation chaining approach, the frame oriented representation and a sophisticated human oriented user interface were demonstrated.

2 Main objective of the development environment

2.1 Introduction

The strategy on which the present work is integrated addresses the flexible development of multisensorial perceptional oriented sub-systems.

In general terms it seems acceptable to correlate sensor based information integration with data/knowledge representation hierarchy needs. Normal starting points are based on the physical layer (transducer signal filtering, analogue processing and digital conversion) and end at some virtual level (semantic aspects associated). This last level should also be supported on additional extensions of conceptual architecture hierarchies. An image derived set of points can be translated to a "line segment" concept, a region can support a "hole" definition, etc. In general, this hierarchic process implies loss of detailed specific information. Gains can be obtained not by destroying the existing collected information, but by making the logical access more context dependent, and at the same time profi the concept structuring virtual power. At the physical level, precise data and algorithms dominate. At more abstract levels, concepts derived from the processed data associated with fuzzy inference represent the main data types.

Integration can be forecasted to be predominantly done on the upper levels. That implies the use of filtering techniques as well as the definition of virtual layers, derived from the initial sensorial, plain, nonstructured information. Massive relevant work has been done mainly considering isolated sensors. The difficulty arises from the inexistence of consistent and sound integration methodologies. Also important is the final purpose definition of such systems. Who is the consumer? Usually, another system. Unless a strong decoupling is necessary, dependencies on common information can be explored, leading to more sophisticated and complex integration possibilities. Being uncommitted to a solution, this work will consider an open field of possibilities, instead of premature compromising. The development environment will reflect the pre-conditioning needs for system's study, implementation, simulation and testing. A preliminary definition led to the following requirements:

- Different levels of sensors have to be supported. There is a need for the description of physical and simulated (CAD) sensors with different virtual description.
- Different representation paradigms for data and knowledge representation.
- A user friendly graphical oriented interface [3].
- Interactive consistent debugging facilities, active either during development or execution phases (Supported by the graphical interface).
- Modification facilities, even at functional level (adding new software or hardware oriented modules), that include the possibility of logically defining new transformations based upon existing ones.
- Open communication facilities between different sub-systems.

Such an ambitious environment has to be considered as a preparation step previous to the definition and development of consistent multisensorial integrated architectures. Modelling, including techniques for fuzzy knowledge representation are of major importance, as well as learning and simulation.

A multisensorial integrated architecture is usually not self-consistent. A more general system aimed to provided a larger degree of autonomy can be the final target to incorporate previous defined sensor based architectures. Autonomy is strongly dependent on the ability to deal with the external world. Gathering information from the outside world represents a common way of building up a feedback loop allowing the possibility of developing adaptation mechanisms. In our case, examples will stick mainly to tactile sensing. That doesn't represent a severe limitation since position detection, contact evaluation and generalized forces sensing have to be integrated. Additionally, limited cooperation with vision and proximity sensing will be considered for initial positioning of the sensor (making the first contact).

The hardware/software architecture supported by the development system integrates the following elements:

- Real or simulated sensors (tactile,force,vision, ultrasonics).
- Manipulator's controller (real or simulated), with adequate sensorial interfaces.
- Logical support for object modelling. Different representations can be made available at tools level (e.g., CSG or B-rep) or at a higher level of abstraction (surfaces, vertex, lines)
- An adequate level of logical communication between the different processing systems.

2.2 Architecture

Our environment is essentiality a "graphical language" with the concept of "transformation". The operations on this concept are essentially: define, select and link. A full description of them will be given later on.

To implement our environment a three level division is proposed: a declarative "graphical language", a functional kernel and a procedural level. The declarative "graphical language" implements the user's interface to the environment, the functional kernel implements the functionality of the "transformation" concept and the procedural level implements the algorithms.

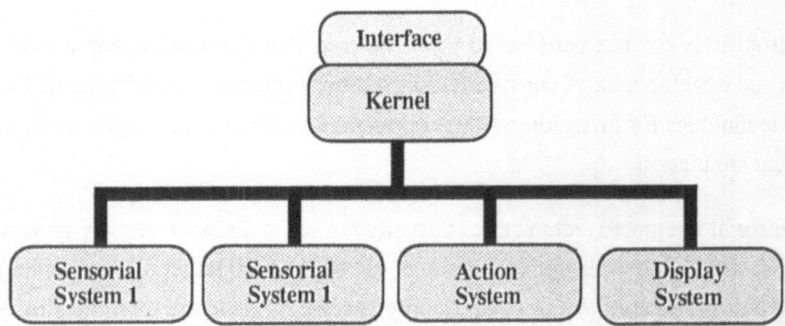

Fig.1 Architecture of the environment

The user interface and the kernel are implemented by the same physical development system. The procedural level can be implemented by the same system or can be developed by specialized hardware/software. The environment's architecture in which different subsystems (sensorial, action and display) are shown in Fig.1.

Table 1 lists the different concepts used in each level.

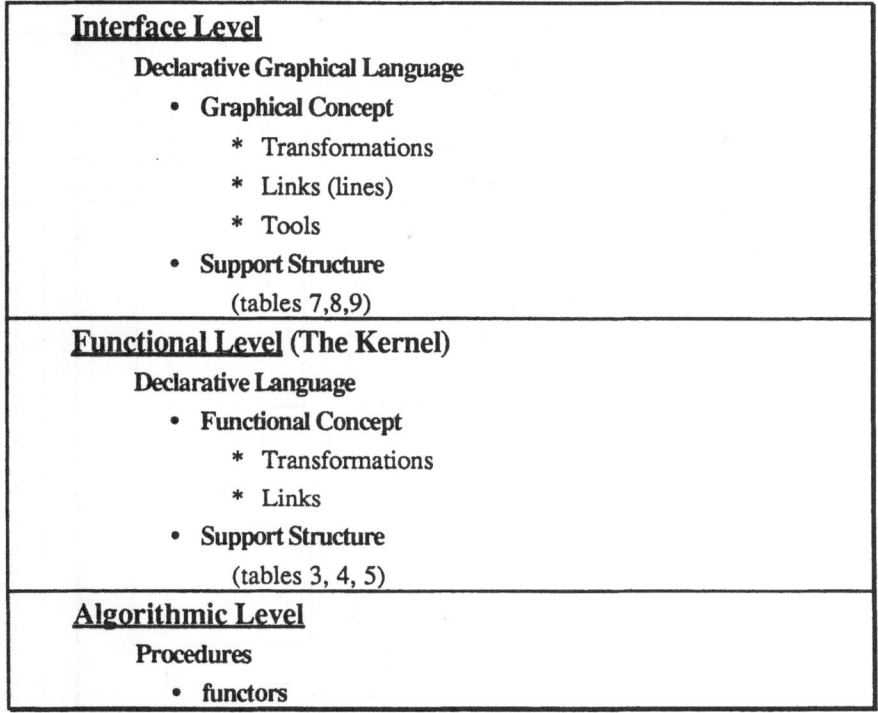

Table 1 Concepts in the environment levels

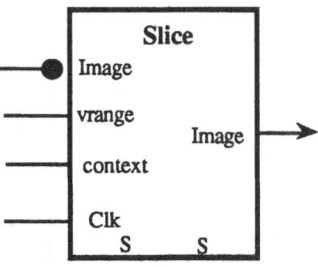

Fig. 2 The transformation's icon

In the declarative graphical language, icons represent the transformations (for instance a slice transformation is illustrated in Fig. 2.) and lines to represent the links between different transformations. Concepts implemented in the three levels have different supporting structures.

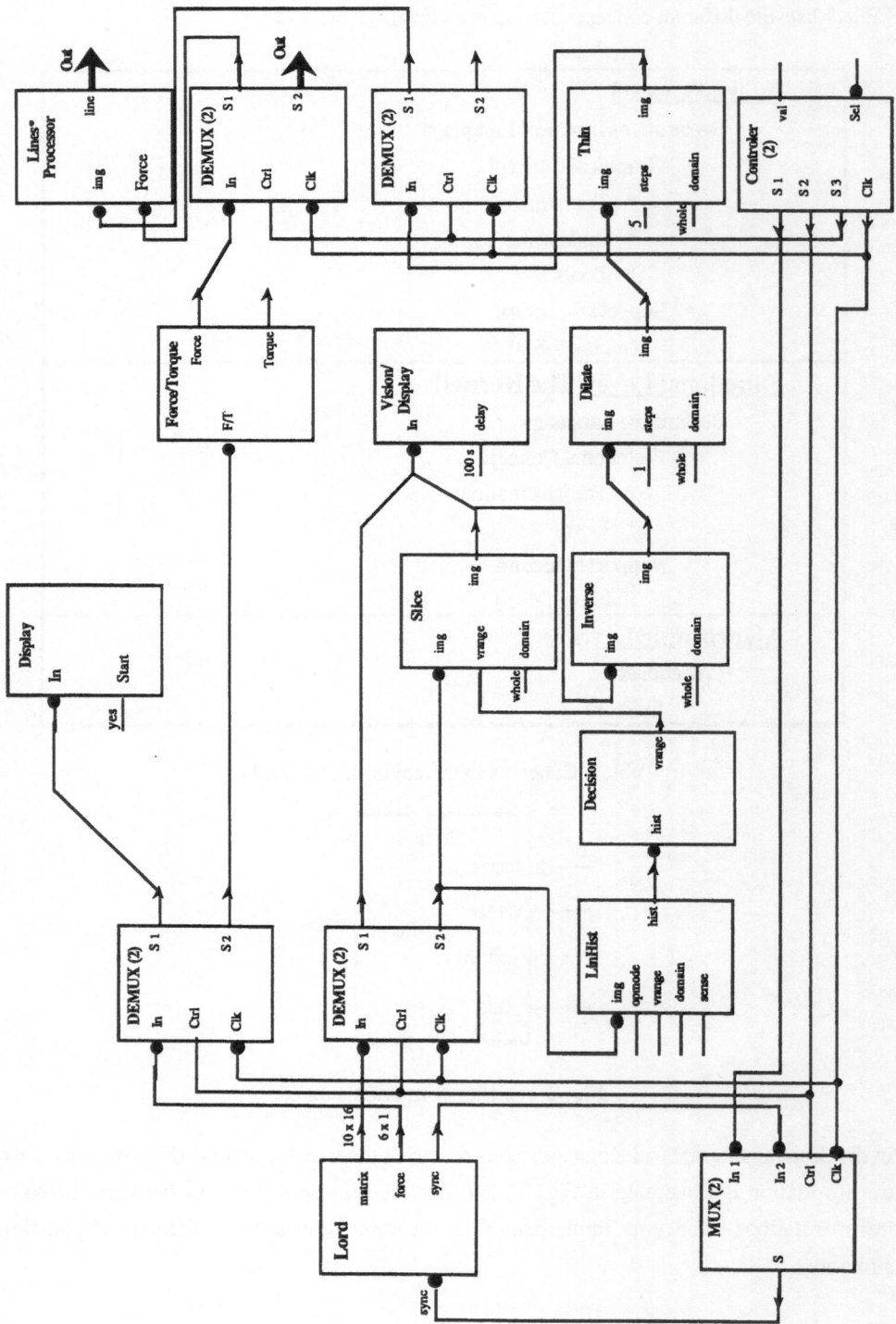

Fig.3 An example of a graphical program

The functional level discussion is the aim of this paper, as well as the implementation details of the transformation concept.

The procedural level implements the transformation's algorithm. As an example we can envisage the existence of hardware and of a specialized library for a vision system which supports the algorithms needed for different transformations - e.g. a Pascal procedure in the form of

    ....
    procedure slice (ImgIn : image, Range: vrange, var ImgOut : image);

but this level is not what we are really addressing in this paper.

To illustrate the previous statements in Fig.3 an example of a graphical program implemented in our environment is documented. Different transformation (icons) and the links (lines) are represented.

This graphical description has to be viewed as a program. In this example icons are representing sensors (for instance a tactile sensor develop by Lord Corporation is represented by a transformation named "Lord" which has as output: a matrix of 10 by 16 points and a force vector (Fx, Fy, Fz, Mx, My, Mz)). In this program we can also visualize icons as filters (slice, inverse, dilate - a binary transformation, thin - another binary transformation), icons as controllers of paths of information (for instances multiplexers, demultiplexers, controllers, and so on) or specialized icons that are application dependent.

2.3 Kernel

All the transformations have a set of common properties. Essentially they have inputs [i1...in], outputs [o1...ok], and the transformation's function $\mathcal{F}$ (the functor), i.e.,

$$[O1...Ok] = \mathcal{F}([I1...In])$$

Every transformation is characterized by:
- **input synchronization condition** - they can be synchronous or asynchronous. If the inputs are synchronous there is a transformation's trigger only when all the input data is available, otherwise the input is asynchronous;
- **output synchronization condition** - they can be synchronous or asynchronous. If the outputs are synchronous only when all output are available, the output values are sent to the transformations' inputs linked to those outputs. Otherwise the outputs are asynchronous, i.e., when each output has its value it is sent to the transformations linked to this output;

- **functor** - is the procedure that implements this function. We consider that it is possible to have more than one functor for each transformation;
- **context** - is one of the possible inputs to define the transformation's functor.

All the inputs are characterized by the parameter:

- **type** - the inputs can be twofold: queue or level. A queue-type input has a functionality similar to a "FIFO" (First In First Out) queue, i.e, one input value produces at most one output and then it is consumed. With a level-type input the value is not consumed by the output and when one writes into a level-type input the new value destroys the previous.

The outputs are characterized by :

- **links** - define what transformations and what inputs of those transformations are linked to each output, i.e., to which transformations this output must send its value when computed.

These items describe the concept of "generic transformation", i.e., all the requirements every transformation must satisfy. However there are different transformation concepts (for instance we have slice, thin, multiplexer transformations, and so on), in which each transformation has a concrete definition; i.e., the number of inputs, the number of outputs, the type of each input and output, the different functors, and so on. A complete definition is a "specific transformation".

However in a program, we can use a concept several times; for instance in Fig.3 we use the concept of multiplexer four times. Each use of this concept is in fact an instance of the same concept, the "multiplexer".

In our model, three basic transformation concepts are supported: the generic transformation, the specific transformation and the instances of specific transformations.

However, the specific transformation concept has two different types: simple or composed.

A simple transformation is the transformation as we have defined; i.e. the transformation defined by the input and output definitions and the functor.

A composed transformation is a transformation that is defined using simple transformations, and (eventually) other composed transformations. As with simple transformations, one has to define:

- The input and output parameters;
- The simple (and composed) transformations used;
- The definition of the existing links among the composing transformations;

- Initialization of the values for some inputs;
- The connections between the inputs and outputs of the composed transformation and the composing transformations.

Now that we have defined the building blocks of the kernel, we shall define (Table 2) the valid operations supporting the basic concepts.

Generic transformation
• Create a transformation
• Delete a transformation.
Transformation
• Create an instance of a transformation
• Delete an instance of a transformation
• Add a functor (context, functor) to the transformation
• Delete the functor from the transformation.
Instances
• Send input values to the instances
• Connect the output of an instance to the input of another instance
• Disconnect the output of the selected instance form the other given instance
• Debug either inputs or outputs of a given instance.

Table 2 The valid operations for transformations

To create a new transformation or an instance of a transformation is to create a new entity. This new entity has all the properties of the transformation concept besides the specific changes the user eventually makes. One can therefore conclude that the use of programming languages that incorporate the inheritance concept can be very useful for the implementation of this kind of systems.

To debug an input or an output of a transformation, is to tell the user (either directly, sending a message to his computer terminal; or indirectly, through the use of a file) of every change that line suffers.

2.4 Interface - short definition of a interface

As known, the interface is fully dependent on the the means available to implement the system. It can either be very simple, such as a computer terminal, or a complex one based on the use of environment including, for instance, menus, tools, icons, etc.

It was assumed that this work would be supported by a good graphical interface. This chapter is devoted to describing general principles of such an interface; more specific information can be found on the description of the first prototype.

The interface translates graphically, to the user, what was defined for the kernel; the following possibilities were made available:
- ability to define simple or composed transformations
- to program the system; i.e. to develop a schema (Fig. 3)
- to endow the user with debugging possibilities

The transformations based on following rules have to be represented by icons as shown in Fig. 4:
- every transformation is represented by a rectangle, with the inputs on its left and the outputs on its right.
- a black dot on an input means that the input is a FIFO queue; the absence of the dot means it is a level-type input.
- on the lower part of the icon appear the letters "S" and "A", meaning that the inputs or the outputs are either "Synchronous" or "Asynchronous".
- on the upper part of the icon is shown the name of the transformation.

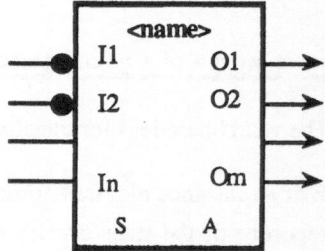

Fig. 4 Generic icon of a transformation

In defining a transformation, the user has to choose between the definition of a simple transformation and a composed transformation.

In the case of a simple transformation, the system is able to ask for the parameters of that transformation and its functor's definition.

If the transformation is to be a composed one, the parameters of the transformation are also to be asked; but instead of asking the user for the definition of the functor, the user is asked to define the meaning of the transformation (composing elements, connections, and so on), in a manner similar to the one used on schema edition.

The connections made during the definition of a composed transformation correspond to the concept of link, whose instances only exist on the definition of instances of that composed transformation.

The definition of a composed transformation can be viewed in one of two ways; either the system keeps the definition of the transformation (sequences), as was defined, or the system is able to "compile" the definition and generate a composed functor for the transformation.

On the edition of schemas, several possibilities to select a transformation, create and delete instances of transformations and to connect and disconnect different transformations between them must exist. Besides these operations, one must be able to set the input values of the used transformations (e.g. in Fig. 3, one can set the "start" input to "yes" in transformation or set the "step" input to "5" on the transformation "thin").

All these operations correspond to the graphical equivalents of the kernel primitives.

Debugging and control facilities can be considered as single item to be discussed, because execution control can be thought as a debugging activity.

Debugging facilities can be divided into three groups of operations: environment debugging, instance analysis and specialized debugging transformations.

On a first group of operations, operations of "trace", "spy" and "line debugging" can be found.

If the "trace" operation is active, whenever an instance of a transformation is changed, the user is informed of the current status of that transformation (input and output values).

The "spy" case is similar to the previous one; however, it only applies to transformations placed under surveillance.

Whenever the user selects the option "line debugging", the flow of data through the schema is signalized; i.e. when data flows through a line, the line's aspect changes.

On a second group of operations, an instance of a transformation can be picked up from the schema; from that moment on, every data flowing through the instance, is visualized on the outputs.

On a third group of operations, two examples are show; usually, this kind of operations is application dependent.

As an example, in Fig. 5 a we have a transformation that allows the visualization of numeric values flowing through that transformation, provided that the "debug" input is set to "yes", otherwise, the value just flows through the transformation, unchanged and untouched.

In Fig.5 b a particular type of display is represented to be used for complex data types, for instance the display of an image.

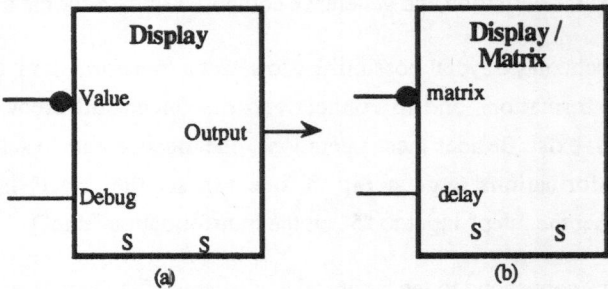

Fig. 5 Examples of specialized transformations

3 Description of the first prototype

3.1 Introduction

The first prototype was developed in a Macintosh SE computer using a "frame kit" developed in Prolog (LPA MacProlog)by our research group. The Macintosh was selected due to the existence a good graphical interface, with built in "easy to use" graphical primitives. The use of a frame kit with inheritance capabilities was essential to fulfill the environment requirements. In the following we assume the reader is familiarized with frame-based knowledge representation [4]

The frame kit supports the notion of **"generic slot"**, i.e., the name of a slot can be any Prolog term, (for instance, the slot name - input(_) - can be used to represent the slot input(1), input(2) or input(3)), etc. This concept, is not implemented in commercially available frame engines, but has been very useful in many in robotic developed applications.

3.2 Kernel

Transformation definition

The "transformation" concept was divided in three parts: "generic transformation", "simple transformation" and "composed transformation".

The "generic transformation" structure is shown in Table 3. In this structure the slots contain: the number of inputs and outputs (nInput, nOutput), the description of input and output synchronization parameters (syncInput, syncOutput), the concepts of generic input and output (input(_), output(_)) and the demons ("if write") for the input and output connections.

The function of the "if write" demon on the generic slot "input(_)" is the following: when we write a new value into an input slot, this slot inherits the demon from the generic slot input(_), and this demon tests if the inputs are synchronous or asynchronous. If the conditions to trigger the functor are met the functor is activated.

```
frame : transformation;
        slot : graphicImage ;
                facet : value ->nil /* perhaps to be implemented by a relation*/
        slot : nInput ;
                facet : value ->1 ; /* default value */
        slot : nOutput ;
                facet : value ->1 ; /* default value*/
        slot : syncInput ;
                facet : value ->S ; /* default value */
        slot : syncOutput ;
                facet : value ->S ; /* default value */
        slot : input(_) ;
                facet : links -> [(nil, nil)] ;
                /*queue (nameOfFrame,output number)*/
                facet : type -> FIFO ; /* FIFO or CONST */
                facet : value -> nil ;
                facet : if_write : f_input ;
        slot : output(_) ;
                facet : links -> [(nil, nil)] ;
                        /*queue (nameOfFrame,input number)*/
                facet : value -> nil ;
                facet : if_write -> f_out
        method : deleteTRF ;
        method : connect ;
        method : cutconnection;
        method : debugTRFin;
        method : debugTRFout.
```

Table 3 Generic Transformation's structure

The function of the "if write" demon for the generic slot "output(_)" is the following: when a new value is writen into an output slot, this slot inherits the demon from the generic slot "output(_)", which tests if the outputs are synchronous or asynchronous. If the conditions are met the demon will send the output value to the transformations linked to this one.

The methods are the primitives defined for the concept of transformation.

The slot graphicImage makes the connection between the functional structure and the graphical structure.

Definition of a simple transformation

A simple transformation is defined in Table 4. This definition inherits part of its structure from the definition of "generic transformation" and adds the concept of context, and the primitives defined for the kernel.

The context is defined by the facets: "domaiOfContex", "functor" and "value". The facet "domaiOfContex" contains a list of possible pairs domain (context, functor), the facet "functor" contains the name of the functor selected, and the facet "value" contains the name of the selected context . This slot has a demon "if write" whose effect is the following: when the slot context is written, the demon verifies if a new context belongs to the domain and then puts the name of the functor is put into the facet "functor" and facet value is updated.

```
frame : simpleTRF;
    isa : transformation
    slot : context ;
        facet : domaiOfContex -> [ ] ;
                /* queue of pair values (context, functors)*/
        facet : functor -> nil ;
                /* select functor */
        facet : value -> nil ;
                /* select context */
        facet : if_write -> f_context ;
    method : functor ;
    method : createTRF;
    method : createTRFInst ;
    method : deleteTRFInst ;
    method : addfunctorTRFInst ;
```

Table 4 Simple transformation's structure

Definition of composed transformation

In the definition of the composed transformation, besides the information available in the case of a "simple transformation", the names of the transformations used in this definition, the information of the links and the initial value for some inputs are considered.Therefore the methods "create a transformation", "create an instance" and "delete a transformation" must be redefined,

Table 5 gives the definition of a composed transformation with a slot, "descOfElement", containing the transformations used in the definition of this composed transformation. This definition is formed by a list of pairs of values (a identifier to represent an instance of a transformation, name of the transformation).

```
frame : ComposeTRF;
        isa : transformation;
        slot : descOfElement ;
               facet : value ->[ ] ; /* [(name,type)] */
        slot : links
               facet : value -> [ ] ; /* [((nameOut,out),(nameIn,In))] */
        slot : initialValue -> [ ] ; /* [(nameIn, In, Value)] */
               facet : value -> [ ] ; /* [((nameOut,out),(nameIn,In))] */
        slot : input() ;
               facet : if_write : new_f_input ;
        slot : aux1() ;
               facet : links -> [(nil, nil)] ;
                          /*queue (nameOfFrame,input number)*/
               facet : value -> nil ;
               facet : if_write -> f1_out ;
        slot : aux2() ;
               facet : value -> nil ;
               facet : if_write -> f2_out ;
        method : createTRF;
        method : createTRFInst ;
        method : deleteTRFInst ;
```

Table 5 Composed transformation's structure

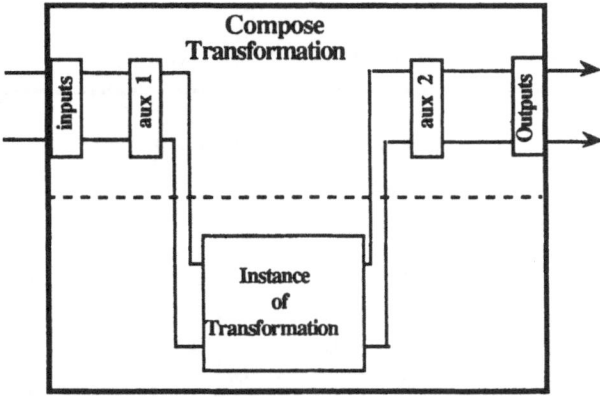

Fig. 6 Architecture of an instance of a compose transformation

A slot named "links" describes all links between the transformations, each one is defined by a list of a pair of pairs of values, i.e., ((transformation identifier, output number), (transformation identifier, input number)).

Finally we have a list containing the initial values, each one defined by a list of triples (transformation identifier, input number, value).

When an instance of a composed transformation is created, the instances of all transformations used in the definition of the composed transformation and all the links between them are established, and initialization of the inputs with the given values is done. This process is transparent, meaning the user does not notice that the system has created the instances of the particular transformations and he believes that he has a single transformation. The structure of composed transformation is represented Fig. 6 where the auxiliary inputs and outputs do the linking between the transformation inputs and outputs and the instance. The "if write" demon on the inputs must be redefined not to trigger the functor but to send the value for the respective auxiliary input.

Definition of the transformation's functor

The functor is implemented in Prolog and using two terms: **getIn**(X, input number, ValueIn), **putOut**(X,input number, ValueOut) to read and write the values in the respective frame. The general structure of a functor is shown in Table 6, where "X" is a variable and represents the name of the frame.

```
name(X) :-
    ...
    getIn(X, input number, ValueIn),
    ...
    putOut(X,output number, ValueOut),
    ...
```

Table 6 Structure of a functor

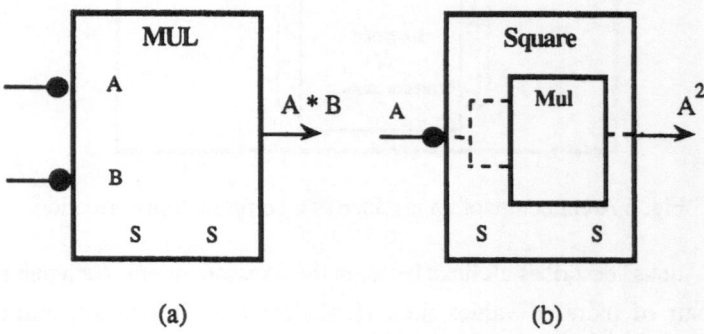

(a)   (b)

Fig. 7 Two examples of transformations

Two examples of transformations

Two very simple examples of transformations are shown in Fig. 7

The "simple transformation" **MUL** (multiplication) is defined by:

- Number of inputs   = 2 ;
- Type of Input 1   = FIFO ;   /* default value */
- Type of Input 2   = FIFO ;   /* default value */
- Name of functor, and context   = (f_mul ,real);
- Prolog term

         f_mul(X) :-

                 getIn(X, 1, V1),

                 getIn(X, 2, V2),

                 V is V1 * V2,

                 putOut(X,1,V).

The composed transformation, **Square** is defined by:

- Number of inputs   = 1 ;   /* default value */
- Type of Input 1   = FIFO ;   /* default value */
- Description of the elements   = [(m1, MUL)]
- Links   = [((input, 1), (m1,1)), ((input, 1), (m1,2))]
- Initial values   = []

3.3 Interface

In this environment the interface is implemented using two options: menus and tools. In "menus" all the options of the kernel which are not directly related to the system's programming exist. With these tools all the needed facilities for programming (schema editor) are defined.

Menus

There are two different types of menus: a menu for concepts and a menu for instances.

In the concept option the operations - "load", "save", "save as...", "create", "delete" and "change" transformation - can be chosen. With the "load" and "save" or "save as..." operations all the definitions of the transformation's concept can be loaded or saved. In the "instance" option selection the following operations can be done: "load", "save", "save as...".

Tools

The following tools were implemented: "select" and "paste" a transformation, "link" and "disconnect" transformations, "initialize" inputs and "debug" the transformations.

The tools recognize the objects: transformation's "icons" and "lines". The definition of the icon concept is made in Table 7, in which we have the information of the window where the picture is drawn and the position in the window where the icon is displayed. This entity has two primitives: "create" and "kill" an icon.

```
frame : icon ;
        slot : window ;
        slot : position; /* (x,y) */
        method : create_pic(mt_create_icon);
        method : kill_icon(mt_kill_icon).
```

Table 7 Picture's structure

The representation of a transformation icon is shown in Table 8. The definition of the icon is: number of inputs and outputs (nOfInput and nOfOutput), the height and the width of the icon, and the drawing functor ("descriptor").

Two pointers are also used: "transformation" and "setOfLines".The "transformation" is the name of the frame which represents the functional transformation and the setOflines represents the set of lines connected to this transformation icon.

```
frame: function_icon;
        isa: icon ;
        slot : transformation ;
        slot : nOfInput ; value :0 ;
        slot : nOfOutput ; value :0 ;
        slot : high ; value : 40 ;
        slot : wide ; value : 60 ;
        slot : setOfLines ;
        slot : descriptor ; value : draw_function;
        method : create_pic(mt_create_fnt);
        method : kill_pic(mt_kill_fnt);
        method : find_input(mt_find_in);
        method : find_output(mt_find_out).
```

Table 8 Transformation Icon's Structure

Finally we have the concept of "line" shown in Table 9. This definition is composed by the line definition and start and end point.

```
frame :line;
        isa: picture;
        slot : draw;
        slot : start ;
        slot : end ;
        method : kill_pic(mt_kill_line).
```

Table 9 Line's Structure

4 Conclusion

The present work, initially tuned to support tactile sensor oriented developments, is in fact much more general. The introduced concepts and facilities are very well adapted to the implementation of architectures in which multisensorial integration can be done. The adequate combination of different representation and programming paradigms also facilitates the integration of those architecture in others environments like CIM systems.

The first implemented prototype demonstrates the feasibility of the proposal and will be followed immediately by new developments including a strong support on simulation aspects. Also expectations exist regarding the adequacy of a new developed frame engine (KRIP) [5] already tailored for robotics applications, to be highly instrumental in near future developments.

Acknowledgements

This work has been supported in part by Junta Nacional de Investigação Científica, Instituto Nacional de Investigação Científica, ECC (ARTIFACTS).

Bibliographic references

[1]   F. Moura-Pires, J. Pimentão; Sistemas de Autónomos: Uma proposta de modelação do Sistema Sensorial. GR-RT-PS-49-88, 1988

[2]   PS300 Document Set, Volume 3a, Programmer Reference, Evans and Sutherland Computer Corparation

[3]   S. P. Reiss: An Object-Oriented Framework for Graphical Programming, SIGPLAN Notices, V21, 10, October 1986

[4]   M. Minsky:; A Framework for Representing Knowledge; Mind Design; MIT Press: 1981: pp 95-128

[5]   J. Moura Pires et al.: KRIP - Knowledge Representation in Prolog a frame based system (ver 0.9), GR-RT-G-01-89, 1989

Video-speed   Triangulation   Range   Imaging

P.W. Verbeek, J. Nobel*, G.K. Steenvoorden* and M. Stuivinga*
Applied Physics Dept.
Delft University of Technology
P.O. Box 5046
2600 GA Delft
The Netherlands

Abstract

When a light plane is projected on a scene, the resulting profile line can
be recorded by a camera system looking from aside. This technique for range
imaging with a 1-D or 2-D video camera is well known. It takes the recording
of an image line (say 256 pixels) to establish the position of one profile
point. If a 1-D video camera (line camera) is replaced by a Position
Sensitive photo Detector (PSD) the position is found from only two current
values, simultaneously measured. A speed gain of 256 is thus achieved.
However, like the line camera the PSD is a 1-D detector that needs a scanning device to measure the different parts of the profile line.

We want the speed gain without the scanning. Therefore, we propose to replace the 2-D videocamera by a novel type of line profile detector: an array
of PSDs that allows recording of a profile in a video line time (64us).

---

* TNO Institute of Applied Physics, P.O. Box 155, 2600 AD Delft,
  The Netherlands

To realize a video-speed range camera a PSD-array chip has been constructed. On the 10 x 10 mm2 large chip a row of 96 single PSDs have been implemented to be read out externally. In the experimental configuration a fan beam strikes the surface of a test object (cross section 100 x 30 mm2). With a 50 mm camera-lens the illuminated profile of the surface is imaged on the array sensor. The arrangement used is of the triangulation type: the height z at a lateral position x_i at the object is sharply imaged at position z' on the corresponding single PSD numbered "i". The position z' and thus the height z follow directly from both currents of the PSD.

Using a 2MHz clock and a multiplexing system all 96 PSDs can be read out in less than a video line time. By moving the object relative to the fan beam (either by a conveyor belt or by a scanning mirror) a real time video image of 256 range profiles is formed.

In this paper preliminary results are given, based on the read-out of a subset of eight PSDs of the array sensor.

1. Introduction

Triangulation is a standard principle for range finding [1]. Stereovision and focusing of a lens are triangulation techniques that use available light; they only work for objects that show detail. The detail may be added artificially by illumination through a projected dot pattern.
Taking the full advantage of controlled illumination one may project well-defined calibrated dot-, line-, or stripe-patterns. Now only one camera is needed to perform the triangulation; together with the projector it is equivalent to a pair of eyes: active triangulation.
Projected dots or lines must be uniquely labeled for identification in the camera image. A slow way is to label by frame number [2]: project one line per camera frame. A sequence of k frames is sufficient to label 2^k lines with a time sequential on/off code [3]. For grey scenes color labeled lines have been proposed. When camera resolution exceeds scene detail dot pattern code can be used [4]. Color and dot codes yield a range image per frame.

The method to be described exploits the knowledge that profile images as recorded by a camera under single line illumination will in general have one
bright pixel at each image line. The method gives a range image in one TV
frame time without color or resolution restrictions.

2. Line profile recording by PSD-array

As mentioned in the introduction projection of a single line (lightplane)
yields a profile of the scene that can be recorded by a camera. The profiles
can be analyzed to construct a range image.
For a traditional TV camera the analysis consists of localizing on each
image line the bright spot where the profile crosses. Recording all dark
pixels and searching for the spot is time consuming.

Localization of a bright spot is the specialized task of a different kind of
detector: the position sensitive device or photo detector, PSD. Several versions exist. The basic form is a long narrow photodiode with two end-
contacts and a common ground (linear PSD). The photocurrents I_1 and I_2 from
the end contacts are a measure for the position z' of the centre of gravity
of the light distribution on the PSD. If the length of the PSD equals L we
have

$$z' = L/2 \quad (I_2 - I_1)/(I_2 + I_1) \tag{1}$$

When the centre of the PSD is illuminated the photocurrent distributes
symmetrically and $I_1 = I_2$ and z'=0.

Measuring the spot position on a PSD takes as much time as measuring the
greyvalue in one pixel of a normal TV camera.
We have therefore proposed to build a camera chip that has a PSD at every
image line: the PSD-array [5]. We have produced a prototype chip of 96
linear PSDs [6], of which eight are in regular use for testing. Each PSD has
a photosensitive area of 6.6 mm x 8 um. The pitch of the array is 28 um.

3. Triangulation geometry

A set-up with standard triangulation geometry was used (fig. 1).
A laser diode plus cylinder optics produces a fan of light that strikes the object. The resulting profile is imaged on the PSD-array.
In order to get the next profile the object will be moved either physically by a conveyor belt or virtually e.g. by means of a rotating mirror. In the present set-up the object can be spun by an electric motor.
All positions in the light plane are imaged in the plane of the PSD-array (Scheimpflug arrangement). Any profile will thus have a sharp image.

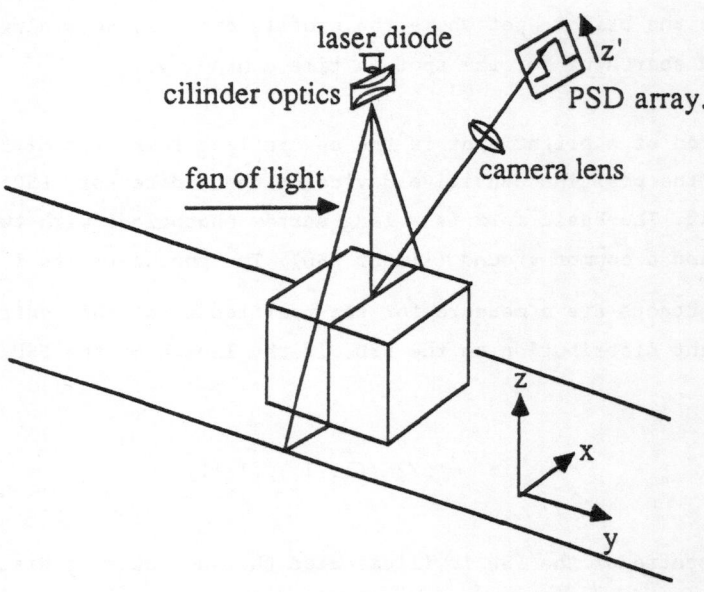

Fig. 1. Triangulation set-up.

The direction of the laser diode is z, the light plane is parallel to the x-z-plane ($y=y_0$), the conveyor belt will move in the y-direction.

The image coordinates are z' and x', where the x'-axis is parallel to the x-axis and the z'-axis is along the PSD strips. In a profile measurement z'is sampled at 96 x'-positions, x' = i. 28um, where i is the index of the linear

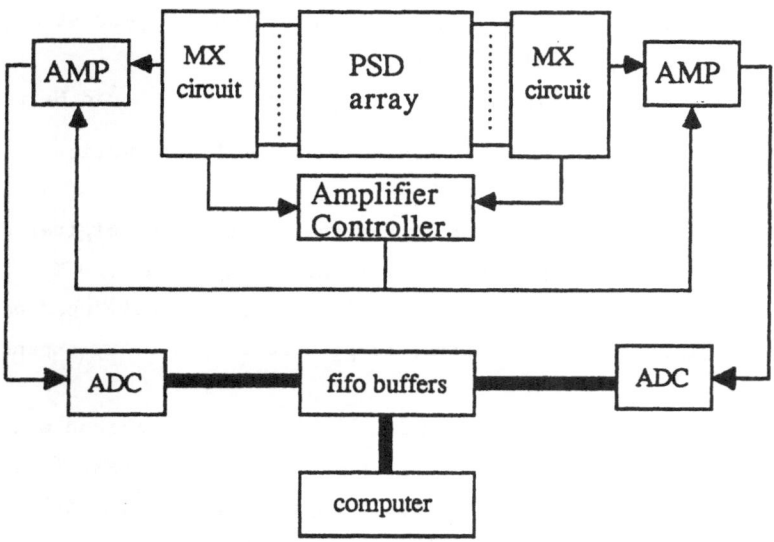

Fig. 2. Schematic diagram of the sensor electronics.

PSD. The value of z' follows from the currents I1i and I2i according to eq.(1). For given x', z' and y_0 the 3-D position xyz must be calculated. The relation between xyz and x'z' coordinates has been described in [1].

In the present set-up the range of the sensor is 50 mm in the z-direction and 24 mm in the x-direction. The range is mainly limited by the power of the laser diode.

4. Signal processing

The two currents I_{i1} and I_{i2} of PSD no i are selected by synchronized analog multiplexing. From I_{i1} and I_{i2} the spot position z_i' is calculated using eq. (1). The depth information is contained in the ratio of I1i and I2i rather than in their absolute magnitudes. The magnitudes depend strongly on the reflective properties of the object.

In fact the sum $I_{i1}+I_{i2}$ is a measure of the greyvalue that would be recorded by a conventional camera and can be used as additional information.

The ratio could be calculated either in the analog or in the digital domain. Analog dividers at video frequency are difficult to construct.

On the other hand an accurate digital calculation of the ratio woud demand a very accurate digital representation, large word length, very expensive A/D converters.

The solution we propose consists of analog block scaling combined with digital division. The schematic diagram is given in fig. 2. The essential components here are two identical steerable amplifiers that multiply both I_{i1} and I_{i2} by one, the same, of a fixed set of factors. The factor is chosen so as to have optimal use of the input range of the (eight bit) A/D converters. The choice is automatically made on the basis of the value of $I_{1i}+I_{2i}$. The eight bit outputs Ri1 and Ri2 of the A/D converters now allow a calculation of z'_i with roughly constant accuracy

$$z'_i = L/2 \quad (R_{2i}-R_{1i})/(R_{2i}+R_{1i}) \tag{2}$$

The A/D converters sample at 2MHz. The conversion results R_{1i} and R_{2i} are shifted into two first-in-first-out (FIFO) buffers. These buffers solve the problem of timing differences between the sensor and the computer. The computer reads the data from the FIFO buffers and uses these values to calculate the profile.

The calculation of a z'_i value from each read-in pair R_{1i} and R_{2i}, followed by the transformation calculation [1] of x_i, y_i-y_0 and z_i from z'_i, i and the triangulation parameters, would be too time-consuming for fast applications.

Therefore a different method is used. Since the number of possible pairs (R_{i1},R_{i2}) is limited to 256x256 combinations, the corresponding z'_i values are precalculated for all these cases and stored in a 64 kbyte memory. Using this look-up table z'_i can directly be found from R_{1i}, R_{2i}.

Next, the number of possible (z'_i,i) pairs is limited to 256x96. Three look-up tables are prepared that contain x_i, y_i-y0 and z_i respectively for each possible z'_i,i combination. The tables are filled on the basis of the trans-formation. The parameters of the transformation are derived from calibration measurements.

5. Results

A set-up as described above was made for testing purposes. Only eight of the 96 PSDs were used in the test. The consequence of the limited number of PSDs is that the profile is sampled at 3 mm intervals, instead of .25 mm

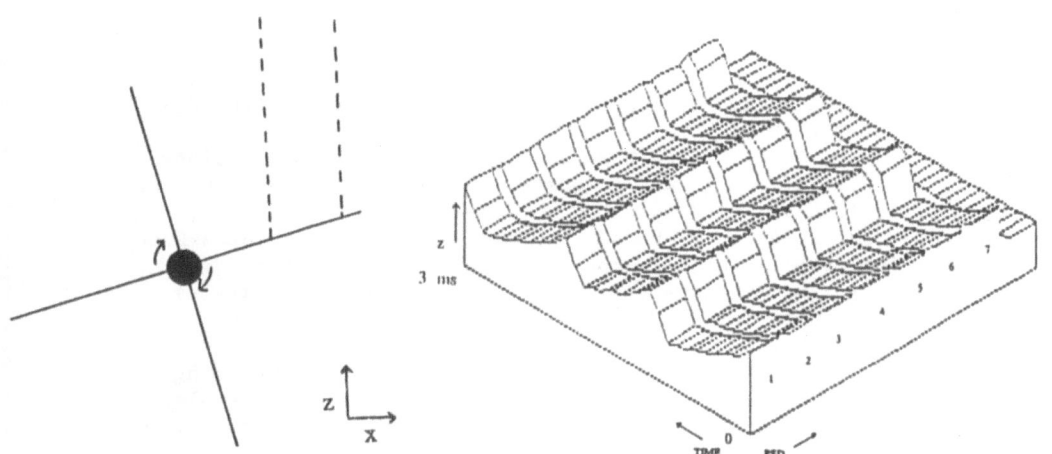

Fig. 3. Four blade rotor rotating at 14000 rpm and its time dependent eight-sample profile (uncalibrated, PSD no 8 has been used as a zero reference).

intervals when using all 96 PSDs. The extension to 96 PSDs only requires a minor adjustment in the multiplexing circuit, which does not influence the overall system operation. Therefore the test with eight PSDs is representative for the overall system behaviour.

The test measurements were done at full operating speed (two million samples per second). This means that a profile using eight PSDs is measured in 4 us and a complete two-dimensional profile with all 96 PSDs would take 48 us to measure. The table look-ups can be performed in real time.

An uncalibrated example, the range image of a four blade rotor, is given in fig. 3. The rotor speed was 14000 rpm implying a blade passage time of about 1 ms. The display shows only 100 us resolution.

An accuracy of 1% was designed for. Quantization noise contributes one least significant bit of the A/D converter (0.4% of the full depth range). Thermal noise is the main physical noise source in PSDs.

The range of the sensor has been kept at 50 mm x 24 mm in order to assure sufficient (0.5%) signal to noise ratio in the light of the limited laser diode power. Repeated calibration measurements should yield a total accuracy of 1% according to the noise contribution.

References

1. Ballard, D.H, C.M. Brown: Computer Vision. Englewood Cliffs, N.J., Prentice-Hall (1982)
2. Technical Arts Corp.: White Scanner. In: Industrial Applications of Image Analysis (N.J. Zimmerman and A. Oosterlinck eds.), Antwerp, p. 209. Oct. 17-19 (1983)
3. Inokuchi, S., K. Sato, F. Matsuda: Range-imaging system for 3-D object recognition. Proc. 7th ICPR, Montreal, pp. 806-808 (1984)
4. Vuylsteke, P.: Een meetsysteem voor de acquisitie van ruimtelijke beelden, gebaseerd op niet-sequentiële belichtingscodering. Thesis, Leuven, 1987
5. Verbeek, P.W., F.C.A. Groen, G.K. Steenvoorden, M.E.C. Stuivinga: Range cameras at video speed based on a PSD-array or on CCD-interpolation. Proc. IASTED Int. Symp. on Robotics and Automation, Lugano, pp.67-69 (1985)
6. Stuivinga, M., J. Nobel, P.W. Verbeek, G.K. Steenvoorden and M.M.Joon: Range-finding camera based on a position-sensitive device array. Sensors and Actuators 17, 255-258 (1989)

Image Sensors for Real–Time 3D Acquisition

Part – 1 –

Three Dimensional Image Acquisition

P.Vuylsteke, C.B.Price, A.Oosterlinck

ESAT–MI2 Departement of Electrical Engineering
Katholieke Universiteit Leuven,
Kardinaal Mercierlaan 94,
3030 Heverlee, BELGIUM

NATO ASI Series, Vol. F 63
Traditional and Non-Traditional Robotic Sensors
Edited by·T. C. Henderson
© Springer-Verlag Berlin Heidelberg 1990

1 Introduction

Development of depth–sensing systems has been influenced to a large degree by discoveries from perceptual psychology. The Human Visual System (HVS) provides one with information about the shape and spatial relationships between objects in an observed scene, even while the scene is changing. The ability of the HVS to function correctly is mainly thanks to a built–in redundancy of operations, several mechanisms being brought into play giving various interpretations of the scene which are combined to give the best most plausible global description of the scene. This process can of course yield inconsistent interpretations, which we call optical illusions.

The most inportant of the biological perception mechanisms are based directly on physical models of image formation. *Stereopsis* and *motion parallax* are based on pure geometrical principles. *Dynamic Focussing* draws on the laws of geometrical optics. Some of these mechanisms for absolute depth–measurement are found in the lower levels of the visual cortex and demand no higher–level input. This is one conclusion from random dot stereogram experiments where it is possible to perceive a three-dimensional image by fusing two random dot patterns, each of which has no structure. However, it has also been demonstrated that other of these mechanisms require implicit information concerning the object surroundings. For example, in the process which derives depth information from the gradient of observed texture size, the actual texture used is homogeneous, but this fact is not explicitly known at any level in the HVS , but is 'assumed'.

While research into the fundamental mechanisms of depth-perception in the HVS (from neurophysiological, psychophysical and computer model studies) has led to better understanding of three-dimensional computer vision, it has not led to any applications in industry. Such research leads to methods which are complex, and inherently parallel demanding new approaches, such as Neural Networks .

We can classify depth–measurement systems into *passive* and *active*. In the former class, the measurement process only extracts information already present in the scene, while active systems first add information to the scene, and observe how the scene modulates this information. Passive methods use simple , usually natural, illumination, while the added information of the active systems takes the form of structured or coded illumination.

Active methods may be furthur classified into two groups, firstly *coaxial light beam* systems, such as optical radar, where depth is obtained from the time–of–flight of a laser pulse reflecing off the object. Secondly, *triangulation* techniques make use of a projected light soure directed in one direction, received by a sensor directed in another. This is the most important technique as far as industry is concerned.

2 Passive Methods

Spatial information can be obtained from two–dimensional projections made from a single viewpoint. The resulting images give information about the orientation of object surfaces. The relevance of these techniques to industry is not great, research here leading to deeper understanding of the mechanisms of visual perception.

2.1 Shape From Shading

This term was introduced by Horn who showed how a recorded intensity image can be used to provide spatial information about a scene. The intensity of light reflected from an object is determined by the illumination, the object's albedo, and most importantly its orientation. A mathematical relation between the reflected image light, and the illumination can be constructed and then used to recover orientation from measured intensity. With a coordinate system centered at the camera with large focal length, it is possible to construct a *reflectance map* which establishes the exact relation between measured image values and object surface orientation. This map can be obtained analytically for simple setups e.g., point illumination, Lambertian or specular surface reflexions, but is ususally obtained experimentally.

Of course the image grey–values are by themselves not sufficient to determine the local surface orientation , which involves the specification of two parameters, while only one is provided by the image intensity values. This problem is solved by assuming the surfaces to be smooth. This assumtption, (which is not always easy to justify where e.g., the scene contains edges), together with the large number of system parameters which need to be known (nature of the reflexion, albedo, illumination distribution) restrict the possible applications of the method.

2.2 Photometrical Stereopsis

Here a relationship is established between the recorded image grey–values and local object surface orientation. Several image recordings are made from a single, fixed, viewpoint. The (misleading) term stereopsis comes from the effective use of several illumination directions. Strictly speaking, this is an *active* method since the nature of the illumination is an essential part of the measurement process.

Equations of reflexion link desired surface orientation to observed recorded image intensity. This relationship is used to build up a reflectance map, countours of constant grey–value in the gradient space for given reflexion and illumination conditions. When the solution in gradient space is sought for locally, the solution is underdetermined, requiring the specification of further constraints such as continuity. Since recordings are repeated with several illuminations, the solution must satisfy several conditions, imposed by each separate illumination. Since the equations are non–linear there is not necessarily one unique solution. Computations can be executed in real time through the use of look–up tables produced from calibration objects.

One method has been constructed to make use of *specular* reflexions, where line-illumination is required (since each surface element only reflects in a certain direction, a point–source would only return information from a small part of the object). Specular reflexion can only come from one object point, a realization which led to the develpoment of a method using four point ligh sources, The solution is computed for each of the four possible combinations of three light sources using a Lambertian reflexion model. This allows specular reflexion resulting from any one light source to be detected and the corresponding solution rejected. In a typical measurement situation, a combination of specular and *isotropic* reflexion occurs.

By virtue of the possible redundancies involved in photometrical stereopsis, it is more robust than 'shape from shading'. Also the strict control of illumination is no great problem in many industrial applications. The most important disadvantages of this method are the dependance on a specific relfexion model, the disturbing influence of scattered light from one object to another and the rather high demands made of the sensor. Grimson has suggested the use of photometrical information as an addition to classical stereopsis, to allow to allow the interpolation within visually 'flat' regions.

2.3   Shape from Texture

Bajcsy and Lieberman have developed a technique to determine relative distance from observed texture size, using distortions of regular texture patches by perspective effects. Texture elements projected onto a background object are observed to be larger than those from a foreground object. This allows measurement of relative object distances, which can then be converted into absolute distances if the height of the camera and its optical parameters and the size of the texture elements are known. The most general form of this procedure is underdetermined, so its use is only suggested for natural scenes where the terrain is flat and the texture pattern is homogeneous.

Ikeuchi defines a distortion factor fo the observed texture, which determines the direction of the surface normals compoatible with the known view direction. This is rather like shape from shading, here too additional restrictions can lead to a simple solution. Ikeuchi suggests the use of a combination of multiple viewpoints.

3   Stereopsis and Parallax

The fundamental underdetermined nature of a three–dimensional reconstruction from a two-dimensional projection can be for the most part solved using two or more images, recorded from different view–points. We consider two approaches; the classical *stereopsis* approach where two or more camera positions are used, or the *parallax* approach where the view–point is made to move continuously.

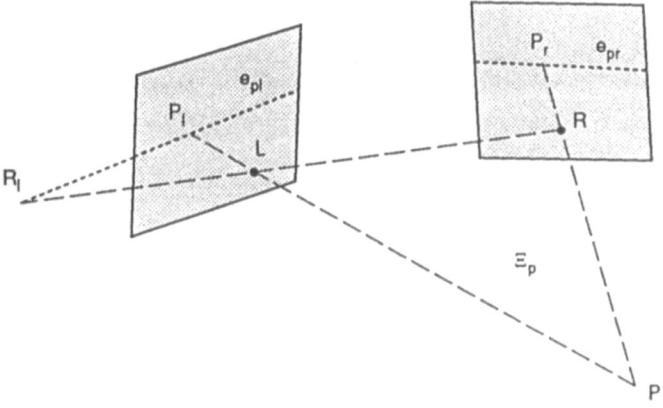

Figure 1. The Geometry of Stereopsis

Consider an arrangement with two cameras, with lenses at L and R with fields of view which overlap to a large extent (Fig.1). An object point P is projected in P_l and P_r. For a given geometry, the coordinates of the projections are sufficient to determine the point P. Stereopsis is at first sight a straightforward geometrical problem. Although sufficient information is present to provide the reconstruction, the problem arises of finding corresponding image points in the two images, required for the reconstruction calculations. This is the so–called *correspondance* problem.

3.1   The Correspondence Problem

Finding corresponding points in a stereo–pair of images is at first glance an enormous combinatorial problem; each point in one image must be associated to one point in the second image. For an image of linear size n, this problem is of order $O(n^4)$. Fortunately, the geometry of the situation allows a reduction in the dimension of the search problem by 1. The plane LRP through the lenses is called the *epipolar surface* through P. The projection of this plane through the left image is the *epipolar line* e_{pl} Epipolar lines e_{pl} belonging to different points P cut each other in a common point R_l (possibly outside the image). An equivalent situation is found in the right image. For a given image point P_r in the right image,the corresponding epipolar plane is LRP_r. Since the object point P also lies in this plane the corresponding point in the left image must lie on the epipolar line e_{pl}. This *epipolar restriction* means that for each point in one image, the corresponding point in the second must be sought for only on the associated epipolar line.

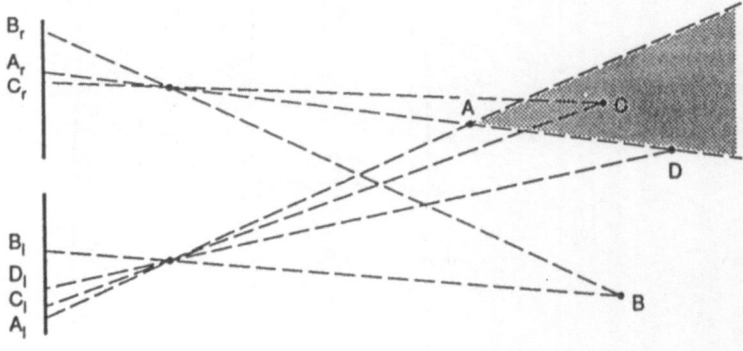

Figure 2. The Correspondence Problem

For a general optical setup these lines can in principle be determined, but the calculations are complex for this general case, so usually a simplified setup is used with coplanar aligned image planes and identical lenses. All epiploar lines are then parallel and corresponding points are found at the same height, i.e., row in the image. The normal distance from an object point to the image plane is then inversely proportional to the relative horizontal distance or *disparity* between the two image points and proportional to the lens focal length and to the length of the baseline LR.

Whereas the epipolar restriction is simply a consequence of the projection geometry, in most applications there is no simple and faultless procedure for matching the corresponding points. To this end, extra constraints are imposed, based on general considerations of the scene. The first steps in this direction were made by Marr in his research into steropsis in the Human Visual System. Marr proposed two basis theorems; (i) each image point has at the most one corresponding point in the other image (principle of unicity) (ii) the disparity changes continuously over the scene. The resulting relaxation algorithm has recently been implemented on the Connexion Machine.

Most of the stereo alogirthms implicity assume that the order of the points along the epipolar line in each image is the same (Fig.2) where point B_r is found on the same side of A_r as B_l to A_l. This assumption is not valid on the grey area behing point A which forms a 'forbidden region'. Object points on the boundary lines of this region do not satisfy Marr's first principle; A_r having two corresponding points A_l and D_l. This situation arises in practice for transparent objects, skeleton–like structures and where a large object is located just behind a small one. The forbidden region has been extended to the three dimensional case by Yuille and Poggio. Likewise the principle of continuity has been revised. The proposal that the disparity changes continuously along an epiploar line is not satisfied if the projections of the object cross.

3.2 Primitives and Strategies for Stereo Correlation

The search for corresponding points in a stereo pair assumes that both points are projections from one physical point in the scene, and not artifacts or noise. Finding corresponding object points in two images requires that in each image an object point can be distinguished from its surroundings. In an image plane, a point can only be distinguished using its value i.e., brightness. Brightness variations in an image can be produced by texture or surface colour from shadow–effects and finally from shape. Where such variations are not present, e.g., surfaces of uniform grey–value, stereopsis is not suited.

The use of raw grey–values as features is itself unsuitable, since, in general, too many points have the same grey–value. So groups of pixels must be used. Groups may be selected over limited neighbourhoods, e.g., local patterns of edge pixels, or else as macroscopic elements e.g., line segments. In general, the higher the level of primitives used, the easier the correspondance problem can be solved. This of course requires more preprocessing and leads to a loss of flexibility of the system which then handles one problem class well, but loses the ability to handle a range of situations.

Many systems find correspondance by direct correlation of the images. The local maxima of cross–correlation over small regions gives a good indication of the most similar positions. Here there are two problems: firstly the computations involved are intensive, and secondly it is unlikely that the correlated regions agree on a point–wise basis.. The same surface is observed from two different angles, so their intensities can vary. But also the *perspective shortening* differs in the stereopair, that is the apparent distortion that results from the sideways view of the object. Thus there are evident distortions between the correlated patterns. These can be reduced by a reduction in size of the correlation window, but this in turn leads to more maxima being detected and an increase in false correlations. There is a tradeoff.

To combat this problem, the correlation terms are often weighted with coefficients which fall off towards the sides of the correlation window. Also the number of candidates for correlation can be restricted, e.g., by the Moravec interest operator , calculated as the difference between squares of neighbouring pixel values, leading to the selection of image regions only with a strong varying content. Ellis suggests that in many industrial scenes, the background is constant and thus contains no relevant information. The image is first segmented into objects and background and only non–background pixels are used in the correlation. Segmented edge–points give good starting points for the correlation opeation. To combat the problem of occlusion, the correlation operation is duplicated; for each point in the left image the corresponding right image point is sought for and vice versa.

In the second category of stereo methods, the agreement is determined using *elementary image primitives* such as edges, the local maxima of the intensity gradient characterized by their direction and magnitude (edge contrast). These primitives are compact, universal and contain the most relevant information required for stereo correlation. The latest version of the Marr–Poggio–Grimson algorithm uses the zero–crossings of the Laplacian as the edges. While the gradient is directional the Laplacian is not and can be computed with a single mask. Also the Laplacian allows greater accuracy in edge detection, since where it crosses zero, it is changing rapidly, while at the maximum of a directional gradient operator, the rate of change lies around zero. Before application of the Laplacian operator, the image is first smoothed with a Gaussian filter. This allows noise reduction, and in combination with the following Laplacian allows selection of primitives' size since the combination Gaussian smoothing plus Laplacian edge filter is in fact a band–pass spatial filter. This suggests the use of several filter sets allowing a set of primitive images at several resolutions to be constructed ,leading to a *hierarchical search strategy* for correspondance. First correspondance is sought for at coarse resolution, and when found can be passed up to the next finer resolution where the search space is then much smaller. This leads to a reduction in computation time, but more importantly, since the search regions becomes progressively smaller, there is less chance of incorrect matching.

Stereocorrelation can also be determined using macroscopic features such as line–segments. This requires a good image segmentation which restricts the method to the small class of polyhedra–like objects. Search for correspondance is dramatically simplified when a high–level structural description of the image is available (e.g., high level primitives combined with relations, such as 'left–of', 'up', 'touching'). Such high–level primitives are parts of a skeleton and have an associated parameter set (such as length, orientation, mass). These structural descriptions are also useful in further stages of processing, e.g., determination of a stable grip position for a robot.

3.3 Multiple Viewpoints

The principles behind stereopsis can be easily extended to more than two viewpoints. Through the increased redundancy so obtained, the reliability of the stereo correlation is increased, and occluded points detected easily. While this technique is not new, it has recently found renewed interest. Three cameras are usually used in colinear, coplanar or arbitrary configurations. Usually the redundant information is used in the last stage of the stereocorrelation, and may serve simply as a check on the results. Ohta processes the stereoimages in pairs, and uses the dynamic programming technique to find the correspondances between each pair, the whole set then fused together using a relaxation method. Tsai has made a different suggestion, where the measurement system consists of eight cameras in a coplanar arrangment on orthogonal axes. For a given point in one basis image, the corresponding points in the other seven images are sought for directed by one disparity parameter (eg the disparity between the basis image and the nearest left image), as a consequence of the epipolar restriction. The correspondance problem remains one-dimensional for several viewpoints. The redundancy emerges from the fact that for each disparity parameter value, all eight image points must agree, since they are all projections of the object point. For any given small object area, the eight projections are similar, and so are their local statistics. Thus the disparity parameter is determined as that which (i) maximizes the combined moment or (ii) eliminimates the variance. This assumes that the image functions are stationary so that the estimate of the parameters can be obtained via local averaging.

3.4 Parallax

The principles of stereopsis can be applied to a sequence of images obtained where the camera makes a known sweep across the scene. The average disparity between successive images is dependent on the ratio of sweep speed to image spatial frequency. By keeping this ratio small, the search area for stereocorrelation can be dramatically reduced.

Thinking of an infintesimal step in camera sweep, it is possible to write down equations of optical flow, which relate the intensity variation at a point in the recorded image to the movement of a passing scene. There are assumptions that the intensity gradient is continuous and non–zero. The solution is under–determined so additional constraints are needed e.g., smooth variation. These conditions are ususally satisfied except at object boundaries.

The optical flow is strongly dependent on the spatial form of the scene. The inverse problem of recovering the geometry of the scene from observed optical flow is for the most part soluble, with the assumption that the scene contents do not change. The motion of the camera does not need to be specified, but can be deduced form the optical flow of a few selected points. Research here is directed towards how many datum points are needed and to the constraints layed upon them.

4   Active Methods

In the *passive* techniques discussed above, no restrictions on lighting were imposed, (with the exception of photometic stereopsis). These techniques can be applied under arbitrary conditions of illumination, but this universality comes at a high price. A profound intermingling of effects of illumination and spatial form yield the recorded image. To invert the process of image formation, it is neccessary that these different factors can be separated. For most passive methods, this problem is ill-conditioned, requiring the addition of extra constraints. Robust passive algorithms are moreover computationally expensive.

Three dimensional image acquisition can in theory be simplified via the *integration of the illumination with the measurement process*. This assumes that the sensor can separate the effects of the specific light source from other background light sources. In practice this is achieved through screening of background illumination or by matching the spectral characteristics of sensor and light source (through band-pass colour filters). The methods can be grouped into *coaxial light-beam* methods such as the laser rangefinder and *triangulation* techniques, both developed in many versions. The following section presents a summary of these methods.

4.1 Optical Radar

When a plane–wave passes through a known medium, the distance travelled can be computed as the product of wavespeed and time to cover the distance. This simple principle is applied in the laser–rangefinder, which comprises a laser source and a photodetector.

In optical radar, the time–of–flight is measured, the time needed for the emitted laser pulse to reach the detector after scattering from the object surface. The optics are designed so that the only scattered light reaching the detector arrives parallel to the emitted light, so that the object distance is half the total distance travelled by the light pulse. The accuracy of the distance measurement is determined by the accuracy of the time–of–flight measurement. Given the large velocity of light, the demands on the detector are extreme (a resolution of 1mm requires an accuracy of 3 *pico*–seconds in time). Moreover, the signal at the detector has a finite rise–time and a variable amplitude precluding the use of simple thresholding to detect the arrival of the signal. Many systems make use of the 'constant fraction timing discriminator', where the trigger level is chosen at a fixed fraction of the peak value of the detected pulse. Also, to reduce the possible dynamic range of the detector signal by several orders of magnitude, variable optical and electronic attenuators are employed.

Two forms of modulation are commonly used; pulse and amplitude modulation. For pulsed systems, the accuracy of the distance measurement is bounded by a compromise between the laser power and the pulse frequency. Repeated measurements are made and averaged to improve the accuracy at the expense of measurement speed . With the addition of a collimator, a complete distance image can be obtained using a scanning procedure.

For shorter distances, continuous modulation is normally used, the distance being computed from the phase difference between detected and emitted signals. The resolution of measurement here is no greater than one half a wavelength of the modulation. The effects of noise are dramatically limited through filtering at the detector with a sharp band–pass filter, tuned to the modulation frequency. Detection is preceeded by a mixing of emitted and received signals with a common intermediate frequency to allow the phase detector to operate at a low frequency.

Optical radar measurements have a very discrete character, and can be obtained without complicated post–processing, in contrast to the passive methods described earlier. Thanks to the coaxial paths of emitted and detected light beams, the problem of partial occlusion, specific to stereopsis (and also triangulation) disappears. The cost of these methods is extremely high, and due to the inherent danger of laser beams, their application is confined to screened workareas.

4.2 Triangulation with Simple Illumination Patterns

The spatial position of a point is geometrically determined by the cutting of a plane by a line, by two lines in a plane, or by three planes. These geometrical facts are used in a series of simple triangulation techniques, which make use of a directional projector and a directional sensor. Projected light, either with a point or line distribution is reflected or scattered from the objects' surfaces and arrives at the sensor, which 'looks' in a particular direction. The reflected component in the direction of the sensor must be sufficiently large which in general requires diffuse reflexion or a degree of scattering by the object. Only the illuminated object points are available for triangulation, which is a very small fraction of the total surface area. Hence a scanning or deflection capability is usually built in to the system allowing the illumination of a wider field of view. This typically takes the form of a mechanical mirror arrangement. A plethora of implementations using this concept have been made, which may be classified according to the following characteristics:

- The dimensionality of the illumination (point or line)

- The number of degrees of freedom of the scanning(0,1 or 2)

- Whether or not the scan position is specified ; the instantaneous scanning parameters are not needed if a redundant sensor arrangement is used

- The dimensionality of the light sensor (line or plane)

- The optics (spherical or cylindrical lens, mirrors, multiple apertures)

- The number of sensors

Typical configurations have been summarized in Fig.3 (details of specific systems have been abstracted). ϕ and θ represent the two degrees of freedom for rotation of the scan.

Fig.3(a) shows a simple but not often used system, where the sensor is two–dimensional so that the projector scanning can have two independent axes of rotation. If two sensors are used, the points on the object can be determined from the two recorded image points as in stereopsis. Here the projector scanning does not have to be calibrated, but serves simply to illuminate one point on the object allowing the correspondence problem to be side-stepped. The acquisition speed is very slow (0.5 sec/point). In another variant, one light sensor is surrounded by six LED's which are illuminated in sequence, which provides enough information to measure the distance and local orientation of the object surface. Yet another variant uses a kaleidoscopic tunnel of mirrors to effectively extend the sensor area, but with a more difficult interpretation of the resulting image.

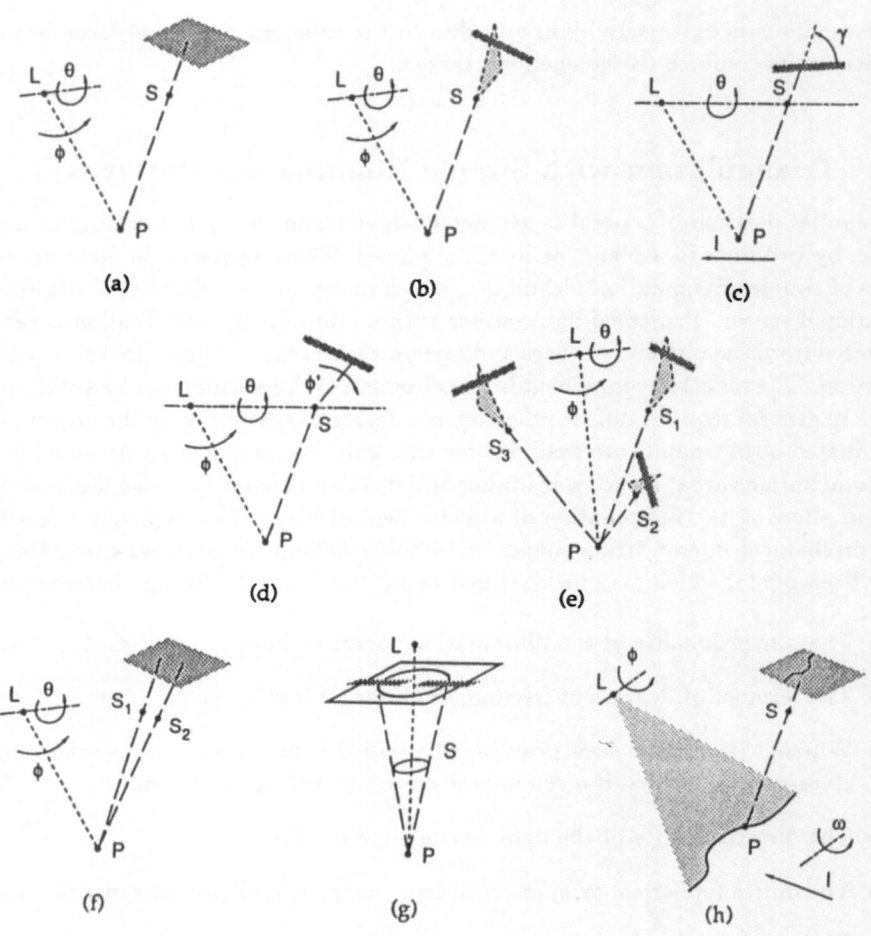

Figure 3. The Various Triangulation Configurations

Fig.3(b) The use of a plane light sensor as described above serves only to ensure that light will actually be received from an object point. This can be done more efficiently by the use of a linear sensor combined with a cylindrical lens. This lens serves to broaden light arriving from a single object point into a line. Its orientation is arranged so that this line always intersects with the line sensor, thus guaranteeing capture of the illuminated object point. Haugen developed this technique by the use of a digital line sensor comprising a series of long, homogeneous photosensors arranged in a parallel matrix, covered with a binary mask (Fig.4). The mask is constructed so that each line of light traversing the array excites a different combination of sensors, yielding a Gray-code. Here, the system speed is also increased by replacing the classical mechanical mirror deflexion system with a rotating hologram, which scans a complete 128x128 raster in one revolution, giving an acquisition speed of $2x10^5$ points/second at a distance resolution of 4-bits.

Fig.3(c) Here the linear sensor is accurately aligned in the same plane as the projection ray, so that the sensor always receives some light from the object. For a certain sensor angle γ, given by the so–called Scheimpflug condition, all points along the sensor line are in focus. Scanning is accomplished by rotating (θ) both projector ray and sensor together (which obviates the need for a cylindrical lens) and through lateral translation (l) of the object. A typical commercial instrument has a precision of $34\mu m$ in a depth of 32mm at a working distance of 10cms, and an acquisition speed of 125 points/second.

Fig.3(d) A further enhancement couples projector and scanner deflexion to allow use of the entire sensor length.
Fig.3(e),(f) In the configurations discussed above, the illumination geometry is an essential parameter and must be known for each illumination position. The scan direction is usually encoded optically. If several light sources are used, then this requirement can be dropped. The position of a point is determined by the cutting of three surfaces through the centres of the linear sensors, perpendicular to the sensors. Nishikawa suggests a coplanar arrangement of two sensors. With the need for a compact and reliable system in mind, the possibility of bringing both projections together at one sensor is being researched **Fig.3(f)**.

Fig.3(g) Idesawa uses a mirror–tunnel in front of the sensor elements which has the effect of projecting an object point into a ring, the diameter of which is proportional to the object distance. This diameter is measured using two linear sensors. Given the small distance between the sensors in these latter techniques, it is hardly surprising that their accuracy is very sensitive to system defects.

Fig.3(h) Illumination with a point source allows only a single measurement to be made from each read–out of the sensor device, resulting in a rather slow measurement procedure. Alternatively, by projecting a line onto the scene, and making the triangu-

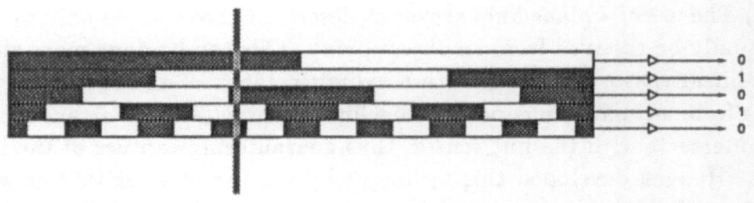

Figure 4. Digital–readout line-stripe sensor

lation calculations for all points on the line profile a speed–up can be obtained, even though it takes longer to read out the two dimensional sensor required. This system is applied in welding where the scanning follows the welding line. Here, specific problems need to be dealt with; smoke and spats are blown off, light emanating from the welding is removed by interferometric filtering. Welding light has a broad spectrum peaking in the ultra–violet, while the projected (usually laser) light is red. This measurement technique is also suitable for the alignment of parts on automatic assembly lines, in contrast to 'bin-picking' situations where scanning is normally required.

One variant of the light–stripe technique uses a linear camera set up so that only light arriving from points of known depth is received. The working distance and view angle are fixed, allowing optimal setting of the optical parameters. The object is placed on a precision translation table, and moved in small steps along the depth direction, building up a depth map across an entire section of the object surface.

The accuracy of depth measurement by triangulation methods is mainly determined by the distance between camera and projector and by the precison of measurement of the imaged light. The projected light must remain in focus over the working distance, a distance limited by the depth of focus of the projector. This can be increased by use of a ring–like diffraction pattern instead of a single spot. Another inportant parameter is the resolution of the sensor. As well as classical CCD or CID sensors, analogue sensors are also used, e.g., lateral effect photodiodes. These devices are in essence, a light–sensitive diode combined with a distributed resistance. Here, injected current flows to the edges of the sensor with a distribution given by the position of the point illumination. The centre of the light patterns can be directly determined from the measured current distribution.

4.3 Periodically Structured Illumination Patterns

Triangulation techniques which use simple illumination patterns find numerous applications due to the simple and unambiguous interpretation of the measurements. However, the available surface of the image sensor is underused, and the acquisition time is high due to the required systematic scanning of the scene.. To improve upon these limitations, systems have been developed which project a pattern onto the entire scene, allowing a distance image to be produced from one single recorded image.

A direct enhancement of the light–stripe patterns takes the form of a wedge of light patches projected on the scene. This allows each illuminated object point to be available for the triangulation calculation. Since the projected pattern is periodic, a correspondance problem arises since it is not known from which particular light patch a recorded pixel originates. For the case of simple scenes where there are no overlapping parts or shadow effects, the ordering principle from stereopsis can be applied, that is the natural order of the light patches in the projected image is maintained in the recorded image, so that the patch number can be obtained by counting. Reference points are obtained from fixed surfaces with known positions, e.g., the floor. All light lines on this surface are known, and can be traced throughout the entire recorded image, if they remain continuous. Nevertheless there is a real danger that the overlapping light lines originating from different patches remains undetected.

The largest research and development effort in this area is directed to improving the precision of depth measurements. One such method involves accurate generation of the light stripes using a laser–illuminted hologram, and a double–aperture image sensor. Depth measurement is obtained from the displacement of the lines measured by the two apertures. The resolution is increased by displacement of the line patterns. In another system, designed to inspect turbine blades to an accuracy of 25μ, subpixel resolution in measurement of the line positions is required. This is achieved using a superposition of two projected line patterns, one half–line out of phase. Each pattern is a different colour allowing separate recordings of the reflected patterns. Depth is obained from the zero–crossings of the difference signal, which do not depend on the variation of reflectivities within the scene, providing any variations apply equally for both colours.

Identification of the light stripe remains a problem for both local orientation or relative depth measurements. The projection of line patterns can be thought of as a the application of a carrier signal to the scene which is then modulated by the spatial form of the scene, recorded and demodulated. Object surfaces with a specific orientation lead to separate delta–functions in the Fourier–domain and can be segmented by the use of band–pass filters. The determination of orientation requires two independent components. These can be obtained from the observed direction of the horizontal and vertical raster lines in the recorded image. For surfaces which vary slowly, relative depth can be obtained by simple integration. It has even been shown that orientation can be obtained from a single parallel–line raster, using the observed directions and the distance between neighbouring lines. Accuracy increases with the line separation. Other alternatives include the use of a regular texture pattern, and several projections from various directions to allow the measurement of absolute position. Here the resolution is determined by the size of the texture elements.

In the last ten years, much research has been directed to potential solutions for the correspondance problem with structured lighting. This should allow the high–speed measurement of depth, without needing to fall back on specifics of particular situations as in the case of the passive methods. For example an illumination pattern can be added to a stereopsis measurement system. Correspondance between the recorded images is sought for using agreement between the observed networks. This does not take into account any partial occlusion which would lead to differences in the struture of the two observed patterns.

Like the correspondance problem, the identification problem associated with active triangulation methods is essentially one–dimensional. The stereopsis geometry of Fig.1 can be applied directly to the methods described here, where one camera is replaced by the projector. The epipolar restriction thus remains valid. An extra limitation has its origins in the discrete nature of the illumination pattern: Consider an orthogonal line raster, defined in the image–plane of the projector, for any observed raster point in the camera image, the corresponding point in the projection image must be sought for on the epipolar line. But the chosen point must lay on fixed discrete positions in the projection image. If the epipolar lines do not lay on the raster rows, then the number of possible correspondances is limited, at least in theory. In practice, the epiploar lines are surrounded by an uncertainty region. In a typical measurement system, the density of raster points is very low (total 30 points), so that an epipolar band only intersects one raster point in the projection image, namely the chosen raster point. For larger rasters, one can see that the average number of possible correspondances is proportional to the linear density of raster points, and to the breadth of the uncertainty band around the epipolar lines.

To further reduce the number of correspondance pairs, extra criteria can be applied. For example (i) for each observed raster point in the recorded image,there is only one corresponding raster point in the projector image; (ii) the relationship between row and columns also holds in the recorded image, (iii) one raster point has a maximum of four neighbours (iv) the local order of rows and columns is maintained.. Using these criteria, the remaining possibilities are submitted to a relaxation network which yields the required consistant correspandances. Experiments show that for a 14x16 point raster, the local relaxation method eliminates only 2–12Following the addition of global limitations, between one and four spatial interpretations remain.

The ability to obtain a simple identification of discrete raster points or lines is provided by the addition of an extra camera, allowing the verification of each suggested solution. Abolute indentification of rasterlines or points cannot be guaranteed since the redundancy limitations must be applied to a controlled extent otherwise the correct solution may also be eliminated.

One suggestion uses pairs of adjacent raster lines. The corresponding pair is sought–

for in the corresponding epipolar line . If these pairs are not parallel, then there is only one place where the distance along the raster direction is exactly equal to the raster period. This position must correspond to the sought for raster position. Such a position is easy to find, but the solution is sensitive to errors, since the epipolar lines corresponding to adjacent points intersect with a very small angle, an effect exaggerated with a high raster-point density.

The epipolar restriction technique can be iteratively applied in a dynamic configuration. For an observed rasterpoint in the recorded image, the corresponding position in the projector image is limited by the epipolar line (Fig.1) If the projector is rotated around its optical axis, and the rasterpoint is tracked in the recorded image, then each point determines a new epiploar line in the projector image. Both epiploar lines belong to the same rasterpoint which is determined by the intersection of the lines. Identification is not affected by possible movements in the scene. This requires correct tracking of the raster points, despite discontinuities in the scene where moving rasterpoints may suddenly disappear from the image and reappear at other places. Again, increased raster-point density aggravates this problem.

4.4 Moire-Contours

Moire patterns are produced by a low pass filtering of a combination of periodic or quasi-periodic structures. Perception of this effect is made possible by the preference of the HVS for low spatial frequencies. The use of moire patterns to code spatial information is a well-established technique, with several variants. In *shadow-moire-topography*, a line screen is placed in front of the object (Fig.5), masking all object points except those at specific depths. As the distance between projector and camera is reduced, the spacing of the depth-levels approaches a constant value. Low-pass filtering produces a percept of light and dark contours of constant depth. This technique was originally developed for photographic recording, allowing a simple, manual interpretation of the photographs. In practice, the use of a large mask screen is too clumsy for many applications, which led to the use of separate mask screens placed in front of the projector and the camera. This technique is known as *projection-moire-topography*. With the use of digital processing, the camera mask can be replaced with an electronic (or software) mask, where the mask is effectively constructed by selective processing of the recorded data. This is known as *scanning-moire*. For scenes with slowly-varying surfaces the order of the fringe can be obtained by simple counting, with one known position obtained by projection of a datum mark. Successive fringes from a single moire pattern give no indication of the direction of surface slope, this direction is easily obtained by changing the phase of the detection mask and observing the direction of shift of the moire contours.

The next stage in the developments of the moire system involves the projection of a moire reference images, obtained by recording an image of the object illuminated with a moire mask. When a similar object (eg the same object type but with a defect, or in vibration) is illuminated with the reference pattern, the moire contours produced reveal

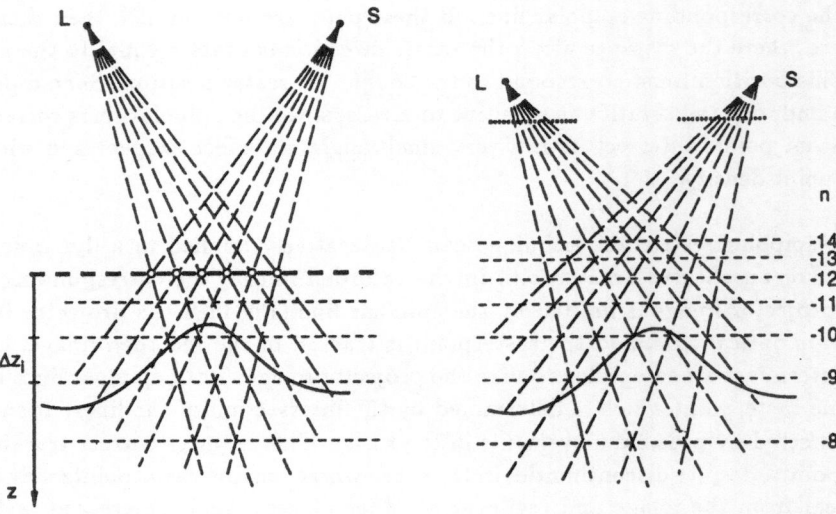

Figure 5. Moire topography

differences between reference and measured object. This leads to defect–detection applications.

The attraction of moire techniques if found in their historical origin , a direct and easily interpretable *visualization* of the depth results in the form of iso–distance lines. No additional information is provided by moire over the previous triangulation methods discussed, indeed the moire equipment is identical to line–raster triangulation. The identification problem still appears, leading to failure to interpret broken contour lines without additional information.

4.5   Coded–Illumination Methods

A solution to the correspondance problem can be directly obtained by enabling each individual position of the projected pattern to be unambiguously identified. Such a pattern must contain more information than periodic lattices, hence the name *coded illumination*. Of course the location code must be easily extractable from the recorded image, despite the unknown modulations of the pattern by the scene in question. The code can be integrated in the projection pattern in an **analogue** fashion, or more often in a **binary** manner. The amount of information coded per unit area can be increased by the use of several colour channels.

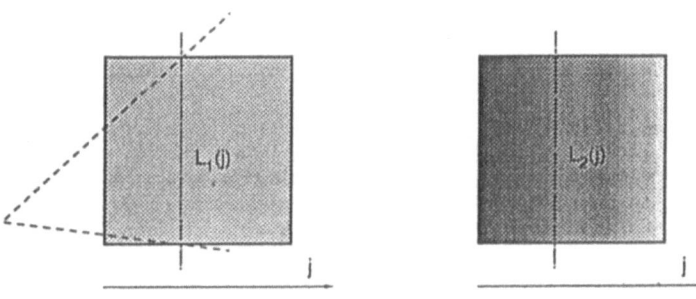

Figure 6. Intensity–ratio depth sensor

4.5.1 Analogue Techniques

The 'intensity–ratio' depth–sensor uses the local image intensity (grey–value) as information carrier. This intensity is not only dependent on the geometry and optics of the system, but also on the unknown form and reflection parameters of the object under investigation. The influence of all these factors is eliminated by the use of two illuminating patterns, and taking the ratio of corresponding pixels in the recorded images. Schwartz, e.g., first applies homogeneous illumination, then a pattern with a continuous intensity variation in the horizontal direction (Fig.6).

The intensity ratio $R(j)$ at each pixel is a known function of the lateral displacement in the projected image, and this relationship is maintained in the recorded images. The corresponding horizontal position j in the projected image can be calculated from the inverse of the function $R(j)$, with the assumption that this function is monotonic.

The analogue nature of this approach places tight constraints on the apparatus, especially stability and linearity. For good depth–resolution, the function $R(j)$ should have a large range, which can be set in the linear region of the camera response–curve. In practice, the actual light levels vary widely, according to the nature of the surfaces and the viewpoint, which implies that the effective dynamic range of $R(j)$ must be correspondingly limited, to allow for a sufficient margin between sensor noise and saturation over a wide range of reflectances.

4.5.2 Binary Techniques

The problem of camera dynamic range discussed above is greatly eased if only two intensity levels are projected. To allow a larger number of regions to be individually characterised, several binary patterns are sequentially proso that every region is represented with a unique n-bit codeword.

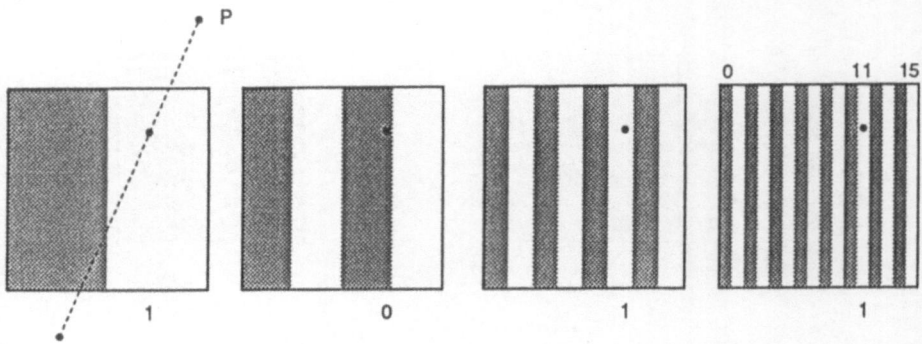

Figure 7. Sequential Binary Coding

This principle is outlined in Fig.7 where an object point P is illuminated by n sequential images. Point P is sometimes in light, sometimes in darkness. This sequence of illuminations is recorded at every pixel in the recorded image, leading to the formation of a binary codeword for each pixel which give the corresponding horizontal position in the projector image. For a sequence of n images, a maximum of 2^n positions can be coded, which implies a good increase in speed compared with the light-stripe method where a sequence of 2^n projected images is needed to code 2^n positions. This **time-sequential** coding technique is protected by US patent 4,175,862 (Matteo, 1979) .

An equivalent method was almost simultaneously developed by Altschuler et al where the projection pattern is organised as 128 columns of 128 pixels. Each column can be switched between light and darkness. Measurements can therefore only be made at known discrete positions in the scene. The projection pattern is generated by interference between laser beams and subsequently modulated using an electro-mechanical shutter matrix based on a liquid crystal display (LCD).

Alternative binary coding techniques have been proposed in an effort to achieve higher robustness, e.g., the Hamming–code provides extra redundancy. This has been researched by Minou et al., who made a study of the noise-sensitivity of several coding techniques. Inokuchi employed Gray-coding in a prototype system, so that neighbouring columns differed by just one bit–position. This prevents the occurrence of large identification faults at image points near transition regions of the binary pattern, or where misalignment of successive illumination patterns occurs, perhaps in the case of mechanical pattern swapping. The Gray code ensures that such errors remain limited to one position of the finest

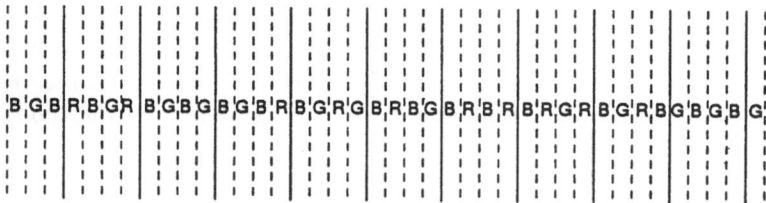

Figure 8. Colour Coding

resolution. The illumination and recording sequence can also be extended with the addition of two recordings; the first is the sum illumination over all columns and the second is the scene without illumination. These extra images are discretized using the average of the reference images as a local threshold, in order to discount the effects of the reflectance variations due to the surrounding light. The robustness can be raised even further by the addition of the negatives of all projection masks to the illumination sequence, the binary information being applied and extracted in a differential way.. At the same time this allows the transitions between light and dark regions to be more sharpy defined—as the zero–crossings of the difference signal. Mechanical mask changing can be replaced by a programmable LCD shutter with contrast 75:1 and a switching time of 8 milliseconds. Despite the rapid growth in use of electro-optical shutters, there is renewed interest in simple mechanical systems, e.g., a rotating Gray-code generator which produces 10+2 patterns at a resolution of 1000 lines.

4.5.3 Scenes Including Motion

For static images, the number of steps in the coding sequence is of little importance. But, recording ten or so images form a scene with a TV–camera requires around half a second. Any movement in this period will cause successive images in the coding sequence to fall out of registration, rendering identification impossible. In this case is it vital that the coding is obtained using a **single illumination pattern**. Boye considers the use of several colour channels to this end. A complete projection pattern is composed from a juxtaposition of subpatterns, each in turn being composed of n lines chosen from a set of γ possible colours. This composition is done in such a way so that each subpattern is given a different set of coloured lines or code, as in Fig.8.

Each subpattern and hence its position is recognizable in the recorded image, assuming that each subpattern is present in this image in its entirety and not broken up by the edges of an object. The actual positions of the colour lines in the recorded image are obtained from peaks in the various colour channels. Neighbouring lines must have different colours, and signal contrast can be further increased by the insertion of a dark line between each colour line. Robustness is obtained by minimizing the code–distance between all subpatterns.

Despite the advantages of colour coding over binary coding, there are several disadvantages worth noting. Firstly, the codes can be identified at *fixed* positions, i.e., the smallest resolvable detail ranges between n and $2n - 1$ neighbouring lines for a n–bit code. In contrast, pseudo–noise sequences, e.g., can be made where identification is possible at *each and every* position of n sequential numbers, so the minimum resolvable detail is 1. Secondly, concerning the claimed redundancy, the code distance is only minimal between *pairs* of subpatterns. The recorded image is not explicitly composed of subpatterns, but is simply a sequence of lines, with the result that for optimal redundancy, a minimum distance should be sought between *all groups* of n (or more) lines.

Closely related to the above method is the "rainbow–depth–measurement" system where the scence is illunimated with a rainbow of light, whose colour continually varies from red to violet, obtained via the dispersion of white light. Directional information is thus obtained via colour. Of course the intensity of coloured light received depends on both the spectrum of the source and the reflectance and colour of objects in the scene. These effects are removed by using two colour channels in the camera, with the requirement that the camera channels' colour dependency is monotonic. Ideally, the response of one channel should increase with wavelength and the other decrease, not usually the case with naked cameras, requiring the addition of colour filters. With these corrections, the camera transforms colour into angle, and can thereby be used in triangulation. The elegance of this methods rests on the need for a single illumination of the scene and a single recording, making the technique applicable to moving scenes. The depth image has a high lateral resolution since all pixels in the recorded image (except the black ones) are explicitly available for triangulation. Object surface reflectivity and colour have in theory no effect.

Both the intensity ratio and the rainbow systems are very sensitive to noise and fluctuations in the camera responses. The relation between projection angle and system response is complex, any error in colour measurement is passed on to the depth values. The dynamic range of the system is limited to the linear region of the camera responses (since the above mentioned dual-channel method involves addition of the camera signals). This method cannot be expected to yield accurate results when the colour of the object is almost complementary to either of the dual colour channels, yielding a very small recorded signal.

Figure 9. Vuylsteke's non–sequential pattern

4.6   Non–Sequential Coding

A novel projected pattern has been developed by Vuylsteke at this laboratory which guarantees good detection of codes and robust identification.

Fig.9 shows part of the checkerboard projection pattern used. A code point is defined at the corners of each checkerboard element. At each code point, a single bit 0 (dark), 1 (light) is assigned. The checkerboard elements are devised so that over each 2x3 window of code points, a *unique* 6-bit code is found, moving over the image columns. The column position can be extracted at once from the pattern. The information required to locate a particular position has been spread over *space* (the 2x3 local region) in this method, contrasting with sequential coding techniques where the information is spread over *time*. The information density in the projected image is low (1 bit per point), yet triangulation can be guaranteed.

Fig.10 shows the results of illuminating a scene with this checkerboard pattern, where scene modulation of both pattern geometry and intensity are clearly visible. The spatial position of the code points is determined as follows;

- The code points in the recorder image are detected and their coordinates stored.

- The codebit is determined at each code point.

- The codeword is constructed using the code bits from the local area, and the column number (in the projected pattern) is found.

- Using the coordinates of the code point in the recorded image, and the column number found above, the spatial position of each point in the scene is found by simple triangulation.

The optical and physical parameters of the system must also be known, and are obtained from a single calibration of the system. The principle of using local information to extract position requires that the neighbouring code points in the projected pattern remain neighbours in the recorded image. This is true unless there are strong discontinuous surfaces in the scene, resulting codewords must be rejected.

Figure 10. Example of Illumination with the Vuylsteke Coded Pattern

Image Sensors for Real–Time 3D Acquisition

Part – 2 –

Back Shape Measurement for Evaluating Scoliosis using a Single Binary Encoded Light Pattern

M.De Groof, P.Suetens, G.Marchal, A.Oosterlinck

ESAT–MI2 Departement of Electrical Engineering
Katholieke Universiteit Leuven,
Kardinaal Mercierlaan 94,
3030 Heverlee, BELGIUM

NATO ASI Series, Vol. F 63
Traditional and Non-Traditional Robotic Sensors
Edited by T. C. Henderson
© Springer-Verlag Berlin Heidelberg 1990

Abstract - In this paper an acquisition system is described allowing an optical contact free three-dimensional modelling of the human back. Surface curvatures representing local shape are used to calculate a median profile on the surface of the back. Using this surface information a three-dimensional reconstruction of the vertebral column inside the body is carried out. Clinical relevant parameters are extracted out of the obtained shape of the spine making a comparison with radiological findings possible.

1   Introduction

In this paper a system is described for three-dimensional reconstruction of the spinal midline. The reconstruction is used to evaluate patients with structural deformities of the back such as scoliosis or kyphosis. Usually such patients are evaluated by means of radiographs. An optical method for analysis of the topography of the human back is now developed to limit x-ray exposure of such patients. The impracticability of fixing the human body during measurement requires that the acquisition is carried out instantaneously. The acquisition system [7] is based on triangulation with periodically structured illumination. After calibration all the information for identifying the three-dimensional position of a scene point is available in a single camera picture. The instantaneous illumination makes the measurement system also suited for moving scenes.

2 The acquisition system

The measurement setup consists of a ccd camera and a slide projector both at a fixed distance from an object. A grid of 64 rows by 63 columns is projected onto the object (see fig. 1) and the scene is instantaneous captured by the camera. The camera (High Technology Holland, MX type) delivers a 625-lines CCIR video signal which is sampled and digitized to a 512 x 512 by 8 bits digital picture.

2.1 The layout of the illumination pattern

A chess-board appears as the basic structure in the projected and hence the captured image as seen in figure 1.

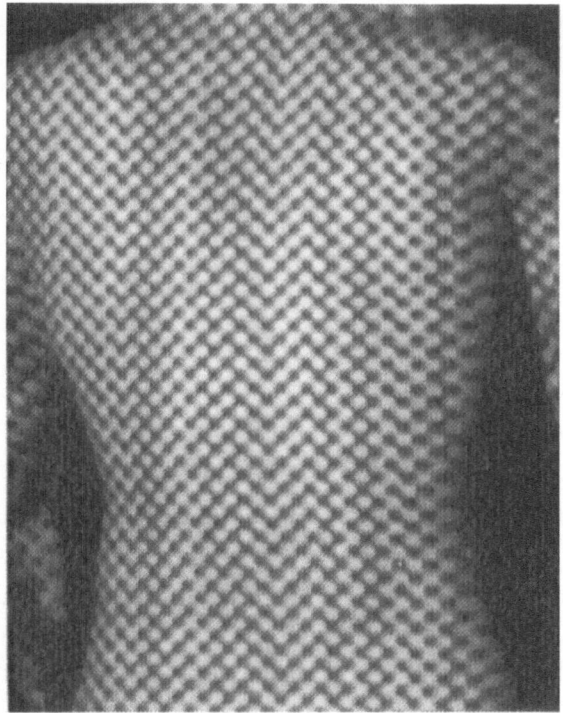

Fig. 1 Projection of the grid on the human back

At the intersection of the horizontal and vertical chess-board edges, i.e. where four squares meet, a grid point is located. So the chess-board pattern is used to mark the place of the grid points. The entire image processing is based on a grid of 63 x 64 grid points. For triangulation (see section 2.2) it is necessary to know the coordinates in the image of each grid point and also which column a specific grid point belongs to. For encoding the number of a column the grid points are individually marked with a bright ("1") or dark ("0") spot. As seen in figure 2 the chess-board also exhibits two basic appearances of the grid points, denoted as "+" or "-". The complete pattern is composed of the four primitives depicted in figure 2. So a single binary illumination pattern with only two intensity levels is used to indicate the place of a grid point and to mark it with a codebit.

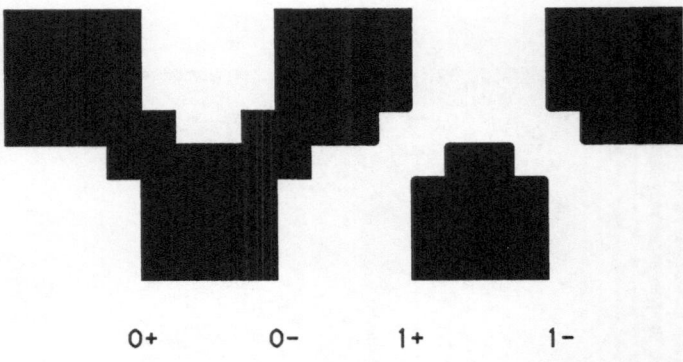

0+   0-   1+   1-

Fig. 2   The four possibilities of codebits

Demodulation of the captured image consists of the following steps : 1) Detecting the grid points in the image and calculating their precise coordinates. 2) Determining the individual codebit ("1" or "0") of each grid point with its appearance in the chess-board ("+" or "-"). The rest of the processing is based only on the detected grid points. Decoding of the image consists of the following steps : 1) Searching for the closest neighbours of every detected grid point using dynamic windows, hence locally recovering the original grid topology in the observed image. 2) Constructing a directed graph with a node for every detected grid point and branches to the immediate four neighbours. 3) Identifying the column number of each grid point using windows of adjacent grid points. For the purpose of identification of the column number the codebits are assigned to each of the 63 grid points in a row in a special way. In fact the assignment is based on two binary sequences of 63 bits length. As a consequence the bit map assignment is repeated every second row. At an arbitrary grid position every window of 2 x 3 points carries a 6 bit codeword which is in a one-to-one correspondence with the underlying column number. This means that the grid points are identifiable only within their local context and an identification window size of 2 x 3 is sufficient for unique identification. The applied binary code is derived from *pseudo-noise sequences* and is optimized to make the identification as fault-tolerant as possible. The degree of fault-tolerance may be expressed as the minimal number of binary positions in which any two windows taken from arbitrary columns in the grid differ (see [7] for a detailed discussion on this subject). The network of projected grid points need to be locally preserved in the camera image. This assumption holds true because the human back contains no jump boundaries.

2.2 Triangulation

Upon projection every column of the grid generates an *indication* plane (as seen in fig. 3) also through the optical centre of the projector.

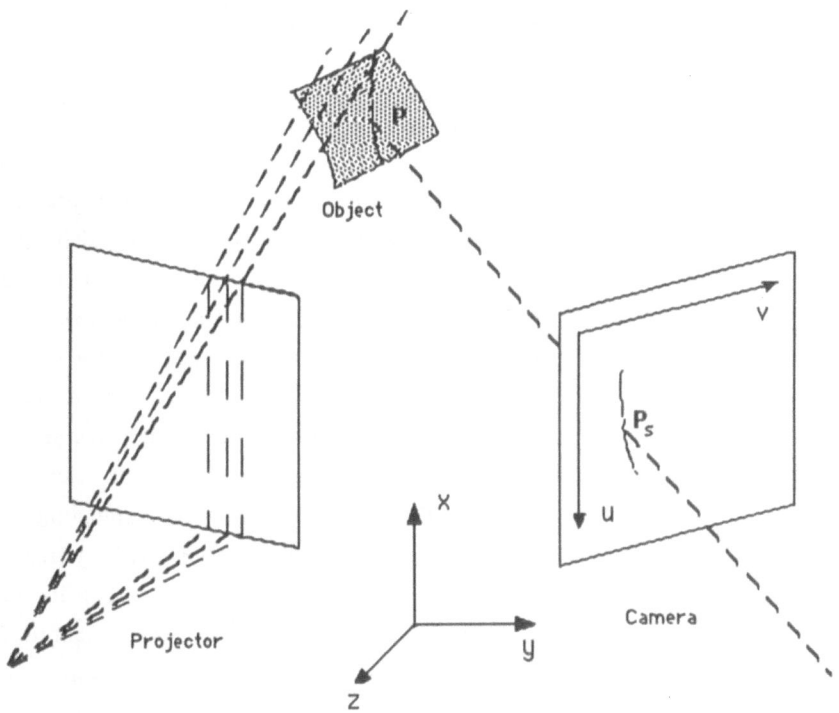

Fig. 3 The measurement setup

The explicite equations of each of the 63 planes through a column of the grid and the lens center of the projector are of the general form :

$$x + a_j\, y + b_j\, z + c_j = 0 \ , \quad j = 0 \ldots 62. \tag{1}$$

These equations are suited to represent any plane that is not parallel to the x-axis. For every plane only three independent parameters a_j, b_j, c_j need to be determined, which is done in a calibration procedure.

The camera is modeled according to a pin-hole model. The mapping of a point in cartesian coordinates (x,y,z) to the corresponding pixel coordinates (u,v) in the image can then be expressed by the following equations :

$$
\begin{bmatrix} t\,u \\ t\,v \\ t \end{bmatrix}
=
\begin{bmatrix}
S_{11} & S_{12} & S_{13} & S_{14} \\
S_{21} & S_{22} & S_{23} & S_{24} \\
S_{31} & S_{32} & S_{33} & S_{34}
\end{bmatrix}
\begin{bmatrix} x \\ y \\ z \\ 1 \end{bmatrix}
\tag{2}
$$

These equations include the transformation from world to camera coordinates and corrections for perspective projection, rescaling and misalignment. The precision of the range measurement still critically depends on the accuracy of the detected grid point coordinates in the camera image. Given the image coordinates of a point (u,v) and the number of the indication plane the point belongs to, the equation of that plane (1) together with the camera equations (2) give a set of three linear equations in the unknowns x,y,z which can be solved. The scaling factor t can be eliminated in (2) yielding the equation of a straight line. In fact equation (2) represents the line through a point on the object and the corresponding point in the camera image. So the spatial position of a point P is uniquely defined as the intersection of a plane and a straight line. The 3x4 matrix S in (2) is called the camera matrix. Given a set of calibration points with known world coordinates and observed image coordinates the components of that camera matrix can be calculated.

2.3   Experimental results

A series of experiments have been carried out to estimate the measurement accuracy of the acquisition system. The appropiateness of the linear camera model is verified by backprojecting the imaged calibration points on the planes from which they originate. For plane positions at a distance of 95-140 cm from the camera-projector base line, the RMS deviation between the actual calibration points and the backprojected points varied from 0.1-0.2 mm. These figures are indicative of a small non-linear lens distortion.

As long as the object surface is nearly perpendicular to the bisector line of the optical axes of camera and projector, and the angle subtended by the axes is not too large, good measurement results are obtained. The range limitation is evaluated quantitatively using a flat test scene. In each experiment a least squares plane is fitted through the obtained range coordinates. Two parameters are considered to be indicative of the measurement performance : the RMS value of the distance d of the range points with respect to the fitted plane, and the total number of range points N_{cop} lying within a distance of less than 2 mm to the plane. Numerical results using a white frontal target plane are presented in Table 1. N_{det} is the number of grid points detected in the image and N_{id} is the number of grid points that have their column number identified.

Depth (cm)	N_{det}	N_{id}	N_{cop}	RMS d (mm)
85	1438	776	-	-
90	3310	3214	3191	0.2
95	3712	3708	3708	0.2
100	4032	4031	4031	0.2
105	4032	4031	4031	0.3
110	3776	3757	3757	0.3
115	3520	3291	3291	0.4
120	3300	2507	2434	0.6
125	2841	1922	1502	0.7

Table 1. Numerical measurements results.

As seen in Tabel 1 the range measurement shows no remarkable degradation within the depth range from 90-105 cm. Although N_{det} is smaller than 4032 outside the range 100-105 cm, we verified that all *visible* points have been effectively detected in these experiments. The fact that many points are missing outside this range is exclusively due to the limited field of view of the camera.

2.4 Constructing an image with local shape information

As a result of the acquisition a list of x,y,z coordinates representing the shape of the human back is obtained. These x,y,z coordinates are related to a fixed coordinate system. However the position and orientation of a patient will change at each acquisition session. For automatic computer processing the algorithm has to be independent of an arbitrary choice of a coordinate system. Therefore intrinsic shape properties of the surface are considered. Only the immediate environment of every recorded point of the back is considered. Surface curvatures [4] are used to describe the local shape at any point. In a 3 x 3 neighbourhood of a point an approximating paraboloid is calculated. As a consequence two curvatures having two independent directions are enough for an unambiguous description of local shape. In addition it has mathematically been proved that two orthogonal directions with extreme curvatures (called the principal directions) always can be found. Formulas of differential geometry are used to calculate the two principal curvatures and directions (see [4] for a detailed discussion). According to the sign of those curvatures four basic surface entities can be distinguished : a planar, parabolic, elliptic or hyperbolic case. Further shape analysis mostly uses the mean and the product (the so-called Gaussian curvature) of the two principal directions. The first step in the processing of the x,y,z coordinates consists in constructing an image of the back containing these surface curvatures as local shape information.

3. Interpretation of the three-dimensional data

The surface of the back is now represented using locally obtained information. For a clinical interpretation further analysis still need to be performed. Among the entities physicians are interested in concerning the human back are pathological curves of the spine. An abnormal curvature in a frontal plane is called scoliosis and in a sagittal plane it is called kyphosis. The parameters involved in measurement of scoliosis and kyphosis routinely used in radiology are : 1) Cobbs angle [1] as a measure of the curvature of the spine in a frontal view for defining scoliosis. 2) The analogue angle in a sagittal view for defining kyphosis. 3) The rotation of the vertebrae. For comparison with radiological data the same parameters need to be extracted from the image containing the local shape information. Therefore it is first necessary to find some kind of relationship between surface shape and underlying skeletal structures like e.g. the vertebral column.

3.1 Extracting a median symmetry line out of the surface data

According to [6] a relation between back surface and underlying spinal shape can be obtained using the spinous processes, the bony landmarks of the vertebra on the surface. First the line connecting the spinous processes has to be extracted out of the image containing the local shape information of the back. Instead of looking for a symmetry plane, a median profile is constructed which divides the back into a left and a right half with minimal asymmetry [3]. For this purpose transversal sections of the back are considered. The parts to the left and to the right of every point in a transversal section are compared. An asymmetry function is used to evaluate the profile points lying symmetrically to the left and the right of a point. This asymmetry function is based on the differences in magnitude and direction of the principal curvatures of symmetrically located points. The points with minimal assymetry in every transversal section are not simply connected to yield an overal symmetry line of the back. Instead a minimal costpath algorithm is used to find the best overal symmetry line for the back.

3.2 Three-dimensional reconstruction of the spinal midline

Out of the position of the symmetry line on the surface (assumed connecting the spinous processes) a three-dimensional reconstruction of the spinal midline can be obtained [5]. A vertebra consists of a vertebral body and a spinous processus. The rotation of a vertebra produces a distortion of the surface covering it and hence the rotation can be estimated from this distortion. In every transversal section of the back the direction of a perpendicular on the surface in the point of the overal symmetry line is calculated. The first derivative of a spline function [2] is used for calculating the perpendicular. As an approximation the direction of the perpendicular on the surface is considered equal to the angle of vertebral rotation. The position of the centre of the vertebral body is calculated using this angle of rotation of the vertebra and an estimation of the length of its spinous process. A three-dimensional location of the vertebral bodies inside the human body is thus obtained. A projection on any two-dimensional plane can be made. In a next step the curvature of the whole vertebral column is analysed yielding relevant shape parameters of the spine like Cobbs angle [1]. The degree of scoliosis is evaluated and a comparison with radiological data can be carried out. An antero-posterior radiograph of the spine is taken at the same moment without moving the patient. At this moment the shape parameters of the spine obtained from the radiograph are provided by a physician. Computer digitization and evaluation of the radiographs will be carried out in the near future so a full-automatic comparison of both methods using Cobbs angle will be possible. The quantitative correlation to radiological measures of the skeletal geometry will then be independent of any subjective contribution.

4. Conclusion

The developed acquisition system allows for a contact-free optical three-dimensional reconstruction of the human back. Using curvature analysis the obtained surface data are converted to a form which makes them independent of an arbitrary choice of a coordinate system. Analysis from a medical point of view of the obtained data is carried out. Cobbs angle is calculated making an easy comparison with radiological data possible. The initial results are very promising, but the method still has to be extensively evaluated in clinical practice.

5. Bibliography

1. Cobb, J.R.: Outline for the study of scoliosis. Instruct. Course Lect. : V, 261-275, 1948.

2. Dierckx, P.: An improved algorithm for curve fitting with spline functions. Report TW 54, K.U.L. : Department of Computer Science, 1981.

3. Drerup, B. and Hierholzer, E.: Automatic localization of anatomical landmarks on the back surface and construction of a body fixed coordinate system. J. Biomechanics 20 ,961-970.,1987.

4. Frobin, W. and Hierholzer, E.: Analysis of human back shape using surface curvatures.J. Biomechanics 15 , 379-390,1982.

5.Hierholzer, E. and Drerup, B.: Three-dimensional reconstruction of the spinal midline from rasterstereographs. Preprint, To be published in : Biostereometrics '88, Proc. SPIE 1030 5, 1988.

6. Turner-Smith, A.R.; Harris, J.D.; Houghton, G.R. and Jefferson, R.J.: A method for analysis of back shape in scoliosis. J. Biomechanics 21 , 497-509, 1988

7. Vuylsteke, P. : Een meetsysteem voor de acquisitie van ruimtelijke beelden, gebaseerd op niet-sequentiele belichtingscodering. Doctoraatsthesis, K.U.L. : Faculteit Toegepaste Wetenschappen (E.S.A.T.), 1987.

ACTIVE SENSING WITH A DEXTROUS ROBOTIC HAND

Peter K. Allen

Department of Computer Science
Columbia University
New York, New York 10027

Abstract

What separates sensing with a hand from other more passive sensing modalities such as vision is its active nature. This paper describes our efforts in building a useful active hand sensing environment that can be used for a number of different tasks including intelligent grasping, manipulation, and haptic object recognition. We outline the system we have built to control the hand/arm system, discuss the tactile sensors we have mounted on the hand's fingers, and elaborate on some exploratory procedures (EP's) we have implemented to allow the hand to do active tactile sensing for object recognition tasks.

1. INTRODUCTION

What separates sensing with a hand from other more passive sensing modalities such as vision is its active nature. A robotic hand can probe, move, and change its environment. If we desire to use robotic hands for tasks such as grasping, manipulation, assembly, inspection and object recognition, then we need to develop *systems* that address many different areas of research including mechanical design of hands, control architectures for real-time concurrent control of multiple degree of freedom devices, software and programming environments for distributed robotic systems, sophisticated sensors, and task-level control. This paper describes our (ongoing) efforts in attempting to build such a system for active sensing tasks using a Utah-MIT hand [15]. This hand is the outcome of a lengthy design process, and it has allowed us to concentrate on the task level control of the hand rather than building a hand itself. Our effort has begun with building a control architecture that will interact well with the existing analog joint-level servos that come with the hand. In addition, we have developed hardware to allow us to interface versatile tactile sensor pads on the fingers of the hand to allow us to explore robotic shape perception using touch alone.

The outline of this paper is as follows: Section 2 is an overview of the hardware/software/sensing environment we have built to perform intelligent hand functions, Section 3 outlines our experience in using the hand system to understand shape, and Section 4 is a summary outlining future work to be done with the hand.

NATO ASI Series, Vol. F 63
Traditional and Non-Traditional Robotic Sensors
Edited by T. C. Henderson
© Springer-Verlag Berlin Heidelberg 1990

2. SYSTEM OVERVIEW

The system we have built consists of a Utah-MIT hand attached to a PUMA 560 manipulator. The hand contains four fingers, each with four degrees of freedom. It resembles the human hand in size and shape, but lacks a number of features that humans find very useful. In particular, it has no palmar degree of freedom (closing of the palm) and the thumb is placed directly opposite the other three fingers, with all fingers identical in size (see figure 1). The hand has joint position sensors that yield joint angle data and tendon force sensors that measure forces on each of the two tendons (extensor and flexor) that control a joint. The PUMA adds 6 degrees of freedom to the system (3 translation parameters to move the hand in space and 3 rotational parameters to orient the hand), yielding a 22 degree of freedom system. Clearly, such a system is a nightmare to control at the servo-level in real-time. Our approach is to use the embedded controllers in each of these systems, controlling and communicating with them through an intelligent, high-level controller that links together the movements of arm, hand and fingers with the feedback sensing of joint positions, tendon forces, and tactile responses on the fingers.

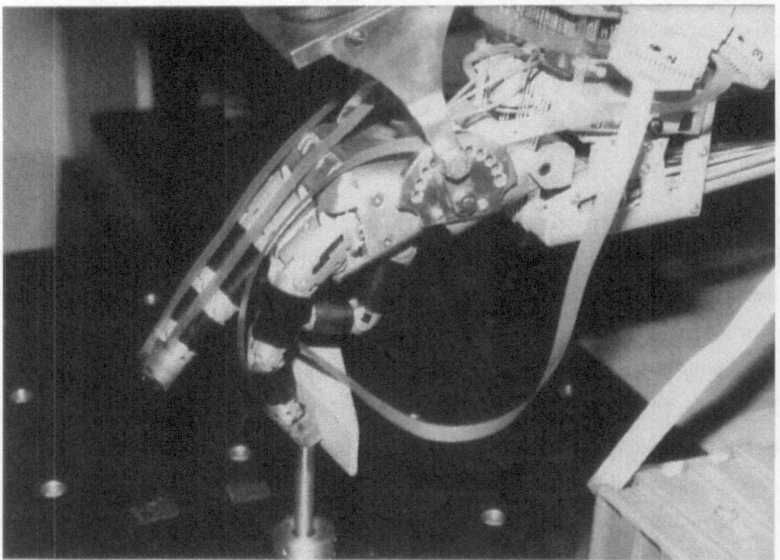

Figure 1. Utah-MIT hand with tactile sensors mounted.

The hardware structure of the system is shown in Figure 2. The high-level control resides in a SUN-3 processor. The SUN serves as the central controller, and has access to a full UNIX based system for program development and debugging as well as a set of window-based utilities to allow graphical output and display of the system's various states. The hand is controlled by an analog controller that is commanded through D/A boards from a dedicated 68020 system. The SUN is capable of downloading and executing code on the 68020 and can communicate with it

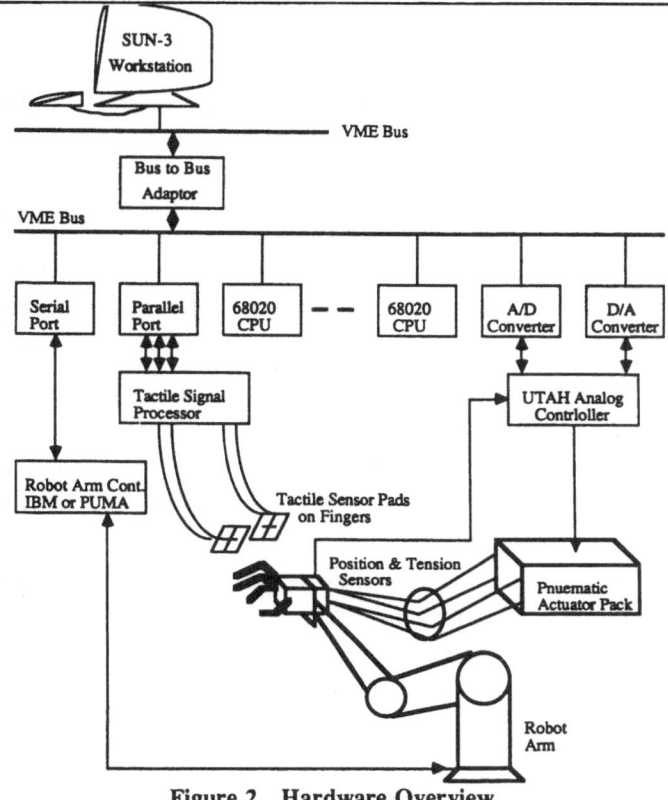

Figure 2. Hardware Overview.

through a shared memory interface [23]. The tactile sensing system is controlled by another dedicated 68020 that monitors the forces on each of the sensor pads. The connection from the SUN
to the PUMA is via the VAL-II host control option over a serial interface. We are currently
changing the interface to the PUMA to RCCL [10] to make the hand-arm interaction more tightly
coupled.

2.1. Low-Level System Primitives

To create a system that is useful at the higher levels of reasoning, a rich set of low-level
primitives is needed. The hand system contains sensors for joint position, tendon force, and tactile
contact, which gives the system user a wide range of frames and data to reason about. The low-
level system system primitives described below are organized along two dimensions, *type* and
domain. *Type* refers to the type of sensing or actuation (or both) which the primitive implements.
Continuous sensing implies a monitoring mode while one shot sensing implies a static sensor
reading. Continuous action implies a synchronized control loop while one shot action implies an
imperative command. *Domain* refers to the coordinate frame or sensor domain that the primitive

operates in. The primitives are:

- *Get Joints:* Reads the joint angles on the hand.

- *Get Forces:* Reads the forces (measured at the wrist of the hand) on the flexor and extensor tendons that control each joint.

- *Get Tactile:* Reads a 16x16 tactile array on each finger.

- *Get Wrist:* Reads the Cartesian position of the PUMA wrist.

- *Goto Move:* A one shot move that is done atomically. Implemented by setting the desired joint position in the analog control system directly to the final position value of the move for each joint. Move commands have the ability to use symbolic names for specified poses. There is a set of standard pose names available for use by the high-level programming system, particularly for hand pre-shaping operations.

- *Trajectory Move:* Allows a single joint or a number of joints to be moved along an interpolated trajectory from a starting joint space vector to an ending joint space vector. There is also an analogous command that interpolates in Cartesian space for fingertip motion.

- *Position Contact:* This is a continuous sensing primitive that is implemented in the Cartesian position domain. When the difference between actual and desired Cartesian fingertip position exceeds a threshold, contact is signaled.

- *Force Contact:* Same as above but tendon forces are monitored for changes. A change in force differentials implies a finger contact.

- *Tactile Contact:* The tactile arrays are monitored for readings above a prespecified threshold.

2.2. Composite Functions

A number of composite functions that have been built out of the low-level system primitives. They are:

- *Grasp with Force:* Used to grasp objects with a desired grip strength. The command closes all joints specified by a joint mask incrementally while monitoring the tendon forces controlling the joints in the tendon mask. When the difference between the flexor and extensor tendon forces on a joint exceeds the specified threshold, movement of that joints stops. This process continues until the forces on all of the tendons in the tendon mask have surpassed the threshold. This primitive is useful for grasping objects when their precise dimensions are unknown.

- *Guarded Move:* This composite function combines one or more of the three low-level contact primitives (*Position Contact, Force Contact, or Tactile Contact*) with the either hand finger motion or motion of the PUMA arm. When a contact is detected, the relevant motion ceases.

- *Limp:* This primitive is very useful in establishing grasps. It allows a human to interact with the hand and position it manually. It is implemented by comparing the actual joint positions with the previous joint positions, and updating the current desired position to reflect the new actual joint positions. For example, a hand can be placed on an object in a known position and the position can then be recorded for future use. It has a feature to allow the masking out of joints that need to be fixed (not *Limp*) and those which should be under LIMP control.

2.3. Tactile Primitives

While the level of sensing provided by the joint position and tendon force sensors on the Utah-MIT hand is better than earlier implemented hands, it still falls far short of the requirements for a dextrous manipulation system. In particular, what is desired is accurate positional contact information between the hand and a target object, and a measure of the forces exerted by the fingers at these contact points. The sensory feedback provided by the hand does not allow for localization of contacts. Hence, a requirement for this system is a robust and accurate tactile sensing capability, utilizing sensors mounted on the links of the fingers. Tactile sensing differs from traditional vision sensing in its active nature. Thus, a robotic system that employs tactile sensors on the fingers of a dextrous hand must deal with three related issues: 1) acquisition and interpretation of tactile sensor data from many sites on multiple fingers, 2) control of the dextrous hand using tactile sensor feedback, and 3) development of sensing strategies using tactile feedback.

To satisfy the first requirement, we have implemented tactile sensors on each of the hand's fingers. The technology being used is a piezo-resistive polymeric material manufactured by Interlink, Inc. [27, 30]. The design of the tactile pads we are using sandwiches the polymer between two pliable sheets of Kapton material that contains electrical etching. The application of forces on the pads provides an increased electrical flow channel between the two sheets as the material within is compressed. The piezoresistive polymer is patterned to form rows on one substrate and columns on the other. The rows and columns form a grid in which each intersection acts as a force-sensitive variable resistance whose value decreases approximately exponentially with normal force. The pads consist of 16 rows by 16 columns, providing a sense resolution of 256 points on a 0.5 x 1.0 inch pad.

The 256 sites of each sensor pad are addressed independently by analog circuitry that cancels current flow in all paths of the grid except the one containing the resistive element being measured using a method developed by van Brussel and Belien [32]. A hardware interface board has been developed to perform this operation at high-speed using analog CMOS components. The interface board performs the analog-to-digital conversion task by means of an 8-bit flash A/D converter and allows up to sixteen sensor pads to be addressed.

The tactile sensor system is site addressable as well as sensor pad addressable. Thus, the low-level control software provides a sensor pad ID and a (row, column) coordinate within the 16x16 matrix to the interface hardware via a parallel port interface. A touch "frame" containing the force values from any subset of the sites on the pads can be obtained in this manner. This method of requesting data/receiving data was chosen for four reasons. First, hardware is simplified since it does not have to generate pad IDs and site coordinates. Second, this method of handshaking maximizes data integrity since "we get what we ask for". Data provided by the hardware interface will never be misidentified because of a timing problem. Third, the touch frame sampling frequency is determined by software which allows for easier testing of the hardware and experimentation to determine the optimal sampling frequency. Finally, between one and sixteen sensor pads can be connected to the hardware interface without making any hardware modifications, easing incremental development.

Results with this sensor have been good, particularly with respect to signal isolation and hysteresis. The sensors are monitored by a separate 68020 that is responsible for low-level tactile processing. Some of the low-level tactile primitives that have been implemented are:

- *Tactile Filters:* A number of useful digital filters have been implemented including averaging and median filters which are very useful in processing noisy tactile data [21].

- *Tactile Moments:* A useful technique for quickly getting contact information is central moment analysis [14]. The contact area and centroid of the contact can be determined using moments. The second moments are useful for determining the eccentricity of the contact region and the principal axes of the contact.

- *Edge Detection:* A number of edge detectors have been developed and used for feature extraction from tactile images.

- *Line Detection:* Lines are detected by using the output of the edge detection procedure in a Hough transform [6].

Figure 3 shows the output of tactile responses filtered with median and averaging filters when a pen cap is placed on the sensor pads and when a credit card is applied across the sensor pad. The signal is very localized (the sensing area is on the order of 0.5 inches by 1.0 inches).

2.4. PUMA Arm Primitives

The arm primitives are already embedded in VAL-II, the programming environment of the PUMA. They include movement primitives, asynchronous interrupt capability, and the ability to establish arbitrary coordinate frames such as the hand coordinate system. It also provides a global coordinate system in which the tasks can take place.

2.5. DIAL: DIagrammatic Animation Language

The high-level control program is called DIAL and was originally developed by Steven Feiner [8] for use as a graphical animation language. DIAL is a diagrammatic language that allows parallel processes to be represented in a compact graphical "time line". It has been used for animation of graphical displays but it has been transported by us to the robotics domain so we may exploit its ability to express parallel operation of robotic devices. It also provides a

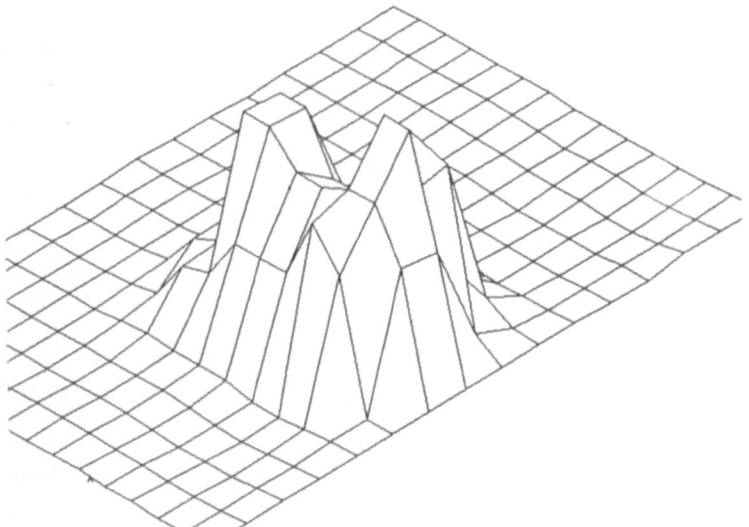

Figure 3a. Tactile impression of a pen cap.

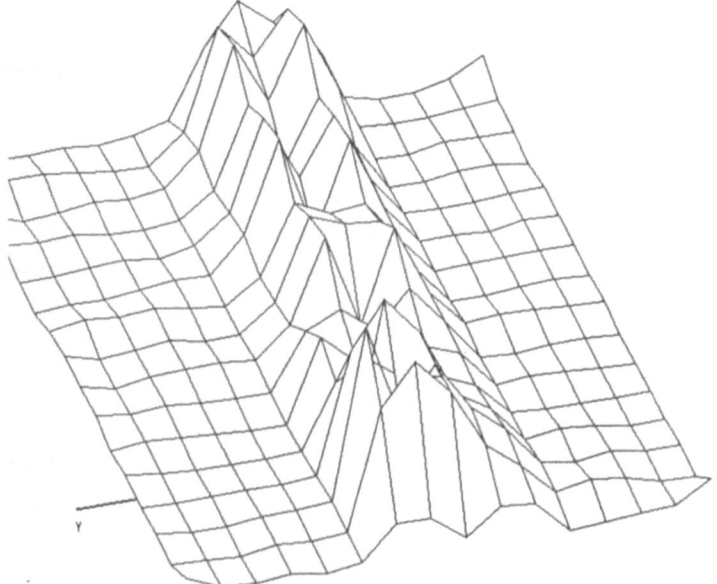

Figure 3b. Tactile impression of the edge of a credit card.

convenient way to implement task-level scripts which can then be bound to particular sensors, actuators and methods for accomplishing a generic grasping or manipulation task. The instruction set that DIAL supports is specified at run-time in a user-supplied backend. We have created a number of DIAL backend's to work in our environment which has the need to express parallel actions in the performance of simple grasping and manipulation tasks. Given that we need to control the hand and arm movement concurrently, as well as integrating information from three different sensor systems (joint position, tendon force, tactile array), we found it natural to program tasks as DIAL scripts.

2.6. SCL: Script Control language

Although DIAL has great readability and writability for robot programming, it tends to support simple and small programming. It doesn't support the higher level control abstractions, communications between processes and asynchronous error handling.

To alleviate this, we have built a level above DIAL called Script Control Language. This is a programming environment that treats DIAL scripts as the objects of interest. In order to develop a robot task in the SCL system, programmers need to prepare three parts:

- A SCL program (high-level): provides control flow and monitoring of DIAL scripts.
- Several DIAL scripts (middle-level): provide timing and synchronization for concurrent, low-level events.
- A set of backend functions (low-level): provide the functionalities of low-level events.

The back-end functions are C functions that contain the low-level system primitives described earlier. These can be built into DIAL scripts for more concise, task level scripts. To embed error-handling and asynchronous sensor control, SCL invokes each script as necessary. Using this environment, we have built task level scripts to perform pick and place operations, pouring of liquids from pitchers, unscrewing of lightbulbs, and haptic exploration procedures which are fully described in [2, 3]

3. ACTIVE HAPTIC SENSING FOR OBJECT RECOGNITION TASKS

A focus of our work has been in the use of the hand system described above to recover the shape of objects in a scene. Object recognition has traditionally been associated with vision sensor systems. However, these systems suffer from a number of inherent problems, not the least of which is occlusion. Any visual system will be limited to a view that obscures all back-facing areas of the object. In robot manipulation tasks, important areas of the work environment are occluded by the end-effector itself. This difficulty is especially acute during the act of acquiring a grasp on an object, when the contact areas will be occluded.

Our work is oriented towards building robotic analogues of the human capabilities of hands, particularly as it applies to object recognition †. A number of interesting properties of the human

---

† We must be careful here, in trying to draw too close a comparison between a human hand and devices such as a Utah-MIT hand. Johansson and Vallbo [16] have reported that there are about 17,000 mechano-receptors in the skin of

haptic ‡ system have been investigated by Lederman and Klatzky and their colleagues [18, 20, 19] This work has shown that an important component of the haptic system is its ability to recognize attributes of three-dimensional objects quickly and accurately. Among these attributes are global shape, hardness, temperature, weight, size, articulation and function.

In order to fully understand the power of haptics, we have first set out to perform a series of object recognition tasks using touch alone. If we can achieve success in object recognition using the haptic approach, then we will be able to extend our system's power greatly by using both touch and vision. Lederman and Klatzky [19] have found a 96-99% success rate in identifying common objects using two-handed, haptic object recognition. An outcome of this research is the identification of hand movement strategies that are used by humans in discovering different attributes of three-dimensional objects. They have labeled these EP's, or Exploratory Procedures, and we have found it natural to extend these human capabilities to our robotic domain.

3.1. Grasping by Containment

The first of these strategies we have implemented is grasping by containment. This is an attempt to understand an object's gross contour and volume by effectively molding the hand to the object. We have chosen to model objects as superquadrics [7, 24, 5] whose surface 3-D vector X is defined below using a latitudinal and longitudinal parameterization expressed in spherical coordinates.

$$X(\eta, \omega) = \begin{bmatrix} a_1 C_\eta^{\varepsilon_1} C_\omega^{\varepsilon_2} \\ a_2 C_\eta^{\varepsilon_1} S_\omega^{\varepsilon_2} \\ a_3 S_\eta^{\varepsilon_1} \end{bmatrix} \qquad (1)$$

$$-\frac{\pi}{2} \le \eta \le \frac{\pi}{2}, \quad -\pi \le \omega \le \pi$$

C_η, S_ω are Cosine(η) and Sine(ω).

ε_1, ε_2 are the superquadric shape parameters.

a_1, a_2, a_3 are scaling factors along the X, Y and Z directions.

Superquadrics form a rich set of shape primitives that allows a wide degree of freedom in

---

the human hand; our robotic hand is more limited with 16 joint sensors, 32 tendon force sensors, and 4 16 x 16 fingertip tactile sensors. In addition, a human hand has two main differences in structure from our robotic hand. The first is a highly flexible, opposable thumb that is mounted to the side of the other digits. The Utah-MIT hand thumb is identical to the other fingers and is mounted directly opposite the other fingers. The second difference is a palmar degree of freedom exists in human hands that is missing in the Utah-MIT hand. Humans find this palmar degree of freedom quite useful, especially for encompassing type grasps where the hand is molded to an object and as a grasping mechanism in its own right, almost independent of the existence of multi-jointed fingers.

‡ An important point to be made in applying hands to robots is that the human perceptual process of interest is *haptic* perception. By this, we mean the interplay of both the cutaneous system (skin, tactile receptors) and the kinaesthetic system (joints, muscle and bone) of the arm [9].

modeling objects. The parameter space is continuous and allows a smooth change from a cuboid to a sphere to a cylinder, with more complex shapes derivable with the addition of bending and tapering parameters. These "lumps of clay" are deformable by the usual linear stretching and scaling operations and can be combined using boolean set operations to create more complex objects.

What makes superquadrics particularly relevant for haptic recognition is the following:

- The models are volumetric in nature, which maps directly into the psychophysical perception processes suggested by grasping by containment.

- The models can be constrained by the volumetric constraint implied by the joint positions on each finger.

- The models can be recovered with sparse amounts of point contact data since only a limited number of parameters need to be recovered. There are 5 parameters related to shape (see equation 1) and 6 related to position and orientation in space. Global deformations (tapering, bending) add a few more.

- In addition to the use of contact points of fingers on a surface, the surface normals from contacts can be used to describe a dual superquadric which has the same analytical properties as the model itself.

- The analytic nature of the model created from sparse data allows searching strategies in the model space to proceed in a hypothesize and test fashion as suggested by the work of Allen [4] and Stansfield [28].

3.2. Recovery Procedure

For this initial work on recognition, we have used a simplified procedure to gather data points. Our intent is to use the tactile sensors mounted on the finger links to generate contact position data. However, during our initial trials, our tactile sensors were not yet mounted on the hand. Instead, we opted for a method that used the hand's internal joint angle readings and tendon forces to generate Cartesian positions of contact based upon fingertip contact.

The Puma arm moves the hand to a position in which it will close around the object. The fingers are spread wide during approach. Then the fingers are closed by position commands until the observed force (estimated by the difference between the flexor and extensor tendon tensions) exceeds a given threshold, which indicates that the finger is in contact with the object. The joint angle positions are read, and kinematic models of the hand and the Puma arm are used to convert them to XYZ positions in world coordinates. Then the fingers are opened wide again, and a second containing grasp is executed, with the fingers taking different approach paths. The fingers are spread once again, and the Puma arm moves the hand to the next position.

The sequence of Puma positions is given in advance. Once the contact points are determined using the forward kinematics of the hand derived from the joint angle sensors, the sparse sets of point data is injected into the recovery algorithm developed by Solina [26]. This algorithm uses a Levenberg-Marquardt non-linear least squares approximation to fit the superquadric "inside-out function". This is an implicit form of equation 1 which records if a sample data point lies inside, outside or on the surface of the superquadric model. By summing the squared distance of each sample data point from the current model, an error of fit measure is generated that is minimized by the algorithm.

Equation 1 is for a canonical superquadric located at the origin. Since our sensor data can exist anywhere in the world coordinate space, the algorithm must recover the 6 rotation and translation parameters in addition to the 5 superquadric shape parameters (a_1, a_2, a_3, ε_1, ε_2). In addition, we allow global deformations to include tapering of superquadric forms. The taper is defined to be a linear tapering with 2 parameters that control the tapering in both the X and Y dimensions. The algorithm must recover a minimum of 11 parameters and 13 if the object is tapered.

Figure 4. Object Database.

We tested this procedure against a database of 6 objects (shown in Figure 4 plus a smaller cylinder). The database included objects that could be modeled as undeformed superquadrics (block, large cylinder, small cylinder) and deformed (tapered) superquadrics (lightbulb, funnel, triangular wedge). The recovered shapes are shown in Figure 5 with the sample data points overlaid on them.

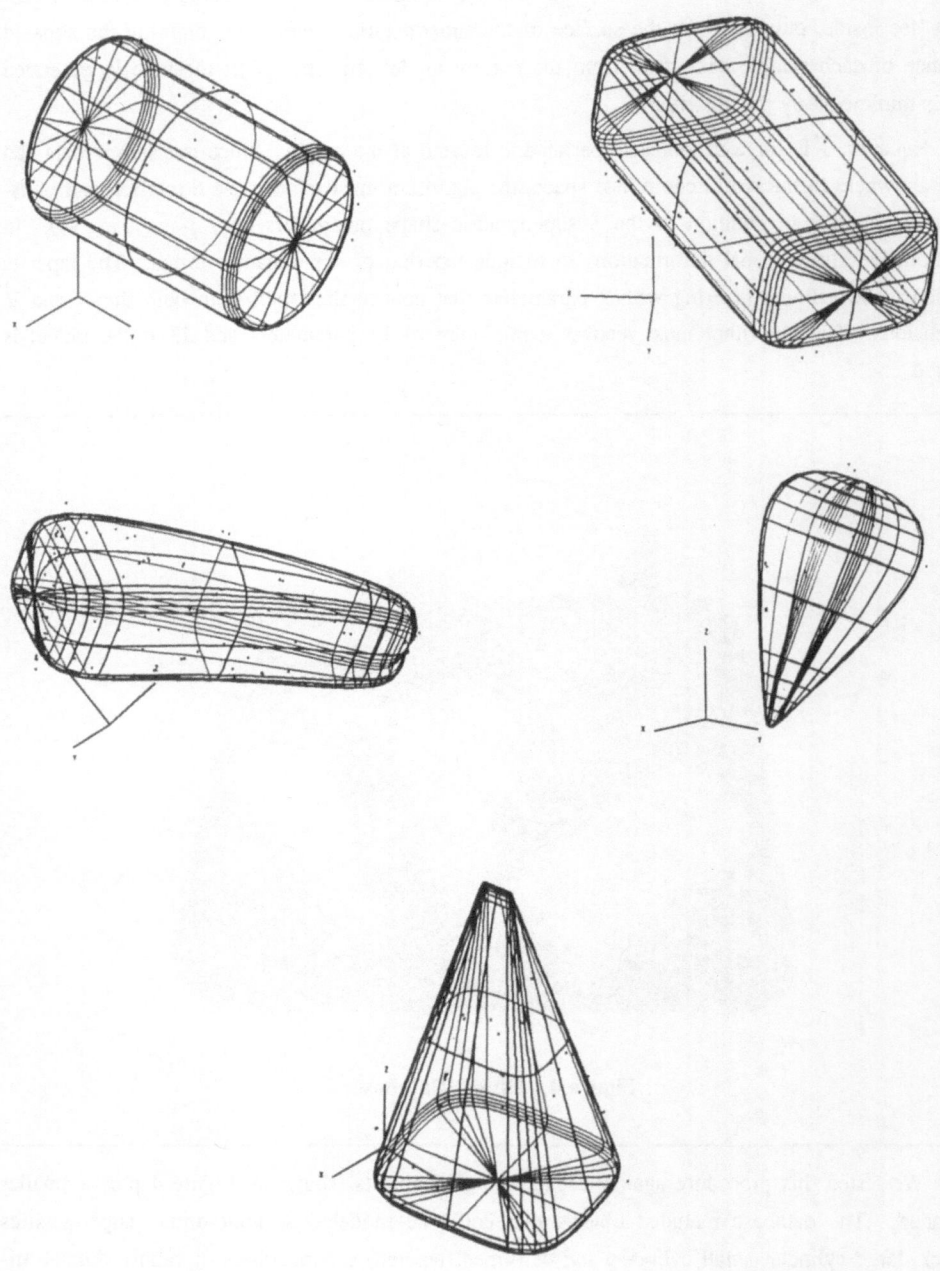

Figure 5. Recovered shape of cylinder, block, wedge, lighbulb and funnel.

The results of these experiments are quite good, especially considering the sparse nature of the data and the errors in the derived contact points. These errors are a function of the accuracy and calibration of the robotic arm, the hand joint position sensors and the kinematic model of the hand itself. In spite of this sensor error, the recovered shapes are an accurate representation of the actual object's shape. The data points are overlayed on the recovered shapes to show the closeness of fit and the sparseness of the data. Each object's shape was recovered with extremely sparse amounts of data; typically 30-100 points, depending on the object. It is important to note that this is about two orders of magnitude less than typical range data images which try to recover shape with denser data, that, unlike touch sensing, is limited to a viewpoint that only exposes half the objects surfaces to the sensor.

3.3. Contour Following

The second EP we have implemented is contour following with a two fingered grasp. This EP will allow us to determine an objects contour which has been shown to be a strong shape cue from previous vision research [29, 25, 22]. The procedure we have developed allows us to do a two-fingered trace of opposing sides of an object to determine the contour.

The problem of using a tactile device to trace a surface on an object is a complicated one. Previous work by Allen [1] using a one fingered tactile sensor mounted on a PUMA traced along an arbitrary surface by calculating a weighted vector of constraint directions that tried to follow the surface curvature while preserving smoothness of the trace and a constraint having to do with creating regions bounded by traces that were equivalent in size. Hor [13] traced contours of planar objects using a planar 4-fingered "chopstick" like manipulator. Strain gauge sensors on the fingers of this device would calculate surface normals and move tangentially along a surface, recording the contour. Stansfield [28] used a planar LORD tactile sensor mounted on a PUMA to trace edges and other features on objects.

Our procedure begins with the hand above the object. The hand moves down, pinching the fingers as it moves down until it contacts the object. Once the object is contacted, the hand is moved along an exploration axis, recontacting the surface as it moves along, recording the point of contact on the tactile sensors mounted on the hand. A least-square cubic polynomial is computed from these points, giving an estimate of the contour.

Determining the exploration axis is a key part of this EP. Knowing in which direction to trace the object is important to higher level recovery procedures which need to use this information in the recognition process. Once the hand makes contact with the object, it explores the contour along a known axis which we calculate apriori. We are currently implementing a vision based technique to determine this axis. Our use of vision is consistent with the multi-sensor approach to perception, which has proven to be quite strong (see Kak and Chen [17]). One of the key findings in using multi-sensor data is that over-reliance on one sensor can cause error. It appears to be an empirical fact that pushing one sensor results in a degradation of results; however, using only the most reliable and highest confidence sensor data allows one to proceed along a path that is known to be correct. We call this principle "less is more", in that reduced amounts of reliable data from a single sensor are more useful than large amounts of data which may be spurious. By combining the data that is most reliable from each of a number of sensors, more accurate results may be computed.

Our method of visual recovery of the exploration axis (currently being implemented) exploits the recent work of Wolff [31] in stereo line matching. Point-based stereo techniques tend to be unreliable in that multiple correspondences between images can cause mismatches and error. More stable matching can occur using larger primitives such as lines [12]. Even using line-based matching, problems can still occur. Matching the endpoints of lines can be prone to errors in the output of the line finder which may break a single line into multiple segments due to differing edge strengths along the line. The problem here is that 3-D depth is being computed, which requires an absolute correspondence of points (whether from point-based or line-based methods).

Our method alleviates this dependence on absolute matching of unstable primitives to generate 3-D depth. All we require of the algorithm is an *orientation* vector in 3-D. We do not need to have its absolute depth, but need to generate a match between a family of parallel lines sharing the same orientation. This orientation can then be used by the active hand as the exploration axis. The 3-D depth has already been determined from the contact of the hand with the object. Given this 3-D depth from tactile contact, we can follow the 3-D axis determined by the line based stereo matcher to continue our exploration.

It is important to note that this method is less sensitive to matching errors and baseline measurement, another common cause of stereo error. In addition, it is also less prone to the effects of physical point mismatches as the baseline increases, since we are still matching a larger entity, the line itself. Intuitively, the method creates a 3-D plane in space from the camera center and any two points on the line. This plane and a similar plane from the other camera are all that are needed to create a 3-D intersection line which we can use as the exploration axis.

As there are many lines in a scene, we have to choose a criteria for deciding which lines constitute the axis of the object. For exploration purposes, we simply want to discover a maximum length line which will serve as an axis. In most cases, this is part of the visual occluding contour of the object, which is exactly the axis we desire for active tactile exploration.

3.4. Planar Surface Explorer

This EP is used to explore planar surfaces with the hand's index finger. While the index finger is held in an extended position, the hand and arm are moved to a specified location. Then the index finger is flexed until the fingertip's tactile sensor detects contact with a surface. (If no contact is detected, the procedure terminates.) After the initial contact, the Cartesian position of the contact point is noted. The hand and arm then begin an iterative search for the boundaries of the surface by performing the following sequence: (a) lift the finger off the surface until tactile contact is lost; (b) move the arm in a direction parallel to the surface; (c) if the finger is in contact after the movement, note the new contact location, otherwise lower the index finger until it makes contact with the surface again; (d) repeat steps (a)-(c) until the finger fails to make contact in step (c). In step (d), if the finger does not contact the surface, then either the finger has moved beyond the edge of the surface, or the surface is too far away from the finger to be detected. To check for the latter case, the arm must be moved toward the surface. After completing the first collection of data points and finding the edge of the surface, the index finger is moved back to the position of initial contact, and a second mapping of the surface is undertaken in a direction 180 ° opposite. This procedure continues until a second surface edge is detected. The search now continues as before but in a direction perpendicular to the first two traces. This procedure then is able to map out a set of contact points on the surface, describing its extent. Each time the finger-tip contacts the surface, the Cartesian coordinates of the contact are retained. Periodically, a linear, least-squares fit to a plane is performed to derive the normal to the plane and a measure of fit. The surface normal is useful not only for the sake of object recognition, but also in speeding the exploratory procedure by refining the hypothesis of where to move the arm. The acquisition of data points in this method is also compatible with the three-point seed method of Henderson and Bhanu for forming planar surfaces from range data [11]. Figure 6 shows a pattern of traces on a planar surface of a rectangular block using this EP. The contact points are all within a single plane with the exception of the end points where the finger has dropped off the edge of the block, thus ending the search procedure.

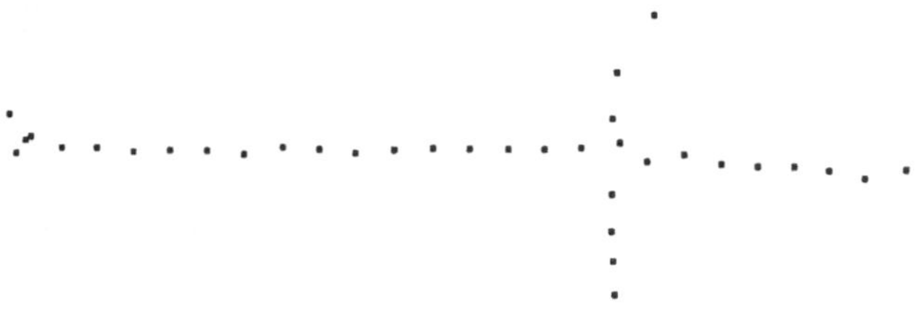

Figure 6. Surface Explorer tactile contacts on a rectangular planar surface.

4. SUMMARY AND FUTURE WORK

We have described an integrated system for active sensing with a dextrous robotic hand. The system is comprised of a rich set of sensing and actuation primitives that are linked by a script like task level command language that can control concurrent robotic actions that include hand and arm movements. Using this system, we have acquired sparse 3-D contact data from a variety of different sensors (joint positions, tendon forces, tactile arrays) to build accurate volumetric descriptions of objects. In addition, we have implemented two other active exploratory procedures, contour following and planar surface exploration, that can be used to discover finer object detail.

In the future, we hope to link all the exploratory procedures into a fully autonomous system that will be able to use gross object structure as a generator of sensing hypothesis for the finer level EP's. In this way we hope to be able to recover the shape of more complex objects using tactile and visual processing.

5. ACKNOWLEDGEMENTS

This work was supported in part by DARPA contract N00039-84-C-0165, NSF grants DMC-86-05065, DCI-86-08845, CCR-86-12709, IRI-86-57151, North American Philips Laboratories, and the AT&T Foundation. Special thanks to Franc Solina for allowing us to use his recovery algorithms; Paul Michelman and Peter Armstrong for acquiring the tactile sensor data; Ken Roberts for setting up and calibrating the Utah hand; and Steve Feiner, Cliff Beshers and Wen-Way Hseush for help with DIAL and SCL.

References

1. Allen, Peter, "Integrating vision and touch for object recognition tasks," *International Journal of Robotics Research*, vol. 7, no. 6, pp. 15-32, 1988.

2. Allen, Peter, Paul Michelman, and Kenneth S. Roberts, "An integrated system for dextrous manipulation," *IEEE Conference on Robotics and Automation*, pp. 612-617, Scottsdale, AZ, May 15-19, 1989.

3. Allen, Peter and Kenneth S. Roberts, "Haptic object recognition using a multi-fingered dextrous hand," *IEEE Conference on Robotics and Automation*, pp. 342-347, Scottsdale, AZ, May 15-19, 1989.

4. Allen, Peter K., *Robotic object recognition using vision and touch*, Kluwer Academic Publishers, Boston, 1987.

5. Bajcsy, Ruzena and Franc Solina, "Three dimensional object representation revisited," *Proceedings International Conference on Computer Vision*, London, June 1987.

6. Ballard, D. and C. Brown, *Computer vision*, Prentice-Hall, 1982.

7. Barr, Alan, "Superquadrics and angle preserving transformations," *IEEE Computer Graphics and Applications*, vol. 1, pp. 11-23, 1981.

8. Feiner, Steven, David Salesin, and Thomas Banchoff, "DIAL: A diagrammatic animation language," *IEEE Computer Graphics and Animation*, vol. 2, no. 7, pp. 43-54, September 1982.

9. Harmon, Leon, "Tactile sensing for robots," in *Recent Advances in Robotics*, ed. G. Beni, pp. 389-424, 1985.

10. Hayward, Vincent and Richard Paul, "Robot manipulator control under UNIX," *Proc. of the 13th ISIR*, pp. 20:32-20:44, Chicago, April 17-21, 1983.

11. Henderson, T. C. and Bir Bhanu, "Three point seed method for the extraction of planar faces from range data," *Proc. of IEEE Workshop on Industrial Applications of Machine Vision*, pp. 181-186, Research Triangle Park, NC, May 1982.

12. Henriksen, Knud, "Line-based stereo matching," MS-CIS-87-52, Grasp Lab 109, Department of Computer and Information Science, University of Pennsylvania, Philadelphia.

13. Hor, Maw-Kae, "Control and task planning for a four finger dextrous manipulator," Ph.D Thesis, Courant Institute, New York University, October 1987.

14. Horn, B. K. P., *Robot vision*, M.I.T. Press, 1986.

15. Jacobsen, S. C., E. K. Iversen, D. F. Knutti, R. T. Johnson, and K. B. Biggers, "Design of the Utah/MIT dextrous hand," *Proceedings of the IEEE Conference on Robotics and Automation*, pp. 1520-1532, San Francisco, April 7-10, 1986.

16. Johansson, Roland S. and Ake B. Vallbo, "Tactile sensory coding in the glabrous skin of the human hand," *Trends in Neurosciences*, vol. 6, pp. 27-32, 1983.

17. Kak, A. and S. Chen editors, *Proceedings of the AAAI Workshop on Spatial-Reasoning and Multisensor Integration*, Morgan Kauffman, Los Altos, CA, Oct. 1987.

18. Klatzky, Roberta, Susan Lederman, and Victoria Metzger, "Identifying objects by touch: An "expert system"," *Perception and Psychophysics*, vol. 37, pp. 299-302 .

19. Lederman, Susan and Roberta Klatzky, "Hand movements: A window into haptic object recognition," *Cognitive Psychology*, vol. 19, pp. 342-368, 1987.

20. Lederman, S. J. and R. A. Browse, "The physiology and psychophysics of touch," *Sensors and sensory systems for advanced robots*, NATO ASI series, Springer-Verlag.

21. Muthukrishnan, C., D. Smith, D. Myers, J. Rebman, and A. Koivo, "Edge detection in tactile images," *Proc. of IEEE International Conference on Robotics and Automation*, pp. 1500-1505, Raleigh, 1987.

22. Nalwa, V., "Line drawing interpretation: Bilateral symmetry," *Proceedings of DARPA Image Understanding Workshop*, pp. 956-967, Morgan-Kauffman Publishers, Los Angeles, CA, February 1987.

23. Narasimhan, Sundar, David M. Siegel, and John M. Hollerbach, "Condor: A revised architecture for controlling the Utah-MIT hand," *IEEE Conference on Robotics and Automation*, pp. 446-449, Philadelphia, April 24-29, 1988.

24. Pentland, Alex P., "Recognition by parts," Technical Report 406, SRI International, December 16, 1986.

25. Ponce, J., D. Chelberg, and W. Mann, "Invariant properties of the projections of straight homogeneous cylinders," *Proc. First International Conference on Computer Vision*, London, 1987.

26. Solina, Franc, "Shape recovery and segmentation with deformable part models," Ph.D Dissertation, Department of Computer Science, University of Pennsylvania, December 1987.

27. Speeter, Thomas, "Flexible piezo-resistive touch sensing array," *SPIE Conference on Optics, Illumination and Image Sensing for Machine Vision III*, Cambridge, November 1988.

28. Stansfield, Sharon, "Visually-guided haptic object recognition," Ph.D. dissertation, Department of Computer and Information Science, University of Pennsylvania, October 1987.

29. Stevens, Kent, "Computation of locally parallel structure," *Biological Cybernetics*, no. 29, pp. 19-28, 1978.

30. Tise, B., "A compact high resolution piezo-resistive digital tactile sensor," *IEEE Conference on Robotics and Automation*, pp. 760-764, Philadelphia, April 24-29, 1988.

31. Wolff, Lawrence B., "Measuring the orientation of lines and surfaces using translation invariant stereo," *SPIE Conference on Sensor Fusion*, vol. 1003, Cambridge, MA, November 1988.

32. van Brussel, H. and H. Belien, "A High Resolutiion Tactile Sensor for Part Recognition," *Proceedings of the 6th International Conference on Robot Vision and Sensory Controls*, Paris, June 1986..

Sensor Integration Using State Estimators

J.G. Balchen   F. Dessen   G. Skofteland

Norwegian Institute of Technology, Division of Engineering Cybernetics
N-7034 Trondheim, Norway

Abstract: Means for including very different types of sensors using one single unit are described. Accumulated data are represented using an updatable dynamic model, a Kalman filter. The scheme easily handles common phenomena such as skewed sampling, finite resolution measurements and information delays. Included is an example where 3D motion information is collected by one or more vision sensors.

1 Introduction

Many robotic systems employ a multiplicity of sensory devices which extract information about the behavior of the system and its environment. For the purpose of deriving specific information useful for control of the system, it is necessary to process the individual type of information in a systematic manner. This task is refered to as *sensor integration (fusion)*. Sensors and sensor systems are established fields of research within robotics as well as within general control and dynamic estimation theory. Within the latter area, the concept of state and parameter estimators (observers), (Jazwinski, 1970) is well established and found useful at least in cases where the process can be considered continuous in the quantities to be estimated. Within robotics (including autonomous vehicles) contributions come from a large number of research communities resulting in a situation where many different, however no general, tools have been described for the fusion and interpretation of sensor data. This partly results since the number of different types of sensors is large and also because of the special research made on specific devices, such as cameras.

Often sensor integration is employed to resolve the problem of sensory redundancy which occurs when the system has more sensors than strictly necessary to determine a state variable of the system, say position. A frequently encountered example of this kind is when the position of an object is measured by a number of different sensors based on different physical principles and with different characteristics with respect to accuracy, noise, availability, etc. under varying operating conditions. If for instance one sensor fails, the other sensors can still be available to fulfill the task provided they have been properly integrated.

This contribution describes sensor systems for robotics in terms of state and parameter estimation schemes. Basic estimation schemes are outlined in § 2. A systematic means which allows the time of appearance of sensor data to be irregular or even stochastic is developed in § 3. The resulting scheme is intended for use in large sensory systems. A example is given in § 4, where the motion of an object is monitored by a number of cameras.

2 Methods for Sensor Integration

In Fig. 2.1, two alternatives are indicated for sensor integration. Part A suggests that the primary information from the sensors is handled directly by some algorithm producing the integrated information whereby proper account must be made for the individual characteristics of the sensors, their measuring uncertainties, etc. No general concept exists for this type of direct integration, but a large number of proposals have been made.

NATO ASI Series, Vol. F 63
Traditional and Non-Traditional Robotic Sensors
Edited by T. C. Henderson
© Springer-Verlag Berlin Heidelberg 1990

A : Direct sensor integration

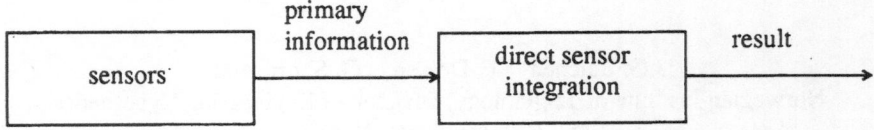

B : Sensor integration using a recursive algorithm

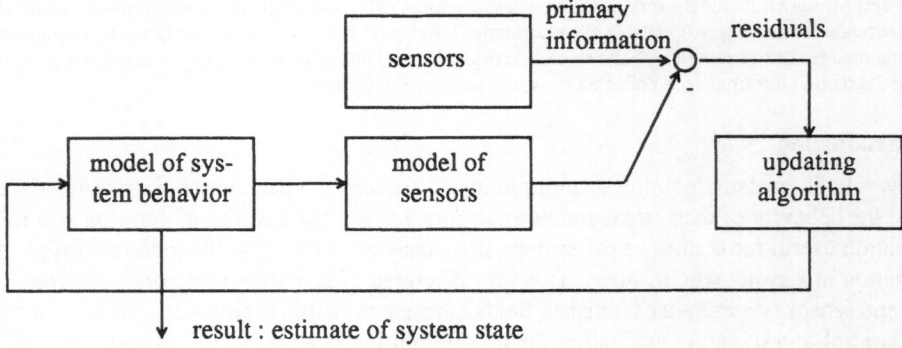

result : estimate of system state

Figure 2.1 Two alternatives for sensor integration

Part B of Fig. 2.1 is shows a block diagram of a system for sensor integration using a recursive algorithm. This reflects a very general approach applicable to a large number of practical problems. It consists of two mathematical models and an updating algorithm. One of the models describes how the primary information is generated by the sensors in relation to the behavior of the system (the motion of robots, objects and so on). The other model describes the behavior of the system (robot arm states, object states, etc.). The updating algorithm takes the residuals which is the difference between the primary information received from the sensors and the estimated sensor information received from the model of the sensors and corrects the behavior of the system model in such a way as to reduce the size of the residuals. The model of the system behavior may assume different levels of complexity. If the model describes the free movement in three dimensional space of a point (very small object), its total state vector may be composed of a position vector **p,** a velocity vector **v** and an acceleration vector **a** which are all three-dimensional and related by a differential equation involving 9 variables,

$$\dot{\mathbf{p}} = \mathbf{v} \quad ; \quad \dot{\mathbf{v}} = \mathbf{a} \tag{2.1}$$

If the object has a finite size, its position and orientation needs to be described by 6 independent variables (6 dof). The same applies to the velocities and accelerations. A number of different coordinate systems may be chosen to describe the equations of rigid body movement yielding equivalent types of information.

One particular, theoretical basis for the realization of the recursive algorithm shown in Fig. 2.1 is the well known Kalman Filter (KF). The updating mechanism of the KF is a *least squares, minimum variance algorithm.* The KF appears in a continuous version applicable when the system model is in the form of differential equations and a discrete version applicable when the dynamic model consists of different equations. The elementary KF is based on linear models whereas the Extended Kalman Filter (EKF) allows for nonlinear models both of the system behavior and of the sensors.

The EKF algorithm is reviewed as follows using standard state space notation:

$$x(k+1) = f(x(k)) + n(k) \tag{2.2}$$

with measurement model

$$y(k) = h(x(k)) + w(k) \tag{2.3}$$

Here

$x(\cdot)$	:	State vector consisting of positions, velocities and accelerations
k	:	discrete time variable
$f(\cdot)$	:	vector of nonlinear functions in behavior model
$n(\cdot)$	:	vector of excitation forces, covariance matrix N
$y(\cdot)$	:	vector of primary sensor variables
$h(\cdot)$	:	vector of nonlinear sensor relations
$w(\cdot)$	:	vector of sensor uncertainties (noise), covariance matrix W

A posteriori state estimate:

$$\hat{x}(k) = \bar{x}(k) + K(k)[y(k) - h(\bar{x}(k))] \tag{2.4}$$

Predicted (a priori) state estimate:

$$\bar{x}(k+1) = f(\hat{x}(k)) \tag{2.5}$$

KF gain matrix :

$$K(k) = \bar{X}(k)D^T(k)[D(k)\bar{X}(k)D^T(k) + W(k)]^{-1} \tag{2.6}$$

Predicted covariance matrix of a priori state estimate:

$$\bar{X}(k+1) = \Phi(k) \hat{X}(k) \Phi^T(k) + N(k) \tag{2.7}$$

Covariance matrix of a posteriori state estimate:

$$\hat{X}(k) = [I - K(k)D(k)]\bar{X}(k) \tag{2.8}$$

where

$$D(k) = \partial h(x)/\partial x^T |_{x = \bar{x}(k)} \quad ; \quad \Phi(k) = \partial f(x)/\partial x^T |_{x = \hat{x}(k)}$$

The KF gain matrix of (2.6) is the contents of the block "updating algorithm" in Fig. 2.1. One of the major advantages of employing the recursive algorithm of the EKF is that it is very general and allows for the application of an arbitrary number of sensors as long as a sensor model, $h(\ )$ has been established. If more sensors are employed than strictly necessary from an observability point of view, it is possible to arrive at more accurate estimates of system states and insensitivity to sensor failure. Above, the basic EKF is described. This is often modified to cope with special situations.

3 Event Driven State Estimators

State estimators are usually described in terms of continuous-time differential equations or by discrete-time difference equations where all measurements are available simultaneously at equally spaced instances in time. Thus, common phenomena such as skewed sampling are neglected. Also, a fixed rate of appearance of data implies that a complete update of the state estimate is made even if little new information is available. The schemes are convenient since their structures are well defined, and conceptually simple means are available for system design and performance analysis. The present section describes means for the systematic treatment of now-and-then appearing sensor data, useful in large-scale sensory systems.

3.1 Basic Concept

To make it simple, the system below is based on a continuous-time kernel state estimator. This also emphasizes the difference between ballistic estimation and sensor-based updates, and makes it possible to take into account continuously available data as well as data available at specific moments. For digital computer implementations, the continuous-time kernel can be replaced by a fixed-rate discrete-time equivalent in the usual manner. It is assumed that the process is represented by the differential equation

$$\mathbf{x}^{\cdot} = g(\mathbf{x}, \mathbf{u}, \mathbf{n}) \tag{3.1}$$

where $\mathbf{x}$: state vector, $\mathbf{u}$: control vector, $\mathbf{n}$: process disturbance. Assuming continuity, the process can be linearized about trajectories $\bar{\mathbf{x}}(t), \bar{\mathbf{u}}(t)$ and $\bar{\mathbf{n}}(t)$ of a state estimator by

$$\dot{\xi} = A\xi + B\eta + C\zeta \tag{3.2}$$

where $\xi = \mathbf{x} - \bar{\mathbf{x}}$, $\eta = \mathbf{u} - \bar{\mathbf{u}}$, $\zeta = \mathbf{n} - \bar{\mathbf{n}}$ and

$$\dot{\bar{\mathbf{x}}} = g(\bar{\mathbf{x}}, \bar{\mathbf{u}}, \bar{\mathbf{n}}) \tag{3.3}$$

Usually $\bar{\mathbf{u}} = \mathbf{u}$ so that $\eta = \mathbf{0}$, simplifying (3.2). ξ is the state estimate error whereas ζ is considered to be the white-noise, zero-mean part of $\mathbf{n}$. Thus the state estimate covariance develops (Jazwinski, 1970) by

$$\dot{\mathbf{X}} = A\mathbf{X} + \mathbf{X}A^T + CNC^T \tag{3.4}$$

where $\mathbf{X} = E(\xi\xi^T)$, $N = E(\zeta\zeta^T)$ are covariance matrices. Continuously available measurements $\mathbf{y}_0 = \mathbf{D}\mathbf{x}_0 + \mathbf{w}_0$ will modify the covariance to

$$\dot{\mathbf{X}} = A\mathbf{X} + \mathbf{X}A^T - \mathbf{X}\mathbf{D}_0^T\mathbf{W}_0^{-1}\mathbf{D}_0\mathbf{X} + CNC^T \tag{3.5}$$

where $\mathbf{W}_0$ is the covariance of measurement vector $\mathbf{y}_0$. These measurements may or may not be present. The important point is that it is possible continuously to keep track of the covariance of the error in the state estimate. Measurements available at discrete points in time are divided into groups $\mathbf{y}_1$ through $\mathbf{y}_R$ such that when presented,

$$\mathbf{y}_1 = h_1(\mathbf{x}) + \mathbf{w}_1 \quad , \mathbf{y}_2 = h_2(\mathbf{x}) + \mathbf{w}_2 \quad , \dots \tag{3.6}$$

and so on. Here $\mathbf{w}_i$ is a white-noise, zero-mean disturbance vector for measurement vector $\mathbf{y}_i$, with covariance matrix $\mathbf{W}_i$. Linearization of an arbitrary measurement group $\mathbf{y}_i$ about $\bar{\mathbf{y}}_i$ and $\bar{\mathbf{x}}$ such that $\bar{\mathbf{y}}_i = h_i(\bar{\mathbf{x}})$, is represented by

$$\mathbf{v}_i = \mathbf{D}_i \xi + \mathbf{w}_i \tag{3.7}$$

where $\mathbf{v}_i = \mathbf{y}_i - \bar{\mathbf{y}}_i$. It is now assumed that measurement vector $\mathbf{y}_i$ is available at time $t(j)$. Vector $\mathbf{y}_i$ may or may not be sufficient for the observability of the dynamic system.

Available from (3.3) is an a priori state estimate $\bar{x}(j)$ with error covariance $X(j)^\dagger$ given by (3.4) or (3.5). The ordinary minimum variance estimate $\hat{x}(j)$ based on $\bar{x}(j)$ and $y_i(j)$ is then

$$\hat{x}(j) = \bar{x}(j) + K_i(j)[y_i(j) - h_i(\bar{x}(j))] \tag{3.8}$$

where

$$K_i(j) = X(j)D_i^T[D_iX(j)D_i^T + W_i]^{-1} \tag{3.9}$$

Similar to (2.8), the error covariance of this estimate is given by

$$\hat{X}(j) = [I - K_i(j)D_i]X(j) \tag{3.10}$$

When these computations have been made, $X(j)$ in (3.4) or (3.5) is reinitiated by $\hat{X}(j)$ whereas $\bar{x}(j)$ is set to $\hat{x}(j)$ before the continuous updating resumes. In this way, means are available for correcting the state estimator at arbitrary instances $t(j)$ based on a part $y_i(j)$ of the complete measurement vector.

3.2 Covariance based Data Selector

Future robots and autonomous vehicles are assumed to have installed large numbers of very different sensors. The number and diversity can, when considering the computation and data transmission capacities, imply that only a lesser number of sensors can be handled at a sufficient regularity. The problem is then to select those measurements which result in a sufficient quality for the continuously updated state estimate. The relevant system states should at least be observable, but a better quality measure can result from the covariance matrix X itself.

$$J = E(\xi^T\xi) = \text{trace}X \tag{3.11}$$

is a simple example. Due to measurements $y_i(j)$ it is altered by

$$\Delta J_i = \text{trace}[\hat{X}(j) - X(j)] \tag{3.12}$$

which can be computed using (3.9) and (3.10).

The covariance based data selector is intended for on-line use in large-scale sensory systems. However, off-line modifications have their use as well, for instance as the means for inferring the necessary sample rates of each group of measurements in medium and small-scale control systems. It is the authors' opinion that, in general, the sample rates for measurements should be considered as a separate problem and without reference to the rates for setting of the control values.

---

† $\bar{x}(j)$ and $X(j)$ are short for $\bar{x}(t(j))$ and $X(t(j))$.

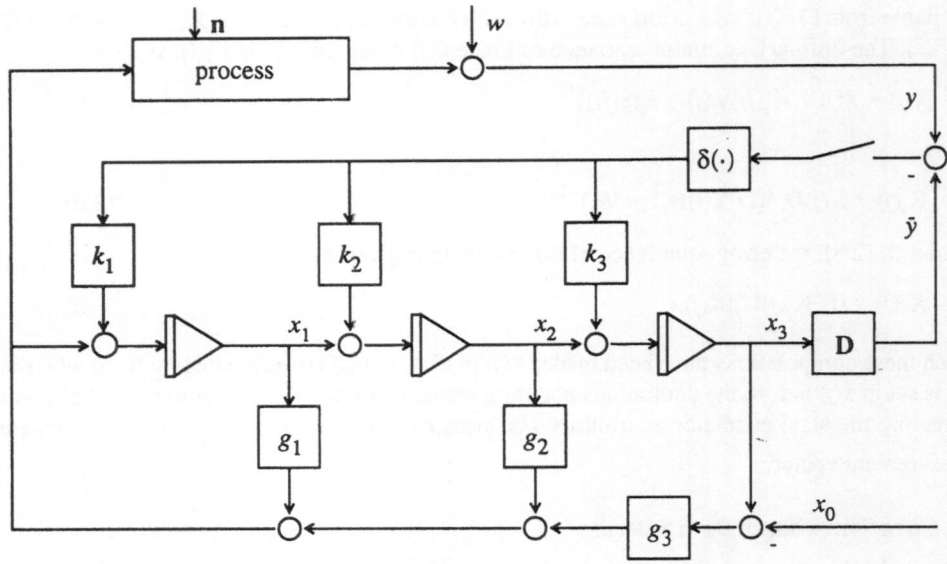

Figure 3.1 Servo control system

3.3 A simple Servo

As an example, a servo is assumed which is controlled using analog computation, or at a high sample rate. However, positional measurements are available at a low rate only. As indicated in Fig. 3.1, the state estimator models the mechanism as three cascaded integrators, and takes feedback from the continuous state estimate. This approach allows very high feedback gains $[\,g_1, g_2, g_3\,]$, resulting in a high-performance servo. Using a traditional controller, where feedback is taken directly from positional measurement, the performance is inherently bounded due to the low rate of appearance of y. The presented estimation and control structure is usable for instance in camera-guided autonomous vehicles as well as in high-performance servo mechanisms. An application within 3D vision is presented below.

4 Camera based Estimation of 3D Motion

The Kalman Filter can be used for sequential estimation of position and motion parameters of a rigid body based on camera measurements (Broida, 1986; Dickmanns, 1985). In this section the stereoscopic camera system of Fig. 4.2 is used for estimating motion in 3D space. The positions of the measured features in the pictures are compared to the predicted positions, and the differences are used to refine the estimates. Only a small set of simple, easy-extractable features, is needed. These can be assumed noisy, and the uncertainty is reduced in a least squares fashion in the Kalman Filter The feature extraction time can also be reduced, because the Kalman Filter not only gives a prediction of the position and motion parameters, but provides a representation of the uncertainty associated with the estimates as well. The method is not restricted to a stereoscopic camera system. More cameras and other sensors can be integrated in the state estimator by just expanding the Kalman measurement vector. The measurements don't even have to come at the same time, which shall be shown later in this section.

One application for this type of camera system is direct and flexible programming of a robot based on monitoring an experienced operator moving the tool (e.g. spray-painting).

4.1 The Measurement Model

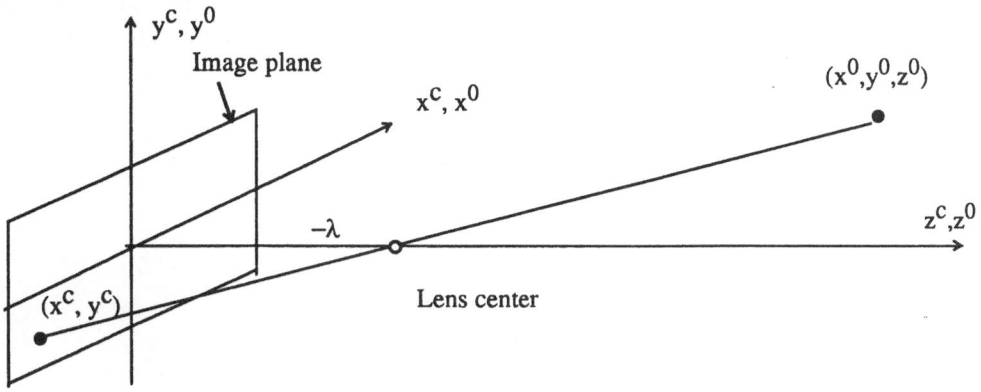

Figure 4.1 Basic model of the imaging process. Camera coordinate system (x^C, y^C, z^C) is aligned with the world coordinate system (x^0, y^0, z^0)

The camera model is that of central projection. The camera coordinate system (x^C, y^C, z^C) has the image plane coincident with the x-y plane and the optical axis along the z-axis. The center of the image plane is at $(0, 0, -\lambda)$, and the center of lens is at the origin. Here it is assumed for simplicity that the camera coordinate system is aligned with the world coordinate system (x^0, y^0, z^0). The image plane coordinates (x^C, y^C) of the projected 3D point (x^0, y^0, z^0) are

$$x^C = \frac{-\lambda x^0}{z^0} \tag{4.1}$$

$$y^C = \frac{-\lambda y^0}{z^0} \tag{4.2}$$

The measurement equations are nonlinear which means that the Extended Kalman Filter must be used. The measurements are linearized about the most recent estimate of the object position. The new D-matrix is

$$D_k = \left. \frac{\partial h(x)}{\partial x^T} \right|_{x = \bar{x}_k} \tag{4.3}$$

The two cameras are placed as in Fig.4.2 at $(\Delta x, 0, 0)$ and $(-\Delta x, 0, 0)$. The two camera axes crosses in the middle of the observation area, (at $z = x_0$) so that depth resolution is increased.

The measured features are the camera coordinates of two *matchpoints* on the observed object, which gives four measurements in each camera, a total of 8 measurements. How these coordinates are deduced from the camera image is not a central point in this paper. It might be extracted simply by the knowledge of the intensity value of the endpoints in the images, or the gradient in the discrete image function around the endpoints. This can be further simplified by placing light points (e.g. LEDs) at the matchpoints. The two cameras don't have to measure the same points on the object, so the correspondence problem is not dealt with in this strategy. As pointed out earlier, the Kalman Filter gives a prediction of the measurements with an associated uncertainty. This should be used to reduce the search area in the image and thus speed up the feature extracting algorithms. The measurements are assumed noisy. The physical sources of noise are quantization error, calibration error and vibrations in the equipment.

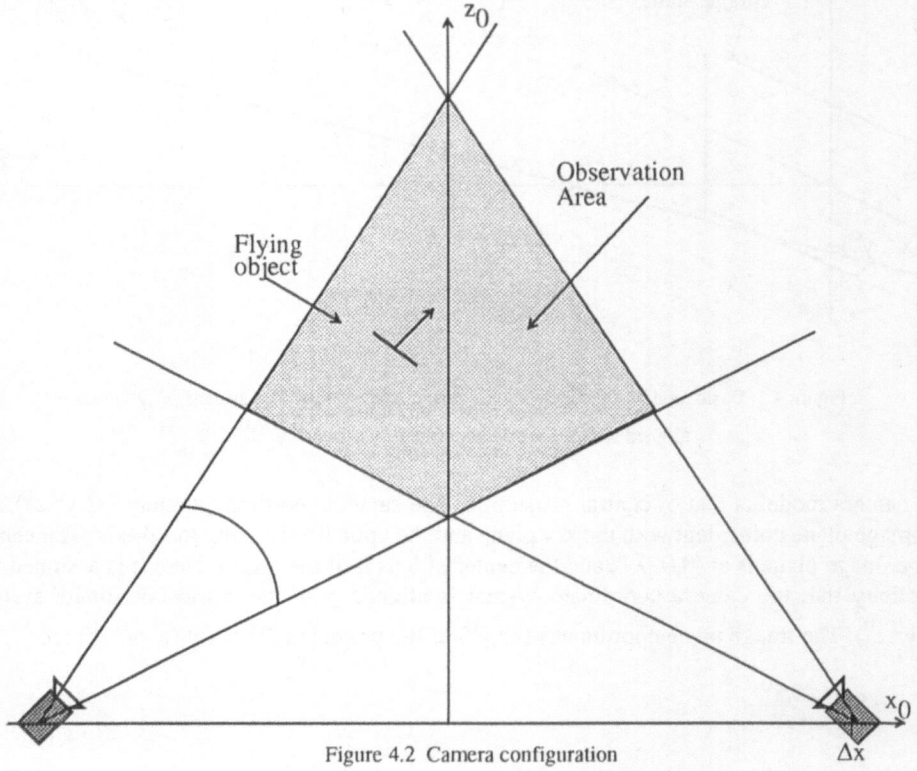

Figure 4.2 Camera configuration

4.2 Motion and Kinematics

The camera system is used to predict the motion of a free flying rigid object. The state-space model can be developed from Newton's law for the translation, and for Euler's law for the rotation around the center of gravity. An inertial base frame 0 is used to describe the motion of the rigid body. The center of mass has the position **p** with respect to the origin in the inertial frame. The velocity of the center of gravity is denoted **v**, while the angular velocity vector with respect to the inertial frame is ω. A superscript 0 is used to denote decomposition of a vector in the base frame , while decomposition in the body frame is denoted by superscript b. The equations of motion are

$$\ddot{\mathbf{v}}^0 = 0 \tag{4.5}$$

In the special case that the body is symmetric around its center of gravity, the equation of angular motion is simply

$$\ddot{\omega}^0 = 0 \tag{4.7}$$

The differential equation for the position vector **p** is

$$\dot{\mathbf{p}} = \mathbf{v}^0 \tag{4.8}$$

The angular velocity, however, cannot be integrated to an orientation vector in the same way.

Classical Euler angles describe orientation through the rotations ψ around the z axis, then θ around the y axis, then finally a rotation ϕ around the rotated z axis. The kinematic differential equation is found from

$$\omega^0 = \begin{bmatrix} \sin\theta\cos\psi & -\sin\psi & 0 \\ \sin\theta\sin\psi & \cos\psi & 0 \\ \cos\theta & 0 & 1 \end{bmatrix} \dot{r}_E \tag{4.9}$$

where

$$r_E = \begin{bmatrix} \phi \\ \theta \\ \psi \end{bmatrix} \tag{4.10}$$

The desired inverse relation is

$$\dot{r}_E = C_E \omega^0 \tag{4.11}$$

where

$$C_E = \frac{1}{\sin\theta} \begin{bmatrix} \cos\psi & \sin\psi & 0 \\ -\sin\theta\sin\psi & -\sin\theta\cos\psi & 0 \\ -\cos\theta\cos\psi & -\sin\psi\cos\psi & 1 \end{bmatrix} \tag{4.12}$$

The description has a singularity at $\sin\theta = 0$.

An alternative set of Euler angles are the roll-pitch-yaw angles: a rotation α around the z axis, β around the y axis, and finally γ around the x axis. The inverse transformation is

$$\dot{r}_R = C_R \omega^0 \tag{4.13}$$

where

$$r_R = \begin{bmatrix} \alpha \\ \beta \\ \gamma \end{bmatrix} \tag{4.14}$$

In this case the singularity appears at $\cos\beta = 0$.

A singularity-free description is obtained by using roll-pitch-yaw angles when $|\beta| <= \pi/4$ and switch to classical Euler angles when $|\theta| > \pi/4$.

When classical Euler angles are used, the state vector is

$$x = [\, p \quad r_E \quad v^0 \quad \omega^0 \quad \dot{v}^0 \quad \dot{\omega}^0 \,]^T \tag{4.15}$$

The state-space model is then

$$\dot{x} = A(x)x \tag{4.16}$$

where

$$A(x) = \begin{pmatrix} 0 & 0 & I & 0 & 0 & 0 \\ 0 & 0 & 0 & C_E & 0 & 0 \\ 0 & 0 & 0 & 0 & I & 0 \\ 0 & 0 & 0 & 0 & 0 & I \\ 0 & 0 & 0 & 0 & 0 & 0 \\ 0 & 0 & 0 & 0 & 0 & 0 \end{pmatrix} \qquad (4.17)$$

When θ becomes $\pi/4$, a switch is made to

$$x = [\, p \quad r_R \quad v^0 \quad \omega^0 \quad \dot{v}^0 \quad \dot{\omega}^0 \,]^T \qquad (4.18)$$

$$A(x) = \begin{pmatrix} 0 & 0 & I & 0 & 0 & 0 \\ 0 & 0 & 0 & C_R & 0 & 0 \\ 0 & 0 & 0 & 0 & I & 0 \\ 0 & 0 & 0 & 0 & 0 & I \\ 0 & 0 & 0 & 0 & 0 & 0 \\ 0 & 0 & 0 & 0 & 0 & 0 \end{pmatrix} \qquad (4.19)$$

4.3 Simulation Results

The simple object of Fig. 4.3 is to be used in the simulation run. It is symmetric about its major axis. The length is 0.3 m. The measurements come from the stereoscopic camera system of Fig. 4.2, where Δx is 2.5 m. The cameras have a resolution of 512 by 512 pixels and a field of view of 30° with a focus length of $\lambda = 0.016$ m. The matchpoints are the two end points of the object. These measurements are distorted with gaussian random noise. The noise level has standard deviation of 5 pixels which corresponds to 5-7 percent of the picture of the rod in the simulation runs.

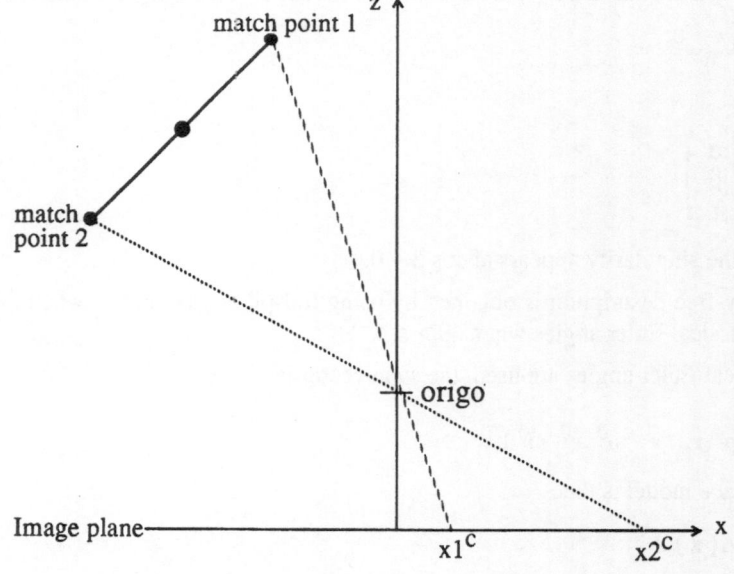

Figure 4.3 Object and measurement model

The initial values of the simulator were: $\mathbf{p} = [0,0,2.5]^T$, $\mathbf{v} = [-1,-1,-1]$, $\dot{\mathbf{v}} = [\,0.75, 0.75, 0.75\,]$. In the simulation run the object is rotating in the x-y plane, with initial angular velocity and acceleration of -1.0 rad/s and 0.75 rad/s^2. The initial estimates are different from the true values and are therefore associated with a low certainty, reflected by a large value in the covariance matrix. Assuming 30 pictures per second, the sample interval is T = 1/30 s. Two different results are presented, differing in the picture sampling. In the first simulation, measurements come from both cameras simultaneously at a rate of 30 pictures per second. In the second simulation the first measurements come from camera 1, the second from camera 2, then again from camera 1 and so on. The data rate is in this manner reduced to the half of that in the first simulation.

Some results from the first simulation run are presented in Figs. 4.4 - 4.6. The measurements come from the two cameras simultaneously. All the state variables converge quickly to their correct values. The position parameters, which are the most important, converge very fast, and their covariances converge to a low value. The error in z^c improves to less than 1 cm which is about 1or2 pixels in a camera. Figure 4.6 shows that the error in the camera measurements are reduced from a standard deviation of 5 pixels to a standard deviation of 1-2 pixels.

The result from the second simulation run, where measurements from only one camera at the time are available, is presented in Fig. 4.7. This is a special case of the theory presented in §3, where the measurements could be available at arbitrary instances $t(j)$. The state estimator is updated according to the available part of the measurement vector, $\mathbf{y}_i(j)$ of the complete vector $\mathbf{y}$.

The result shows that the state estimator still works well. Slower convergence is expected since the number of measurements are only half of the number used in the first simulation.

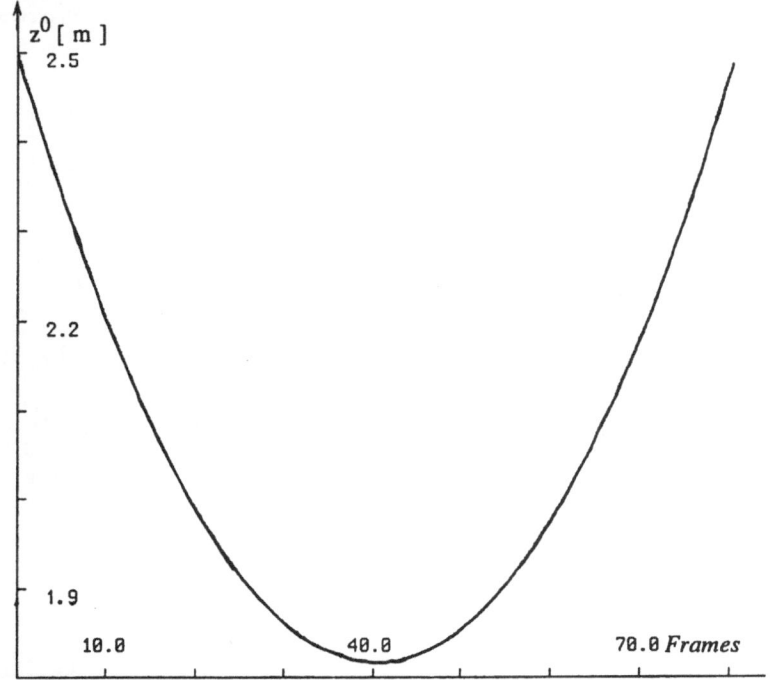

Figure 4.4 Translation in z-direction (depth) of the objects center of mass in both simulation runs. Prediction error is shown in Fig.4.5. 30 image frames are provided per second..

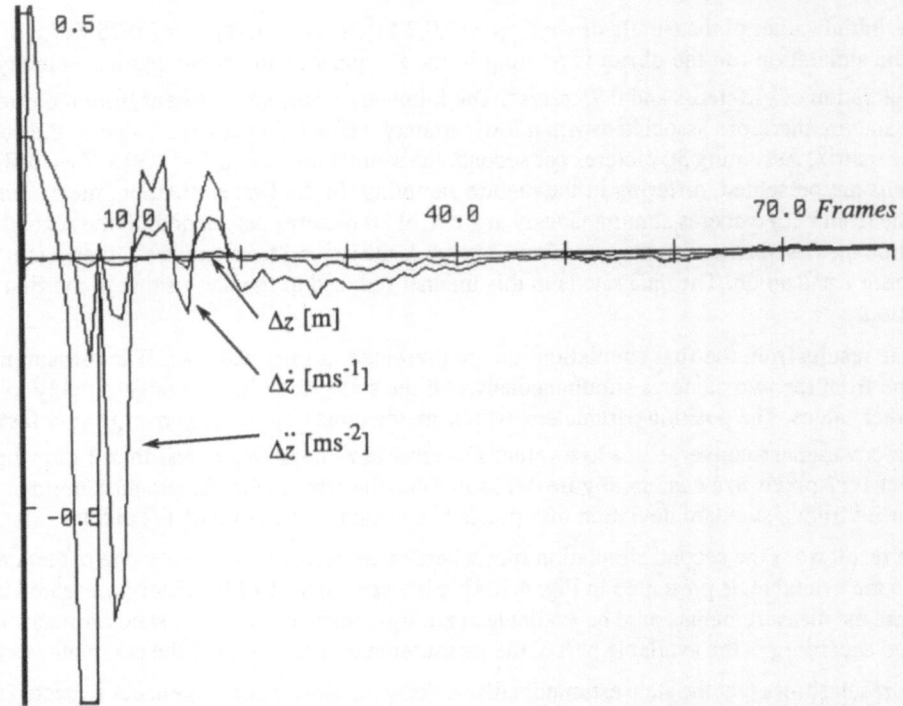

Figure 4.5 Error in predicted position, velocity and acceleration of the center of mass of the object (the z-component) in the first simulation run. All estimates converge. Position convergesthe fastest. 30 picture frames/sec.

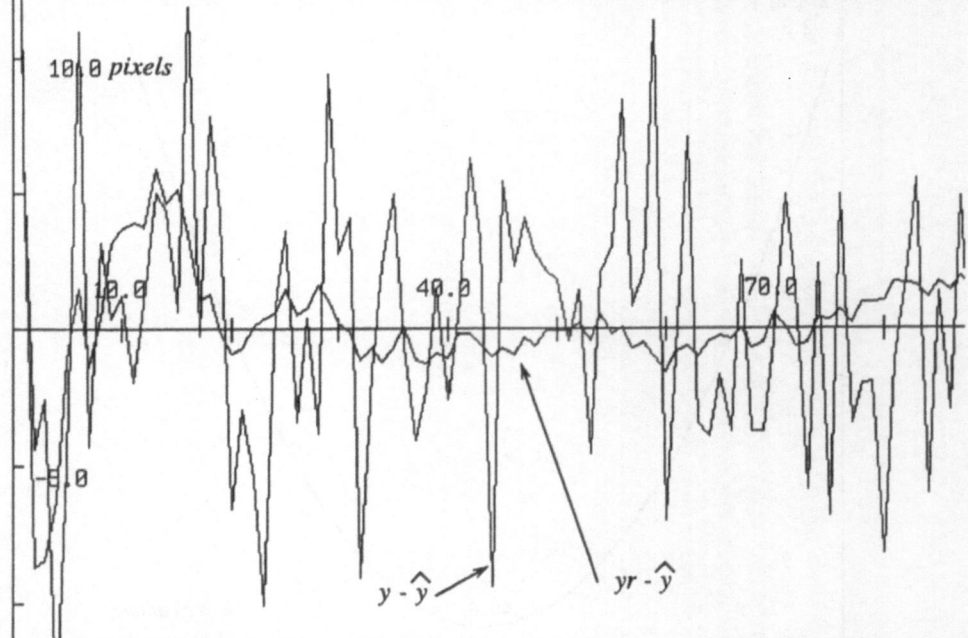

Figure 4.6 From simulation 1. Innovation process $\varepsilon = \hat{y} - y$ and $y_r - \hat{y}$, where $y = y_r + w$ (w: the noise term). Predicted measurement $\hat{y}$ is very close to the noise-free measurement y_r. It differs with only 1-2 pixels. Difference between the noisy measurements y and the predicted measurements $\hat{y}$ is almost uncorrelated.

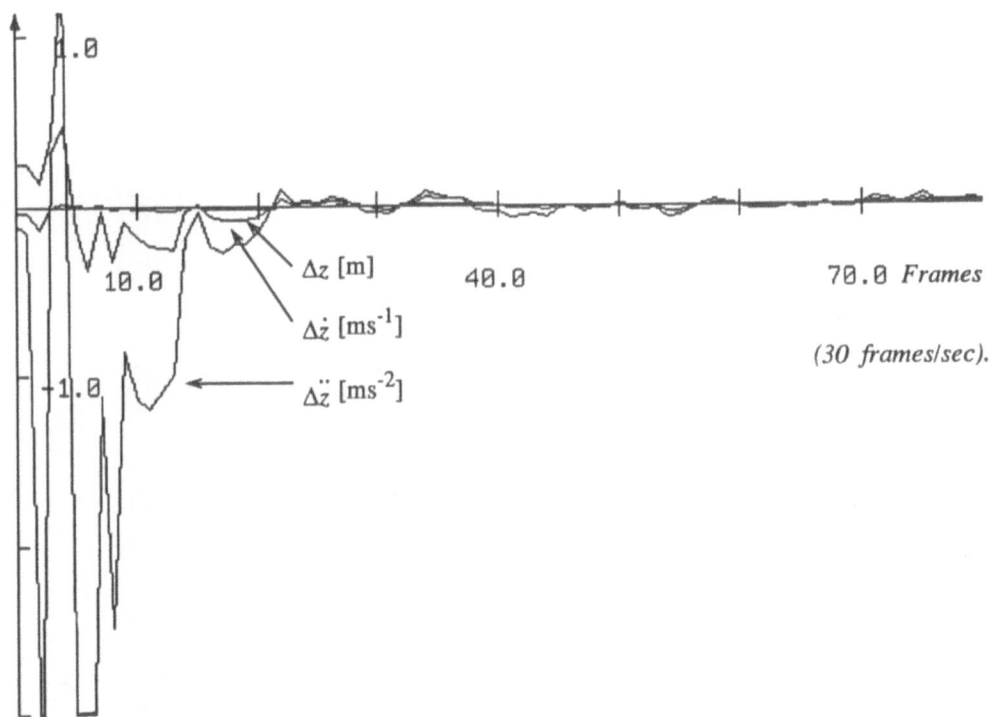

Figure 4.7 Error in predicted position, velocity and acceleration of the center of mass of the object (z-component) when measurements from only one of the two cameras are available at each instance (simulation 2). The figure corresponds to Fig.4.5. The state estimator needs a few more frames to converge, compared to simulation run 1, Fig.4.5.

5 Conclusion

It has been shown that applying the Kalman Filter algorithm is a powerful and general technique for integrating sensors. This has been demonstrated for the task of inverting primary camera data in order to establish estimates of coordinate values, as well as of the velocity and acceleration of a moving object in 3D space. The KF can, using simple means, integrate data from any number of possibly very different sensors. Also, sample rates and the time of appearance of the information need not be synchronized. Feedback to a control system is made available without direct reference to the sensors.

Parts of this work was sponsored by the Royal Norwegian Council for Scientific and Industrial Research, on contract TU 22301.

REFERENCES

Broida, T.J. , Chellappa, R. (1986): Estimation of Object Motion Parameters from Noisy Images. *IEEE Trans. on Pattern Analysis and Machine Intelligence*, PAMI-8(1):90-98

Dickmanns, E. D. , Zapp, A. (1985): Guiding Land Vehicles along Roadways by Computer Vision. *Proc. Congres Automatique 1985, The Tools for Tomorrow*, Toulouse, 233-244

Jazwinski, A. H. (1970): *Stochastic Processes and Filtering Theory*. Academic Press, New York

Multisensory Telerobotic Techniques

J. Dietrich, G. Hirzinger, J. Heindl, J. Schott

D L R

German Aerospace Research Establishment

Institute for Flight Systems Dynamics

Oberpfaffenhofen, D-8031 Wessling, FRG

Abstract

The paper outlines the telerobotic concepts as presently developed for a small multisensory robot to fly with the next spacelab mission D2; the robot is supposed to work in an autonomous mode, teleoperated by astronauts, and teleoperated from ground. Its key feature is a recently developed multisensory gripper with highly integrated, miniaturized sensor technology including stiff and compliant six-axis force-torque sensing, 9 laser range finders (one of them realized as a rotating laser scanner), tactile arrays, grasp force control and a stero camera pair. Perfect modularity in hard- and software with all preprocessing electronics realized in the gripper was one the major design goals. This multisensory information is a key issue when teleoperating the robot from ground. Sensory simulation on ground computers using advanced stereo graphics is supposed to predict the sensorbased path refinement as induced by the real sensors on board. A particularly interesting situation occurs in the experiment "grasping a floating object from ground" with overall delays of more than 4 seconds. Predictive simulation using realtime fusion of stereo images and laserscan information is the challenging technique envisioned here.

1 Introduction

Automation and robotics (A+R) will become one of the most attractive technologies in space, as it will allow experiment-handling, material processing, assembly and servicing with a very limited amount of highly expensive manned missions, and the expectation of an extensive technology transfer from space to earth seems to be much more justified than in many other areas.

This is one of the main reasons why several activities towards space robotics have started now in Germany with the long-term goal of a major contribution to the space station, e.g. to the subsystem "man tended free flyer" (MTFF). In particular we will fly a robot technology experiment ROTEX with the next "German" spacelab-mission D2 in 1991.

NATO ASI Series, Vol. F 63
Traditional and Non-Traditional Robotic Sensors
Edited by T. C. Henderson
© Springer-Verlag Berlin Heidelberg 1990

ROTEX is kind of a starting shot for a German participation in space automation and robotics.

In this paper - after an outline of the experiment and the robot's multisensory gripper - the telerobotic concepts as envisioned in ROTEX are explained. They include teleoperation from ground, where long delay times have to be compensated by predictive estimation schemes using advanced 3D-computer graphics.

The ROTEX experiment shows up several major features:

- A small, six-axis robot (working space $\approx$ 1m) flies inside a space-lab rack (fig. 1). Its gripper will be provided with a number of sensors, especially two 6-axis force-torque wrist sensors, grasping force control, an array of 9 laser-range finders and a pair of optical fibers to provide a stereo image out of the gripper; in addition a fixed pair of cameras will provide a stereo image of the robot's working area.

- The robot is able to perform automatic, preprogrammed motions as well as teleoperated motions via an astronaut or an operator on ground (fig.1), including experiment handling and servicing operations.

- The main goals of the experiment are:

  - To verfy joint control (including friction models) under zero gravity as well as μg-motion planning concepts based on the requirement that the robot's accelerations while moving must not disturb any μg-experiments nearby.
  - To demonstrate and verify the use of DLR's sensorbased 6 dof-handcontrollers ("control balls") under zero gravity.
  - To demonstrate the combination of a complex, multisensory robot system with powerful man-machine-interfaces (as are 3D-computergraphics, control-ball, stereo imaging), that allow teleoperation from ground, too,

In order to demonstrate servicing and repair capabilities by teleoperation three basic tasks are envisioned (for all operational modes):

a) assembling a mechanical grid structure

b) connecting/disconnecting an electrical plug

c) grasping a floating object.

2 A really multisensory gripper

There are a number of reasons why sensorcontrolled and thus more intelligent robots are still hardly to find in industrial applications despite of all the predictions that have been made already years ago.

Sensors in general are

- too expensive

- too voluminous for an elegant integration into smart grippers

- not reliable enough, with a "black box" and a considerable amount of wiring coming with each sensor, so that multisensory integration became more a pure laboratory concept than a practical item.

- difficult to integrate into robot control systems, from the data technical view point as well as from the algorithmic one.

In our laboratory these observations have led to the design of a multisensory gripper which (as critical end-mass of a robot) had to be of light-weight and in which we tried to avoid all the above-mentioned problems; i.e. the design goals were:

- minimal size and weight for all sensors, with overload protection - as necessary in case of force sensors - integrated, i.e. not introducing extra weight.

- fabrication costs considerably lower than was usual in the past

- all analog electronics and digital preprocessing integrated in the sensors or at least in the wrist

- sensors exchangeable and mountable mechanically in a simple way; but more important they should be easily integrable electrically, that is from the hardware point of view, as well as from the software point of view in a completely modular way.

Before describing the different systems in a more detailed way, let us briefly focus on the above topics of modular integration. Fig. 2 shows that each sensor as well as the gripper drive contains analog electronics, digital electronics and power supply on tiny SMD–boards inside the sensor or as an additional circular slice in the wrist. The one important item is power supply, as in general the sensors need different voltages each and they have to be galvanically decoupled - main reasons for the excessive wirings in the past. In our design there is only one 20 kHz – 50 V power supply providing a rectangularly alternating voltage source; each of the sensors has its own tiny transformer via which it derives the individually needed voltages. Thus galvanic decoupling is assured automatically. Concerning data transmission in the past most sensors were provided with slow serial RS232 interfaces and the robot control system hat to care for the data protocols with each sensor. In our new concept there is now a fast serial bus (RS485) with 375 kBaud signal rate, to which each sensor is connected; a bus master requests data from the active sensors (thus we have a master slave system) and collects the data into a dualport-RAM, from where the higher–level robot control system can simply read them without caring for protocols or interrupt handling. This concept meanwhile is going to influence German robot industry towards creating much more efficient sensor interfaces than were available in the past.

The modularity concept is pertained in the software of all sensors (fig. 3). There is a measurement module that differs with the physical principles used in the different sensors but uniquely supplies digitized voltages into a switch buffer from which the preprocessing module (e.g. using calibration laws) issues physically relevant values. From a second switch buffer the most recent values are fetched by a bus master request.

Fig. 4 shows a block diagram, fig. 5 the real arrangement of the different components integrated into our prototype gripper. The gripper's sensory components are (see fig. 4):

a) an array of 9 very small laser range finders based on triangulation, one of them being somewhat bigger (half the size of a match box, switchable into a rotational scanning mode) for the medium range of $\approx$ 5–30 cm (fig. 6), and 4 tiny ones in each finger for lower ranges of 0–3 cm. The range finders are the result of many years development aiming at a precise performance over a remarkable range, independent of the slant angle and surface of the measured object. One of the main problems to be solved in this context was to design a nonlinear digital control system that adapts the light transmitter's intensity depending on the reflected light intensity in a range of 1 to 4000 within 10 μsec, for each measuring pulse independently. This indeed enables the sensors to measure distances with respect to surfaces that show up strongly and quickly changing reflection characteristics.

The light source for all range finders is a laser diode with integrated power monitor diode. The divergent laser light is focused by a lens to form a parallel light beam, the diameter of which is $\approx$ 1 mm with divergence in the range of 2 mrad. The laser's output power is 3 mW at 790 mm wavelength.

All receiver systems are built up with linear position sensitive detectors (PSD). The received environment light is suppressed by different techniques, e.g. by optical daylight cut-off filters, filtering amplifiers adapted to the laser pulse characteristics, and sample-hold techniques that registrate and subtract the environment signal from the reflected light pulse measurements (see fig. 7 und 8).

Resolution for the medium range/scanner sensor as well as for the finger sensors varies from 0,1 % (near) to 3 % (far) of measurement value.

The medium distance range finder weighs only 0,15 N. So in order to operate it optionally as a scanner we decided to rotate the whole system using a tiny stepper motor instead of implying some oscillating mirror that would yield smaller scan angles and might induce vibrations onto the instrumented compliance. The problem of wiring power and data to such a rotating sensor head is solved by using the scanner's ball-bearings as conductive transducers.

b) a tactile array of 4 x 8 sensing elements (conductive rubber "taxels") in each finger. The dimensions of the tactile area is 32 x 16 mm. The binary state of each taxel is serially transmitted through the analog multiplexers without additional wiring.

c) a "stiff" 6 axis force–torque sensor based on strain–gauge measurements and a compliant optical one. Originally it seemed necessary to make a final decision between these two principles, but as indicated in fig. 3 and fig. 4 they finally were combined into a ring–shaped system around the gripper drive, the instrumented compliance being lockable and unlockable electrically. Shaping these sensors as rings around the gripper drive shows up different advantages:

- it does not prolong the axial wrist length
- it brings the measurement system nearer to the origin of forces–torques and yields a better ratio of torque range to force range than achievable with a compact form.

The "compliant" optical force–torque sensor consists of an inner and an outer part (fig.9). The basic measuring arrangement in the inner ring is composed of a LED, a slit and perpendicular to it a linear position sensitive detector (PSD). The slit/LED combination is mobile against the remaining system. Six of such systems (rotated by 60 degrees each) are mounted in a plane, whereby the slits alternatingly are vertical and parallel to the plane. The ring with PSD's is fixed inside the outer part and connected via springs with the LED–slit–basis. The springs bring the inner part back to neutral position when no forces/torques are exerted. There exists a particularly simple and unique transformation from PSD–signals $U_1...U_6$ to the unknown displacements.

The stiff, strain–gauged force–torque sensor is a completely new design and will be described in another paper. It performs automatic temperature compensation based on the temperature characteristic as stored during the calibration process and continues operating reliably with reduced accuracy if one of the strain–gauges is damaged.

d) A pair of tiny stereo CCD-cameras, the CCD's plus optics plus minimum electronics in a first realization taking a volume smaller than a match–box, too. Of course the camera's wiring is independent of the rest of sensors and signal processing is not done in the gripper.

e) An electrical gripper drive, the motor of which is treated like a sensor with respect to the data bus and the 20 kHz power supply connections. As the whole gripper is part of a light–wheigt robot concept [9], the electrical gripper drive developed in our lab was aiming at a performance improvement by a factor at least 5 compared to available designs.

The drive motor concept is based on our own redesign of commercially available two–phase stepping motors of the ESCAP type (see fig. 5). These motors were basically attractive for our light–weight goals because they show up

- very low rotor inertia (flat disc)

- small size, low weight (1,5 N) but relatively high torque (0,33 Nm), the rotor permanent magnets being integrated into a rare earth disc with **large** diameter

- high positioning accuracy (25 pole pairs).

What we wanted to arrive at however was not really a stepping motor, but an electronically commutated dc motor with the option of switching into a micro–step mode for fine–positioning. Due to the motor's mechanical construction it was fairly easy to integrate hall sensors that allow for the electronic commutation and at the same time serve as position sensors; thus extra resolvers were not needed. Some efforts have been made to raise the torque–speed characteristics at higher speeds (e.g. 6000 – 16000 r.p.m) by a speed–dependent phase–lead circuitry.

In the gripper the problem is to transform the motor's high–speed rotational motion into a fairly slow axial motion to move the fingers (fig. 5). For this type of transmission a new mechanical spindle concept has been designed as shown in fig. 10 [5]. The spindle fixed to the motor axis wears a very fine thread with extremely small pitch (e.g. 0,2 mm/revolution). Six planetary rollers arranged concentrically between spindle and the finger–driving nut show up fine grooves (no pitch) that fit precisely into the grooves of the spindle thread. However they also show up much coarser grooves that mate with corresponding coarse grooves inside the nut, so the rollers do not move axially relative to the nut nor do they change their mutual distance. Why not? The real trick has been to provide the rollers' fine grooves with a "phase shift" from one roller to the next (fig. 10), so that without using a multiple thread (**not** realizeable with this pitch anyway) they simulate a nut with the desired low pitch. And, as the other crucial factor, by using rollers of this type sliding friction as prevailing in a pure spindle–nut concept has been replaced by much smaller rolling friction (1 % as the presently envisioned goal).

What we have gained with this motor–gearing combination is a small prismatic drive (applicable also in a robot joint), which used as gripper drive allows to exert grasp forces of more than 400 N with a gripper weight of 5 N and a grasp speed of about 15 cm/sec; without measuring and feeding back grasp forces we arrived at a feedforward grasp force control resolution of $\approx$ 4 N (1 % of max force) with high repeatability. Reduction rate referring to the finger rotation is $\approx$ 1 : 1000.

With the enormous amount of tiny SMD–electronics and the excessive number of mechanical parts involved (more than 200) we are fairly sure that the gripper is the most complicated multisensory robot end-effector that was built so far.

3 A telerobotic concept with sensory feedback and predictive control

The basics of our telerobotic concept are briefly explained by hand of fig. 11. "Rudimentary" commands are derived either on–line from a human teacher operating the DLR control ball (a non–force–reflecting 6dof–ball–shaped–hand–controller, see fig. 12 using the compliant force–torque–sensor as described above inside, see [10, 11, 12]), or from a path–generator connecting preprogrammed points. They are interpreted in a dual way as force/torque or positional/orientational commands. Loosely speaking if the robot moves in free space, the ball forces are transformed into translational commands, however if the robot senses contact with the environment, it takes the ball inputs as nominal force values and by closing the sensory loop at the robot's site (see fig. 11) it always exerts only those forces which are given by the human operator [11]. Thus although we are not feeding the forces back to the human arm (as does "bilateral" force control with the well–known stability problems in case of delays), the operator is sure that the robot is fully under his control and he easily may lock up doors, assemble parts or plug in connectors. In other words, the human operator (via stereovision and 3D-graphics i.e. the visual system) is enclosed in the feedback loop on a very high level but low band-width, while the low level sensory loops are closed on–board at the robot directly with high bandwith. Thus we try to prepare a supervisory control technique that will shift more and more autonomy to the robot while always offering real–time human interference. In case of large transmission delays stereovision is just replaced by predictive graphics leaving the basic structures untouched.

Looking a little bit more into details (e.g. [7]) shows that the above mentioned rudimentary teach commands Δp_T (i.e. the ball commands here) are in case of sensory contact projected into a position controlled subspace to yield the position feedforward component $\Delta p_{T,cp}$, and into the orthogonal sensorcontrolled subspace ($\underline{f}_{T,cf}$); there they are either neglected in case nominal sensor values have to be kept constant autonomously, or they are counter-balanced with the wrist sensor data as mentioned above yielding a robot that is fully under human control (fig. 13). We treat the range finders as pseudo–force–sensors, too, thus arriving at a unified treatment of completely different sensors. The position–controlled subspace may further be subdivided into a subspace in which motion is generated automatically, and a subsace executing the operator's best projection. Thus any kind of distributing force and position control between robot and human operator or gross path planner is realizable. For the generation of subspaces we use kind of a simple knowledge–base, which depending on the recent task together with the actual sensor data generates these subspaces automatically (fig. 13).

As mentioned above the loops closed around the human operator are of low bandwith with the control ball as input control device and visual displays as feedback devices. When teleoperating on board of spacelab, visual display is fairly restricted using a small colour TV-monitor, where in the present state of planning the (black–white) stereo images produced by the gripper or the global cameras are displayed alternatingly, the operator using shutter glasses as developed in German nuclear power facilities with only 15 V power supply and switching frequencies up to 1 kHz. The sensory information will be added in simple bar–like form at the monitor's edges.

For teleoperation from ground the situation is different: much more powerful equipment is avalilable there but the communication link restrictions are obvious. Indeed it turned out that the "normal" spacelab uplinks as used until now are not at all adequate for tele-presence ideas. Varying up-link delays of up to 15 seconds partly caused by data checks in Houston seemed to destroy the ground teleoperation concept completely. But with the present base–line using the Text and Graphics Channel (TAG) for the up–link these difficulties seem to be solved, although this channel using the TDRS satellite could not be tested until now. Uplink command rates are in the range of 2 kbit/sec, assuming a sampling rate of 20 Hz. Nevertheless we have to take into account an overall delay of 4 sec in the loop closed via the ground station. In order to get exact knowledge about this delay we will provide the ball commands with a code, which when arriving at the robot, is packed into the downlink information.

This down–link information comprises a sequential RGB–video signal; the left and right black–white stereo images are packed into the red and green channel, they are super-imposed and displayed on a polarized screen on ground. The down–link data channel contains all the internal (position encoders) and external sensory signals so that on a 3D–graphics monitor the robot's position is displayed as well as all sensor data. Prefer-ably a stereo graphics system is used with real–time volume–shaded representation of the work-cell. But of course the big problem for teleoperation from ground are the delays. The only way to compensate for them is predictive computer graphics. We want to make extensive use of it in ROTEX. Fig. 14 is to outline that the human operator at the remote workstation handles the control ball by looking at a "predicted" graphics (e.g. wire frame) model of the robot. The control commands issued to this instantaneously reacting robot simulator are sent to the remote robot as well, using the time-delayed transmission links. Now the ground–station computer and simulation system contains a model of the uplink and downlink delay lines as well as a model of the actual states of the real robot and its environment (especially moving objects). Note that we have several alternatives to superimpose the predicted robot model (augmented by predictions of any other moving parts) with other information representations:

- the presently received (of course delayed) TV–Stereo or mono image in case that the globally fixed camera pair is active

- the "delayed" graphics image derived from this delayed TV-image (including the case of hand–mounted cameras) and other sensory data

- the actual graphics image as derived from the state space model of robot and envi-ronment.

We do not have a final conclusion on what the most efficient way of superposition would be, but there is of course evidence that the most crucial problems lie in the derivation of "output data" (e.g. positions/orientations of moving objects) from say stereo images and range finders. As real time is required, this is an extremely challenging preprocessing problem not discussed in more detail here.

For the robot we assume a linearized cartesian state space model $\underline{x}_{k+1} = A\underline{x}_k + \underline{b}u_k$.

Thus the left part of fig. 14 and 15 e.g. contains just a prediction of the robot's present estimated state $\hat{\underline{x}}_k$ robot to the future state $\hat{\underline{x}}_{k+n_u}$; n $_u$ is the uplink–delay–time expressed as a multiple of the sampling period, that makes up one delay d. This predicted state is just the state to which the presently issued ball command has to refer. But the more interesting part is the estimator on the right half of fig. 14 and 15. It compares the measured, but downlink–delayed output data y_{k-n_d} (e.g. the robot's and the environment's positions and orientations) to the output data $\hat{\underline{y}}_{k-n_d}$ from the robot and environment model running through the downlink–delay computer model (n_d is the number of sampling periods in the down link-delay).

Now the important discrimination is between a static and a dynamic environment (fixed workcell and objects on one side and free floating objects on the other side). In both cases knowledge about the environment has to be used in the predictive simulation. In the static case this knowledge is contained in a world model that allows sensory simulation (including feedback simulation), in the dynamic case it is contained in a dynamic model used by a Kalman filter.

a) **the static case**

Sensory simulation and display on the stereographics system is strongly supported by a collision detection scheme working with a real time data base (fig. 16). A special task observes the simulated movement of the robot through its workspace and calculates in several steps if any collison might occur. For this all objects in the workspace as well as the robot links and gripper parts are additionally modelled in the global geometrical data base with simple polytopes such as cubes, balls, cylinders or ellipsoids. Another way of checking for collision is to divide the workcell into a three dimensional grid structure and either assigning the names of the objects – lying inside the grid cubes – to the corresponding elements of the data structure or assigning the numbers of each grid cube – touched by an object – as special parameters to the corresponding list in the database.

In the first step the collision detection task checks if the simple polytopes of the robot arm are in collision course with any polytopes of objects in the workcell. In this case the control system has to decide if the motion of the robot should be stopped or not. The movement should be continued if there are active sensors observing the environment in the direction of the motion. Therefore the collision detection task maps the geometrical and physical data of the colliding objects into the local real time data base with which the corresponding sensor simulator (e.g. force–torque, range finder) is working. That means, in the second step the sensor simulator using the known motion direction vector checks which normal vectors of the approaching object patches indicate potential collision.

In the somewhat simpler case of non-tactile simulated rangefinders only the direction vector of the laserbeam is checked against the normal vectors of the approaching object planes.

The more difficult problem occurs when simulating a force/torque–sensor or an instrumented compliance. Here the normal vectors of the endeffector parts and patches must be checked whether they indicate that the corresponding patch is in collision course with the approaching object part. To get exact simulated sensor data the skin of the possibly colliding objects should be modelled with cubic spline functions.

In the third step of the sensor simulation based on collision detection the selected surface parts of the colliding objects are checked against intersection. In case of rangefinders the simulated sensor beam is hitting the close object plane and the difference between the intersection point and the origin of the sensor beam delivers the corresponding sensor data. But in case of the force/torque sensor the stiffness of the robot and the colliding parts of its environment must be considered, too. Because of the slow motion of the space robot and the limited computer capacity the dynamic model is out of interest for sensor simulation.

Sure, there are errors expected in the sensor simulation and therefore not only the gross commands of the TM command device (control ball) are running into the simulation system , but also the real sensor data coming back from space (including the real robot position) are used to improve the robot's and its sensor's models. Fig. 17 shows the envisioned structure for telemanipulation from ground with simulated sensory path refinement. It emphasizes that sensory simulation for us indeed has a threefold application:

- collision detection and avoidance
- presimulation of sensorbased path refinement
- update of the (static) world model

b) **the dynamic case**
It becomes relevant when grasping a free-floating object via the ground telerobotic station. A successively linearized dynamical model of the free-flyer using differential angles or quaternions and an extended Kalman-Filter as an observer is the adequate technique for estimating and predicting the relative position of free-flyer and robot. We have chosen a symmetric object for getting more grasp capabilities, at the same time the motions become simpler. The crucial item here is deriving 6dof-measurements of the free–flyer's position/orientation in **realtime** (e.g. at least with 5 Hz sampling rate) using combined stereovison range images and laserscans. The high computational power needed is supplied by a massively parallel transputer system; conceptually the aspect of sensor fusion using sparse range images together

with precise but only two–dimensional laserscans ist of main focus now. Concerning the observer equations it has been shown in [13] how in case of large–delay–time–systems it is possible to design first the observer or Kalman-filter for the origitnal dynamical system and afterwards introduce - in a very straight forward manner - a structure taking account of the delays. Especially for systems where the delay is say 20 or 30 times the sampling period, this helps avoiding the design of high-order, artificially "blown up" systems.

4   Conclusion

ROTEX as a first step of Germany's engagement in space robotics aims at the demonstration of a fairly complex system with a multisensory robot on board and human telerobotic interference that makes use of sensor–based 6 dof handcontrollers, new concepts for predictive 3D-computergraphics and stereo display. For us teleoperation from ground is a very challenging technique that forces us to move even stronger into on-board autonomy. The control structures are thus that the human operator may step more and more towards supervisory control without changing the control loop structures. However we feel that for a number of years remotely operating robots will show up limited intelligence only, so that human "anytime" interference will remain important for a long time. Advanced real-time stereo graphics seems to become one of the most powerful man-machine interfaces in the attempts to overcome large transmission delay times, using appropriate estimation schemes in connection with modelling and real-time update of robot environment and sensory perception.

References

[1] Khatib, O., *A unified approach for motion and force control of robot manipulators: the operational space formulation*. IEEE J. Robotics and Automation. RA–3; pp. 43–53.

[2] An, Ch.H., Atkeson, Ch.G., Hollerbach, J.M., *Model–based control of a robot manipulator*, MIT Press 1988.

[3] Asada, H. and Lim, S.K., *The joint torque feedback control of a direct drive arm* in Proc. IFAC World Congress, Munich, 1987.

[4] Hirzinger G., Heindl, H., Landzettel, K., *Predictive and Knowledge–based Telerobotic Control Concepts*, IEEE Conf. on Robotics and Automation, Scottsdale, 1989.

[5] Dietrich, J., Gombert, B., *Vorrichtung zur Umwandlung einer Drehbewegung in eine Axialbewegung*, European Patent No. 0320621 pending.

[6] Dietrich, J., Gombert, B., *Getriebeanordnung in Form eines Umlaufraedergetriebes*, Patent No. 39011070 pending.

[7] Hirzinger, G., Landzettel, K., *Sensory feedback structures for robots with supervised learning*, IEEE Int. Conf. on Robotics and Automation, St. Louis, Missouri, March 1985.

[8] Hirzinger, G., Dietrich, J., *Multisensory robots and sensorbased path generation*, IEEE Int. Conference on Robotics and Automation, San Francisco, April 7-10, 1986.

[9] Dietrich, J., Hirzinger, G., Gombert, B., Schott, J., *On a unified concept for a new generation od light–weight robots*, First Int. Sympl. on Experimental Robots, Montreal Canada, June 19-21, 1989.

[10] Dietrich, J., Plank G., *Optoelectronic system housed in a plastic sphere* US–Patent: No. 4,785,180, Nov. 15, 1988.

[11] Hirzinger, G. Heindl, J., *Sensor programming, a new way for teaching a robot paths and forces torques simultaneously* 3rd Int. Conf. on Robot Vision and Sensory Controls, Cambridge, Massachusetts/USA, Nov. 7-10, 1983.

[12] Hirzinger, G., *The telerobotic concepts of ROTEX – Germany's first step into space robotics* 39th IAF Congress, Bangalore, India, Oct. 8-15, 1988.

[13] Hirzinger, G., Heindl, J., Landzettel, K., *Predictive and knowledge–based telerobotic control concepts* 1989 IEEE Int. Conf. on Robotics and Automation, May 14–19, 1989, Scottsdale, Arizona, USA

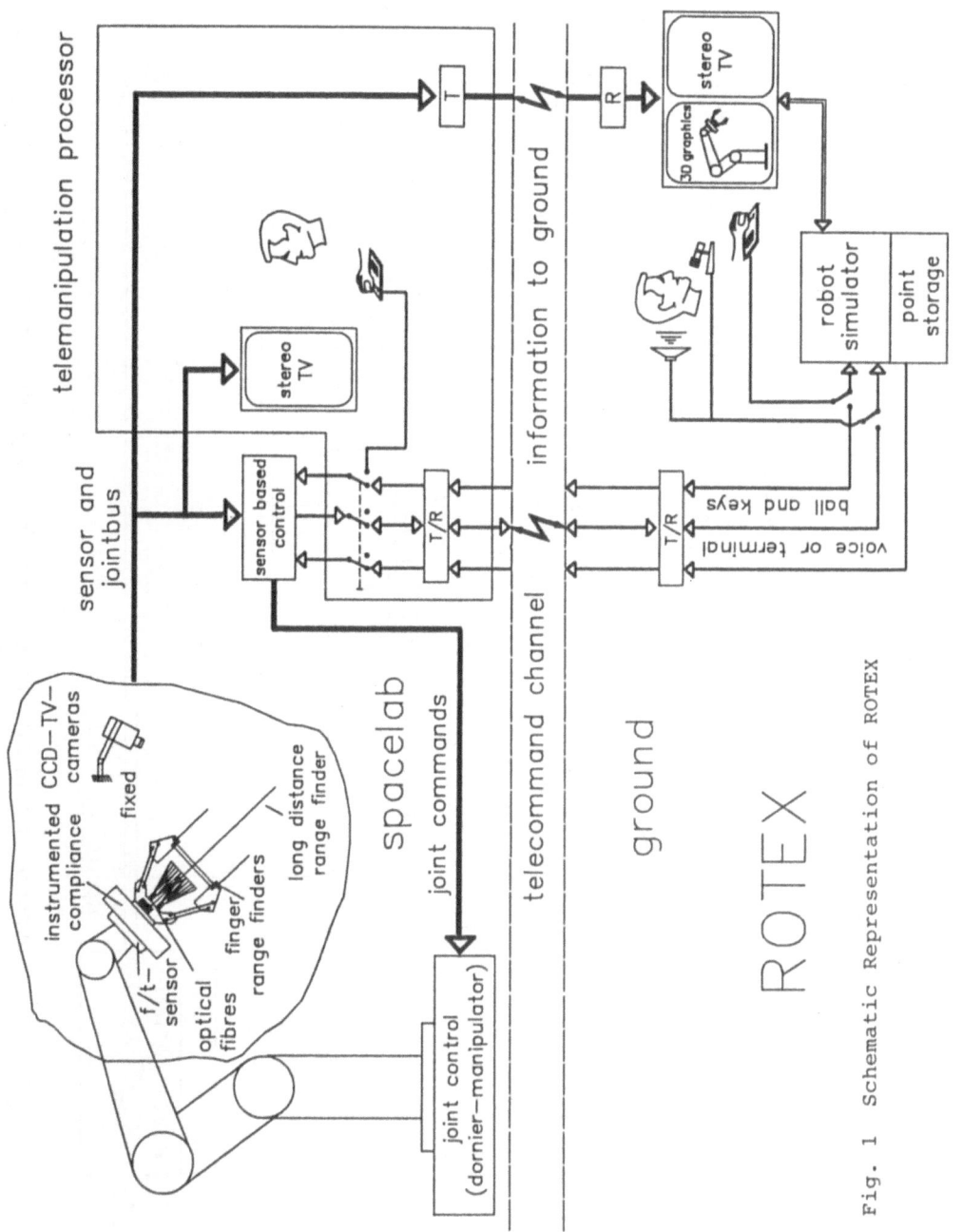

Fig. 1 Schematic Representation of ROTEX

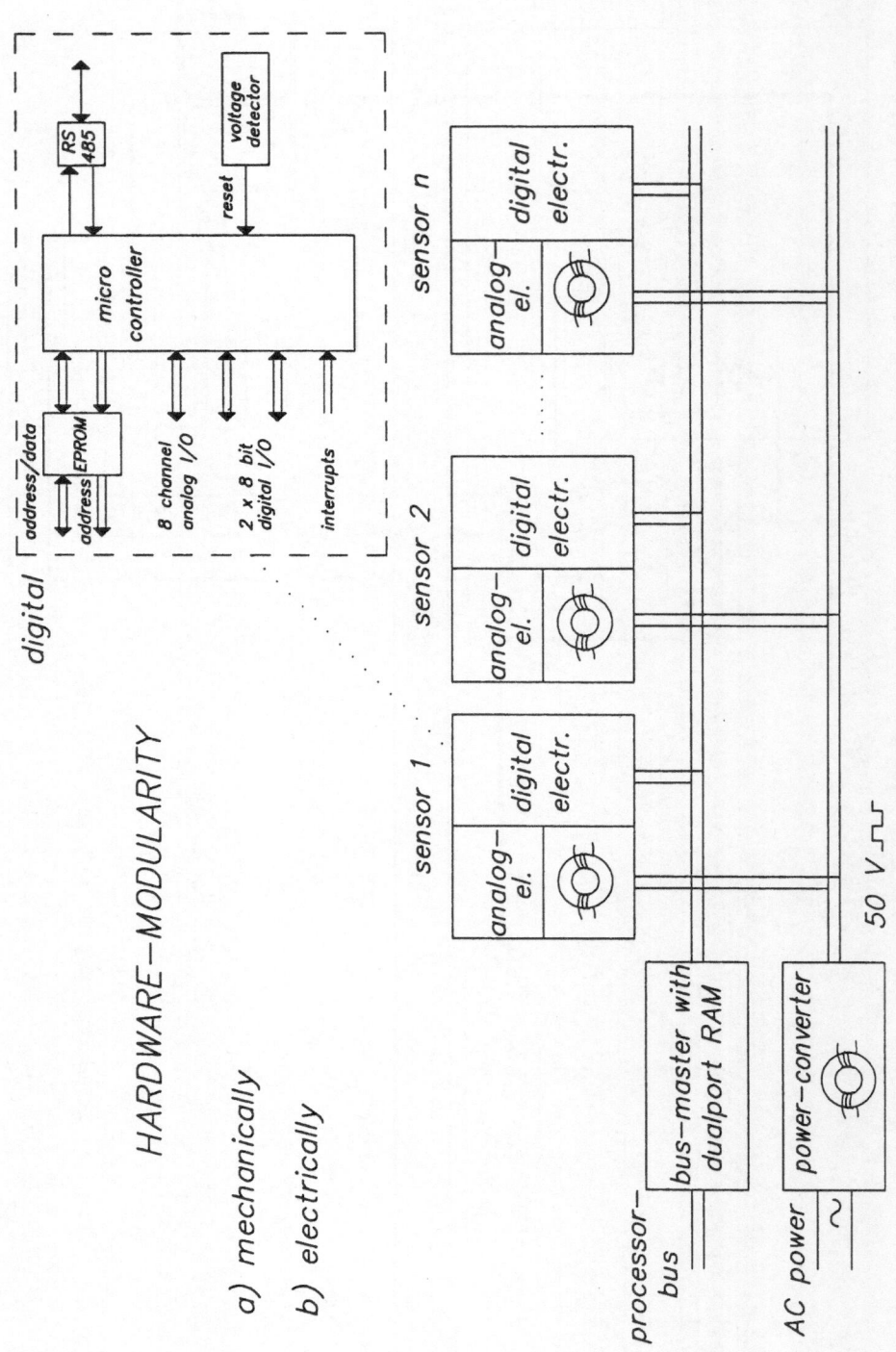

Fig. 2   Hardware modularity in our multisensory gripper

Generation of Data

Generation of Status Data

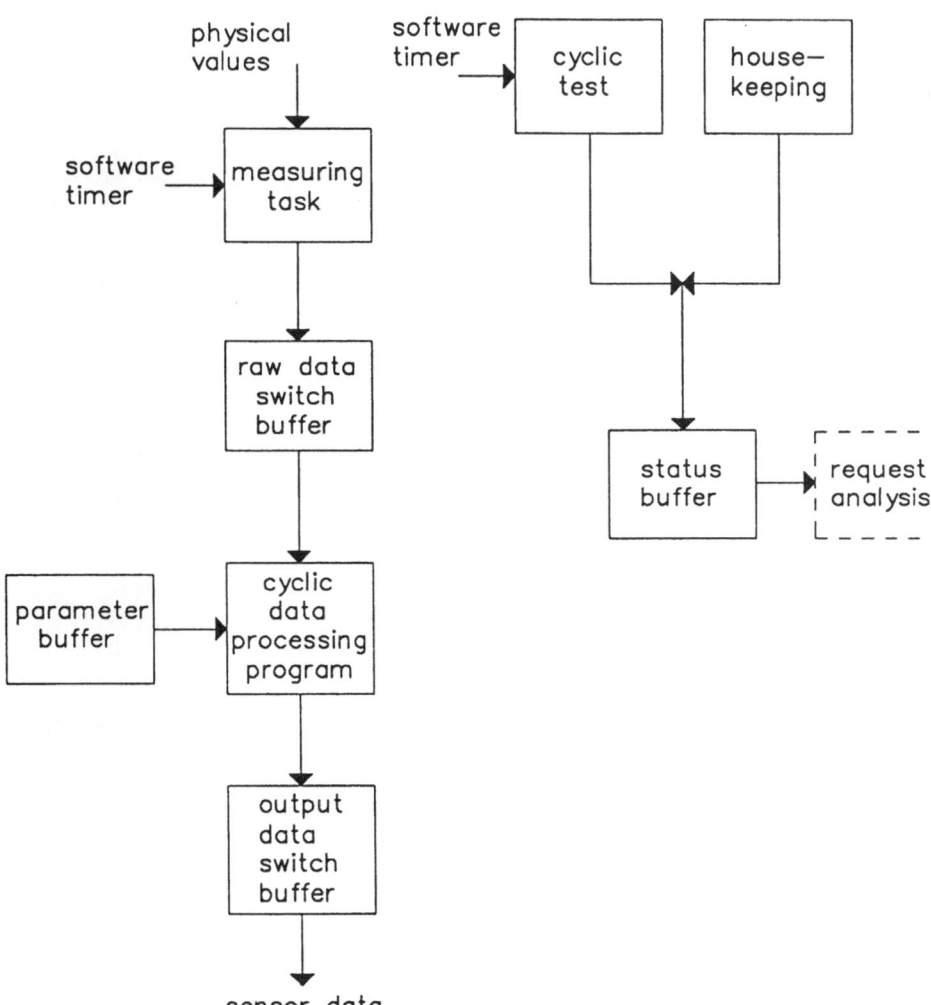

Fig. 3 *Software modularity in the sensor processors*

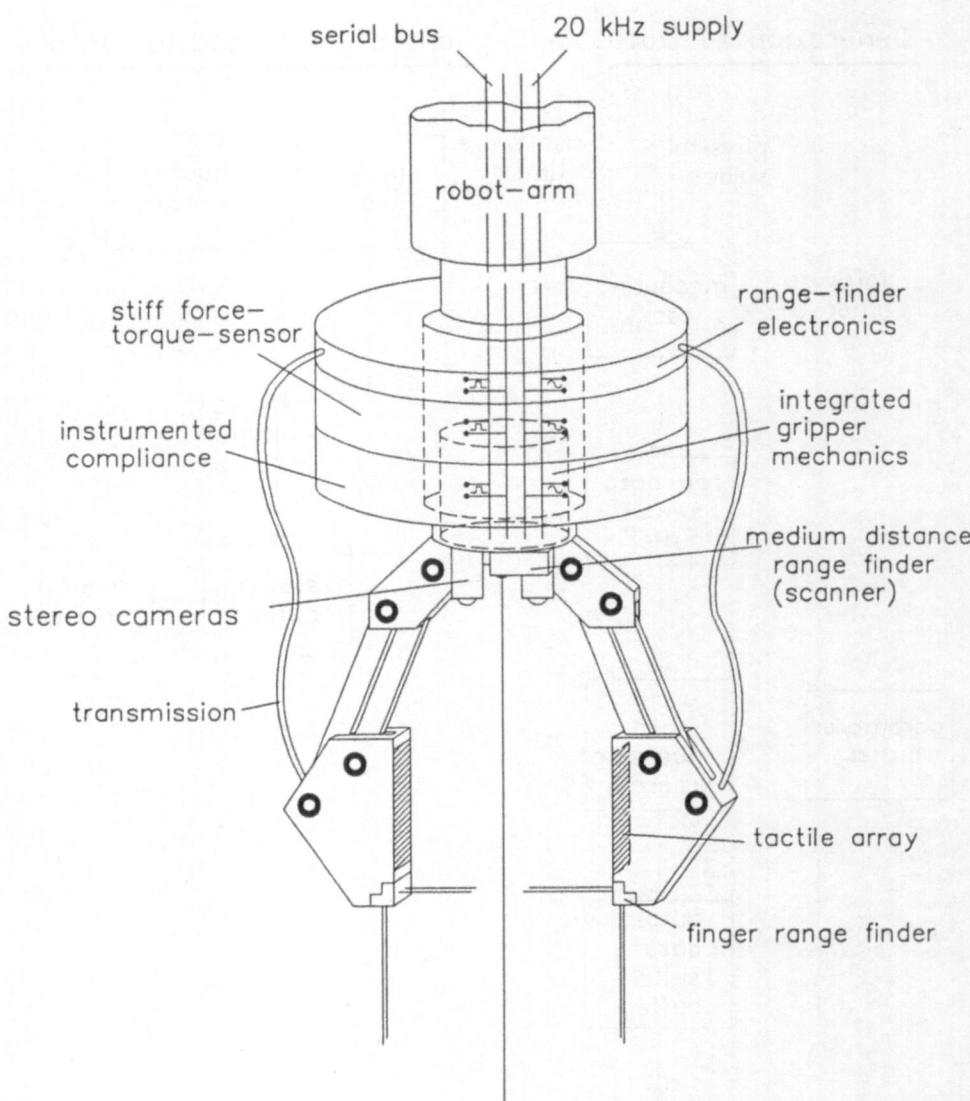

serial bus   20 kHz supply

robot—arm

stiff force—
torque—sensor

range—finder
electronics

instrumented
compliance

integrated
gripper
mechanics

medium distance
range finder
(scanner)

stereo cameras

transmission

tactile array

finger range finder

Fig. 4 *Schematic arrangement of sensors*
in our prototype gripper

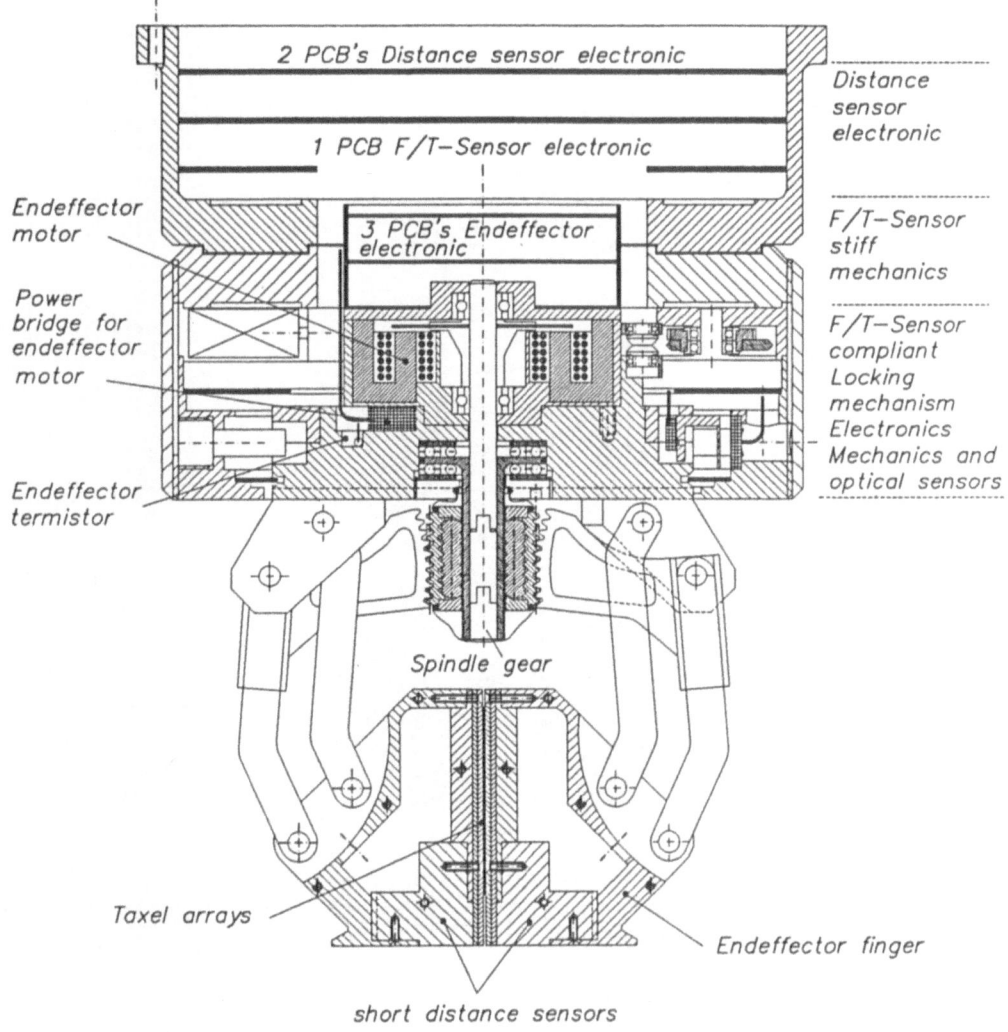

2 PCB's Distance sensor electronic

1 PCB F/T−Sensor electronic

3 PCB's Endeffector electronic

Distance
sensor
electronic

F/T−Sensor
stiff
mechanics

F/T−Sensor
compliant
Locking
mechanism
Electronics
Mechanics and
optical sensors

Endeffector
motor

Power
bridge for
endeffector
motor

Endeffector
termistor

Spindle gear

Taxel arrays

Endeffector finger

short distance sensors

Fig. 5 *Gripper (endeffector) section drawing*

Fig. 6   Medium range finder (5 - 30 cm), based on laser
triangulation and operated as a rotational scanner

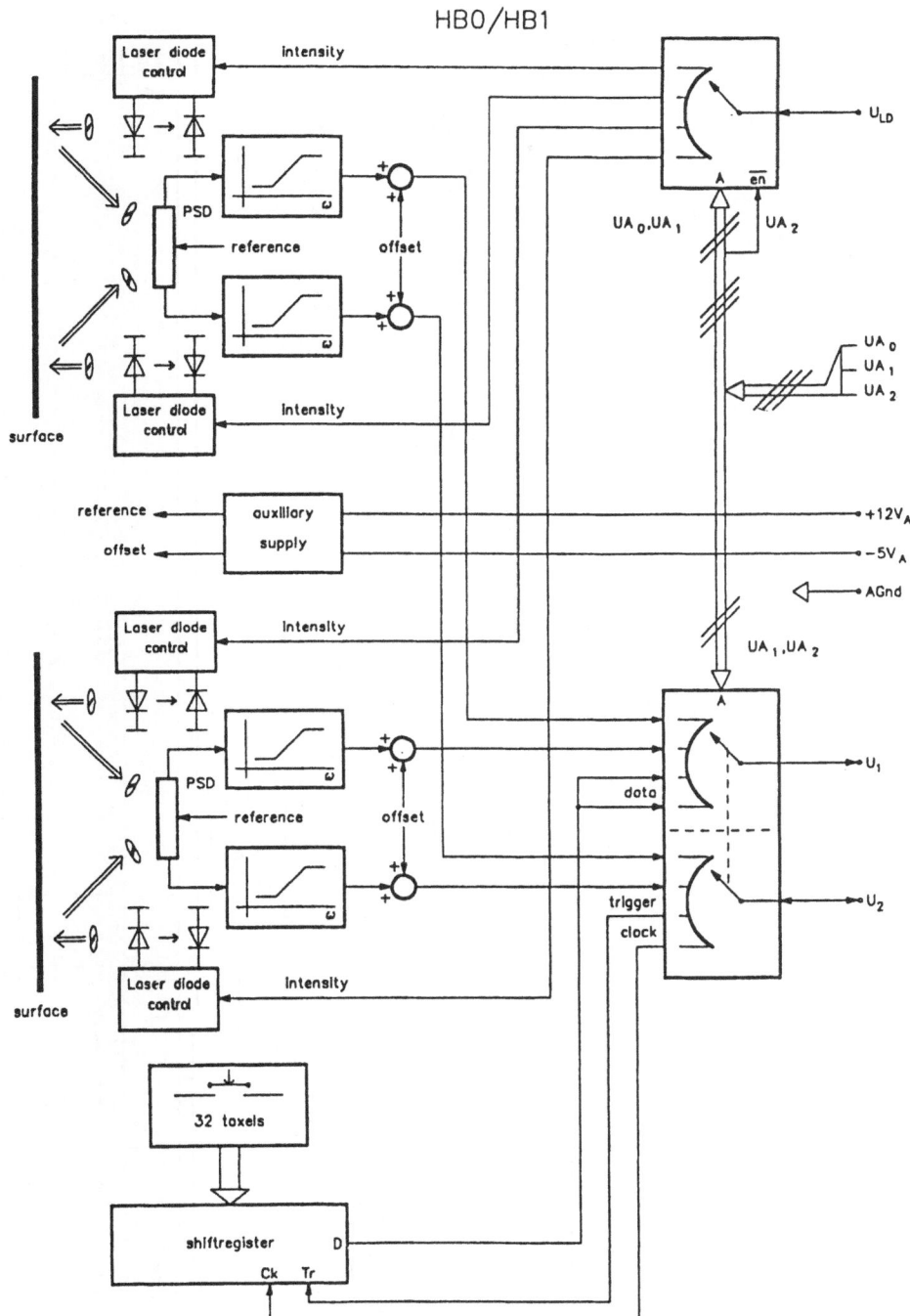

Fig. 7   Short distance sensors / taxel array

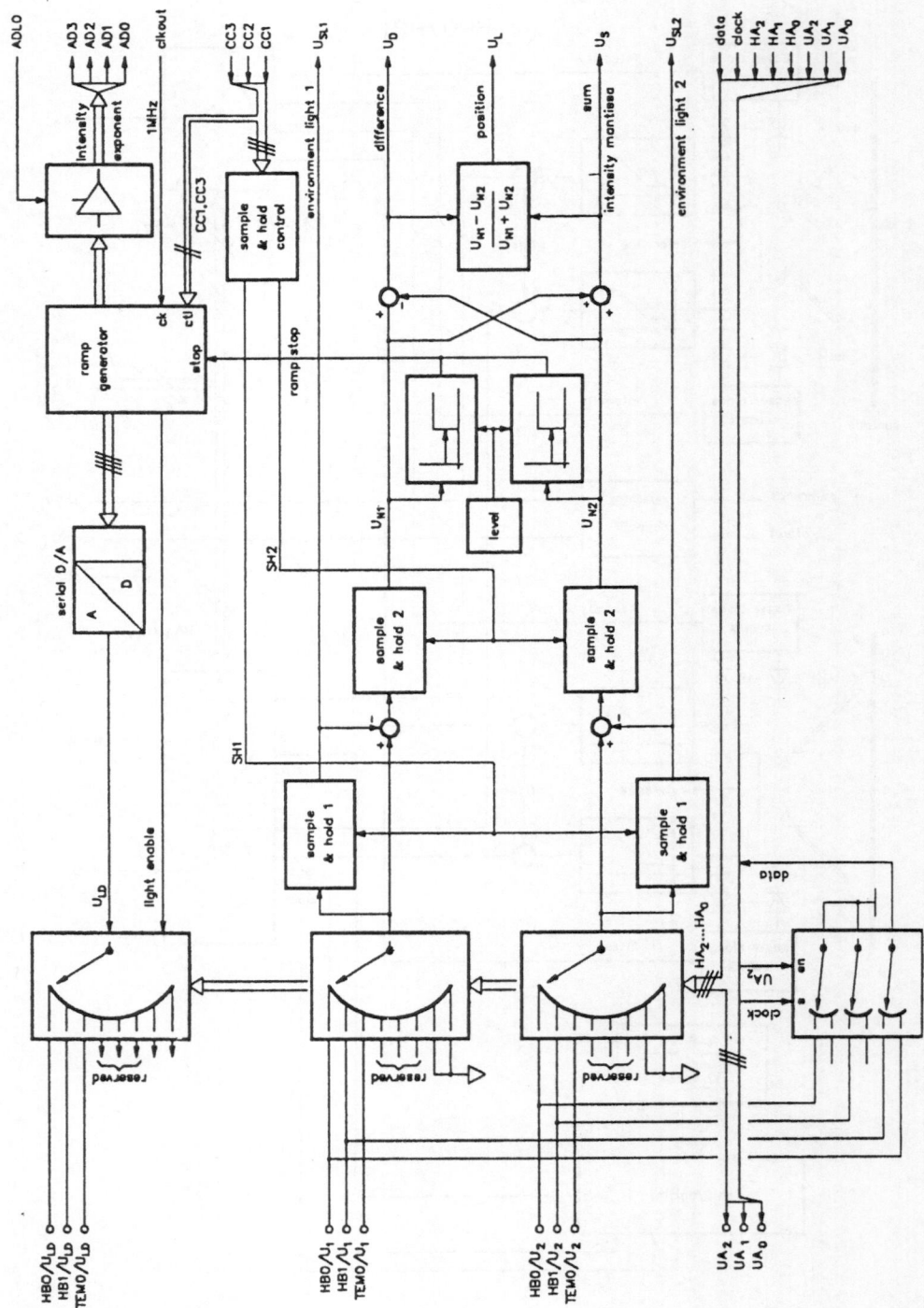

Fig. 8 Distance sensor assembly analog signal conditioning

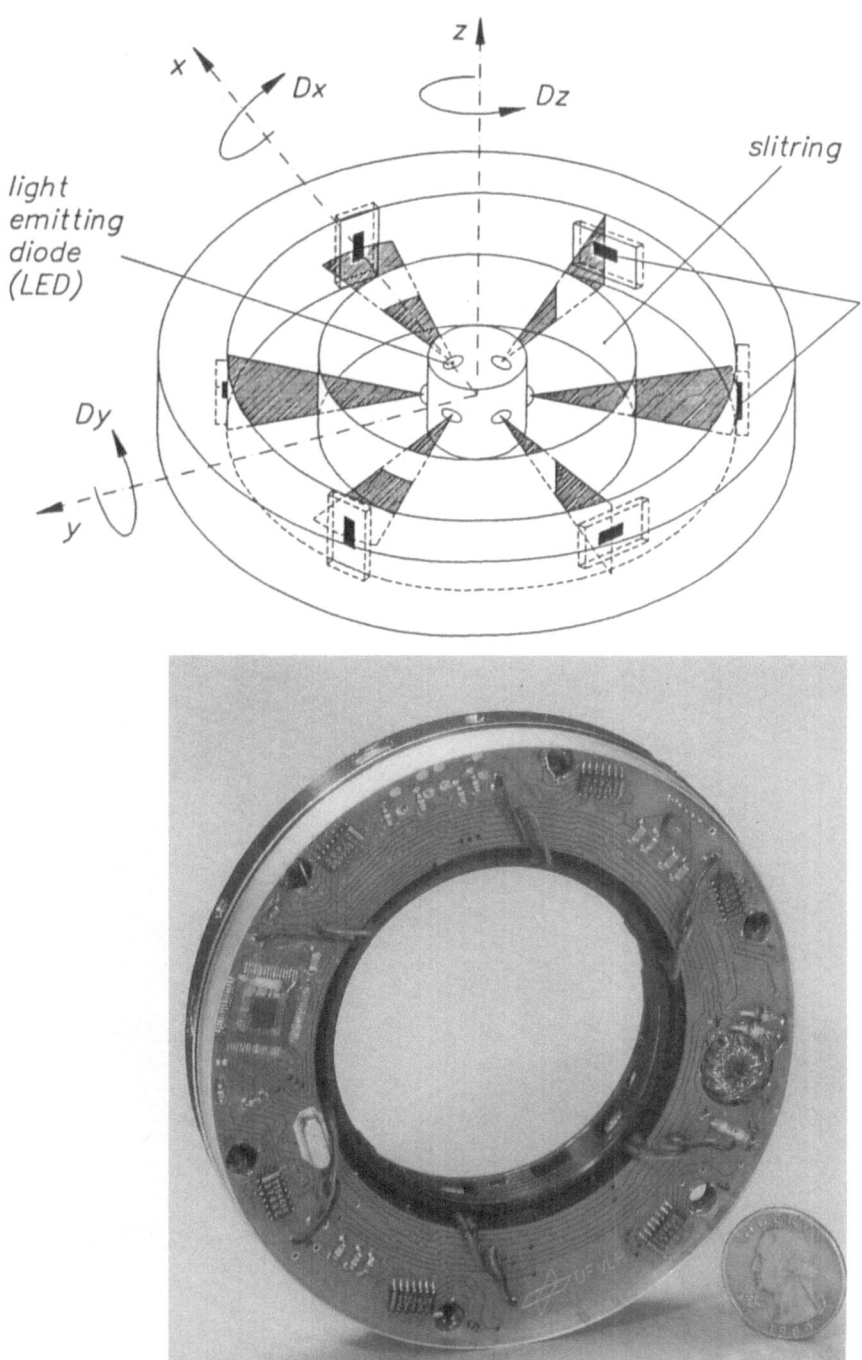

Fig. 9 *Schematic diagram and ring—shape*
realization of the instrumented compliance

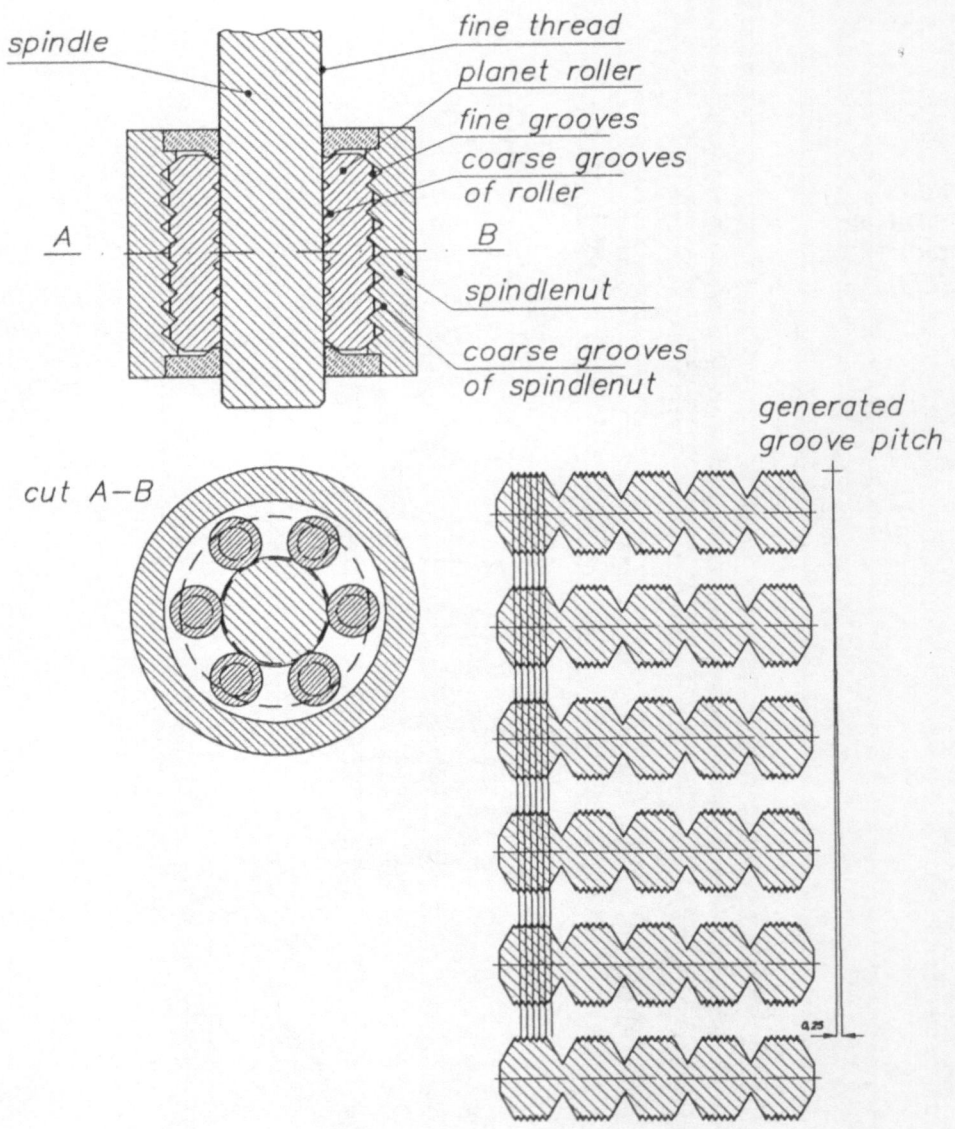

Fig. 10 *Our new spindle concept for the gripper drive*

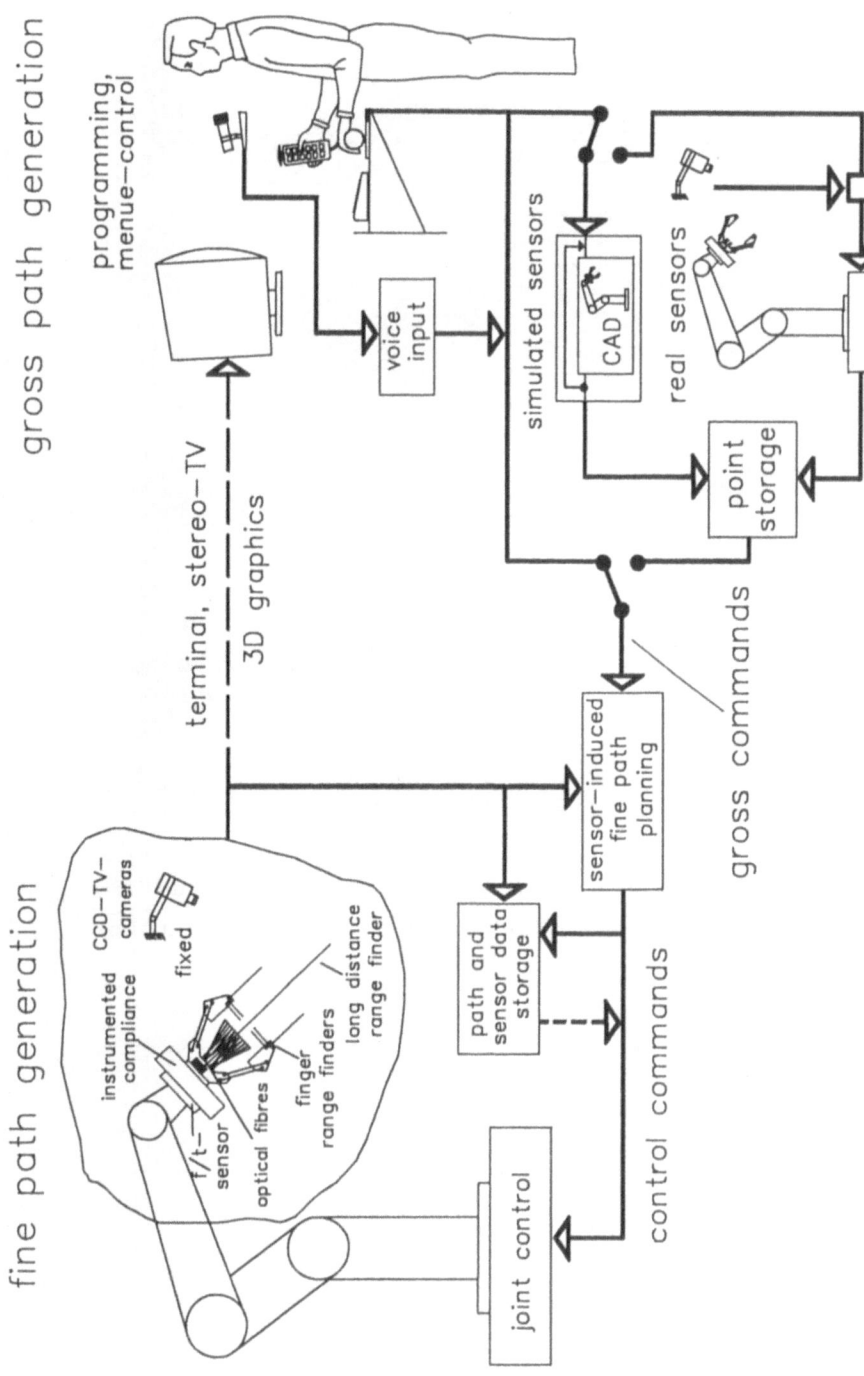

Fig. 11 Overall loop structures for the sensor-based fine path planning

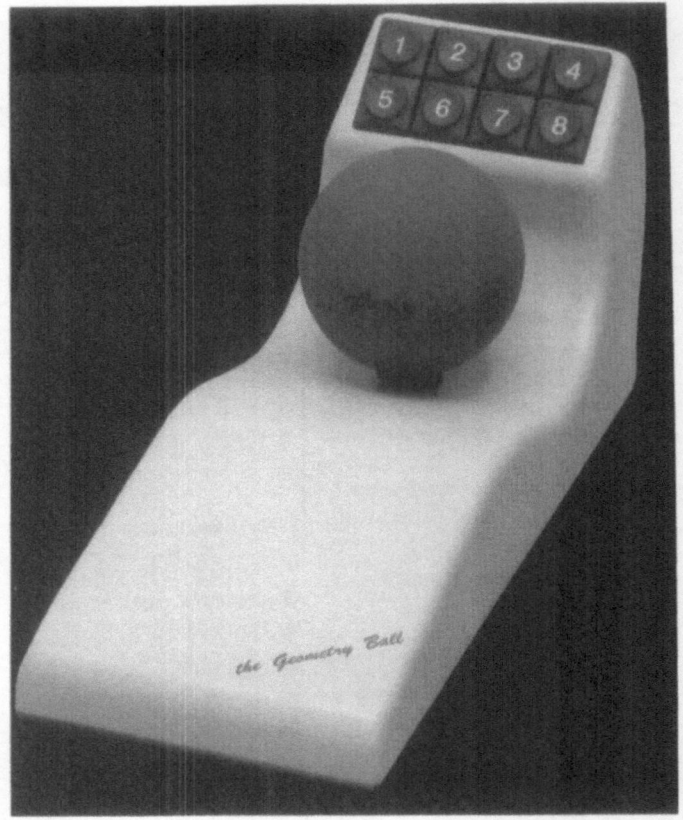

Fig. 12 The DLR 6 DOF hand-controller, used for 3D-computergraphics animation and robot teleoperation

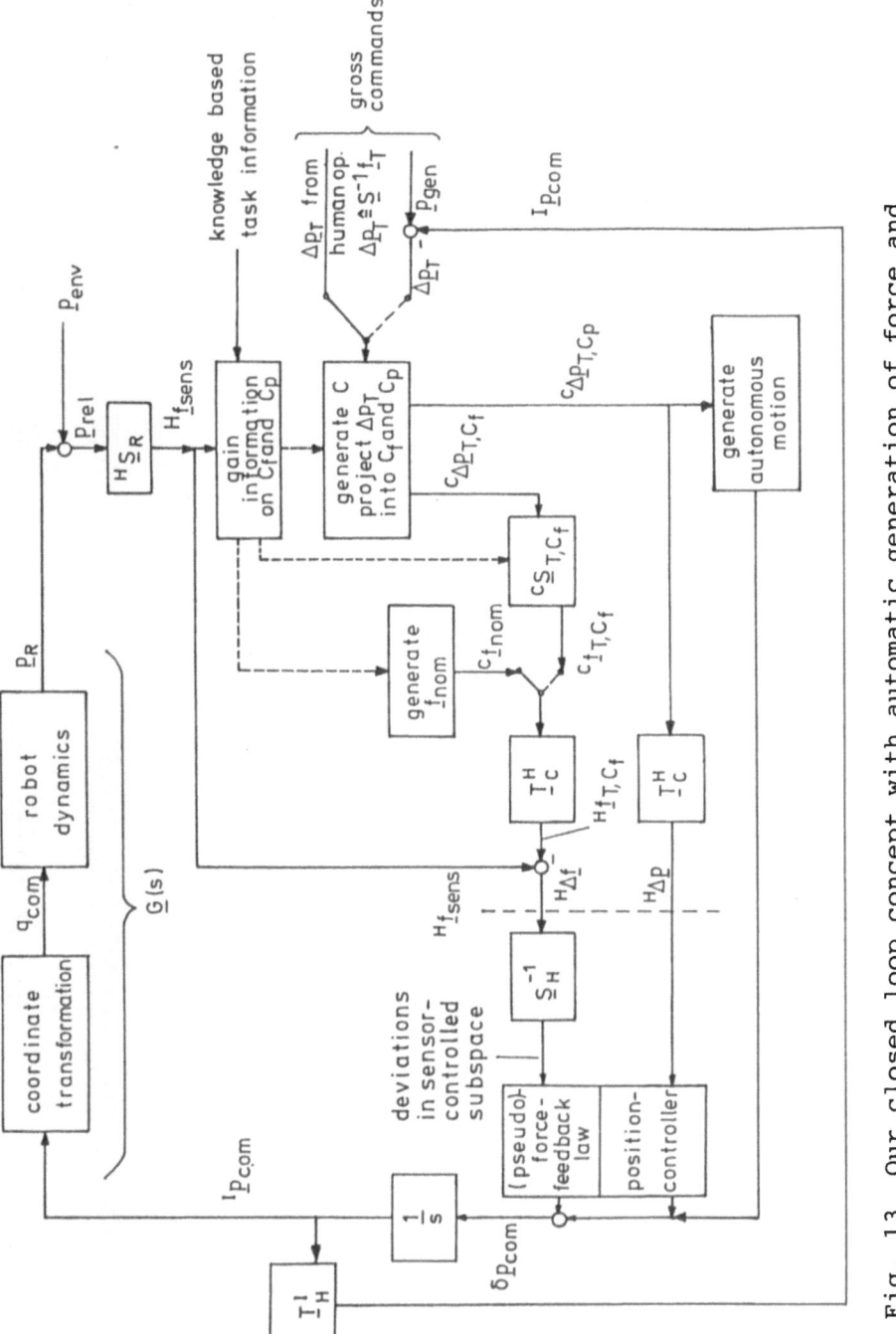

Fig. 13 Our closed loop concept with automatic generation of force and position controlled directions and artificial robot stiffness

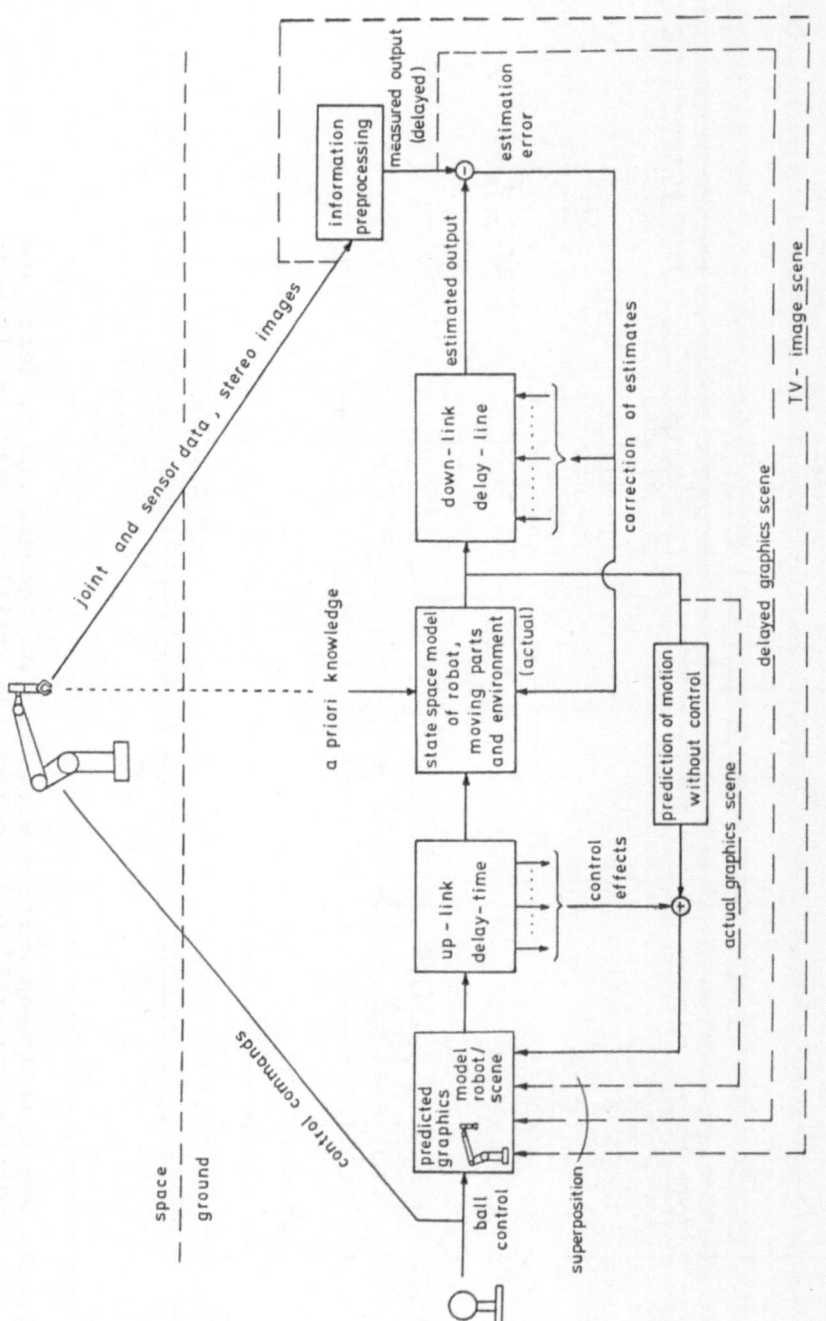

Fig. 14 Block structure of predictive estimation scheme

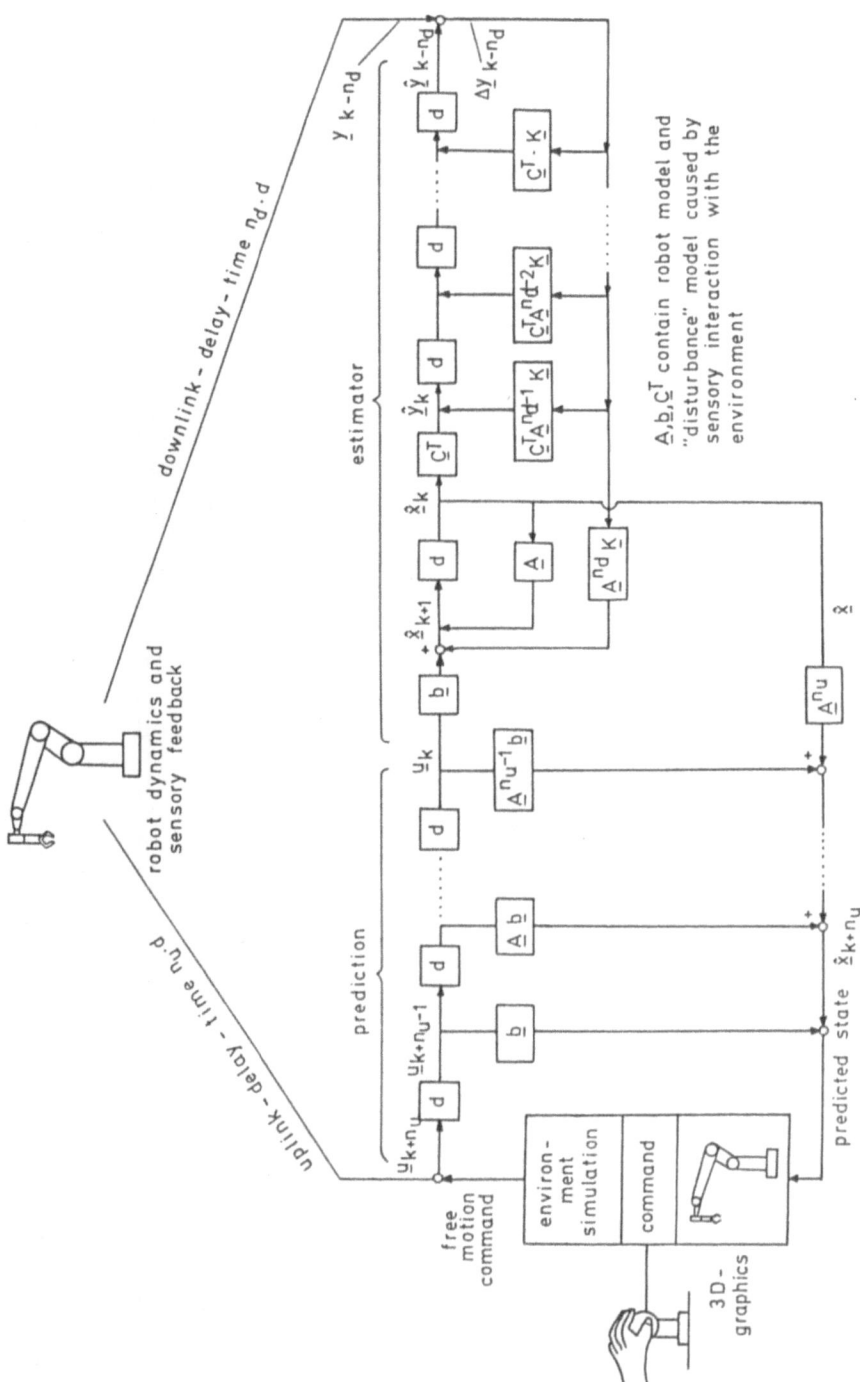

Fig. 15 Detailed predictive estimation scheme for the graphic simulation of a teleoperated robot that has sensory feedback on board and a large time delay in the transmission links.

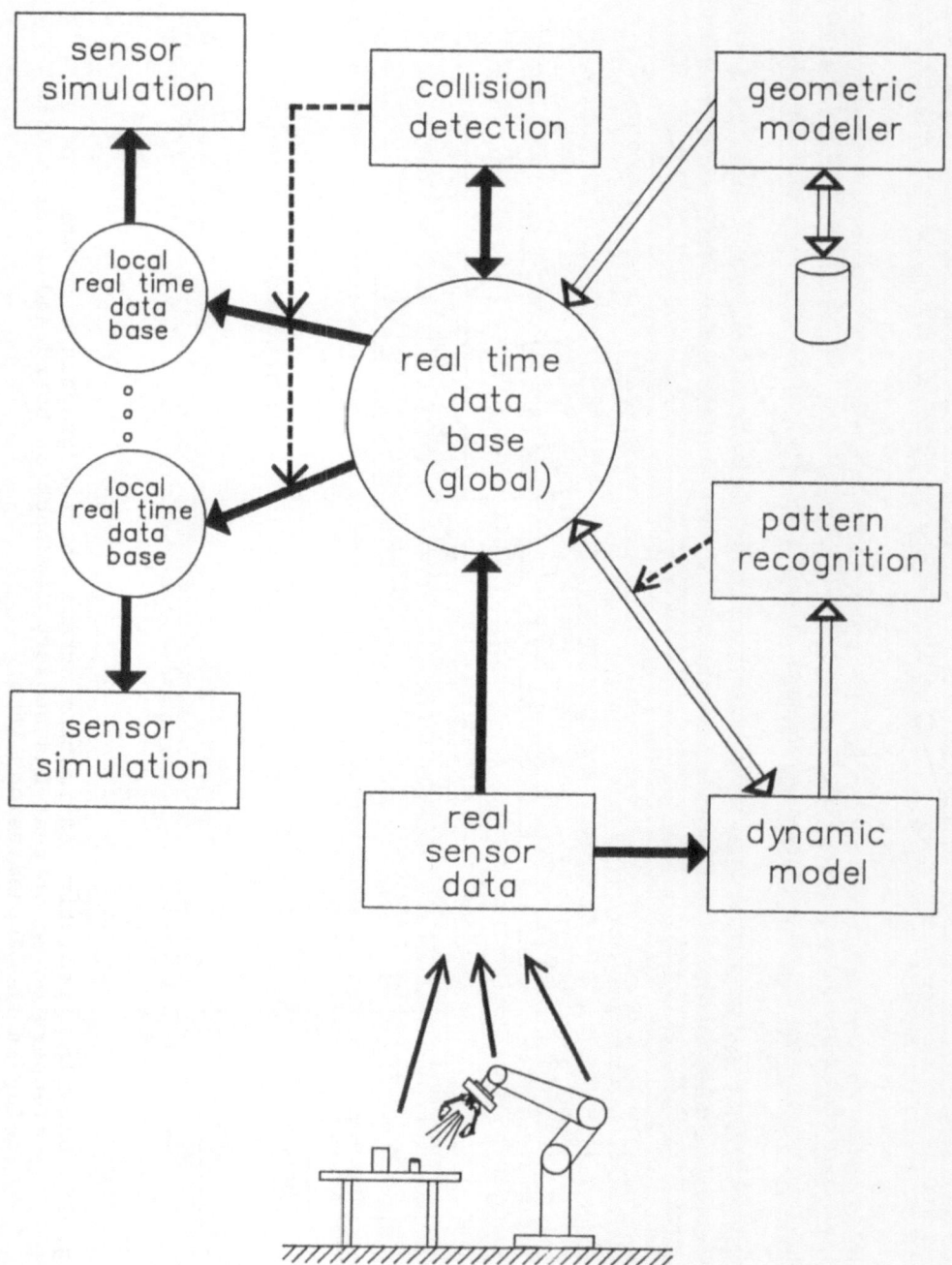

Fig. 16   Sensor simulation scheme

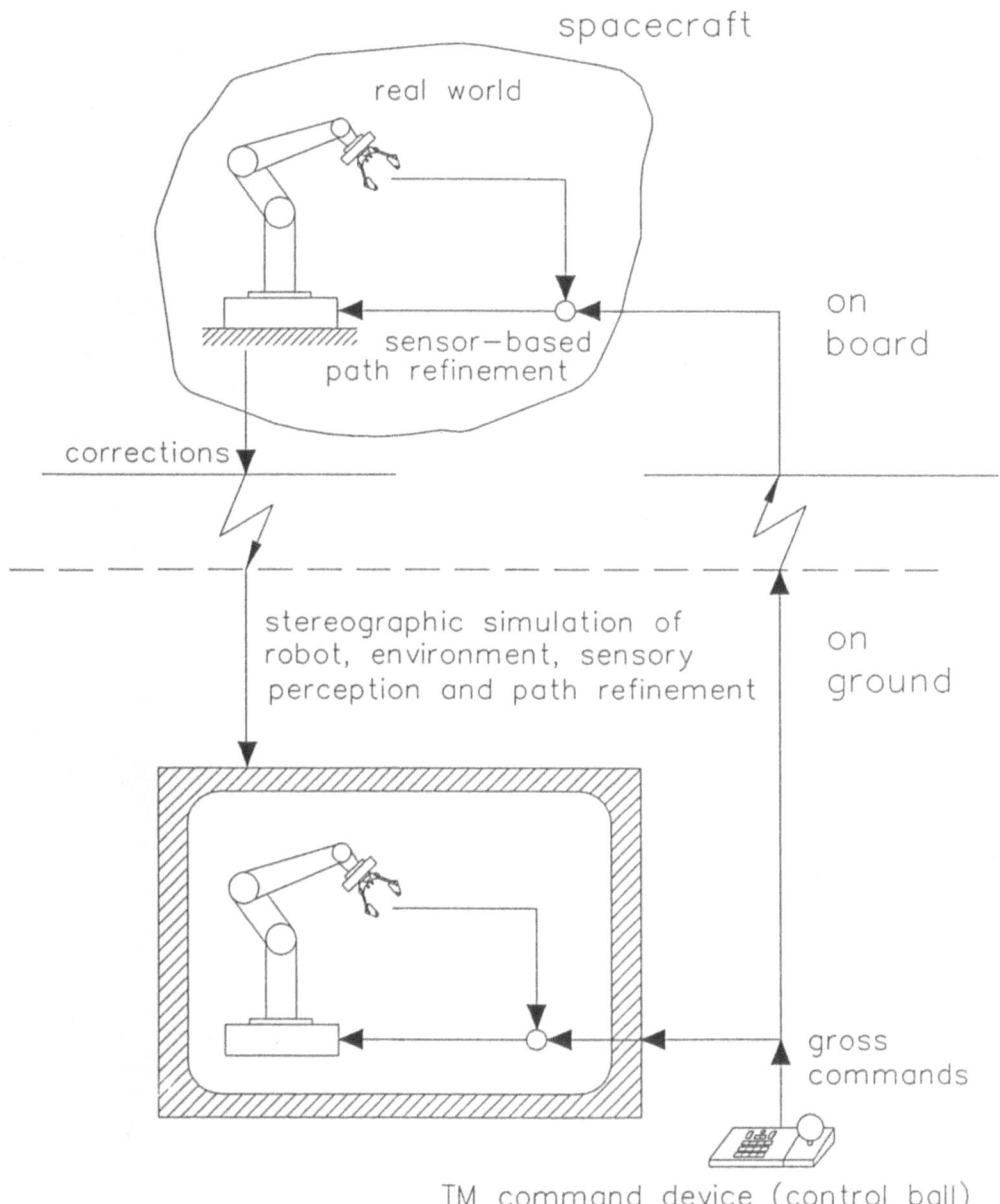

Fig. 17 Presimulation of sensory perception and path
refinement in case of teleoperation from ground

Autochthonous Behaviors - Mapping Perception to Action

Rod Grupen and Thomas C. Henderson*
COINS
University of Massachusetts, USA

Abstract

In this paper we describe an approach to high-level multisensor integration organized around certain egocentric behaviors. The task itself determines the sequence of sensing, the sensors used, and the responses to the sensed data. This leads to the encapsulation of robot behavior in terms of logical sensors and logical actuators. A description of this approach is given as well as some examples for dextrous manipulation and mobile robots.

---

*Department of Computer Science, University of Utah, USA. This author would like to thank Olivier Faugeras and the Robotics and Vision Group at INRIA (Sophia-Antipolis), France for the great sabbatical year there in 1988-89. This work was supported in part by NSF Grant IRI-8802585 and DARPA Contract DAAK1184K0017. All opinions, findings, conclusions or recommendations expressed in this document are those of the author.

NATO ASI Series, Vol. F 63
Traditional and Non-Traditional Robotic Sensors
Edited by T. C. Henderson
© Springer-Verlag Berlin Heidelberg 1990

1 Introduction

Figure 1 depicts the universe as a quantifiable state space. Within this space, subspaces are illustrated which represent the perceivable and actuatable spaces within the universe. The perceivable subspace represents those states which may be measured within the universe, the actuatable subspace represents those states which may be altered by employing a behavior of the system. In the discrete domain, the perceivable space represents states which are fully measureable by available (logical) sensors and the actuatable space are those states which may be transformed to other states by employing (logical) actuators. The dimensionality of the perceivable and actuatable subspace may not necessarily be the same. Moreover, both must be proper subsets of the universe. We will hypothesize an ecological niche in state space for such a system as the task subspace. These are a collection of states which, for whatever reason, identify crucial events in the universe. Such events in biological systems usually correspond to survial – such as the states abstractly indicative of food or danger.

We present such a depiction of an abstract system to discuss properties of systems which directly effect the theoretical limits of the transform from perception to action. Systems employing logical sensors and logical actuators whose state space description consists of disjoint perceivable and actuatable spaces do not have the ability – regardless of training, coaching, teaching, or otherwise cajoling – to effectively map perception to action. The measurable states, in this case, will not correlate to the states from which meaningful behavior can be derived. But, consider the case when the intersection of the perceivable and actuatable spaces is not empty. This condition suggests that certain properties of the universe can be measured, and that this perceived environment can be used to stimulate action such that perceivable state changes occur. Our system now has a region in state space which, in theory, permits abstract goals to be expressed. The proximity of these goals to perceived states can be determined and state changes can be selected which minimize the *distance* to the goal.

At this point in the discussion – rather than digressing into the role of random mutation and evolution in optimizing the task space niche of biological systems – we will instead turn our attention to the design of robotic systems. A (presumabley) human designer typically selects appropriate logical sensing and actuation mechanisms with which to express the desired task domain in an uncertain and partially unpredictable universe. We will call the process the design of *logical behaviors* and will draw on research in robotic control which is cognizant of the principles illustrated above.

Many approaches to multisensor integration have been proposed ranging from low-level descriptions of geometric data sensors[28,29,30] to high-level schemes[1,51,77]. Alternatively, one can focus on the sensors[48] or particular applications[2,6].

Our recent work has focused on a mid-level problem: the organization and integration of sensing in terms of intermediate level types of behavior – that is, activities which are not reflex, but which for the most part are not directly coupled to high-level "intelligent" behavior[45]. An example of such behavior is obstacle avoidance in a mobile robot. Here, data must be integrated from cameras, sonars, and perhaps other sensors as well. However, this function must be performed in an ongoing and automatic way. This is a learned behavior.

For the most part, the types of behavior involved are egocentric, i.e., they maintain spatio-temporal relations between the robot and the world. Our analysis is organized in terms of robot goals and behavior. This is accomplished by the use of what we call: **logical behaviors**. This approach allows for active control and integratoin of multisensor information in the framework of a specific task. We provide examples of the application of these ideas to impedance control, dextrous manipulation and mobile robots.

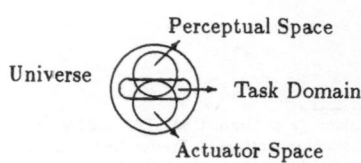

Figure 1. The Task-Perceptor-Actuator Trilogy

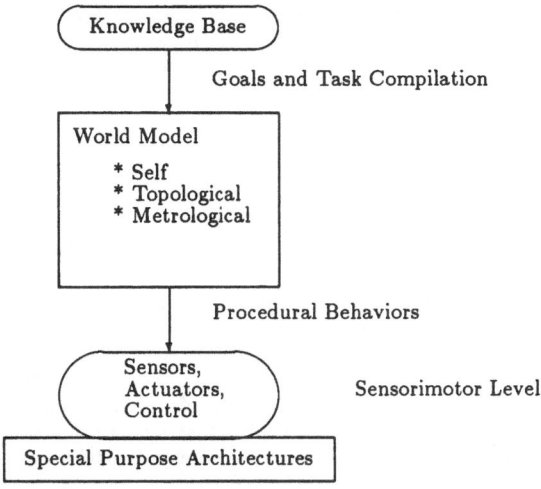

Figure 2. Autonomous Robot Research

2 Background

Multisensor integration has received a good deal of attention in recent years due to the availability of sensors, actuators, and processors. Two major testbeds for such work are:

- robotic workcell automation, and

- mobile autonomous robots.

The first of these involves applying strong knowledge-based techniques to the manufacturing environment, while the second concerns integrating several levels of information processing into a single autonomous system. We restrict our attention here to dextrous manipulation and mobile robots.

Autonomous mobile robots have been studied in a wide range of contexts. Figure 2 imposes an organization on most of the typical keywords. Obviously, the problem of navigation is basic to mobile robots and consequently has been studied by many people on specific implementations [20,22,40,44, 54,59,70,84,85,89,92]. Most such systems must use sensors (e.g., sonar or cameras [26,31,33,91]) and actuators and must control them [32,52,69,75,94,95]. The use of sensors requires the study of uncertainty management [43,78,83,86] and multisensor integration [21,30,50,65,67,72]. More global approaches to the sensorimotor problem can be found in [1,11,39], and special purpose architectures are being planned [4,88,97].

One level up, the mapping of procedural behaviors onto the sensorimotor control structure is of interest [3,5,14,46,47,48,49,93]. The world representations also exist at this level: both the metrological [7,8,73,80], where precise measurement is paramount, and topological [10,12,19,35,36,37,38,41,55,63,68, 87,90,98], where adjacency relations are useful for path planning, etc. It is even possible to study primitive forms of learning in this context [82,90].

Broader studies are usually oriented towards particular applications (e.g., the nuclear industry [17, 96], road following [23,24,74]) or towards well-defined, but limited goals (e.g., indoor [13,34] or outdoor [60,64] navigation).

Finally, the 'highest' level involves the specification and representation of the knowledge appropriate to a given task [18,57,58,62,76] and its compilation into executable robot behavior (or programs) [27,

51,56,61]. The literature is quite large on most of these subjects, and these references are intended as representative of the work in this area. It should be pointed out that most system designers use a central blackboard and some form of direct production system or a compiled version (i.e., a decision tree) to represent knowledge.

From this short summary, it can be seen that the scope of autonomous robot research is indeed vast, but the difficult problems found here are yielding to the steady advance of technical and theoretical developments. In the remainder of this paper, we describe current work on the mobile autonomous robot at INRIA.

3 Behavior Based Sensing and Control

In the most general sense, a robot interacts with its environment by applying operators to the perceived state of the environment. The state and operator may be cognitive - effecting the composition of state parameters without physically altering the environment; on the other hand, elements of the robot's surface may actually be applied to the geometry of the environment. In the latter case, the contacts may be derived from the robot's wheels or bumpers in the case of a mobile cart, or from the fingertips, proximal phalanges, palm or arm of a dextrous manipulator. Characteristics of the environment, the task, and the robot kinematics may be used to construct goal oriented behaviors.

The development of controllers for robotic systems is typically a generalization of the approach used in low level feedback controllers. Elements of the system state are measured and used to quantify the error of the system with respect to a desired state. The operation of the system tends to reduce the state error to zero. The nature of the feedback variables determines the nature of the response. Adaptable control schemes can optimize the response over uncertain inputs by varying the weighting of the feedback variables; however, types of behaviors which are not defined *a priori* cannot be expressed. This suggests that a single control law is not sufficient to manage the complexity of the general purpose robot systems. Control methodologies have been developed which partition the state space of complex systems into disjoint regions, each with an associated control law [79]. The operation of these systems is represented by a finite state automata where state transitions are triggered by sensory events. This approach produces sequences of behaviors in the system.

Behavior based control schemes generalize this approach. Elemental behaviors are instantiated which span the problem domain (see Braitenberg[14]). Braitenberg's work was the precursor of many similar systems, including the subsumption architecture proposed by Brooks[15]. The subsumption architecture is an approach which was developed to construct systems which require composite behaviors[16,15]. Concurrent control laws are defined, each of which acquiring the sensory data necessary for that particular behavior. The so-called *activity producing subsystems* are integrated in a hierarchy in which primary behaviors reside at the lowest levels. Higher level behaviors are used to modulate the output produced by low level behaviors. The work demonstrated a hardwired system tuned to perform a particular task, the navigation of an autonomous vehicle.

These approaches are generalized in the society structures proposed by Minsky[71]. This proposed structure is motivated by observed problem solving behaviors in humans. Agencies are postulated which serve as *proto-specialists* over limited problem domains. The society of such agents is capable of goal oriented behaviors with dynamic priorities based on the current state of the composite system.

The logical behavior systems proposed in this paper are based on a similar perspective. Independent, elemental behaviors are defined which span the required problem domain. We generalize the notion of a behavior to any process which maps information abstracted from (logical) sensors to state transitions which may be mapped onto (logical) actuators. Once again, a logical sensor need not be linked directly or indirectly to a physical sensor, but may represent any hypothetical state from which a state transition is desired. Likewise, the logical actuator need not employ a DC motor, for example, but will transform a state in the actuatable space to some other state. This property provides a mechanism for cognitive and reflexive mappings from sensors to actuators. We also note that the distinction between planning and execution is in some sense a function of whether the logical sensor is in fact terminated at a physical sensor, and whether the logical actuator terminates at a physical actuator.

We describe below the application of this approach to dextrous manipulation and to mobile robots. The example of the design of logical behaviors for multifingered manipulator control includes complex kinematics, multi-functional mechanisms, and complex tasks. The other application is the design of an obstacle avoidance behavior for a mobile robot. We will make an effort to keep our primary focus on the even larger problem domain describing general transformations from sensors to actuators.

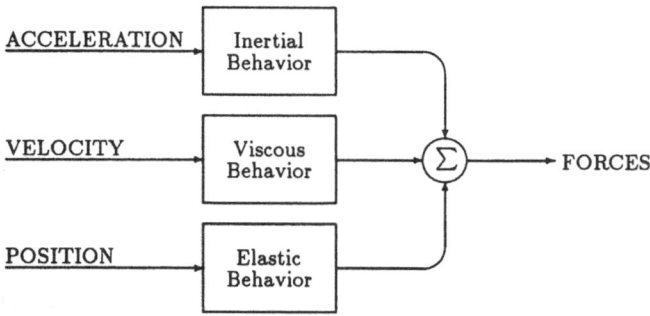

Figure 3. Three Logical Behaviors Which Constitute the Impedance Controller

4 Logical Behaviors for Generalized Impedance Control

Drake designed a passively compliant device which automatically compensates for uncertainty in certain classes of assembly operations[25]. Salisbury employs a stiffness controller to support the construction of stable grasps[81], and Lozano-Pérez *et al.* discuss the use of the generalized damper to plan fine motion assembly strategies[66]. These instances of manipulator control map sensor data to action by modeling the manipulator as an impedance relative to an environmental admittance. As such, the manipulator measures deviations from nominal positions or velocities and applies a correcting force.

To illustrate the use of logical behaviors for multisensor integration, consider an implementation of Hogan's impedance control[53]. Hogan argues that to:

> ensure physical compatability with the environmental admittance, something has to give, and the manipulator should assume the behavior of an impedance.

Others have noted the usefulness of impedance control - loosely defined for our purposes as behaviors which map errors in position, velocity, or acceleration to forces. The terms in the impedance controller are linearly independent functions of separable state variables, $\vec{x}, \dot{\vec{x}}$, and $\ddot{\vec{x}}$. Figure 3 illustrates how the impedance controller may be considered to be a superposition of three separate impedance behaviors: an inertial behavior, a viscous impedance behavior and an elastic impedance bhavior.

The remainder of the presentation will consider only the visco-elastic components of the generalized impedance controller. The structure of each of these logical behaviors consists of a logical sensor, an (optional) reference input, and a logical actuator. The logical sensor is any combination of hardware and software which measures and/or hypothesizes the state of the system[48]. The logical actuator is likewise, any combination of hardware and/or software which transforms the state representation. The logical actuator may simply transform the abstracted state representation, or it may actually employ hardware actuators to physically change the state of the system. To understand the utility in an abstract notion of the logical impedance behaviors, consider the various incarnations of the visco-elastic impedance controllers presented in Figure 4.

The figure defines the constraints which govern the construction of a logical behavior. In this case the behavior describes a transformation from the position/velocity domain into the force/torque domain. The behavior is represented as a combination of a logical sensor and a logical actuator. The generic logical behavior on the top of Figure 4 defines the data type consistency which must be maintained during the construction of the logical impedance behavior. In essence, all data entering or leaving the summation block in Figure 4 must be consistent. We have thus defined the type of the: characteristic output vector (cov) for the logical sensor, the reference input vector (riv), and the characteristic input vector (civ) of the logical actuator.

Figure 4(a) depicts the commonplace joint impedance controller. We illustrate two logical sensors which yield the joint space position and velocities required. The logical actuator simply computes

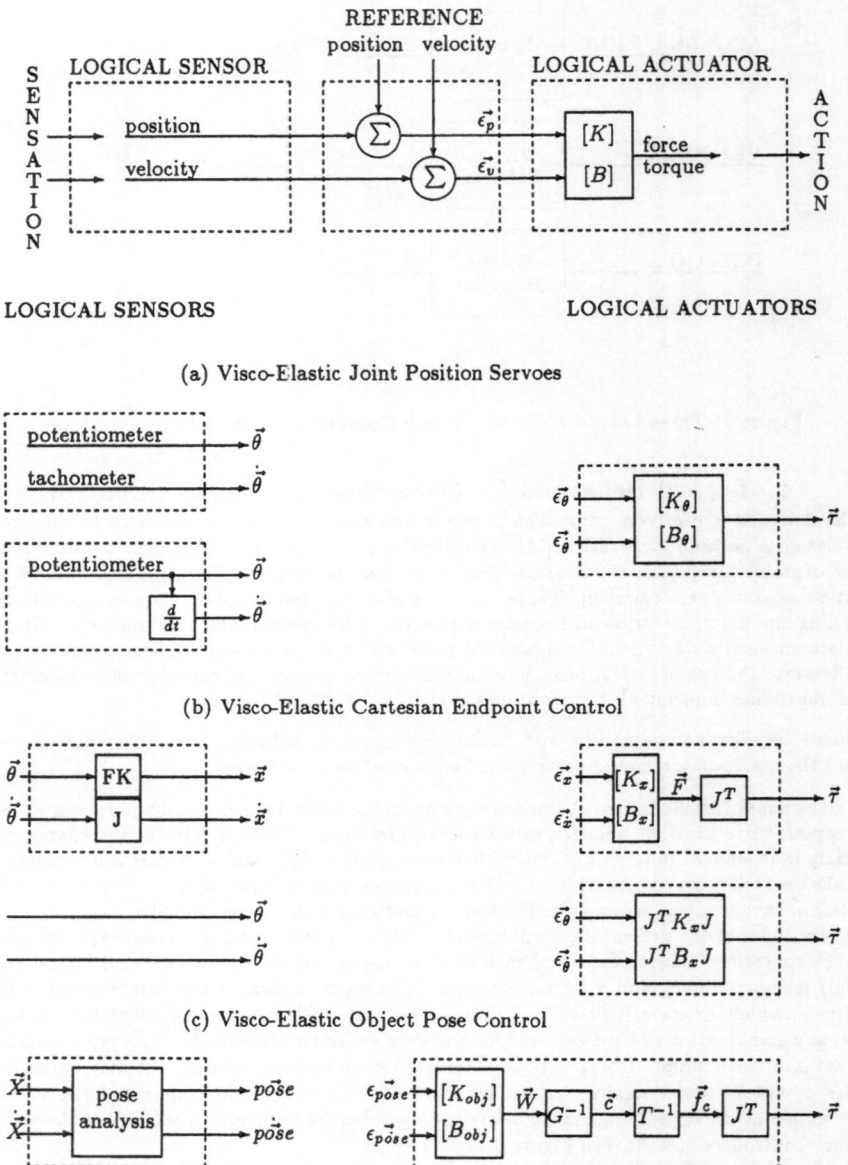

Figure 4. Implementations of Logical Impedance Behaviors

torques which suppress errors in joint space. Figure 4(b) presents logical behavior expressions of two commonly used Cartesian endpoint controllers. The second simply employs the logical sensors presented in Figure 4(a) and a logical actuator which suppresses joint space errors using Cartesian impedance parameters as suggested by Salisbury[81]. The first Cartesian controller shown employs a somewhat different logical sensor which expresses the position and velocity directly in Cartesian space and a logical actuator which supresses errors in Cartesian space. Figure 4(c) presents an object frame Cartesian pose impedance behavior. Another stage is added to the logical sensor which transforms the set of Cartesian contact positions and velocities into an object frame pose and velocity in 6-space. The corresponding logical actuator maps errors in the pose parameters relative to the reference into restoring wrenches, $\vec{W}$. This wrench is in n-space and contains internal (null space) components as well. The inverse of the Grip Jacobian is used to map the restoring wrench into individual wrenches applied through each contact location. If the transform T is defined to map contact forces into contact wrenches, such that:

$$\vec{w}_i = T\vec{f}_i$$

then,

$$\vec{f}_i = T^{-1}\vec{w}_i$$

and, the force commands at each contact are expressed as:

$$\vec{f}_i^c = T^{-1}(\vec{c}^T\vec{w}_i).$$

Finally, the forces applied at each contact location are transformed into actuator torques by way of the appropriate manipulator Jacobian.

Beyond the relative complexity of the impedance controllers presented in Figure 4, each controller is an instance of the same logical behavior. Each instance represents an effective assemblage of logical sensor and logical actuators for some combination of task objectives and constraints in the context of multisensor control.

5 Logical Behaviors for Grasping and Manipulation

A primary motivation for developing the logical behavior formalism is the modularity and data abstraction that is provided in the design of complex robotic controllers. To illustrate this feature, we will first suggest several behaviors which are understood to be useful in the context of grasping and manipulation. The resulting behaviors may be used as a programming language for manipulation, or as a basis for autonomous behavior composition. These behaviors are in fact supported by the Cartesian impedance behaviors described in Section 4. We will employ two principal behaviors in the discussion that follows. These behaviors and others useful in grasping and manipulation are described in detail in [42].

The **wrench closure behavior** is defined to construct 6 dimensional constraint envelopes about the equilibrium position of the object. The logical sensors for this behavior must determine:

1. the geometry of the graspable surface,

2. the type(s) of contact(s) delivered to the object's surface by the manipulator and the environment, and

3. the pose of the object.

This *sensor* data may be derived from a variety of physical sensors, or derived from basic principles and knowledge of the manipulator and the object. In the simulations presented, the geometry of the object is expressed analytically, the contact type is uniformly a point contact with friction model, and the pose of the object is known a priori. The reference input to the logical behavior is a 6-dimensional volume in wrench space which expresses both positive and negative sense wrench magnitudes in each of the object's six degrees of freedom. This *goal* represents a stable grasp objective. The error submitted to the logical actuators represents the difference between the current state of the wrench volume generated by manipulator contacts and the reference input. The logical actuator in this case performs a gradient descent in the error space toward suitable wrench closure states.

The **isotropic manipulator behavior** is designed to condition the manipulator to be compliant to the wishes of the wrench closure behavior. This behavior employs a logical sensor to compute a scalar manipulability index for the hand. This state description is derived from the kinematic configuration of all the fingers in a grasp. The scalar index is normalized so that the reference goal for this behavior is implicitly to generate a unity index for the hand. This behavior performs a gradient ascent in the manipulability index toward isotropic hand states.

Logical Sensors	Logical Actuators	
Object Geometry	Closure Gradient Descent	contact positions
Contact Positions	Hand Index Gradient Ascent	position/orientation

Examples are presented for cylindrical and spherical test objects. All examples employ the Utah/MIT hand geometry. A top view and a side view of the hand/object system are presented. Intermediate positions for the hand frame y- and z-axes are shown. The initial hand frame position is identified in bold print; for clarity, only the final finger configuration is shown. Typical computation time for the four fingered grasps is approximately 10 *seconds* of CPU time on a VAX 750.

The original task stack submitted to the system for the initial grasp task is illustrated in Figure 5.

```
type:              approach
screw task:        NA
contact set:       all fingers
nomadic contacts:  all fingers
↓
type:              condition
screw task:        hypercubic wrench volume
contact set:       all fingers
nomadic contacts:  all fingers
↓
NIL
```

Figure 5. The Task Stack Representation of the Initial Grasp Task

Figures 6, 7 and 8 present the grasps of the cylinder produced when 2, 3 and 4 fingertip contacts are applied to the task, respectively. In Figure 8, the third finger (the ring finger) does not oppose the thumb, but goes to a position where it more effectively complements the wrench subspaces produced by the other contacts. In (b), the task is modified dynamically by adding an incremental task which is the wrench subspace produced by the index finger. The trajectories of the other fingers are then tethered elastically to the index finger. The fingers are essentially grouped into a virtual finger. Which grasp is better for the robot hand is not clear, but (b) appears more anthropomorphic.

Figures 9, 10 and 11 present the grasps of a sphere produced when 2, 3 and 4 fingertip contacts are applied to the task, respectively. In Figure 11 the super symmetric object and the functionally redundant manipulator produced convergence difficulties in the geometry synthesis. There appear to be closely spaced meta-stable states in the contact geometry solution. The convergence problem was controlled by designating a virtual finger over the middle and ring fingers of the hand; the result is not intuitively satisfying. This example suggests the application of additional constraints in the form of virtual fingers and/or geometrical restrictions limiting the portion of the object surface which may be used to address a particular task.

6 Logical Behaviors for Obstacle Avoidance

In this section we consider the following problem: suppose that our mobile robot is wandering through an unknown indoor environment. The robot must:

- **incrementally build a 3-D representation of the world** (i.e., determine its motion and integrate distinct views into a coherent global view),

- **account for uncertainty in its description** (i.e., explicitly represent, manipulate and combine uncertainty), and

- **build a semantic representation of the world** (i.e., discover useful geometric or functional relations and semantic entities).

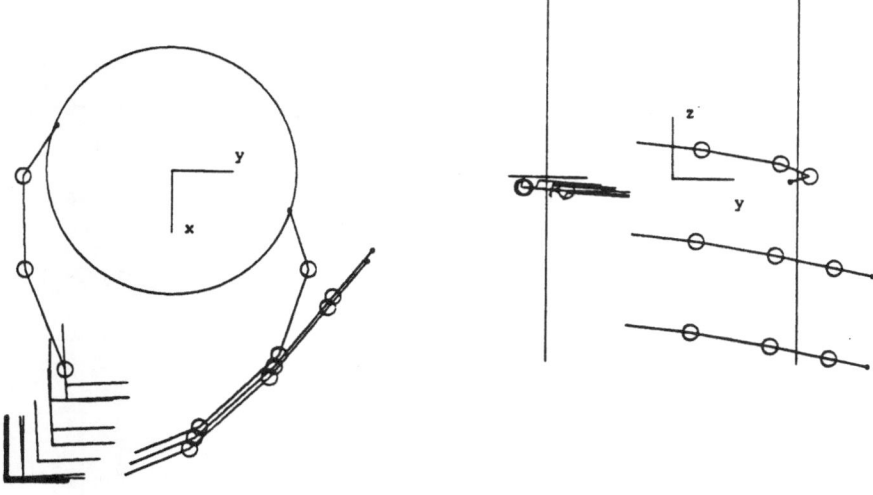

Figure 6. A Two Fingered Grasp on a Cylinder 4 *inches* in Diameter; the Task is a Hypercubic Wrench Volume in Six DOF

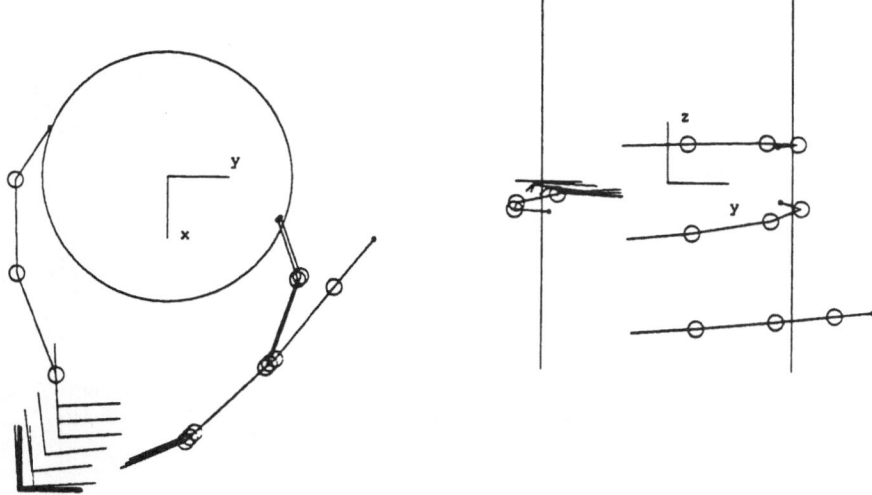

Figure 7. A Three Fingered Grasp on the Cylinder; the Task is a Hypercubic Wrench Volume in Six DOF

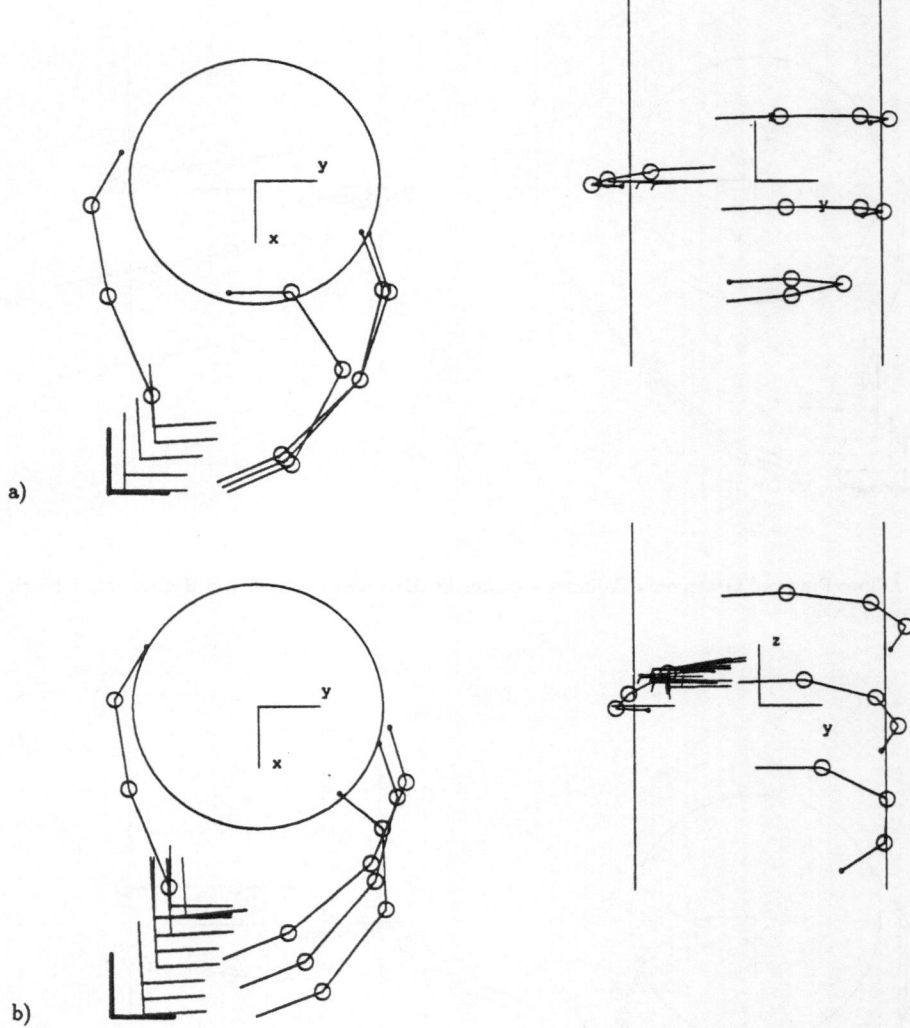

Figure 8. (a) A Four Fingered Grasp on the Cylinder for a Hypercubic Wrench Volume Task; (b) the Same Task with the Virtual Finger Designation over the Index, Middle and Ring Fingers

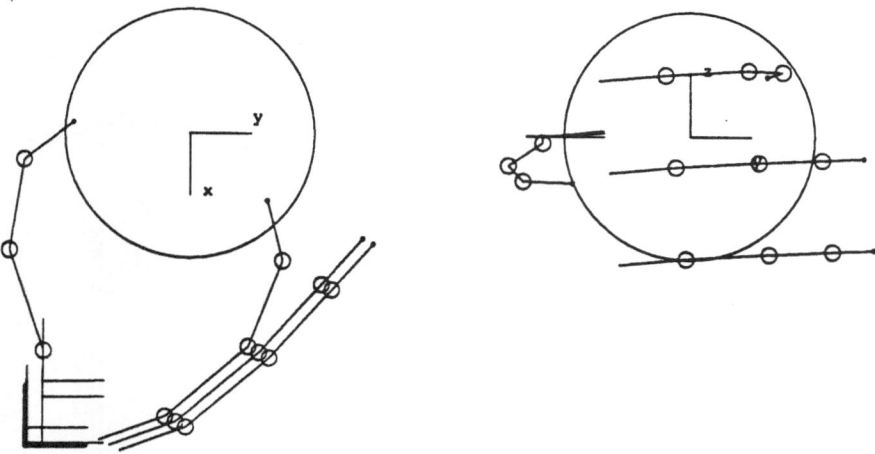

Figure 9. A Two Fingered Grasp on a Sphere 4 *inches* in Diameter; the Task is a Hypercubic Wrench Volume in Six DOF

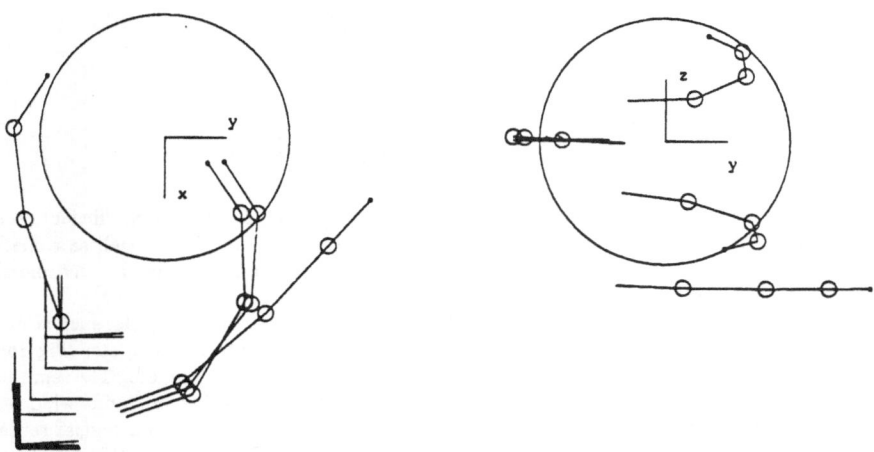

Figure 10. A Three Fingered Grasp on the Sphere; the Task is a Hypercubic Wrench Volume in Six DOF

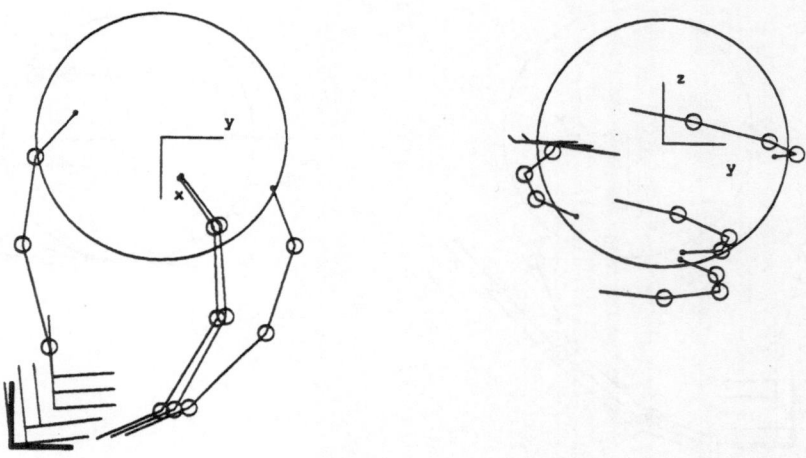

Figure 11. A Four Fingered Grasp on the Sphere

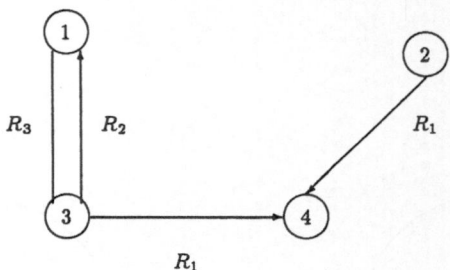

Figure 12. Semantic Net Defining World Model

Here we describe an approach to solving the third problem. (See [6] for details on efficient techniques for producing a local 3-D map from stereo vision and structure from motion as well as a method for combining several viewpoints into a single surface and volume representation of the environment and which accounts for uncertainty.)

The mobile robot must use the 3-D representation to locate simple generic objects, such as doors and windows, and eventually more complicated objects like chairs, desks, file cabinets, etc. The robot can then demonstrate task-based behavior such as going to a window, finding a door, etc. The representation should contain semantic labels (floor, walls, ceiling) and object descriptions (desks, doors, windows, etc.).

The proposed approach is straightforward and exploits our previous work on logical sensors, the Multisensor Knowledge System, and multiple semantic constraints. The World Model is defined in terms of a semantic network (e.g., see Figure 12). The nodes represent physical entities and the relations are (currently) geometric. "Behind" each node is a logical sensor which embodies a recognition strategy for that object. The relations are simply tabulated.

A goal for the robot is defined by adding a node representing the robot itself and relations are added as requirements (see Figure 13). This method permits the system to focus on objects of interest and to exploit any strong knowledge that is available for the task. The added relations are satisfied (usually) by the robot's motion.

As an example, consider the world model in Figure 14 which represents a specific office. The addition of the robot and the "Next_to" relation fires the "Find_door" logical sensor. This in turn causes the strategy for door finding to be invoked. Such a strategy may attempt shortcuts (quick image cues) or may cause a full 3-D representation to be built and analyzed. Logical behaviors are then the combined

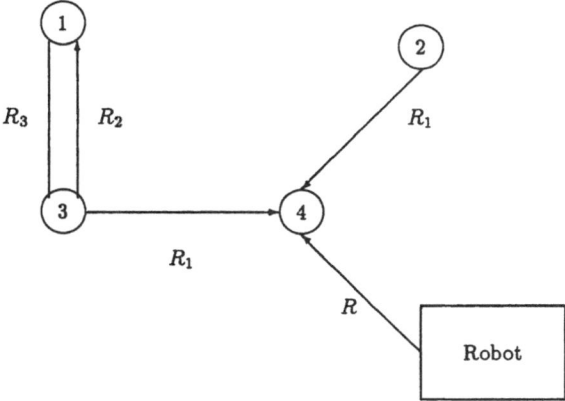

Figure 13. Defining the Robot Task

logical sensors and motion control required to satisfy the "Next_to" relation.

Note that it is in the context of such a strategy that high-level multisensor integration occurs in goal-directed behavior. We are currently implementing a testbed for experimentation.

6.1 Robot Behavior as Real-time Programming

Robots must maintain a permanent interaction with the environment, and this is the essential characteristic of *reactive programs*. Other examples include real-time process controllers, signal processing units and digital watches.

We have selected the Esterel synchronous programming language[9] as the specification language for the reactive kernel of the robot's behavior. A reactive system is organized in terms of three main components:

- **reactive kernel**: specified in Esterel and compiled into C or CommonLisp for execution,

- **interface code**: handles drivers and sensors, and

- **process or data handling code**: routine calculations.

The programs produced are:

- **deterministic**: produce identical output sequences for identical input sequences,

- **concurrent**: cooperate deterministically, and

- **synchronous**: each reaction is assumed to be instantaneous.

Interprocess communication is done by instantly broadcasting events, and statements in the language take time only if they say so explicitly; for example:

every 1000 MILLISECOND do emit SECOND end

In this example, a SECOND signal is sent every thousand milliseconds.

Thus, Esterel provides a high-level specification for temporal programs. Moreover, the finite state automata can be analyzed formally and give high performance in embedded applications. They help encapsulate the specification of sensing and behavior from implementation details. This simplifies simulation, too.

Other advantages include the fact that synchrony is natural from the user's viewpoint; e.g., the user of a watch perceives instant reaction to pushing a control button on the watch. Synchrony is also natural to the programmer. This reconciles concurrency and determinism, allows simpler and more rigorous

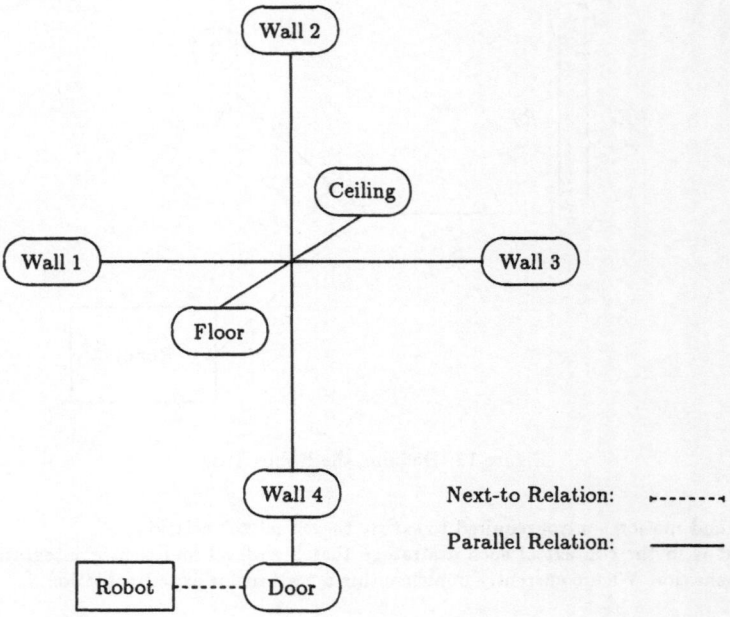

Figure 14. A Representation of the Command: "Go to the office door"

Prior Knowledge	Robot World	Sensors	External Windows

* Source
 	
  - POP11 Code
  - Lisp Code
  - Prolog Code
 	
* CAD Tool
 	
* Other
	* Behavior
	
	  - Automaton
	  - Trace
	  - Robot Dump
	
	* Goals, State
	
	* Maps, Objects
		* Sonar
		
		  - Range
		  - Direction
		
		* Motors
		
		* 3D Segments
			* Camera Images
			
			* Edge Images
			
			* Etc.

POPLOG			*Suntools*

Figure 15. The Debugging System Organization

programs and separates logic from implementation. Finally, such automata are easily implemented in standard programming languages.

Details of the language are not given here; however, a brief summary is in order:

- **variables**: not shared; local to concurrent statements.

- **signals**: used to communicate with environment or between concurrent processes; carry staus (present or absent) and value (arbitrary type).

- **sharing law**: instantaneous broadcasting; within a reaction, all statements of a program see the same status and value for any signal.

- **statements**: two types:

  1. standard imperative style, and
  2. temporal constructs (e.g., **await** *event* do).

An extremely useful output from Esterel is a verbose description of the automaton. This can be used for debugging purposes. Esterel also produces a C program which implements the automaton.

Another useful output is a CommonLisp version of the automaton. This makes simulation straightforward, so long as reasonable functions can be written which simulate the world and the physical mechanisms of the robot. But these, too, can be specified in Esterel and then combined.

6.2 Robot Behavior Debugging Environment

In developing Esterel specifications for robot behavior and sensor control, we are faced with the problem of integrating diverse kinds of knowledge and representations. In particular, debugging robot behaviors requires knowledge of the world model, the robot's goals and states, as well as the behavior specification, and sensor data (intensity images, sonar data, 3-D segments, etc.). Figure 15 shows the current implementation organization. We use POPLOG (an interactive environment which combines CommonLisp, Prolog and Pop11) to support manipulation and display of prior knowledge, the robot world, and some sensor data, while other Suntool-based utilities support display of the trinocular stereo camera images, etc.

Figure 16 shows a representative collection of windows which provide:

- POPLOG source code (window management, etc.)

- Prolog source (semantic entity definition; e.g., walls, doors, etc.)

- sensor data display (e.g., sonar range data, 3D segments)

- Esterel generated automaton (e.g., COMBINE.debug)

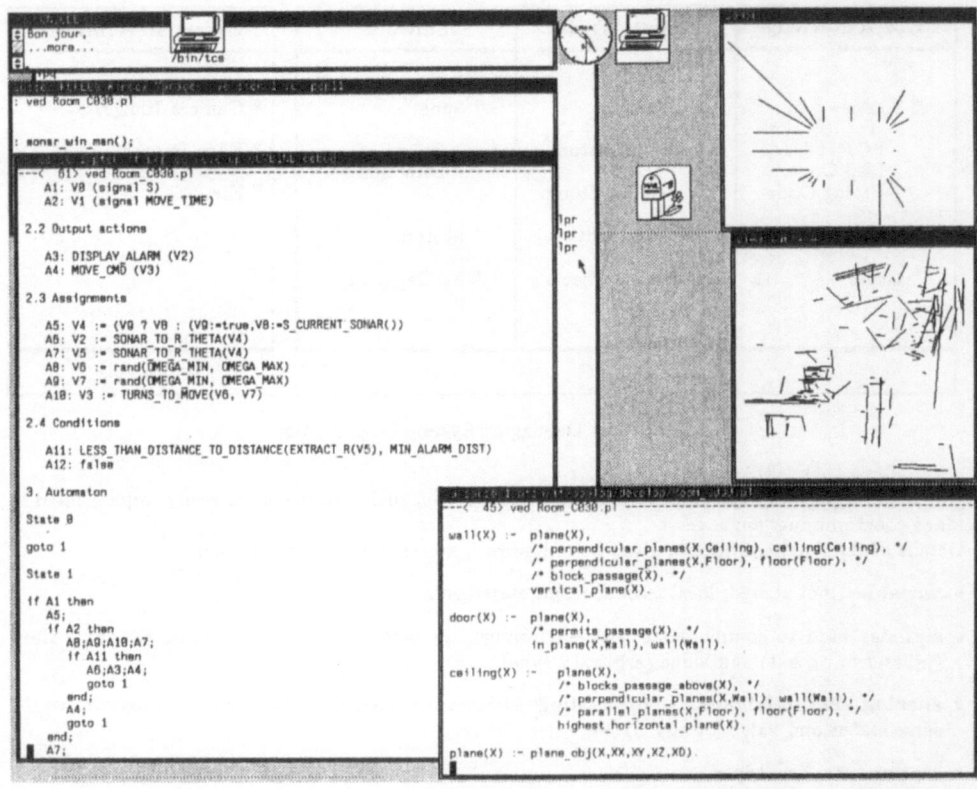

Figure 16. Collection of Windows for Debugging

Esterel permits state tracing during execution, and this combined with access to the robot's sensory data permits rapid and accurate debugging. In Appendix A we give the details for the specification of a wandering robot which must avoid colliding with objects in the world. This specification has been compiled and loaded onto the robot and successfully executed.

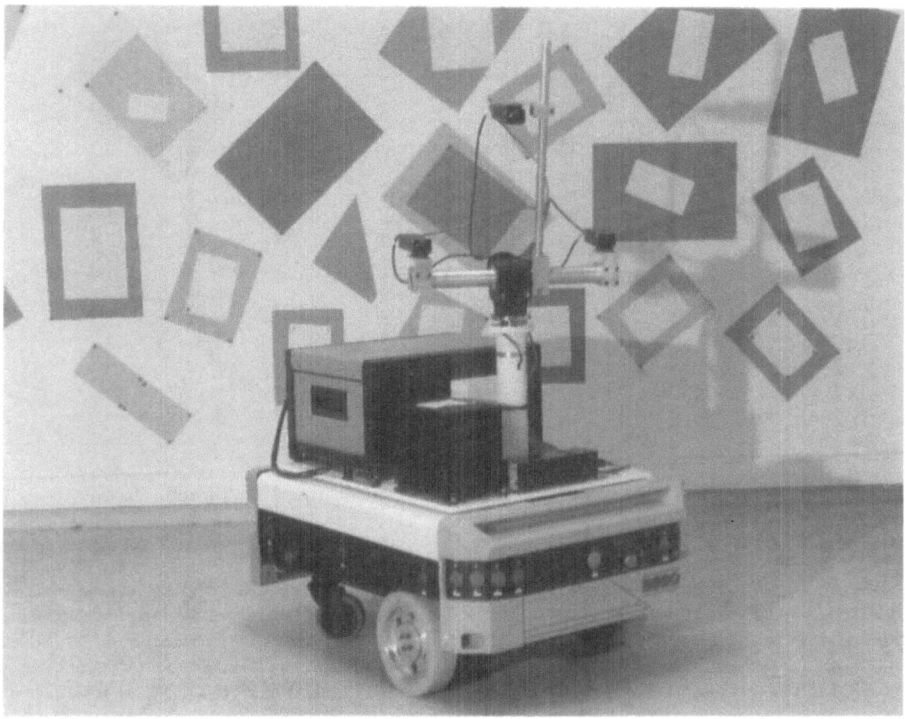

Figure 17. The INRIA Mobile Robot

6.3 Mobile Robot

There is an operational mobile robot at INRIA (Sophia-Antipolis). It is similar to other mobile robots (e.g., like those at CMU or Hilare at LAAS). Figure 18 shows the geometry of the robot (length: 1.025m, width: .7m, and height: .44m) and the locations of the sonar sensors. The two rear wheels drive the robot.

The onboard processing consists of two M68000 series microprocessors on a VME bus; one controls the sonar sensors, and the other runs the real-time operating system, Albatros. The two main wheels are controlled separately, and the system has an odometer.

High-level multisensor integration must be investigated in the context of real-world problems. We have described current work on an autonomous mobile vehicle under development at INRIA. We propose "logical behaviors" as an approach to robot goal representation and achievement.

We intend to continue development of algorithms, architectures and systems for multisensor robotic systems. Moreover, we are currently investigating the simulation of such systems; this involves embedding the reactive kernel in a modeled robot world. Finally, as can be seen by the rough nature of the definitions of walls, doors, etc., we must develop a suitable formal model of the world in which the robot finds itself. We intend to exploit optimized refinements of conceptual clusters defined in first order predicate calculus.

7 Summary

We have proposed **logical behaviors** as a technique for organizing multisensor integration and control. Such an approach allows encapsulation of sensing and actuation and relates those activities to a specific task. Several examples have been presented ranging from impedance control for a robot manipulator to obstacle avoidance for a mobile autonomous robot. We are currently extending the environment to include better debugging and simulation environments.

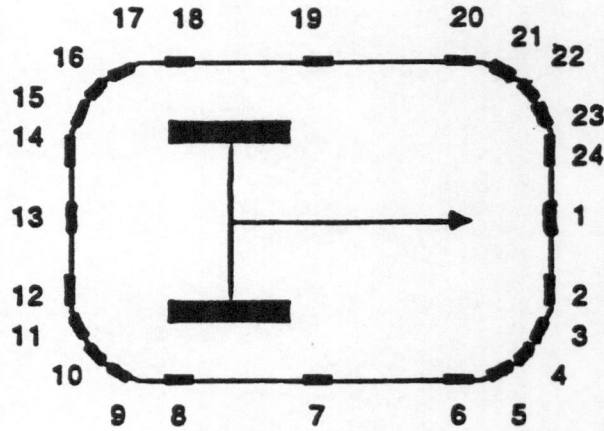

Figure 18. The Geometry and Sensor Placement on the INRIA Mobile Robot

A   Wandering Robot Example

In this appendix, a system is developed which combines several ESTEREL modules (ALARM, GET_MIN_DISTANCE, WANDER and COMBINE) with the on-board robot command routines to generate random robot movement. The robot generates a random move every 10 seconds, and executes it; if there is any obstacle closer than the predefined threshold, the robot makes an emergency stop.

The COMBINE.strl, ALARM.strl, GET_NEAREST_OBJ.strl and WANDER.strl modules are as follows:

```
% $Header: COMBINE.strl,v 1.1 88/12/07  tch Locked $
%%%%%%%%%%%%%%%%%%%%%%%%%%%%%%%%%%%%%%%%%%%%%%%%%%%%%%%%%%
% MODULE TO COMBINE READING WITH WATCHING THE SONAR SENSORS%
%%%%%%%%%%%%%%%%%%%%%%%%%%%%%%%%%%%%%%%%%%%%%%%%%%%%%%%%%%

module COMBINE:

type DISTANCE,
     PING,
     R_THETA,
     MOVE;

constant OMEGA_MIN, OMEGA_MAX : integer;

input S;
input MOVE_TIME;

relation MOVE_TIME => S;

sensor CURRENT_SONAR (PING);

output DISPLAY_ALARM(R_THETA);
output MOVE_CMD(MOVE);

[
signal GET_SONAR(PING), NEAREST_OBJ(R_THETA) in

  every S do
```

```
    emit GET_SONAR(?CURRENT_SONAR)
  end
||
  copymodule ALARM
||
  copymodule GET_MIN_DISTANCE
||
  copymodule WANDER
end
]
.
```

```
% $Header: ALARM.strl,v 1.1 88/12/07 tch Locked $
%%%%%%%%%%%%%%%%%%%%%%%%%%%%%%%%%%%%%%%%%%%%%%%%%%%%%%%%
% MODULE TO SOUND ALARM IF OBJECT TOO CLOSE       %
%%%%%%%%%%%%%%%%%%%%%%%%%%%%%%%%%%%%%%%%%%%%%%%%%%%%%%%%

module ALARM:

type DISTANCE,
     PING,
     R_THETA;

constant MIN_ALARM_DIST : DISTANCE;

input NEAREST_OBJ(R_THETA) ;
input GET_SONAR(PING) ;

output DISPLAY_ALARM(R_THETA);

function LESS_THAN_DISTANCE_TO_DISTANCE(DISTANCE,DISTANCE) : boolean;
function EXTRACT_R(R_THETA) : DISTANCE;
function SONAR_TO_R_THETA(PING) : R_THETA;

every immediate NEAREST_OBJ do
  if LESS_THAN_DISTANCE_TO_DISTANCE(EXTRACT_r(?NEAREST_OBJ),MIN_ALARM_DIST)
    then emit DISPLAY_ALARM(SONAR_TO_R_THETA(?GET_SONAR))
  end
end
```

```
% $Header: GET_MIN_DISTANCE.strl,v 1.1 89/1/17 tch Locked $
%%%%%%%%%%%%%%%%%%%%%%%%%%%%%%%%%%%%%%%%%%%%%%%%%%%%%%
% MODULE TO GET MINIMUM DISTANCE FROM SONARS       %
%%%%%%%%%%%%%%%%%%%%%%%%%%%%%%%%%%%%%%%%%%%%%%%%%%%%%%

module GET_MIN_DISTANCE:

type PING,
     R_THETA;

input GET_SONAR(PING) ;

output NEAREST_OBJ(R_THETA);
function SONAR_TO_R_THETA(PING) : R_THETA;
```

```
every immediate GET_SONAR do
  emit NEAREST_OBJ(SONAR_TO_R_THETA(?GET_SONAR))
end


% $Header: WANDER.strl,v 1.1 88/12/22 tch Locked $
%%%%%%%%%%%%%%%%%%%%%%%%%%%%%%%%%%%%%%%%%%%%%%%%%%%%%%%%%
%  MODULE TO GENERATE RANDOM MOVES          %
%%%%%%%%%%%%%%%%%%%%%%%%%%%%%%%%%%%%%%%%%%%%%%%%%%%%%%%%%

module WANDER:

type MOVE;

constant OMEGA_MIN, OMEGA_MAX : integer;

input MOVE_TIME;

output MOVE_CMD(MOVE);

function rand(integer, integer) : integer;
function TURNS_TO_MOVE(integer, integer) : MOVE;

every MOVE_TIME do

 var left_wheel_turns, right_wheel_turns : integer in

  left_wheel_turns := rand(OMEGA_MIN, OMEGA_MAX);
  right_wheel_turns := rand(OMEGA_MIN, OMEGA_MAX);
  emit MOVE_CMD(TURNS_TO_MOVE(left_wheel_turns,right_wheel_turns))
 end
end
```

The finite state machine produced for COMBINE is:

Automaton COMBINE   (Debug Format)

1. Memory allocation

```
V0: boolean (boolean of signal S)
V1: boolean (boolean of signal MOVE_TIME)
V2: R_THETA (value of signal DISPLAY_ALARM)
V3: MOVE (value of signal MOVE_CMD)
V4: PING (value of signal GET_SONAR)
V5: R_THETA (value of signal NEAREST_OBJ)
V6: integer (variable left_wheel_turns)
V7: integer (variable right_wheel_turns)
V8: PING (value of sensor CURRENT_SONAR)
V9: boolean (boolean of sensor CURRENT_SONAR)
```

2. Actions

2.1 Present signal tests

```
A1: V0 (signal S)
A2: V1 (signal MOVE_TIME)
```

2.2 Output actions

```
A3: DISPLAY_ALARM (V2)
A4: MOVE_CMD (V3)
```

2.3 Assignments

```
A5:  V4 := (V9 ? V8 : (V9:=true,V8:=S_CURRENT_SONAR()))
A6:  V2 := SONAR_TO_R_THETA(V4)
A7:  V5 := SONAR_TO_R_THETA(V4)
A8:  V6 := rand(OMEGA_MIN, OMEGA_MAX)
A9:  V7 := rand(OMEGA_MIN, OMEGA_MAX)
A10: V3 := TURNS_TO_MOVE(V6, V7)
```

2.4 Conditions

```
A11: LESS_THAN_DISTANCE_TO_DISTANCE(EXTRACT_R(V5), MIN_ALARM_DIST)
A12: false
```

3. Automaton

```
State 0

goto 1

State 1

if A1 then
   A5;
   if A2 then
      A8;A9;A10;A7;
      if A11 then
         A6;A3;A4;
         goto 1
      end;
      A4;
      goto 1
   end;
   A7;
   if A11 then
      A6;A3;
      goto 1
   end;
   goto 1
end;
goto 1
```

Multiple processes can be added to the robot by using the add_process command in the Robuter C interface software. However, a send with APRO works better:

```
sprintf(cmd,"MOVE P RC=%d,%d P=%d \n",move.left_wheel_turns,
                        move.right_wheel_turns,
                            move.period);
send(cmd);
```

The program must be loaded into the robot memory, and the go command issued to start it. The program then requests the user to enter a delay which corresponds to how long the program is to run (independently monitored). The robot then generates random moves (the number of turns for each wheel is independent) of not more than 20 centimeters a move every ten seconds and stops if an object is detected closer than two centimeters.

References

[1] J. Albus. *Brains, Behavior and Robotics*. BYTE Books, Peterborough, New Hampshire, 1981.

[2] Peter Allen. Integrating Vision and Touch for Object Recognition Tasks. *International Journal of Robotics Research*, 7(6):15–33, 1988.

[3] Russell L. Andersson. Investigating Fast, Intelligent Systems with a Ping-Pong Playing Robot. In Robert C. Bolles and Bernard Roth, editors, *Proceedings of the Fourth International Symposium on Robotics Research*, pages 15–22, MIT Press, Cambridge, Massachusetts, 1988.

[4] Ronald C. Arkin. Motor Schema Based Navigation for a Mobile Robot: An Approach to Programming by Behavior. In *Proceedings of the International Conference on Robotics and Automation*, pages 264–271, IEEE, Raleigh, North Carolina, 1987.

[5] Ronald C. Arkin, Edward M. Riseman, and Allan R. Hanson. AuRA: An Architecture for Vision-Based Robot Navigation. In *Proceedings of the DARPA Image Understanding Workshop*, pages 417–431, Morgan Kaufmann, Inc., Los Altos, California, 1987.

[6] N. Ayache and O.D. Faugeras. Building, Registrating, and Fusing Noisy Visual Maps. *International Journal of Robotics Research*, 7(6):45–65, 1988.

[7] Nicholas Ayache and Olivier D. Faugeras. Building a Consistent 3D Representation of a Mobile Robot Environment by Combining Multiple Stereo Views. In *IJCAI-87*, pages 808–810, Munich, RFA, August 1987.

[8] Nicholas Ayache and Olivier D. Faugeras. Maintaining Representations of the Environment of a Mobile Robot. In Robert C. Bolles and Bernard Roth, editors, *Proceedings of the Fourth International Symposium on Robotics Research*, pages 337–350, MIT Press, Cambridge, Massachusetts, 1988.

[9] Gérard Berry and Georges Gonthier. *The Esterel Synchronous Programming Language: Design, Semantics, Implementation*. Research Report 842, INRIA, Sophia Antipolis, France, May 1988.

[10] Bir Bhanu and Wilhelm Burger. DRIVE - Dynamic Reasoning from Integrated Evidence. In *Proceedings of the DARPA Image Understanding Workshop*, pages 581–588, Morgan Kaufmann, Inc., Los Altos, California, 1987.

[11] R. Bhatt, D. Gaw, and A. Meystel. A Real-Time Guidance System for an Autonomous Vehicle. In *Proceedings of the International Conference on Robotics and Automation*, pages 1785–1791, IEEE, Raleigh, North Carolina, 1987.

[12] J.D. Boissonnat, Olivier D. Faugeras, and E. LeBras-Mehlman. *Representing Stereo Data with the Delauney Triangulation*. INRIA Research Report 788, INRIA, Roquencourt, France, February 1988.

[13] Michael Brady, Stephen Cameron, Hugh Durrant-Whyte, Margaret Fleck, David Forsyth, Alison Noble, and Ian Page. Progress toward a System that can Acquire Pallets and Clean Warehouses. In Robert C. Bolles and Bernard Roth, editors, *Proceedings of the Fourth International Symposium on Robotics Research*, pages 359–374, MIT Press, Cambridge, Massachusetts, 1988.

[14] Valentino Braintenberg. *Vehicles: Experiments in Synthetic Psychology*. MIT Press, Cambridge, Massachusetts, 1987.

[15] R.A. Brooks. A Robust Layered Control System For A Mobile Robot. *IEEE Journal of Robotics and Automation*, 2(1):14–23, March 1986.

[16] Rodney Brooks. A Robust Programming Scheme for a Mobile Robot. In *Languages for Sensor Based Control in Robotics*, pages 509–522, Springer-Verlag, Berlin, 1987.

[17] Robert E. Carlton and Stephen J. Bartholet. The Evolution of the Application of Mobile Robotics to Nuclear Facility Operations and Maintenance. In *Proceedings of the International Conference on Robotics and Automation*, pages 720–726, IEEE, Raleigh, North Carolina, 1987.

[18] B. Chandrasekaran. Towards a Functional Architecture for Intelligence Based on Generic Information Processing Tasks. In *Proceedings of IJCAI-87*, pages 1183–1192, Munich, RFA, August 1987.

[19] Raja Chatila. Mobile Robot Navigation: Space Modeling and Decisional Processes. In Olivier D. Faugeras and George Giralt, editors, *Proceedings of the Third International Symposium on Robotics Research*, pages 373–378, MIT Press, Cambridge, Massachusetts, 1986.

[20] R. Chattergy. Some Heuristics for the Navigation of a Robot. *International Journal of Robotics Research*, 4(1):59–66, Spring 1985.

[21] S.L. Chiu, D.J. Morley, and J.F.Martin. Sensor Data Fusion on a Parallel Processor. In *Proceedings of the IEEE Conference on Robotics and Automation*, pages 1629–1633, San Francisco, CA, April 1986.

[22] James L. Crowley. Coordination of Action and Perception in a Surveillance Robot. In *IJCAI-87*, pages 793–796, Munich, RFA, August 1987.

[23] Larry S. Davis, Daniel Dementhon, Ramesh Gajulapalli, Todd R. Kushner, Jacqueline LeMoigne, and Phillip Veatch. Vision-Based Navigation: A Status Report. In *Proceedings of the DARPA Image Understanding Workshop*, pages 153–169, Morgan Kaufmann, Inc., Los Altos, California, 1987.

[24] E.D. Dickmann. 4D-Dynamic Scene Analysis with Integral Spatio-temporal Models. In Robert C. Bolles and Bernard Roth, editors, *Proceedings of the Fourth International Symposium on Robotics Research*, pages 311–318, MIT Press, Cambridge, Massachusetts, 1988.

[25] S.H. Drake. *Using Compliance in Lieu of Sensory Feedback for Automatic Assembly*. PhD thesis, Massachusetts Institute of Technology, Cambridge, MA, September 1977.

[26] Bruce A. Draper, Robert T. Collins, John Brolio, Joey Griffith, Allan R. Hanson, and Edward M. Riseman. Tools and Experiments in the Knowledge-Directed Interpretation of Road Scenes. In *Proceedings of the DARPA Image Understanding Workshop*, pages 178–193, Morgan Kaufmann, Inc., Los Altos, California, 1987.

[27] Bruno Dufay and Jean-Claude Latombe. An Approach to Automatic Robot Programming Based on Inductive Learning. In Michael Brady and Richard Paul, editors, *Proceedings of the First International Symposium on Robotics Research*, pages 97–115, MIT Press, Cambridge, Massachusetts, 1984.

[28] H. F. Durrant-Whyte. Consistent Integration and Propagation of Disparate Sensor Observations. *International Journal of Robotics Research*, 6(3):3–24, 1987.

[29] H. F. Durrant-Whyte. Sensor Models and Multisensor Integration. *International Journal of Robotics Research*, 7(6):97–113, 1988.

[30] H.F. Durrant-Whyte. Consistent Integration and Propagation of Disparate Sensor Observations. In *Proceedings of the IEEE Conference on Robotics and Automation*, pages 1464–1469, San Francisco, CA, April 1986.

[31] Alberto Elfes. Sonar-Based Real-World Mapping and Navigation. *IEEE Journal of Robotics and Automation*, 3(3):249–265, June 1987.

[32] Bernard Espiau. Closed Loop Control of Robots with Local Environment Sensing: Principles and Applications. In Hideo Hanafusa and Hirochika Inoue, editors, *Proceedings of the Second International Symposium on Robotics Research*, pages 147–154, MIT Press, Cambridge, Massachusetts, 1985.

[33] O.D. Faugeras, N. Ayache, and B. Faverjon. Building Visual Maps by Combining Noisy Stereo Measurements. In *Proceedings of the IEEE Conference on Robotics and Automation*, pages 1433–1438, San Francisco, CA, April 1986.

[34] Olivier D. Faugeras. Artificial 3D Vision. In *IJCAI-87*, pages 1169–1171, Munich, RFA, August 1987.

[35] B. Faverjon and P. Tournassoud. The Mixed Approach for Motion Planning: Learning Global Strategies from a Local Planner. In *IJCAI-87*, pages 1131–1137, Munich, RFA, August 1987.

[36] Bernard Faverjon and Pierre Tournassoud. Planification et calcul de trajectoires pour Robots Manipulateurs en présence d'obstacles. In *Jounees Geometrique et Robotique*, pages 1–9, INRIA, Toulouse, France, 1988.

[37] William T. Gex and Nancy L. Campbell. Local Free Space Mapping and Path Guidance. In *Proceedings of the International Conference on Robotics and Automation*, pages 424–431, IEEE, Raleigh, North Carolina, 1987.

[38] Malik Ghallab, Rachid Alami, and Raja Chatila. Dealing with Time in Planning and Execution Monitoring. In Robert C. Bolles and Bernard Roth, editors, *Proceedings of the Fourth International Symposium on Robotics Research*, pages 431–443, MIT Press, Cambridge, Massachusetts, 1988.

[39] Georges Giralt. Research Trends in Decisional and Multisensory Aspects of Third Generation Robots. In Hideo Hanafusa and Hirochika Inoue, editors, *Proceedings of the Second International Symposium on Robotics Research*, pages 511–520, MIT Press, Cambridge, Massachusetts, 1985.

[40] Georges Giralt, Raja Chatila, and Marc Vaisset. An Integrated Navigation and Motion Control System for Autonomous Multisensory Mobile Robots. In Michael Brady and Richard Paul, editors, *Proceedings of the First International Symposium on Robotics Research*, pages 191–214, MIT Press, Cambridge, Massachusetts, 1984.

[41] M. Goldstein, F.G. Pin, G. de Saussure, and C.R. Weisbin. 3D World Modeling based on Combinatorial Geometry for Autonomous Robot Navigation. In *Proceedings of the International Conference on Robotics and Automation*, pages 727–733, IEEE, Raleigh, North Carolina, 1987.

[42] R.A. Grupen. *General Purpose Grasping and Manipulation with Multifingered Robot Hands*. PhD thesis, University of Utah, Merrill Engineering Building, Salt Lake City, UT 84112, August 1988.

[43] Gregory D. Hager. *Active Reduction of Uncertainty in Multisensor Systems*. PhD thesis, University of Pennsylvania, Philadelphia, Pennsylvania, July 1988.

[44] Scott Y. Harmon. The Ground Surveillance Robot (GSR): An Autonomous Vehicle Designed to Transit Unknown Terrain. *IEEE Journal of Robotics and Automation*, 3(3):266–279, June 1987.

[45] T. Henderson and Rod Grupen. Autochthonous Behaviors: Mapping Perception to Action. In Thomas C. Henderson, editor, *NATO ASI on Traditional and Non-Traditional Robotic Sensors*, page to appear, Springer-Verlag, Maratea, Italy, August 28 - September 2 1989.

[46] T.C. Henderson, C.D. Hansen, and Bir Bhanu. The Specification of Distributed Sensing and Control. *Journal of Robotic Systems*, 2(4):387–396, 1985.

[47] T.C. Henderson, Chuck Hansen, and Bir Bhanu. A Framework for Distributed Sensing and Control. In *Proceedings of IJCAI 1985*, pages 1106–1109, Los Angeles, CA, August 1985.

[48] T.C. Henderson and E. Shilcrat. Logical Sensor Systems. *Journal of Robotic Systems*, 1(2):169–193, 1984.

[49] T.C. Henderson, E. Shilcrat, and C.D. Hansen. A Fault Tolerant Sensor Scheme. In *Proceedings of the International Conference on Pattern Recognition*, pages 663–665, August 1984.

[50] Thomas C. Henderson. *Workshop on Multisensor Integration for Manufacturing Automation*. Technical Report UU-CS-87-006, University of Utah, Department of Computer Science, Feb. 1987.

[51] Thomas C. Henderson, Eliot Weitz, Chuck Hansen, and Amar Mitiche. Multisensor Knowledge Systems: Interpreting 3D Structure. *International Journal of Robotics Research*, 7(6):114–137, 1988.

[52] G. Hirzinger. Robot Learning and Teach-In Based on Sensory Feedback. In Olivier D. Faugeras and George Giralt, editors, *Proceedings of the Third International Symposium on Robotics Research*, pages 155–163, MIT Press, Cambridge, Massachusetts, 1986.

[53] N. Hogan. Impedance Control: An Approach to Manipulation: I – Theory, II – Implementation, III – Applications. *ASME Journal of Dynamic Systems, Measurement, and Control*, 107:1–24, March 1985.

[54] Can Işik and Alexander M. Meystel. Pilot Level of a Hierarchical Controller for an Unmanned Mobile Robot. *IEEE Journal of Robotics and Automation*, 4(3):241–255, June 1988.

[55] R.A. Jarvis and J.C. Byrne. An Automated Guided Vehicle with Map Building and Path Finding Capabilities. In Robert C. Bolles and Bernard Roth, editors, *Proceedings of the Fourth International Symposium on Robotics Research*, pages 497–504, MIT Press, Cambridge, Massachusetts, 1988.

[56] Leslie P. Kaelbling. An Architecture for Intelligent Reactive Systems. In M.P. Georgedd and A.L. Lansky, editors, *Proceedings of the Workshop on Reasoning about Plans*, pages 395–410, Timberline, Oregon, June 30 - July 2 1986.

[57] A.C. Kak, B.A. Roberts, K.M. Andress, and R.L. Cromwell. Experiments in the Integration of World Knowledge with Sensory Information for Mobile Robots. In *Proceedings of the International Conference on Robotics and Automation*, pages 734–740, IEEE, Raleigh, North Carolina, 1987.

[58] A.C. Kak, A.J. Vayda, R.L. Cromwell, W.Y. Kim, and C.H. Chen. Knowledge-Based Robotics. *Int. J. Prod. Res.*, 26(5):707–734, 1988.

[59] David J. Kriegman, Ernst Triendl, and Thomas O. Binford. A Mobile Robot: Sensing, Planning and Locomotion. In *Proceedings of the International Conference on Robotics and Automation*, pages 402–408, IEEE, Raleigh, North Carolina, 1987.

[60] B. Kuipers and T. Levitt. Navigation and Mapping in Large-Scale Space. *AI Magazine*, 9(2):25–43, Summer 1988.

[61] J.-C. Latombe and C. Laugier. Systèmes de programmation pour la robotique. In *Journees geometrique et robotique*, pages 223–235, INRIA, Toulouse, France, 1988.

[62] Douglas B. Lenat and Edward A. Feigenbaum. On the Thresholds of Knowledge. In *Proceedings of IJCAI-87*, pages 1173–1182, Munich, RFA, August 1987.

[63] Paul Levi. Principles of Planning and control Concepts for Autonomous Mobile Robots. In *Proceedings of the International Conference on Robotics and Automation*, pages 874–881, IEEE, Raleigh, North Carolina, 1987.

[64] Tod S. Levitt, Daryl T. Lawton, David M. Chelberg, and Philip C. Nelson. Qualitative Navigation. In *Proceedings of the DARPA Image Understanding Workshop*, pages 447–465, Morgan Kaufmann, Inc., Los Altos, California, 1987.

[65] John D. Lowrance and Thomas D. Garvey. *Evidential Reasoning: An Implementation for Multisensor Integration*. Technical Note 307, SRI, Inc., December 1983.

[66] T. Lozano-Pérez, M. Mason, and R. Taylor. Automatic Synthesis of Fine-Motion Strategies for Robots. *International Journal of Robotics Research*, 3(1):3–24, 1984.

[67] R.C. Luo, M.-H. Lin, and R.S. Scherp. Dynamic Multisensor Data Fusion System for Intelligent Robots. *IEEE Journal of Robotics and Automation*, 4(4):386–396, August 1988.

[68] Mark B. Metea and Jeffery J.-P. Tsai. Route Planning for Intelligent Autonomous Land Vehicles using Hierarchical Terrain Representation. In *Proceedings of the International Conference on Robotics and Automation*, pages 1947–1952, IEEE, Raleigh, North Carolina, 1987.

[69] A. Meystel. Nested Hierarchical Controller for Intelligent Mobile Autonomous System. In L.O. Hertzberger and F.C.A. Groen, editors, *Intelligent Autonomous Systems*, pages 416–448, North-Holland, Amsterdam, The Netherlands, 1987.

[70] J. Milberg and P. Lutz. Integration of Autonomous Mobile Robots into the Industrial Production Environment. In *Proceedings of the International Conference on Robotics and Automation*, pages 1953–1959, IEEE, Raleigh, North Carolina, 1987.

[71] M. Minsky. *The Society of Mind*. Simon and Schuster, NY, NY, 1986.

[72] A. Mitiche and J.K. Aggarwal. An Overview of Multisensor Systems. *SPIE Optical Computing*, 2:96–98, 1986.

[73] Hans Moravec. Sensor Fusion in Certainty Grids for Mobile Robots. *AI Magazine*, 9(2):61–74, Summer 1988.

[74] Hatem Nasr, Bir Bhanu, and Stephanie Schaffer. Guiding an Autonomous Land Vehicle Using Knowledge-Based Landmark Recognition. In *Proceedings of the DARPA Image Understanding Workshop*, pages 432–439, Morgan Kaufmann, Inc., Los Altos, California, 1987.

[75] James L. Nevins, M. Desai, E. Fogel, B.K. Walker, and D.E. Whitney. Adaptive Control, Learning, and Cost Effective Sensor Systems for Robotic or Advanced Automation Systems. In Michael Brady and Richard Paul, editors, *Proceedings of the First International Symposium on Robotics Research*, pages 983–994, MIT Press, Cambridge, Massachusetts, 1984.

[76] Nils J. Nilsson. *Triangle Tables: A Proposal for a Robot Programming Language*. Technical Report Technical Note 347, SRI International, February 1985.

[77] K. Overton. *The Acquisition, Processing, and Use of Tactile Sensor Data in Robto Control*. PhD thesis, University of Massachusetts, Amherst, Massachusetts, May 1984.

[78] Jocelyne Pertin-Troccaz and Pierre Puget. Dealing with Uncertainty in Robot Planning Using Program Proving Techniques. In Robert C. Bolles and Bernard Roth, editors, *Proceedings of the Fourth International Symposium on Robotics Research*, pages 455–466, MIT Press, Cambridge, Massachusetts, 1988.

[79] M.H. Raibert. *Legged Robots that Balance*. MIT Press, Cambridge, MA, 1986.

[80] Nageswara S.V. Rao, S.S. Iyenfar, C.C. Jorgensen, and C.R. Weisbin. On Terrain Acquisition by a Finite-Sized Mobile Robot in Plane. In *Proceedings of the International Conference on Robotics and Automation*, pages 1314–1319, IEEE, Raleigh, North Carolina, 1987.

[81] K. Salisbury. *Kinematic and Force Analysis of Articulated Hands*. PhD thesis, Stanford, Stanford, California, May 1982.

[82] Steven Salzberg. Heuristics for Inductive Learning. In *IJCAI-85*, pages 603–609, San Francisco, California, August 1985.

[83] R. Schott. On Mobile Robots: A Probabilistic Model for the Representation and Manipulation of Spatial Uncertainty. In *Proceedings of the International Conference on Robotics and Automation*, pages 409–415, IEEE, Raleigh, North Carolina, 1987.

[84] Oliver G. Selfridge and Richard S. Sutton. Training and Tracking in Robotics. In *IJCAI-85*, pages 670–672, San Francisco, California, August 1985.

[85] Grahame B. Smith and Thomas M. Strat. Information Management in a Sensor-Based Autonomous System. In *Proceedings of the DARPA Image Understanding Workshop*, pages 170–177, Morgan Kaufmann, Inc., Los Altos, California, 1987.

[86] R. Smith and P. Cheeseman. On the Representation and Estimation of Spatial Uncertainty. *International Journal of Robotics Research*, 5(4):56–68, 1987.

[87] Ralph P. Sobek. A Robot Planning Structure using Production Rules. In *IJCAI-85*, pages 1103–1105, San Francisco, California, August 1985.

[88] Anthony Stentz and Yoshimasa Goto. The CMU Navigational Architecture. In *Proceedings of the DARPA Image Understanding Workshop*, pages 440–446, Morgan Kaufmann, Inc., Los Altos, California, 1987.

[89] Charles Thorpe, Steven Shafer, Takeo Kanade, and the members of the Strategic Computing Vision Lab. Vision and Navigation for the Carnegie-Mellon Navlab. In *Proceedings of the DARPA Image Understanding Workshop*, pages 143–152, Morgan Kaufmann, Inc., Los Altos, California, 1987.

[90] Pierre Tournassoud. Motion Planning for a Mobile Robot with a Kinematic Constraint. In *Jounees Geometrique et Robotique*, pages 1–26, INRIA, Toulouse, France, 1988.

[91] Ernst Triendl and David J. Kriegman. Stereo Vision and Navigation within Buildings. In *Proceedings of the International Conference on Robotics and Automation*, pages 1725–1730, IEEE, Raleigh, North Carolina, 1987.

[92] Ernst Triendl and David J. Kriegman. Vision and Visual Exploration for the Stanford Mobile Robot. In *Proceedings of the DARPA Image Understanding Workshop*, pages 407–416, Morgan Kaufmann, Inc., Los Altos, California, 1987.

[93] Saburo Tsuji and Jiang Yu Zheng. Visual Path Planning by a Mobile Robot. In *IJCAI-87*, pages 1127–1130, Munich, RFA, August 1987.

[94] F.M. Tuijnman, W. Beemster, W. Duinker, L.O. Hertzberger, E. Kuipers, and H. Muller. A Model for Control Software and Sensor Algorithms for an Autonomous Mobile Robot. In L.O. Hertzberger and F.C.A. Groen, editors, *Intelligent Autonomous Systems*, pages 610–615, North-Holland, Amsterdam, The Netherlands, 1987.

[95] Volker Turau. A Model for a Control and Monitoring System for an Autonomous Mobile Robot. *Robotics*, 4:41–47, 1988.

[96] J.R. White, H.W. Harvey, and K.A. Farnstrom. Testing of Mobile Surveillance Robot at a Nuclear Power Plant. In *Proceedings of the International Conference on Robotics and Automation*, pages 714–719, IEEE, Raleigh, North Carolina, 1987.

[97] William L. Whittaker, George Turkiyyah, and Martial Hebert. An Architecture and Two Cases in Range-Based Modeling and Planning. In *Proceedings of the International Conference on Robotics and Automation*, pages 1991–1997, IEEE, Raleigh, North Carolina, 1987.

[98] Dit-Yan Yeung and George A. Bekey. A Decentralized Approach to the Motion Planning Problem for Multiple Mobile Robots. In *Proceedings of the International Conference on Robotics and Automation*, pages 1779–1784, IEEE, Raleigh, North Carolina, 1987.

Interpreting 3D Lines[†]

Amar Mitiche,   Robert Laganière

INRS Télécommunications
3, Place du Commerce, Ile des Soeurs, Qc, Canada H3E 1H6

Abstract

Computational methods to interpret straight line correspondences in images necessitate the observation of a large number of lines, and introduce systems of equations the numerical solution of which often exhibits several ills. Man-made environments, however, contain many special configurations such as parallel lines, perpendicular lines, known angular configurations in general, etc.; if the existence of such configurations can be ascertained from image data, the problem of 3D interpretation becomes drastically simpler. Moreover, if part of the environment is interpreted, this interpretation can be propagated to other parts of the environment. We propose a consistent labelling formulation which allows the interpretation of general configurations of lines by taking into account the occurrence of special configurations and using propagation.

1.   Introduction

Given image line correspondences over several views, the problem is to determine the position (equation) of the observed lines in space and the motions (displacements) between the viewing positions. The problem, in its generality, requires a minimum of three views [1].

Several 'direct' formulations of this problem have been proposed : the *angular invariance*, the *intersecting planes*, and the *linear* formulations.

The angular invariance formulation [2] uses explicitly the fact that the angles between the lines of a rigid configuration of lines do not change when the configuration undergoes a (rigid) motion. The resulting system of equations involves line orientations only. When

---

[†] This work was supported in part by the Natural Sciences and Engineering Research Council of Canada under grant NSERC-A4234

these orientations are determined, the rotational parameters of the motions are obtained. The final step consists of computing the translational parameters of the motions, and the complete position of the lines.

The intersecting planes formulation [3], [4] translates explicitly the fact that an observed line in space is the intersection of three projective planes. The resulting equations contain only the rotational parameters of the motions. When these are determined, line orientations are obtained. Finally, as with the angular invariance formulation, the translational parameters of the motions, and the complete position of the lines, are recovered.

The intersecting planes formulation can lead to a linear set of equations following a change of variables [5], [6]. The original unknowns are recovered from the solution to the linear system.

The difficulty in the problem is with the computation of line orientations/rotational components of motions [1]. Algorithms associated with these direct formulations share a number of problems. First, a large number of observations is required: at least three views and, with three views, at least six lines for the nonlinear formulations and thirteen lines for the linear formulation. It is difficult to observe such a large number of lines in practice. When it is possible to do so, it is difficult to establish correspondence. Second, assuming we observe a sufficient number of lines for which correspondence is established, distinct views are generally not sufficiently different. This amplifies the problems observed to occur with the algorithms: instability, sensitivity to noise and initial approximations, convergence, etc. [2]-[6]. As a result, these algorithms are impractical.

To overcome these problems, we advocate the use of the constraint satisfaction formulation. First we note that if, instead of general configurations of lines in space, we consider special (but commonly found in man-made environments) configurations such as orthogonal lines, parallel lines, configurations of known angular relations, etc., the problem is simple. Moreover, if a part of space is interpreted, this interpretation can be propagated to interpret other parts of space. One way to proceed is then to have computational units tuned to special configurations of lines in space, instantiating them when the occurrence of such configurations is either directly verified or is hypothesized. When a computational unit is instantiated, the immediate result is the interpretation of the lines involved in the special configuration to which the unit is tuned; propagation will then yield the position of lines in other parts of space; lines reached by propagation are not necessarily part of any special configuration.

Results from a given hypothesis are accepted until they are proven inconsistent with results from another hypothesis (or ground truth if any). Now, of course, *inference control* and *truth maintenance* are the real problems. Taking two hypothesized special configurations to be compatible over a period of time if, during that time, they do not lead to contradictory interpretation of any part of space, the process of 3D interpretation can be formulated efficiently as a search for a globally consistent solution in a network of competing hypotheses. Consistent labelling resolution [7] (e.g. relaxation) can then be used to perform such a search.

These are the basic ideas incorporated in 3D-KNOBIS (3D KNOwledge-Based Interpretation System) described in this paper.

The remainder of this paper is organized as follows : Section 2 gives examples of computational units and explains propagation, Section 3 describes the general structure of 3D-KNOBIS, while Section 4 presents an application to the interpretation of a sequence of image line correspondences. Section 5 contains a summary.

2. Computational units

The following computational units will be used to illustrate the consistent labelling formulation of the problem of computing 3D line orientations (Section 4). Their individual performance has been reported in [2]. The special configurations of lines to which they are tuned are ubiquitous in space. Indeed, a variety of objects and architectural elements have known angular relations; orthogonal lines occur at corners of walls-floor, walls-ceiling, corners of objects such as cabinets and housing boxes; parallel lines also abound. Man-made environments which contain such special configurations are of prime importance in autonomous robot navigation.

The viewing system is modeled by a cartesian reference system and central projection (Figure 1). Line L in space has (unit) orientation $\mathbf{A} = (A_1, A_2, A_3)$. Its projection is ℓ. According to projective relations $\exists \lambda > 0$ such that:

$$A_1 = \frac{\lambda x_2 - x_1}{a}$$
$$A_2 = \frac{\lambda y_2 - y_1}{a} \tag{1}$$
$$A_3 = \frac{(1 - \lambda)f}{a}$$

Where f is the focal length, (x_1, y_1) and (x_2, y_2) are *any* two distinct points on ℓ and $a = \sqrt{(\lambda x_2 - x_1)^2 + (\lambda y_2 - y_1)^2 + (1 - \lambda)^2 f^2}$ is a normalizing factor. Equation (1) indicates that line L is in plane Π (projective plane).

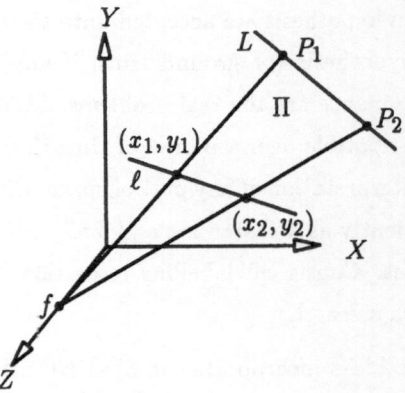

Fig. 1 The viewing model.

2.1 Configurations with known angular relations

The equation for line i and line j is:

$$\mathbf{A}^i \cdot \mathbf{A}^j = cos\theta_{ij} \tag{2}$$

where $\mathbf{A}^i$, $\mathbf{A}^j$ are the orientations of lines i and j (as in Equation (1)), and θ_{ij} is the (known) angle between them. Three or more lines in only one view are needed as three lines will contribute three unknowns and three equations.

2.2 Orthogonal lines.

With orthogonal lines, only one view is necessary as in the case of known angular configurations. The equation for lines i and j is:

$$\mathbf{A}^i \cdot \mathbf{A}^j = 0 \tag{3}$$

The equation above is simplified by dropping the normalizing factors appearing in the denominator of the expressions of $\mathbf{A}^i$ and $\mathbf{A}^j$ (Equation (1)). Its expanded form with three lines is, assuming unit focal length :

$$(\lambda x_{21} - x_{11})(\mu x_{22} - x_{12}) + (\lambda y_{21} - y_{11})(\mu y_{22} - y_{12}) + (1 - \lambda)(1 - \mu) = 0$$
$$(\lambda x_{21} - x_{11})(\gamma x_{23} - x_{13}) + (\lambda y_{21} - y_{11})(\gamma y_{23} - y_{13}) + (1 - \lambda)(1 - \gamma) = 0 \qquad (4)$$
$$(\mu x_{22} - x_{12})(\gamma x_{23} - x_{13}) + (\mu y_{22} - y_{12})(\gamma y_{23} - y_{13}) + (1 - \mu)(1 - \gamma) = 0$$

where (x_{ij}, y_{ij}) are the coordinates of (arbitrary) point i ($i = 1, 2$) on image line j ($j = 1, 3$). The system above can be solved analytically, leading to two solutions. In the absence of any other cue, these two solutions are both acceptable. To disambiguate the interpretation, an additional line with known angular relation to the three lines can be used. Another way is to use a second view and propagation as explained subsequently.

2.3 Parallel Lines

With a viewing system as in Figure 1, the direction of two (or more) parallel lines is the direction of the line through the center of projection and the focus of expansion.

2.4 Propagation

Propagation allows the use of context to spread interpretation from interpreted parts of space. Lines reached through propagation are not necessarily part of special configurations. There are two cases.

(1) *Propagation from one line.*

The equation involving interpreted line L^0 and uninterpreted line L^i is :

$$\mathbf{A}^i \cdot \mathbf{A}^0 = \mathbf{B}^i \cdot \mathbf{B}^0 \qquad (5)$$

where $\mathbf{A}^i$, $\mathbf{B}^i$ are the orientations of L^i in the first view and second view respectively, and $\mathbf{A}^0$, $\mathbf{B}^0$ are those of interpreted line L^0. With $m \geq 2$ views of $n \geq 2$ lines there are $(2n - 3)(m - 1)$ independent angular invariance equations [1]. Therefore two views are not enough. With three views, two uninterpreted lines and one interpreted line, there are four equations involving the known orientation, and two equations involving only the uninterpreted lines, making a total of six equations. We also have six unknowns (one unknown per line, per view). Therefore propagation can be applied given three or more views of one interpreted line and two or more uninterpreted lines.

(2) *Propagation from two or more lines.*

Here, a count of unknowns and equations indicates that only two views are needed. With two views, two or more interpreted lines are needed. An alternative would be to solve for the rotation between the two views using the interpreted lines, and then compute the position of the other lines.

Propagation can disambiguate the case of perpendicular lines (for which two solutions are possible). If we have two views of three perpendicular lines we compute the two interpretations for each view; then, given an additional line, we use propagation to compute its orientation according to each interpretation and each view. Finally, angular invariance between the two views will disambiguate the interpretation as only the actual line configuration will occur in the two views.

3. 3D-KNOBIS: A 3D KNOwledge-Based Interpretation System

Several researchers have recognized the need for vision systems to use prior knowledge in the analysis of images [10]-[19]. Although 3D-KNOBIS uses hypotheses, it is not a traditional hypothesize-test system (as in [15]-[17], for example). Indeed, 3D-KNOBIS allows explicitly the use of uncertain data in the process of inference itself. Also, the structure of the system is such that the process of interpretation is completely separated from the one of inference as truth maintenance is realized through the resolution of a constraint satisfaction problem.

3D-KNOBIS is composed of three knowledge bases: a *rule base*, a *data base* and a *constraint base*. The rules constitute the knowledge about the problem domain while the data form the knowledge about the specific application domain. Constraints are used to define the conditions under which the mapping between the problem and the specific application should be done. These knowledge bases are supervised by two distinct schemes, a *controller* that manages the strategy of inference, and an *interpreter* that can deduce possible interpretations of the current data base. This structure is shown in figure 2.

Several rule forms are possible for the rules in the **rule base**. The rule form: **If** *(antecedent)* **Then** *(consequent)* is quite powerful and convenient. Both antecedent and consequent are propositions which can be expanded straightforwardly into a tree structure using the following expansion rules:

(proposition) ::= (clause)

(proposition) ::= ((proposition) **AND** (proposition))

(proposition) ::= ((proposition) **OR** (proposition))

(proposition) ::= (**NOT** (proposition))

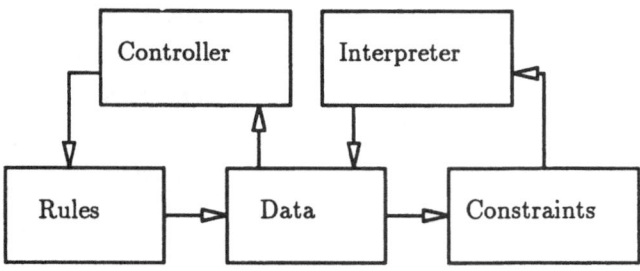

Fig. 2   Structure of 3D-KNOBIS.

A proposition therefore has an AND/OR/NOT tree structure. Although either (*disjunction, negation*) or (*conjunction, negation*) are formally equivalent to (*conjunction, disjunction, negation*), allowing all three operators simplifies greatly transfer of domain knowledge to the system. Parsing such a decision tree is straightforward [9]. A clause is a primitive proposition. This structure allows parallel operations in which the satisfaction of a given condition is verified independently on each branch of the tree.

In addition, we generalize the consequent part of the rule by allowing the presence of several consequents associated with the same antecedent (Figure 3). These multiple consequents represent alternative actions due to the satisfaction of an antecedent (in this sense, the link between each of these consequents can be seen as an exclusive or). In a given situation, only one of the consequents can be valid, but we assume that there is no direct method to determine which one. Rules can also be *certain* or *uncertain*. In an uncertain rule, the consequents do not apply with certainty i.e. they are not a logical consequence of the antecedent. This is translated by the addition of a null consequent as explained subsequently.

In summary, the structure of rules allows two levels of ambiguity: multiplicity of consequents and uncertain rules.

The rule structure leads to a corresponding representation for the **data base**. The application of a multiple consequent rule yields information: a *datum*. Each datum is represented by a node. Several labels are associated with this node, each of them corresponding to one of the possible conclusions of the rule which produces it. A particular label does not corresponds to only one proposition but rather includes all the propositions that would be available if the associated conclusion were true. Furthermore, the same proposition could be part of more than one label (on one or more nodes), each of these different labels representing different ways of infering the same proposition. There-

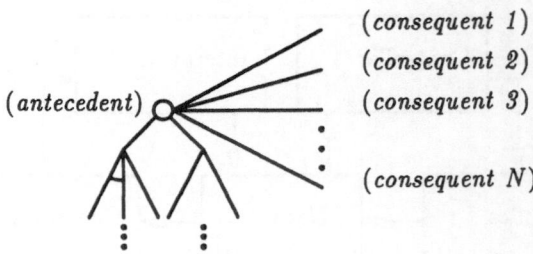

Fig. 3   Rule structure.

fore, this node representation constitutes a non-disjoint segmentation of the data base rather than a simple collection of individual propositions.

Interpreting this nodal structure consists of choosing a label for each node of the data base. All the propositions included in the chosen labels constitute the 'theory' associated with the interpretation (labelling). We say that a label is valid when it is the label retained by the current interpretation. Note that for the analysis to be complete, we must add a null label to nodes to take into account the case where none of the suggested labels of the node can be valid. This can happen if the suggested conclusions of an uncertain rule become false, or if the conditions under which the node has been created are no longer valid. A proposition which is known to be true (fact) is represented by a node with only one label, the one that contains the proposition itself.

The process of inference is managed by the **controller**. If the the propositions contained in a number of labels satisfy the condition expressed by the antecedent of one rule, then a new node is created. There are two restrictions associated with node creation. First, no more that one label per node can be used to satisfy a given antecedent. Second, the different labels used in the derivation must not include contradictory propositions, in order to avoid inconsistency.

Each time a node is created, the **constraint base** must be updated. The role of the constraint base is to keep track of the dependency between a deduced node and its ancestors and also to avoid inconsistent interpretations. Consequently, there are two types of constraints: *dependency* and *compatibility* constraints. Dependency constraints express the fact that the non-null labels of a given node can be valid only if the labels that have been used to create the node are also valid. Compatibility constraints are used to prevent the simultaneous validation of two incompatible labels. Two labels are incompatible if they contain contradictory propositions. The constraint base is updated each time a new node is created.

At any moment during the node creation process, the recorded constraint base can be used to obtain current consistent interpretations. This task is given to the **interpreter**. The interpreter obtains these interpretations by solving the corresponding constraint satisfaction problem [8]. Each consistent labelling found corresponds to an interpretation. Obviously, there may be more than one consistent labelling, in which case one has to use additional information to decrease the number of interpretations, or find a measure of 'plausibility' to classify these interpretations.

4. Interpretation of image line correspondences

This section shows how the concepts introduced in the section above can be applied to interpret a sequence of images using the computational units of Section 2. In this particular application, the analysis will be performed on a sequence of images representing lines in space. The goal here is to recover the 3D orientation of the lines (which is the main difficulty in the line interpretation problem). By hypothesizing some plausible special configurations in the scene, it will be possible to suggest a number of interpretations. Ideally, we end up with a single interpretation. The role of 3D-KNOBIS is to control the process of inference under these uncertainties and to propose possible interpretations.

A proposition designates, here, either a relation existing between a number of lines or a particular numerical assignment for a given attribute of a line. We use orientation as the attribute. In this case, the negation of a proposition would be the assignment of a different orientation (within some tolerance) to a given line.

The following rules are used. Two rules infer the orientation of parallel and orthogonal lines. Another rule is used to propagate the results obtained by their application. In order to allow the application of these rules, the particular configurations to which they refer must be detected or hypothesized. Unfortunately, there is no certain way of detecting such configurations. To cope with this problem, uncertain rules are used to hypothesize the occurrence of these configurations. Obviously, it is possible that configurations predicted by these rules come out to be inconsistent, but it is the role of 3D-KNOBIS to manage this eventuality. Two hypothesizing rules are used: one to hypothesize the occurrence of parallel lines and one to hypothesize the occurrence of orthogonal lines. Parallel lines can be hypothesized by observing that their images often appear nearly parallel in the image. Orthogonal lines, which are often the result of the presence of a 'corner' in the scene, typically appear Y-shaped where the three lines meet at a single point.

Briefly, the rule base that will be used for our particular example is:

Hypothesizing parallel lines rule: if two lines are nearly parallel in at least one image, then these lines are hypothesized to be parallel in the scene.

Hypothesizing orthogonal lines rule: if three lines meet at one point in both images, then these three lines are hypothesized to be orthogonal.

Orthogonal lines rule: if three lines are orthogonal, then their orientation can be computed by the corresponding computational unit (Section 2.2).

Parallel lines rule: if two lines are parallel, then their orientation can be computed by the corresponding computational unit (Section 2.3).

Propagation rule: if the orientation of two non-parallel lines is known over two views, then the orientation of all the other lines can be found by propagation (Section 2.4).

Note that the application of the orthogonal lines rule gives rise to four possible solutions (two per view). As a result, a node created by this multiple consequent rule will have four non-null labels.

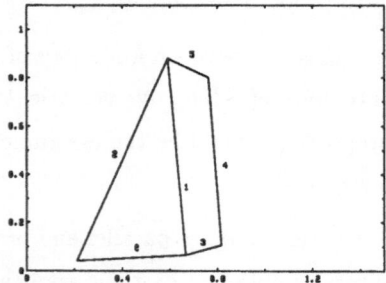 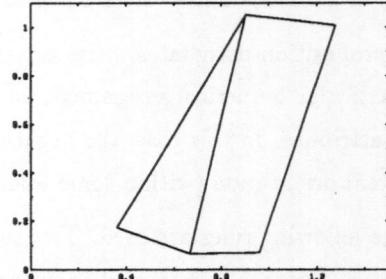

Fig. 4   Images of a wedge.

Figure 4 shows two views of a wedge. With these images as input, the following relations are hypothesized:

```
(parallel    0   5)
(parallel    1   4)
(parallel    3   5)
(parallel    2   4)
(orthogonal  0   1   3)
(orthogonal  1   2   5)
```

To each of these hypotheses corresponds a node; the hypothesis is one label of this node and the null label is another. This latter will be valid if the hypothesis is rejected.

Once a relation is known, the corresponding computational unit can be applied. For example, the fact that lines labelled 0, 1 and 3 are hypothesized to be orthogonal, allows the computation of four solutions:

```
(orthogonal  0  1  3)
                    image 1                      image 2
solution 1:

  0:    ( 0.990,  0.049, -0.131)   ( 0.930, -0.240,  0.279)
  1:    (-0.039,  0.996,  0.081)   ( 0.245,  0.969,  0.017)
  3:    (-0.134,  0.075, -0.988)   ( 0.275, -0.052, -0.960)

solution 2:

  0:    ( 0.990,  0.049, -0.131)   ( 0.117, -0.321, -0.940)
  1:    (-0.039,  0.996,  0.081)   ( 0.019,  0.947, -0.321)
  3:    (-0.134,  0.075, -0.988)   ( 0.993,  0.019,  0.117)

solution 3:

  0:    ( 0.500,  0.000, -0.866)   ( 0.930, -0.240,  0.279)
  1:    (-0.150,  0.985, -0.087)   ( 0.245,  0.969,  0.017)
  3:    ( 0.853,  0.174,  0.492)   ( 0.275, -0.052, -0.960)

solution 4:

  0:    ( 0.500,  0.000, -0.866)   ( 0.117, -0.321, -0.940)
  1:    (-0.150,  0.985, -0.087)   ( 0.019,  0.947, -0.321)
  3:    ( 0.853,  0.174,  0.492)   ( 0.993,  0.019,  0.117)
```

This result is also represented in the data base as a node with four non-null labels. The creation of this node also causes the updating of the constraint base. Dependency constraints are added to state that the results proposed by this node are valid only if the (orthogonal 0 1 3) hypothesis is also valid. Compatibility constraints are also added to record incompatibilities that can arise with results suggested by other nodes.

From two distinct orientations it is possible to propagate interpretation to other lines. For example, the propagation of the first solution gives the following results:

```
                    image 1                      image 2

  0:    ( 0.990,  0.049, -0.131)   ( 0.930, -0.240,  0.279)
  1:    (-0.039,  0.996,  0.081)   ( 0.245,  0.969,  0.017)
  2:    (-0.439, -0.860, -0.259)   (-0.554, -0.741, -0.379)
  3:    (-0.134,  0.075, -0.988)   ( 0.275, -0.052, -0.960)
  4:    (-0.099,  0.995, -0.024)   ( 0.231,  0.968, -0.103)
  5:    ( 0.155,  0.715,  0.682)   ( 0.103,  0.748,  0.656)
```

Once inferencing is completed, the resulting constraint satisfaction problem is solved. A total of eleven consistent labellings are thus found. A measure of plausibility must then be used in order to classify these interpretations. This will be done, here, by considering the number of hypotheses independently confirmed, i.e., the number of hypotheses that lead to the same results. Each interpretation will therefore be ranked according to the number of nodes created by a hypothesizing rule and that have received a non-null assignment. An interpretation is considered only if this number is greater than one. Four interpretations finally remain in our example:

Interpretation #0

```
(parallel 1  4)
(parallel 3  5)
(orthogonal  0  1  3)
```

	image 1	image 2
0:	(0.500, 0.000, -0.866)	(0.117, -0.321, -0.940)
1:	(-0.150, 0.985, -0.087)	(0.019, 0.947, -0.321)
2:	(-0.215, -0.743, 0.634)	(-0.091, -0.504, 0.859)
3:	(0.853, 0.174, 0.492)	(0.993, 0.019, 0.117)
4:	(-0.150, 0.985, -0.087)	(0.019, 0.947, -0.321)
5:	(-0.853, -0.174, -0.492)	(-0.993, -0.019, -0.117)

Interpretation #1

```
(parallel 0  5)
(parallel 1  4)
```

	image 1	image 2
0:	(0.907, 0.061, 0.417)	(0.892, -0.382, -0.240)
1:	(-0.150, 0.985, -0.087)	(0.019, 0.947, -0.321)
2:	(-0.282, -0.829, 0.483)	(-0.003, -0.563, 0.783)
3:	(0.136, 0.147, -0.980)	(0.558, 0.000, 0.806)
4:	(-0.150, 0.985, -0.087)	(0.019, 0.947, -0.321)
5:	(-0.907, -0.061, -0.417)	(-0.892, 0.382, 0.240)

Interpretation #2

```
(parallel 0  5)
(parallel 2  4)
```

	image 1	image 2
0:	(0.907, 0.061, 0.417)	(0.892, -0.382, -0.240)
1:	(-0.175, 0.977, -0.124)	(-0.228, 1.517, -0.028)
2:	(-0.425, -0.693, -0.583)	(-0.555, -0.770, -0.315)
3:	(-0.410, -0.008, -0.912)	(0.707, -0.237, 0.722)
4:	(0.425, 0.693, 0.583)	(0.555, 0.770, 0.315)
5:	(-0.907, -0.061, -0.417)	(-0.892, 0.382, 0.240)

Interpretation #3

```
(parallel 3  5)
(parallel 2  4)
```

	image 1	image 2
0:	(0.192, -0.020, -0.981)	(-0.077, -0.146, -1.192)
1:	(-0.018, 0.994, 0.111)	(-0.042, 1.219, -0.067)
2:	(-0.425, -0.693, -0.583)	(-0.555, -0.770, -0.315)
3:	(0.853, 0.174, 0.492)	(0.993, 0.019, 0.117)
4:	(0.425, 0.693, 0.583)	(0.555, 0.770, 0.315)
5:	(-0.853, -0.174, -0.492)	(-0.993, -0.019, -0.117)

Note that the hypotheses valid in each interpretation are only shown here for plausibility evaluation. Recall that 3D-KNOBIS is not an hypothesize-test system, as nodes are treated uniformly whether they represent certain or uncertain data.

In the absence of any other information, each of these interpretations is acceptable at this point. However, if another view is available (figure 5), the application of the same

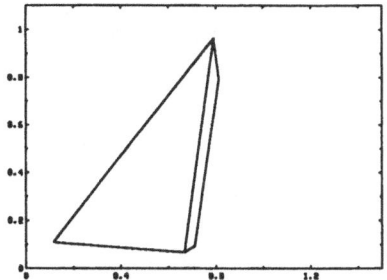

Fig. 5 A third image of the wedge.

process on the second and third views can disambiguate the problem. The possible interpretations obtained with views 2 and 3 are:

Interpretation #0

```
(parallel  1  4)
(parallel  3  5)
(orthogonal  0  1  3)
```

```
              image 1                      image 2

0:   ( 0.117, -0.321, -0.940)    ( 0.754, -0.133, -0.643)
1:   ( 0.019,  0.947, -0.321)    ( 0.004,  0.980, -0.198)
2:   (-0.091, -0.504,  0.859)    (-0.498, -0.652,  0.571)
3:   ( 0.993,  0.019,  0.117)    ( 0.656,  0.147,  0.740)
4:   ( 0.019,  0.947, -0.321)    ( 0.004,  0.980, -0.198)
5:   (-0.993, -0.019, -0.117)    (-0.656, -0.147, -0.740)
```

Interpretation #1

```
(parallel  0  5)
(parallel  1  4)
```

```
              image 1                      image 2

0:   ( 0.892, -0.382, -0.240)    ( 0.671,  0.036,  0.741)
1:   ( 0.019,  0.947, -0.321)    ( 0.004,  0.980, -0.198)
2:   (-0.095, -0.508,  0.856)    (-0.656, -0.758,  0.321)
3:   ( 0.843, -0.021, -0.538)    ( 0.811,  0.441,  0.538)
4:   ( 0.019,  0.947, -0.321)    ( 0.004,  0.980, -0.198)
5:   (-0.892,  0.382,  0.240)    (-0.671, -0.036, -0.741)
```

Interpretation #2

```
(parallel  0  5)
(parallel  2  4)
```

```
              image 1                      image 2

0:   ( 0.892, -0.382, -0.240)    ( 0.671,  0.036,  0.741)
1:   ( 0.350,  0.916,  0.198)    ( 0.440,  0.610,  0.363)
2:   (-0.555, -0.770, -0.315)    (-0.507, -0.621, -0.598)
3:   ( 0.988,  0.022,  0.155)    ( 0.693,  0.343,  0.959)
4:   ( 0.555,  0.770,  0.315)    ( 0.507,  0.621,  0.598)
5:   (-0.892,  0.382,  0.240)    (-0.671, -0.036, -0.741)
```

Interpretation #3

```
(parallel  2  4)
(parallel  3  5)
```

```
              image 1                      image 2

0:   (-0.538, -0.064, -0.840)    (-0.122, -0.270, -0.824)
```

```
1:  ( 0.368,  0.901,  0.231)  ( 0.442,  0.667,  0.434)
2:  (-0.555, -0.770, -0.315)  (-0.507, -0.621, -0.598)
3:  ( 0.993,  0.019,  0.117)  ( 0.656,  0.147,  0.740)
4:  ( 0.555,  0.770,  0.315)  ( 0.507,  0.621,  0.598)
5:  (-0.993, -0.019, -0.117)  (-0.656, -0.147, -0.740)
```

The acceptable interpretation becomes the only one that assigns the same attributes (orientations) to lines of the second view, i.e.

```
Interpretation #0

(parallel   1   4)
(parallel   3   5)
(orthogonal   0   1   3)
```

	image 1	image 2	image 3
0:	(0.500, 0.000, -0.866)	(0.117, -0.321, -0.940)	(0.754, -0.133, -0.643)
1:	(-0.150, 0.985, -0.087)	(0.019, 0.947, -0.321)	(0.004, 0.980, -0.198)
2:	(-0.215, -0.743, 0.634)	(-0.091, -0.504, 0.859)	(-0.498, -0.652, 0.571)
3:	(0.853, 0.174, 0.492)	(0.993, 0.019, 0.117)	(0.656, 0.147, 0.740)
4:	(-0.150, 0.985, -0.087)	(0.019, 0.947, -0.321)	(0.004, 0.980, -0.198)
5:	(-0.853, -0.174, -0.492)	(-0.993, -0.019, -0.117)	(-0.656, -0.147, -0.740)

These are, indeed, the actual orientations.

5.   Summary

We have considered the problem of interpreting 3D lines from image line correspondences. Direct formulations of this problem have led to numerically unreliable algorithms. We have proposed and described a more robust consistent labelling formulation which exploits the existence of special, yet commonly found, configurations of lines in space.

References

[1]   A. Mitiche, O. Faugeras, J. Aggarwal, 'Counting the Line,' *Proc. 9th Int. Conf. on Pattern Recognition,* pp. 693-695, Rome, Italy, November 1988.

[2]   A. Mitiche, G. Habelrih, 'Interpretation of Straight Line Correspondences Using Angular Relations,' *Pattern Recognition,* 22, No.3, pp. 299-308, 1989.

[3]   Y. Liu, T.S. Huang, 'Estimation of Rigid Body Motion Using Straight Line Correspondences,' *Proc. Workshop on Motion: Representation and Analysis,* Charleston, S.C., pp. 47-51, 1986.

[4]   O.D. Faugeras, F. Lustman, G. Toscani, 'Motion and Structure from Motion from Point and Line Matches, *Proc. ICCV,* pp. 25-34, june 1987.

[5]   M.E. Spetsakis, and J. Aloimonos,'Closed Form Solution to the Structure from Motion Problem from Line Correspondences,' *Proc. of AAAI'87,* Seattle, WA, pp. 738-743, 1987.

[6]   Y. Liu, T. S. Huang,'A Linear Algorithm for Motion Estimation Using Straight Line correspondences,' *Proc. 9th Int. Conf. on Pattern Recognition,* pp. 213-219, Rome, Italy, November 1988,

[7] R.M. Haralick, L.G. Shapiro, 'The Consistent Labelling Problem: Part I,' *IEEE PAMI-1*, no 2, pp. 173-184, 1979.

[8] R.M. Haralick, G.L. Elliot, 'Increasing Tree Search Efficiency for Constraint Satisfaction Problems,' *Artificial Intelligence*, vol. 14, pp. 263-313, 1980.

[9] A. Mitiche, A. Mansouri, C. Meubus, 'A Knowledge-based Interpretation System,' *Proc. 9th Int. Conf. on Pattern Recognition*, pp. 992-994, Rome, Italy, November 1988,

[10] R. Jain, 'Dynamic Vision,' *Proc. 9th Int. Conf. on Pattern Recognition*, Rome, pp. 226-235, 1988.

[11] E.L. Walker, M. Herman, 'Geometric Reasoning for Constructing 3D Scene Description from Images,' *Artificial Intelligence*, vol. 37, pp. 275-290, 1988.

[12] R.A. Brooks, 'Symbolic Reasoning among 3D Models and 2D Images,' *Artificial Intelligence*, vol. 17, pp. 285-348, 1981.

[13] T. Kanade, 'Recovery of the 3D Shape of an Object from a Single View', *Artificial Intelligence*, vol. 17, pp. 409-460, 1981.

[14] S.T. Barnard, 'Interpreting Perspective Images,' *Artificial Intelligence*, vol. 21, pp. 435-462, 1983.

[15] P.G. Mulgaonkar, L.G. Shapiro, R.M. Haralick, 'Shape from Perspective: A Rule-based Approach,' *CVGIP*, vol. 36, pp. 298-320, 1986.

[16] D.G. Lowe, T.O. Binford, 'The Recovery of 3D Structure from Image Curves,' *IEEE PAMI-7*, no 3, pp. 320-326, 1985.

[17] V.S. Hwang, L.S. Davis, T. Matsuyama, 'Hypothesis Integration in Understanding Systems,' *CVGIP*, vol. 36, pp. 321-371, 1986.

[18] D.M. Mckeown, W.A. Harvey, J. McDermott, 'Rule-based Interpretation of Aerial Imagery,' *IEEE PAMI-7*, no 5, pp. 570-585, 1985.

[19] M. Herman, T. Kanade, 'The 3D MOSAIC Scene Understanding System,' *Image Understanding Workshop*,' New Orleans, pp. 137-148, 1984.

SENSOR DATA INTEGRATION FOR THE CONTROL OF AN AUTONOMOUS ROBOT

Jörg Raczkowsky and Ulrich Rembold
Institute for Realtime Computer Systems and Robotics
University of Karlsruhe
Federal Republic of Germany

ABSTRACT

The Institute for Realtime Computer Systems and Robotics of the University of Karlsruhe is currently developing the autonomous mobile assembly robot KAMRO. The system consists of a mobile platform with an omnidirectional wheel drive and a superstructure on which two assembly robots are mounted. The sensor system of KAMRO is divided into three functional parts which support the assembly of a product by two arms, the docking of the vehicle and the autonomous navigation. To solve the three tasks, the use of a multisensor system is necessary, combining cameras, distance, approach and tactile sensors. The concept of elementary operations forms the framework for the integration of the sensors into the control system of the mobile robot. In this paper the sensor data processing and various elementary operations of the autonomous mobile assembly robot are described in more detail.

1. Introduction

Robot systems are currently being introduced in many industries. Mostly, they perform complex movements to do their work assignments. Sensors and sensor systems play an increasingly important part for controlling the operation by gathering sensing information about the environment; thus the motion sequences may be changed depending on changes in the workplace. The third generation robots [5] will be able to react intelligently at their work place; research institutes all over the world are working towards their realization. Their characteristics may be briefly described as follows:

- Goal-oriented input of instructions (implicit programming)
- Automatic planning of actions
- Automatic error recovery
- Extensive use of sensor information
- Application of a world model

NATO ASI Series, Vol. F 63
Traditional and Non-Traditional Robotic Sensors
Edited by T. C. Henderson
© Springer-Verlag Berlin Heidelberg 1990

This enumeration shows the complexity of the advanced robot system concept. A control system for these capabilities is usually structured hierarchically [1] [4].

The functional blocks of this hierarchy can be classified in on- and off-line modules, that is, some parts are able to perform in realtime, and other ones are not. The sensor system is an important part of the hierarchy acting on-line. The manipulator system consists of two PUMA 260 robots in a "hanging" configuration. Their relative position was chosen to enable them to perform coordinated two-arm movements. For assembly purposes, electrically driven grippers with inegrated sensors are used.

The sensor system supports the operation of the planning, execution and supervision modules which are incorporated in the robot. The vehicle contains a path planner, a navigator and a docking module. The navigation is done with the help of cameras and sonic sensors in connection with a map of the working environment under the direction of the navigator. The docking of the platform at the work station is controlled by approach sensors and the docking supervisor [12]. The assembly of the two robot arms is planned by the assembly planner [8] and the execution module. The execution of the motion commands is controlled by the effector control and directed by a supervision module [11]. The robot actions are planned and controlled by several expert systems.

To perform an assignment the robot will be given an order for an assembly task. After inspecting the instructions, the robot will drive to a part storage, pick up the workpieces and bring them to an assembly station. Here, the assembly is planned and the manipulator arms will perform the work bench for assembly.

2. The sensor system of KAMRO

KAMRO is a testbed for investigating the possibility of integrating a planning, execution and supervision module into one control system using AI techniques. The apparatus needs a sensor system to understand its world and to react to unforeseen events. For this purpose, it is necessary to perform sensor planning, scene monitoring, sensor data processing and fusion. The sensor system is hierarchically structured in three levels of data processing:

- preprocessing of raw sensor data
- error recognition (comparison with reference data of the world model)
- error diagnosis (state of the real environment)

Preprocessing uses procedural knowledge which is implicitly defined by the algorithms and their control structures. A new approach to speed up the processing time of sensor data are our future work is an investigation of these techniques for the integration into the KAMRO system.

For error recognition, an access mechanism is assigned to the world model for each sensor unit to compute reference data for the current measurement. If deviations are found when comparing the measured data with the reference data, the current action must be slowed-down, possibly stopped and, if necessary, be replaced by another sequence. If a complete local analysis of the situation is not possible, the acquired data must be passed on to the diagnosis module.

For error diagnosis, the fusion of data from different sources is done. In contrast to the two lower control levels, the realtime demands at the fusion level are not so strict, as the manipulation sequence must be completely changed depending on the disturbance. A comprehensive analysis of the situation is more important in order to be able to plan a new manipulation sequence. We employ an expert system based on a blackboard architecture. This concept is frequently applied when AI-methods are used for technical applications.

The sensors are used for three working modes, corresponding to the above described modules, which are closed loop control, error recognition and diagnosis. The selection of the working mode is done by the action control of the system. The actions of KAMRO are composed of a set of elementary operations. These elementary operations define how to use the sensors and how to interpret the results of the sensor operations.

The use of a single sensor is the simplest case. However, the information available from just one sensor often does not sufficiently describe the scenario. In this case, multisensor information must be processed and sent to the control modules as fused data. Since closed loop control and error recognition impose high realtime requirements, the methods for sensor data fusion must be rather simple. However, for the diagnosis mode more time consuming and complex algorithms can be used, e.g. the algorithms of KAMRO sensor system contain methods based on the Fuzzy-set [15] [16] [17] and the Dempster-Shafer theories [14].

The following physical sensors are studied for the integration into the KAMRO system:

- An overhead camera system
- A hand-eye system with two cameras
- A sonic sensor system (200 kHz)
- A sonic sensor system (40 kHz)
- A force-torque sensor in the hand wrist
- A force torque sensor in a robot finger

- A one dimensional force sensor in a finger
- An optical position sensor for the docking

The complex sensor systems, e.g. cameras etc., are divided into a set of virtual sensors. Only a specific part of the information available of a sensor is used for the actual data processing. Table 1, for example, shows the virtual sensors belonging to the overhead camera system.

Sensor	Virt. sensor	Data Type	Interpretation
Camera 1	HISTO1	VECTOR	Basis binarization; existence of parts
	MARK1	INTEGER	Separating parts; number of parts and holes; relations between them
	GRAVI1	INTEGER	Position of parts/holes
	CONTUR1	VECTOR	Recognition of parts/holes
	ORIENT1	INTEGER	Orientation of parts/holes

Table 1. Virtual Sensors of the Overhead System

Particularly, due to the hard realtime requirements, the concept of virtual sensors is very useful with the closed loop and error recognition tasks. Only the absolutely necessary information of the sensors are required. The time consuming preprocessing of the sensor data can be reduced to a minimum.

The specific data acquisition of the real world state is important for a fast diagnosis. Depending on the interim diagnosis results, new sensor data may be requested. Splitting the physical sensors into various virtual sensors is very helpful to speed up the system.

3. Elementary Operations

The agents of the KAMRO system are controlled by basic functions called elementary operations (EO) [7]. A sequence of EOs forms an action plan. Each EO is executed by a specific elementary operation module (EOM). The desired goal is specified by the EO, and the EOM specifies how to achieve this goal. For the navigation of the mobile platform the following EOs are established:

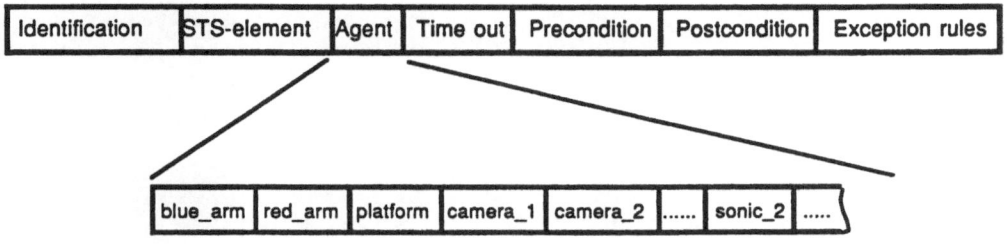

Figure 1.   Specification of the Elementary Operations (EO) [7]

MOVE	specifies the trajectory of the vehicle between a start and a goal position
DOCK	defines the fine motion for the approach of the vehicle to a work station
UNDOCK	defines the fine motion for the departure of the vehicle from a work station

For assembly more elementary operations are necessary:

CALIBRATE	aligns the robot coordinate system with the coordinate system of the work station
TRANSFER	defines the gross motion of the end effector from a start to a goal position along a defined trajectory
APPROACH	defines the fine motion of the end effector to an object
DEPART	describes the fine motion of the end effector from an object
GRASP	describes the action of the end effector for gripping an object
DETACH	describes the action of the end effector to release an object
JOIN	specifies the fine motion of the end effector to connect two objects
DISJOIN	defines the fine motion of the end effector to separate two objects
CONFIGURATE	directs the manipulation of the joints of the robot

Each EO contains a description of the sensor usage to control or supervise an action of the effector system. Fig.1 shows the specification of an EO. The agents in this specification could be the effectors and the sensors of the system. Each action of the system has a specific need for supervision. This will be defined by the EO belonging to the action. The effector control and the supervision module will be loaded with the corresponding parts of the EO. A synchronization signal from the effector control sent to the supervision module starts or stops the supervision task to schedule an effector task.

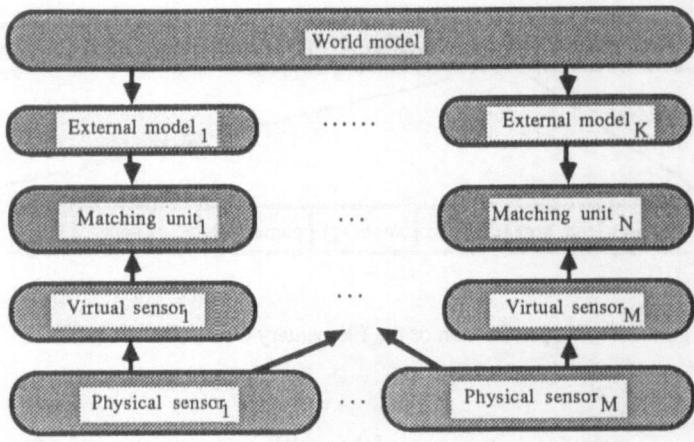

Figure 2. Structure of the Error Recognition

4. Error recognition

Errors must be recognized in realtime. To solve this problem the above described concept of the virtual sensors is used. The lower the amount of the data to be processed, the better the realtime requirements of the error recognition mode can be met. The sensor task planning has to determine the necessary information to detect the potential errors of an action.

Fig. 2 shows the structure of the error recognition module. Each virtual sensor is connected to a matching unit. The reference data from the global world model are transformed through a specific external model for each matching unit. The access to the global data of the world model is done by special modules according to the view concept [9] [13]. Our first approach to implement the world model is done with a relational data base model.

Going from the global model to the external model results in an enormous data reduction. Table 2 shows an example for the overhead camera. The assumed operation is a projection of the object from the 3D-CAD model to the 2D-sensor space. Mostly, the information requested from the world model is geometrical, as shown in Table 2. But the object in the world model is also described with additional information, e.g. weight, reflection parameters etc.

Global model	External model
Surfaces: 209 Directed edges: 1061 Undirected edges: 618 Points: 412	Surfaces:7 Directed edges: 2 Undirected edges: 26 Points: 26

Table 2. Data Reduction in the Global and External Model

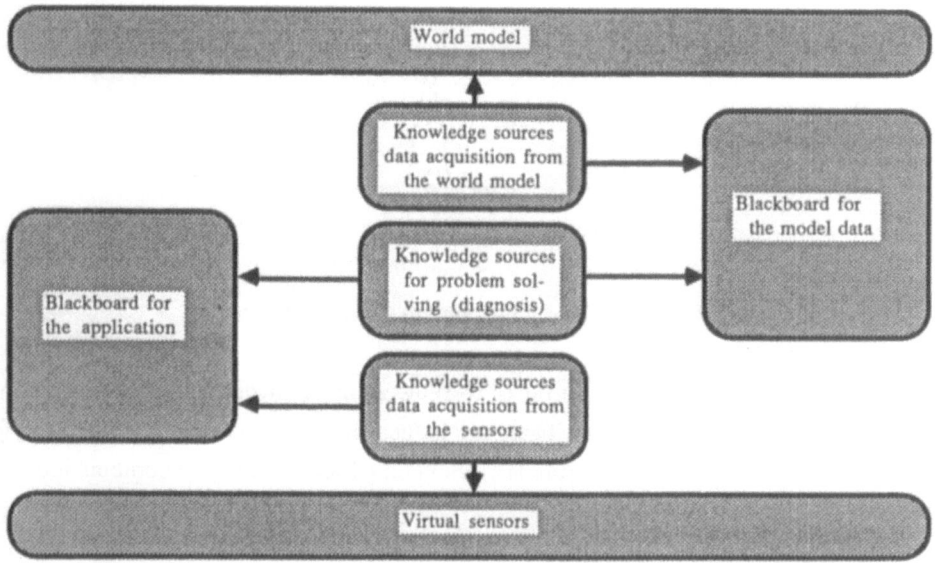

Figure 3.   Structure of the Diagnosis Section

The action of the effector system has to be stopped or slowed down as soon as an intolerable deviation between the actual measured data and the reference data is recognized . In trivial cases the observed data describe the error situation well enough. More complex error states, however, need more intensive data processing. Very often the fusion of data of other sensors is necessary. This task of the error diagnosis is described in the next chapter.

5. Diagnosis

The diagnosis expert system is based on the blackboard concept suggested by Hayes [6]. This architecture is well suited for the use of knowledge based methods in technical applications. It was implemented using the development shell GBB [2]. Fig. 3 shows the internal structure of the diagnosis module.

The blackboard is divided into two parts: the application blackboard is used for problem solving and the model blackboard contains a partial copy of the world model with a specific structure. The knowledge sources are split into two functional groups. One is used for the acquisition of data from the sensor modules and the world model. The other group is used to solve the specific application problem.

The hierarchically structured blackboards contain different information on various levels. The data of the virtual sensors are the basic information on the board. We investigated as our first

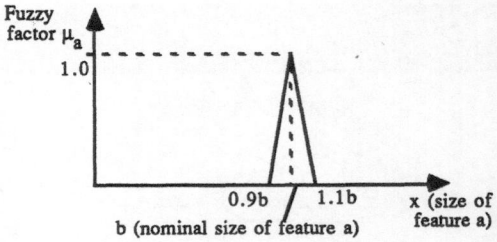

Figure 4. Fuzzy Function to Compare Feature a

application the recognition and measuring of workpieces in a crowded scene with overlapping workpieces. The contour feature attributes are judged by fuzzy membership functions (see Fig.4) and fused by the Dempster´s rule [14]. Equation 1 and 2 show the combination of the evidences:

$$m(\emptyset) = 0 \qquad\qquad [1]$$

$$m(A) = \frac{\underset{\substack{i,j \\ A_i \cap B_j = A}}{\sum} m_1(A_i) \cdot m_2(B_j)}{1 - \underset{\substack{i,j \\ A_i \cap B_j = \emptyset}}{\sum} m_1(A_i) \cdot m_2(B_j)} \qquad\qquad [2]$$

m: $2^\Theta \rightarrow [0,1]$ *basic probaility assignment*
Θ: *frame of discernment*
A,B: *subsets of* Θ

The result of the combination process is a hypothesis that a defined workpiece exists in the scene. To verify the hypothesis a graph matching algorithm is used. For the representation of the workpieces a minimal spanning tree is chosen to decrease the complexity. This method works very fast under the assumption that characteristic features can be found. They allow a well defined entry in the model graph of the workpiece to verify the hypothesis.

The 2D-position calculation is done by partial graph matching. The matching process succeeds if all features of the hypothesized workpieces are found in the scene. Whenever the matching process terminates unsuccessfully, the endnodes of the partial graph provide information about the disturbed workpiece features. This information is used to plan additional sensor tasks.

Distance sensors and tactile sensors can be used to determine the depth information. With their help, the 3D-position of a workpiece surface can be measured with 3 points and an additional measurement to verify the result. All this information is sent to the executive module of the system and further to the planning modules.

In our current research work, we extend the application area to other parts of the assembly process where a multisensory environment is necessary. Another goal is to improve the used

6. Conclusion

Many researchers presently are working on the subject of ´intelligent robots´. Tools of Artificial Intelligence are suitable for controlling these technical applications. However, the strong realtime demands of robotics can not yet be fulfilled using the current AI tools. In sensor technology the knowledge-based systems has mainly been used for picture processing. Multisensor systems have appeared in the last five years, because picture processing on its own unable is to handle complex scenes, e.g. in assembly tasks.

When a multisensory environment is used for a robot, information redundancy may occur. Especially in difficult scenes sensor data can be verified by using data of another sensor. In the future, an advanced robot system must be able to handle redundant information of a complex task.

Our research efforts have first concentrated on specific problem areas. The present work aims at the integration of the various parts to form a complete multisensory system. The concept presented in this paper has been implemented on several computers, depending on the complexity of the tasks. These are VME-systems for the lower level and SUN- and micro VAX-computers for the higher levels.

7. Acknowledgement

This research work was performed at the Institute for Realtime Computer Systems and Robotics, Prof. Dr.-Ing. U. Rembold and Prof. Dr.-Ing. R. Dillmann, Faculty for Informatics, University of Karlsruhe, 7500 Karlsruhe, Federal Republic of Germany.

8. References

1.   Albus, J. S. et. al.: Theory and Practice of Hierarchical Control, 23.IEEE, Washington, 13.-17. Sept. 1981

2. Banzhaf W., Haken, H.: A network for recognition and classification of continuos patterns, INNS, Boston/Mass., September 1988

3. Corkill, D. D., Gallagher, K.Q., Johnson, P. M.: GBB: A generic blackboard development system, Proceedings National Conference on Artificial Intelligence, pp. 1008-1014, 1986

4. Dillmann, R., Rembold, U.: Autonomous robot of the University of Karlsruhe, 15th ISIR, Tokyo, 13.-17. Sept. 1985

5. Giralt, G.: Research trends in decisional and multisensory aspect of third generation robots, Robotics Research: The Second International Symposium, 1985

6. Hayes-Roth, B.: Blackboard architecture for control , Artificial Intelligence 26: 3, 251-321, 1985

7. Hörmann, A., Hugel, Th., Meier, W.: A concept for an intelligent and fault tolerant robot system, Journal of Intelligent and Robotic Systems 1 (1988), pp. 259-286, Kluwer Academic Publishers

8. Hörmann, K., Rembold, U.: A robot action planner for automatic parts assembly, IEEE International Workshop on Intelligent Robots and Systems, Tokyo, Japan Oct. 31-Nov. 2, 1988

9. Kaltenbach, J.: Datenbankzugriffe für wissensbasierte Sensorsysteme in der Robotik, Master thesis, University of Karlsruhe, Faculty for Informatics, July 1987

10. Pearson, J. C. et. al.: Neural network approach to sensory fusion, Nato advanced workshop on highly redundant sensing in robotic systems, Il Chioco, Italy, May 1988

11. Raczkowsky, J.: Ein Konzept für die Multisensordatenverarbeitung in der Robotik, Dissertation, University of Karlsruhe, Faculty for Informatics, June 1989

12. Rembold, U., et. al.: Mobile robot activity model for autonomous free flying platforms, First European In-Orbit Operations Technology Symposium, Darmstadt, ESA SP-272, November 1987

13. Schlageter,G.,Stucky, W.: Datenbanksysteme. Konzepte und Modelle, 2.Auflage, Teubner, Stuttgart, 1983

14. Shafer, G.: A mathematical theory of evidence, Princeton University Press, 1975

15. Zadeh, L. A.: The Role of Fuzzy Logic in the Management of Uncertainty in Expert Systems, Approximate Reasoning in Expert Systems, Elsevier Science Publishers B. V. (North-Holland), 1985

16. Zadeh, L. A.: Fuzzy Sets, Information and Control 8, pp 338- 353, 1965

17. Zadeh, L. A.: Fuzzy Logic and Approximate Reasoning, Synthese 30, pp 407-428, 1975

Cooperation of the Inertial and Visual Systems

Thierry VIÉVILLE   Olivier D. FAUGERAS

INRIA, Sophia-Antipolis, BP 109, 06561 Valbonne, FRANCE

Abstract

This paper introduces a number of issues concerning the use of an inertial system in cooperation with vision. We first present applications of inertial information in a visual system, and then attack the problem of determining motion and orientation of the robotic system from inertial information. An iterative algorithm is finally given, and studied in detail.

1 Introduction

The aim of this study is to introduce the use of inertial information in a visual system. Inertial information is provided by *accelerometers* and *gyrometers*. Such industrial sensors exist, at a price close to the cost of a visual sensor, with a precision of 0.1 m/sec^2 and 0.04 deg/sec, and a bandwidth of $0 - 50\ Hz$.

They provide a partial estimation of the **robot self-motion** and the **robot absolute orientation**, precisely **angular velocity** and **specific forces**.

Although only gravity, the rigid body acceleration, and the angular instantaneous velocity can be measured, it is possible to derive from these quantities various other parameters. In this paper the following parameters are analysed :

- **Instantaneous angular velocity**, which is directly given by the gyrometers output.
- **Linear acceleration**, which is not directly output by the accelerometers but added to gravity. There is no possibility, *a priori*, to separate linear acceleration and gravity, and additional internal hypotheses will be used.
- **Vertical orientation**, which is directly given by the direction of the gravity vector, and which is available as soon as linear acceleration and gravity are differentiated.
- **Angular position**, which is computed by the integration of gyrometers output, or measured using gyroscopes. Angular position evaluation is subject to a drift, linear with respect to time, and has to be periodically recalibrated.
- **Linear translation**, which is computed by the integration of linear acceleration, and is also subject to a drift.

Internal representations of instantaneous self-motion and angular orientation can be based on these parameters. We are going to analyse how such quantities can be computed from sensors output, and study the use of these quantities in cooperation with odometric and visual informations.

Application of inertial informations in a visual system

Visual cues also provide informations about motion and orientation. In order to cooperate with inertial information, either a differential approach (like "optical flow") of visual information, or a "token-tracker" based approach ([TNF88]) analysis is used.

Cooperation between visual and inertial information can be :

- *Estimation of instantaneous motion.* In a rigid, almost stationary background the linear velocity (up to a scale factor) and the angular velocity can be estimated, from the image

NATO ASI Series, Vol. F 63
Traditional and Non-Traditional Robotic Sensors
Edited by T. C. Henderson
© Springer-Verlag Berlin Heidelberg 1990

irradiance variations, on pixels with high contrast. Combined with the inertial estimations of angular velocity and linear velocity time derivative, this estimation can be improved and sped-up.

- *Estimation of angular orientation.* Orientation with respect to the vertical is provided by inertial cues while the variation of orientation in the horizontal plane can be integrated using visual cues, providing a robust estimation of absolute angular orientation.
- *3D-Vision.* When motion is known, the structure of the environment can be inferred from visual information. Using inertial cues in order to estimate motion, it is possible to build an instantaneous depth map of the environment. Using several frames the precision of this estimation is improved.
- *Intrinsic calibration of the visual sensor.* In the case of rotations around the optical center of the visual sensor, the variations of irradiance are no more related to the environment structure, but only depend on the sensor geometry. Using inertial cues, it is possible to generate such rotations and to perform the visual sensor calibration in any, structured, visual environment.
- *Segmentation between a moving object and the background.* The background being stationary and its relative movement with respect to the visual sensor being known through visual cues, it is possible to determine which part of the image belongs to a moving object, the segmentation being based on motion disparity.

All of these applications require an estimation of the angular orientation and of the linear velocity time derivative, computed from sensors output, as discussed.

Integration of inertial informations

Let us briefly describe how to recover ego-motion and angular orientation from the inertial sensor signals ([VF89]). First, vector quantities are expressed in *an orthogonal frame of reference attached to the set of sensors*. Then, *angular velocity* and *accelerations* are computed from gyrometer and accelerometer signals.

Two other additional quantities related to ego-motion and orientation are derived here : *angular position* and *linear translation*

Their computations are interdependent, and raise three problems :

- How to "integrate" the instantaneous angular velocity in order to recover angular orientation ?
- How to separate the gravity field from linear acceleration, given that these two quantities cannot be physically separated ?
- How to use the estimation of gravity (related to the vertical) in order to improve the estimation of angular orientation ?

These three points are analysed in this paper and an iterative method of estimation of absolute angular orientation, including vertical orientation, and of linear acceleration and velocity, is developed.

2 Integration of angular position

We now deal with the problem of integrating $\vec{\omega}$ in order to obtain an estimate of the angular position. We first quickly justify why such an estimation must be done numerically, and why quaternions may constitute a good representation of the angular orientation. We then develop an iterative equation and finally qualitatively estimate the related error.

2.1 Relation between $\vec{\omega}$ and the angular position

The angular velocity $\vec{\omega}$ can be related to the infinitesimal instantaneous rotation by (see for example [Aya88] page 215-217 for details) :

$$\frac{dR}{dt} R^t \cdot \vec{x} = \tilde{\omega} \cdot \vec{x} = \vec{\omega} \wedge \vec{x}$$

if we note for a vector $\vec{X}$: $\tilde{X} = \begin{pmatrix} 0 & -X_z & X_y \\ X_z & 0 & -X_x \\ -X_y & X_x & 0 \end{pmatrix}$ which is an antisymmetric matrix

with $\tilde{X} \cdot \vec{u} = \vec{X} \wedge \vec{u}$. This defines a linear homogeneous differential equation with non-constant coefficients for $R(t)$. It is well known [?] that *such an equation has a unique solution, function of the initial value $R(0)$.* This result is to be interpreted as follow : "One cannot define absolute angular orientation, but only relative angular orientation with respect to a reference". Then, an initial orientation being given, the angular orientation can be expressed as *the rotation to be performed to get from the initial orientation to the actual orientation.* There is no explicit solution for such an equation, in the general case [1]

This numerical integration is, however, particular. Let us note for example :

$$\Delta R(t) = 1 + \tilde{\omega}(t)$$

which represent an infinitesimal rotation, we must, in the case of a discretization, combine these rotations. We have :

$$R((n+1) \cdot \Delta t) = \Delta R(n \cdot \Delta t) \cdot R(n \cdot \Delta t) = \prod_{p=0}^{n} (1 + \tilde{\omega}(p \cdot \Delta t))$$

This combination is *not additive* but multiplicative, and the product is *not commutative* [2] . In addition, because of numerical errors, this product of orthogonal matrixes may generate a non-orthogonal matrix, and this computation *is not stable.*

In order to avoid these problems, we use the *unitary quaternion representation of rotations,* and their product to combine infinitesimal rotations. Although the product of two quaternions requires more operations than a matrix product [3], it will be possible to avoid the problem of unstability.

It is well known [Stu64] that unitary quaternions constitute an efficient way to represent rotations especially in the case where the rotation is related to its derivative. If $\lambda(t)$ is the unitary quaternion representing the rotation $R(t)$, we have :

$$R(t) \cdot \vec{x} = \lambda(t) \times \vec{x} \times \lambda(t)^{-1}$$

3-D vectors being represented as the imaginary component of a quaternion with no real component. The symbol $\times$ denotes the product of two quaternions.

Using these notations one can easily compute [M87] :

$$\frac{d\lambda}{dt} = \frac{1}{2} \cdot \vec{\omega} \times \lambda \tag{1}$$

This last equation will be numerically integrated in the next but one section. Before, since this choice is the result of a comparison of different representations of a rotation, a brief discussion will be given in the next section.

---

[1] There is an explicit solution in the case where

$$\forall t_1, t_2 \, , \; \tilde{\omega}(t_1) \; commutes \; with \; \tilde{\omega}(t_2)$$

This is true if and only if the rotation is performed around a fixed axis, that is if $\vec{\omega}(t) = \theta(t) \cdot \vec{u}$, with $\vec{u}$ fixed. In this case we get :

$$R(t) = e^{\left(\int_0^t \theta(t) \, dt \right) \cdot \tilde{u}} \cdot R(0)$$

[2] One might wish to take only the first terms of this last product, and use :

$$R((n+1) \cdot \Delta t) \simeq 1 + \sum_{p=0}^{n} \tilde{\omega}(p \cdot \Delta t) + o(\| \tilde{\omega}(t) \|^2)$$

which is additive and commutative in terms of $\tilde{\omega}(t)$. Such formula give bad results, since the matrix $R(t)$ computed with such a formula is not orthogonal because (among other things) its diagonal coefficients are equal to one. An improved method has therefore to be used.

[3] The product of two 3x3-matrix requires 9 multiplications and 6 additions, while the product of two quaternions requires 16 multiplications and 12 additions.

2.2 Choosing a representation for the angular position

We list the different usual representations of a rotation and briefly discuss which one would be optimal in our case :

- Using **Matrix representation** is on one hand easy, since directly related to $\tilde{\omega}$. However, based on nine parameters it is rather heavy to compute at each step. In addition, we will need to keep the rotation matrix orthogonal from one step to the next; this is not trivial as opposed to the case of quaternions, for which the similar constraint is to normalize.
- **Euler representation** could have been used, especially if one of the axis of the representation was aligned with $\vec{g}$. This is however not possible, since $\vec{g}$ is not known but to be determined. And all the problems inherent to this representation (use of a lot of trigonometric functions, use of four maps in the atlas) remain.
- **Vector representation of rotation** in terms of its direction and its angle could have been used since we need to describe rotations around specific directions ($\vec{\omega}$ or $\vec{g}$). In this case, if $\vec{u}$ is a unitary vector aligned with the direction of the rotation, while θ is the angle of the rotation :

$$R(\vec{u}, \theta)(\vec{x}) = (1 - \cos(\theta)) \cdot <\vec{u}, \vec{x}> \cdot \vec{u} + \cos(\theta) \cdot \vec{x} + \sin(\theta) \cdot \vec{u} \wedge \vec{x}$$

However this formula is not simple to use when two rotations are combined, as in our case, and is in fact directly derived from quaternion calculus.

- **Representation of rotations as an exponential** of matrix or quaternion have not been used since the combination of two rotations is not commutative and the product of two exponentials is then not equal to the sum of their components. Combining two general rotations is then a complicated problem solved by the *Campbell and Hausdorff* formula ([Hoc68]). Using previous notations, a rotation can be expressed as :

$$R = e^{\tilde{u}}$$

where the exponential the matrix is used here. The direction of the rotation is given by the direction of the vector $\vec{u}$, while the angle of the rotation, taken as positive, is the norm of the vector $\vec{u}$. In our case we have :

$$R(t + \Delta t) = e^{\tilde{u}(t + \Delta t)} = e^{\tilde{\omega}(t)} \cdot e^{\tilde{u}(t)}$$
$$1 + \tilde{\omega}(t) \simeq e^{\tilde{\omega}(t)}$$

The *Campbell and Hausdorff* formula gives an explicit expression of the exponential of the product of two exponentials in a Lie group, in term of a series. In our case, using the third order of the development of this matrix of series, we have :

$$exp(\vec{u}(t + \Delta t)) \simeq exp(\tilde{\omega}(t) + \tilde{u}(t) + \frac{1}{2} \cdot \left[\tilde{\omega}(t), \tilde{u}(t) \right] + \frac{1}{12} \cdot (\left[\tilde{\omega}(t), \left[\tilde{\omega}(t), \tilde{u}(t) \right] \right] + \left[\tilde{u}(t), \left[\tilde{u}(t), \tilde{\omega}(t) \right] \right]))$$

which is valid since the operation :

$$[\tilde{x}, \tilde{y}] = \tilde{x} \cdot \tilde{y} - \tilde{y} \cdot \tilde{x}$$

transforms two 3-D antisymmetric matrix into a 3-D antisymmetric matrix, which can be associated with the skew of a 3-D vector.

This method is the only method which allows a direct computation of the combination of two rotations. The representation of the rotation is minimal (3 parameters), and the operation directly maps two rotations onto a rotation, without the necessity of finding the closest unitary element (as in the case for matrix and quaternions). This operation is a polynomial function of the components of the rotation vectors.

However this method has two drawbacks. First, the number of operations is very high. For example, for the 3rd-order expansion, we have :

$$\begin{vmatrix} \omega_x + u_x + \frac{\omega_y u_z}{2} - \frac{\omega_z u_y}{2} - \frac{\omega_x u_y^2}{12} + \frac{\omega_y u_x u_y}{12} + \frac{u_z \omega_z u_x}{12} - \frac{\omega_x u_z^2}{12} \\ \omega_y + u_y + \frac{\omega_z u_x}{2} - \frac{\omega_x u_z}{2} - \frac{\omega_y u_z^2}{12} + \frac{u_z \omega_z u_y}{12} + \frac{\omega_x u_y u_x}{12} - \frac{\omega_y u_x^2}{12} \\ \omega_z + u_z + \frac{\omega_x u_y}{2} - \frac{\omega_y u_x}{2} - \frac{\omega_z u_x^2}{12} + \frac{\omega_x u_z u_x}{12} + \frac{u_y \omega_y u_z}{12} - \frac{\omega_z u_y^2}{12} \end{vmatrix}$$

which is computed using 21 additions and 45 multiplications. Second, this is only the 3rd-order expansion, and the calculation has to be carried on, since $\vec{u}$ corresponds to the angular position, and is *a priori* small. This method is thus not used here.

- Finally, using a **unitary quaternion**, it is very simple to combine rotations (using the quaternion product), while such a quaternion is easily related to the geometric components of the rotation. Using previous rotations, the quaternion associated to a rotation is simply : $\cos(\frac{\theta}{2}) \oplus \sin(\frac{\theta}{2}) \cdot \vec{u}$, where $\oplus$ denotes the combination of the real and imaginary part of a quaternion.

2.3 Using quaternions to integrate $\vec{\omega}$

In the case of regular samples at a rate of Δt, one can assume [4] the rotation between t and $t + \Delta t$ to be a rotation of direction $\vec{\omega}(t)$ and of angle $\| \vec{\omega}(t) \| \cdot \Delta t$. The corresponding quaternion $\Delta\lambda(t)$ is, for small Δt :

$$\Delta\lambda(t) = \cos(\frac{\|\vec{\omega}(t)\| \cdot \Delta t}{2}) \oplus \sin(\frac{\|\vec{\omega}(t)\| \cdot \Delta t}{2}) \frac{\vec{\omega}(t)}{\|\vec{\omega}(t)\|}$$

$$\Delta\lambda(t) \simeq (1 - (\frac{(\|\vec{\omega}(t)\| \cdot \Delta t)^2}{8})) \oplus \frac{\Delta t}{2} \cdot \vec{\omega}(t)$$

while the residual is of order : $o((\frac{\|\vec{\omega}(t)\| \cdot \Delta t}{2})^3)$.

We get after N samples, an estimation of λ by :

$$\tilde{\lambda}(N \cdot \Delta t) = \Delta\lambda((N-1) \cdot \Delta t) \times \tilde{\lambda}((N-1) \cdot \Delta t) = \prod_{n=0}^{N-1} \Delta\lambda(n \cdot \Delta t)$$

assuming the system is at its initial orientation when $n = 0$.

2.4 Errors in the estimation of the angular rotation

If we denote by $\epsilon_{\vec{\omega}}$ the error on ω, and ϵ_λ the resultant error on λ, we obtain from equation (1) :

$$\| \frac{d\epsilon_\lambda}{dt} \| = \| \frac{1}{2} \cdot \epsilon_{\vec{\omega}} \times (\lambda + \epsilon_\lambda) \|$$

The previous approximation of $\cos()$ and $\sin()$ functions by polynomial expansions induces an error in the norm of $\tilde{\lambda}$, and we get :

$$\| \Delta\lambda \| = 1 + \frac{(\| \vec{\omega} \| \cdot \Delta t)^4}{128} + \theta((\| \vec{\omega} \| \cdot \Delta t)^8)$$

which is expected to be very close to one, but always higher than one.

It is important to recompute after each step the norm of λ and to normalize the result, since our approximate formula tends to increase the norm of λ. Doing this will force the quaternion to be unitary, while the accumulative error will remain only linearly dependent upon time. We thus correct λ by :

$$\tilde{\tilde{\lambda}}(N \cdot \Delta t) = \frac{\tilde{\lambda}(N \cdot \Delta t)}{\| \tilde{\lambda}(N \cdot \Delta t) \|}$$

Then, the norm of $(\lambda + \epsilon_\lambda)$ is not higher than one, even using approximation formulas, and :

$$\| \frac{d\epsilon_\lambda}{dt} \| \leq \| \frac{1}{2} \cdot \epsilon_{\vec{\omega}} \|$$

---

[4] This in fact a "zero order" approximation of $\Delta\lambda(t)$ since $\vec{\omega}(t)$ is assumed to be constant between two samplings. It should be compared to the "rectangle formula" used when numerically computing an integral, while higher order models could have been used, in which $\vec{\omega}(t)$ could have been interpolated using linear ("first order" approximation or "trapezium formula") or quadratic ("second order" approximation or "Simpson formula") methods.

However, a pragmatic argument is in favor of the present method : since Δt must be as short as possible, it is important to compute *quickly* using a simple but efficient method, than using a more sophisticated method, slowing up the calculation and finally degrading performance.

Since the error on $\vec{\omega}$ can be bounded, the time derivative of the error on λ can be bounded too. Then although the errors in $\vec{\omega}$ induce an cumulative error in the computation of λ, this error increase is linear as a function of time, but neither quadratic nor exponential.

3 Separation between gravity and linear acceleration

Gravity and linear acceleration are two quantities which cannot be differentiated with any physical experiment. However, in a robotic environment, special assumptions can be used. Based on these assumptions, additional constraints can be used to separate gravity and linear acceleration. We first motivate two general assumptions on these quantities, and then propose two related methods of separation.

3.1 Using special assumptions on the environment

The accelerations measured by the linear accelerometers are the sum of the object acceleration and gravity. There is no possibility, *a priori*, to separate linear acceleration and gravity, but several additional hypotheses can be used, depending of the context. For example, for a mobile robot running on an floor, one can assume that acceleration is in the plane of the wheels, except if in a lift, or during a free fall... In addition, the direction of the movement is given by the wheel direction, assuming skid does not occur.

However there are two physically realistic hypotheses, not dependent upon the object-specific movement :

1. $\vec{g}$ **is a constant uniform 3-D vector field.** Then, as already formulated :

$$\frac{d\vec{g}}{dt} = \vec{\omega} \wedge \vec{g}$$

2. **It is not physically possible to keep acceleration constant.** Then, for a relatively large window of time, and in an absolute frame of reference, the object acceleration belongs to the higher part of the acceleration frequency spectrum, while gravity is related, in theory, to the lower part of the acceleration frequency spectrum. This means that the DC component of $\vec{\gamma}(O)$ is close to 0 if the time-window is large enough, and we have :

$$\vec{g} = (\vec{\gamma}(O) + \vec{g}) * \int_{-\infty}^{+\infty} I \cdot \delta_0(f) \cdot e^{-2\pi j f t} df$$

where $*$ denotes the convolution operator, I the identity matrix, and vector coordinates are expressed in an absolute frame of reference.

We now describe two methods only based on these hypotheses, and which should be rather general, and not only restricted to, for instance, vehicle movements.

3.2 Method 1 : Estimating $\vec{\gamma}(O)$ in an absolute frame of reference

Using the estimation of the angular rotation through $\lambda(t)$ we can also express $\vec{a} = (\vec{\gamma}(O) + \vec{g})$ with respect to the initial orientation. More precisely, a vector attached to the initial frame of reference is related to a vector attached to the actual frame of reference by :

$$\vec{a}_{absolute} = R(t)^t \cdot \vec{a}_{inertial} = \bar{\lambda}(t) \times \vec{a}_{inertial} \times \lambda(t)$$

since $R(t)$ is the matrix of the rotation from the initial frame to the actual frame, both frames being orthogonal. One should, of course, remember that for an orthogonal matrix $R : R^t = R^{-1}$, while for a unitary quaternion $\lambda : \bar{\lambda} = \lambda^{-1}$ (see for example [Aya88], [Stu64] or [M87] for details).

In this absolute frame of reference we have :

$$\vec{a}_{absolute} = R(t)^t \cdot \vec{\gamma}(O) + \vec{g}(0)$$

where $\vec{g}(0)$ is the fixed value of the gravity field. If $R(t)^t$ had been estimated with no errors, one could write :

$$\vec{\gamma}(O)_{inertial} = \vec{a}_{inertial} - R(t) \cdot \vec{g}(0)$$

This is however not the case, and since the estimation of $R(t)$ is subject to a drift, a filter has to be used.

The second hypothesis provides also an estimation of the object acceleration $\vec{\gamma}(O)$. It is just necessary to filter $\vec{a} = (\vec{\gamma}(O) + \vec{g})$ measured by the accelerometers, in order to eliminate the "zero-frequency" component assumed to be $\vec{g}$, if the vectors are taken in an absolute frame of reference. We have :

$$\vec{\gamma}(O) = \lambda(t) \times (F(t) * (\bar{\lambda}(t) \times \vec{a} \times \lambda(t))) \times \bar{\lambda}(t)$$
$$\vec{g} = \vec{a} - \vec{\gamma}(O) \tag{2}$$

while :

$$F(t) \simeq \int_{-\infty}^{+\infty} I \cdot (1 - \delta_0(f)) \cdot e^{-2\pi j f t} df$$

is the approximation of the impulse response of "frequency-zero" rejection filter. The shape of such a filter is shown in Fig. 1.

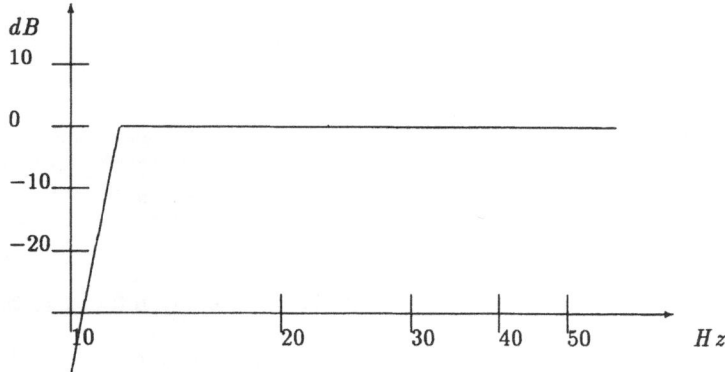

Figure 1. Profile of $F(f)$ gain as a function of frequency (see text for details)

The error on $\lambda(t)$ is mainly a slow drift, that is, a low frequency component of $\lambda(t)$, and is filtered by $F(t)$ in equation (2). Then, contrary to what happens when integrating $\vec{\omega}$, this estimation of $\vec{g}$ for which very low frequencies are filtered out is not subject to drift.

3.3 Method 2 : Estimating $\vec{\gamma}(O)$ using the jerk

The jerk $\vec{j}$ is the time derivative of the accelerations and we have :

$$\vec{j} = \dot{\vec{a}} = \dot{\vec{\gamma}}(O) + \dot{\vec{g}} = \dot{\vec{\gamma}}(O) + \vec{\omega} \wedge \vec{g} = \dot{\vec{\gamma}}(O) + \vec{\omega} \wedge (\vec{a} - \vec{\gamma}(O))$$

which provides a differential equation in term of $\vec{\gamma}(\dot{O})$:

$$\dot{\vec{\gamma}}(O) = \vec{\omega} \wedge \vec{\gamma}(O) + \dot{\vec{a}} - \vec{\omega} \wedge \vec{a}$$

Such an equation can be solved using different methods, either used in the internal model of a Kalman filter, or in a simple feedback as given in figure (2). This feedback is a non-linear filter and will be denoted by :

$$\vec{\gamma}(O)(t) = G(\vec{a}(t), \vec{\omega}(t))$$

In the rest of this paper, only the second method will be used.

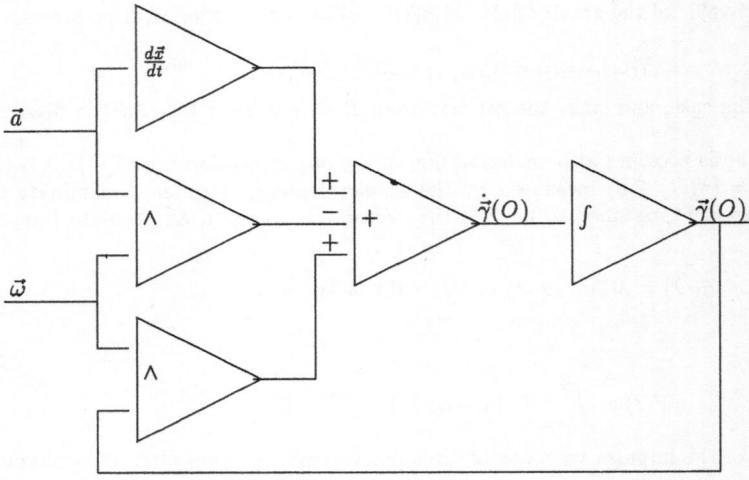

Figure 2. Feedback mechanism for the estimation of $\vec{\gamma}(O)$

4   Correcting angular orientation using $\vec{g}$ estimation

The previous methods provide a way to predict the direction of $\vec{g}$. Then, starting at an initial position for which the system is not moving, while $\vec{g}$ is the only acceleration, we can estimate the vertical direction which is also the direction of $\vec{g}$. If $\tilde{\vec{g}}(t)$ is the estimation of $\vec{g}$ at time t and if at $t = 0$ the system is at an initial position and not moving, we simply have :

$$\tilde{\vec{g}}(0) = \lambda(t)^{-1} \times \tilde{\vec{g}}(t) \times \lambda(t) \tag{3}$$

However, this equation determines only the "vertical orientation", since any rotation in the horizontal plane does not modify $\vec{g}$.

We then first define *horizontal* and *vertical* angular orientations, in other words, the angular orientation with respect to the vertical direction and the orientation in the horizontal plane. We then show how to recover the vertical orientation, knowing $\vec{g}$. We also propose a method to combine the integration of $\vec{\omega}$ developed in a previous section, and the present estimation, in order to improve the angular orientation estimation. We finally briefly describe the kind of errors produced by such a method.

4.1   Decomposition in terms of horizontal and vertical rotation

Let us decompose $\lambda(t)$ in two rotations in the following order :

$$\lambda(t) = \lambda_v(t) \times \lambda_h(t)$$

where :

- $\lambda_h(t)$ corresponds to a rotation in the horizontal plane, around an axis given by $\vec{g}(t)$:

$$\lambda_h(t) = \cos(\frac{\Theta_h}{2}) \oplus \sin(\frac{\Theta_h}{2}) \cdot \vec{v}$$

   where $\vec{v} = \frac{\vec{g}(t)}{\|\vec{g}(t)\|}$ is a unitary vector in the vertical direction.
- $\lambda_v(t)$ corresponds to a rotation orthogonal to $\vec{g}$.

$$\lambda_v(t) = \cos(\frac{\Theta_v}{2}) \oplus \sin(\frac{\Theta_v}{2}) \cdot \vec{u}$$

   with $\| \vec{u} \| = 1$, $\Theta_v > 0$ and $< \vec{u}, \vec{v} > = 0$.

It first should be noted that :

For a given unitary quaternion $\lambda = r \oplus \vec{\eta}$, and a given direction $\vec{v}$ the previous decomposition is unique.

Remembering that if $\lambda_1 = r_1 \oplus \vec{\eta}_1$ and $\lambda_2 = r_2 \oplus \vec{\eta}_2$ are two quaternions, their product is $\lambda_1 \times \lambda_2 = (r_1 \cdot r_2 - <\vec{\eta}_1, \vec{\eta}_2>) \oplus (\vec{\eta}_1 \wedge \vec{\eta}_2 + r_1 \cdot \vec{\eta}_2 + r_2 \cdot \vec{\eta}_1)$, we get :

$$r \oplus \vec{\eta} = \cos(\frac{\Theta_v}{2}) \cdot \cos(\frac{\Theta_h}{2}) \oplus \sin(\frac{\Theta_v}{2}) \cdot \sin(\frac{\Theta_h}{2}) \cdot \vec{u} \wedge \vec{v} + \cos(\frac{\Theta_v}{2}) \cdot \sin(\frac{\Theta_h}{2}) \cdot \vec{v} + \sin(\frac{\Theta_v}{2}) \cdot \cos(\frac{\Theta_h}{2}) \cdot \vec{u}$$

and computing $< \vec{\eta}, \vec{v} >$ and $\vec{\eta} \wedge \vec{v}$ one can easily verify that :

$$\sin(\frac{\Theta_u}{2}) \cdot \vec{u} = \cos(\frac{\Theta_h}{2}) \cdot (\vec{\eta} - < \vec{\eta}, \vec{v} > \cdot \vec{v}) + \sin(\frac{\Theta_h}{2}) \cdot \vec{\eta} \wedge \vec{v}$$
$$\cos(\frac{\Theta_h}{2})^2 = r^2 + < \vec{\eta}, \vec{v} >^2$$
$$\cos(\frac{\Theta_h}{2}) \cdot < \vec{\eta}, \vec{v} > = \sin(\frac{\Theta_h}{2}) \cdot r$$

which determines for $\Theta_h \in [-\pi, \pi]$, $\Theta_v \in [0, \pi]$ a unique representation of the rotations related to λ_h and λ_v. One can also verify that $< \vec{u}, \vec{v} >= 0$ while $< \vec{u}, \vec{u} >= 1$, as requested.

4.2 Using $\vec{g}$ to compute vertical orientation

Since :

$$\tilde{g}(0) = \lambda(t)^{-1} \times \tilde{g}(t) \times \lambda(t)$$
$$\Leftrightarrow \quad \tilde{g}(0) = \lambda_v(t)^{-1} \times \lambda_h(t)^{-1} \times \tilde{g}(t) \times \lambda_h(t) \times \lambda_v(t)$$
$$\Leftrightarrow \quad \tilde{g}(0) = \lambda_v(t)^{-1} \times \tilde{g}(t) \times \lambda_v(t)$$

since the rotation axis of $\lambda_h(t)$ is aligned with $\vec{g}$. We have then to estimate $\lambda_v(t)$ representing a rotation which moves $\vec{g}(0)$ to $\vec{g}(t)$.

The rotation related to $\tilde{\lambda}_v$ is defined by

$$\cos(\tilde{\Theta}_v) = \frac{1}{\|\vec{g}\|^2} \cdot < \vec{g}(t), \vec{g}(0) > \,, \quad \tilde{\Theta}_v > 0$$
$$\vec{u} = \frac{\vec{g}(t) \wedge \vec{g}(0)}{\|\vec{g}(t) \wedge \vec{g}(0)\|}$$

It is the unique rotation with its axis orthogonal to $\vec{g}(t)$ and mapping $\vec{g}(t)$ on $\vec{g}(0)$. Furthermore, $\tilde{g}(0) = \tilde{\lambda}_v^{-1} \times \tilde{g}(t) \times \tilde{\lambda}_v$

4.3 Iterative computation of angular orientation and linear velocity

We can finally correct the estimation of $\lambda(t)$ given by the integration of $\vec{\omega}$ by :

- Decomposing $\lambda(t)$ in term of $\lambda_h(t)$ and $\lambda_v(t)$
- Computing $\lambda_v(t)$ in order to have $\tilde{g}(t)$ vertical
- Computing $\lambda_h(t)$ from $\lambda(t)$
- Recombining the two components

Combining the previous equations, we can compute the angular orientation represented by $\lambda(t)$, the linear velocity $\vec{V}(t)$, and the gravity vector $\vec{g}(t)$. Computation is made iteratively, at each sampling. All equations to be used are reviewed now.

We get for the initial orientation, where the system is assumed to be motionless.

$$\lambda(0) = 1 \oplus \vec{0}$$
$$\vec{V}(O) = \vec{0}$$
$$\vec{g}(0) = \vec{a}(0)$$

where $\vec{a} = \vec{\gamma}(O) + \vec{g}$ denotes the accelerations, and $\vec{V}$ the linear velocity.

Then, from one sample to the next :

- The new angular orientation is given by :

$$\tilde{\lambda}((n+1) \cdot \Delta t) = \Delta \tilde{\lambda}(n \cdot \Delta t) \times \lambda(n \cdot \Delta t)$$
$$\Delta \tilde{\lambda}(n \cdot \Delta t) = (1 - \frac{(\|\vec{\omega}(n \cdot \Delta t)\| \cdot \Delta t)^2}{8}) \oplus \frac{\Delta t}{2} \cdot \vec{\omega}(n \cdot \Delta t)$$
$$\tilde{\tilde{\lambda}}((n+1) \cdot \Delta t) = \frac{\tilde{\lambda}((n+1) \cdot \Delta t)}{\|\tilde{\lambda}((n+1) \cdot \Delta t)\|} = r \oplus \vec{\eta}$$

- Estimating $\vec{\tilde{\gamma}}(O)$ and $\vec{g}$ using G :

$$\vec{\tilde{\gamma}}(O) = G(\vec{a}(t), \vec{\omega}(t))$$
$$\vec{\tilde{g}} = \vec{a} - \vec{\tilde{\gamma}}(O)$$

- The linear translation is simply obtained by integrating $\vec{\tilde{\gamma}}(O)$, that is, if using a trivial integration formula :

$$\vec{V}((n+1) \cdot \Delta t) = \vec{V}(n \cdot \Delta t) + \vec{\tilde{\gamma}}(O)(n \cdot \Delta t)$$

- The orientation as defined by $\tilde{\tilde{\lambda}}$ is constrained to be compatible with the vertical orientation. We then define a quaternion $\tilde{\lambda}_v$ as :

$$\tilde{\lambda}_v = \cos(\frac{\tilde{\Theta}_v}{2}) \oplus \sin(\frac{\tilde{\Theta}_v}{2}) \vec{\tilde{u}}$$

where $\tilde{\Theta}_v$ and $\vec{\tilde{u}}$ are given by :

$$\cos(\tilde{\Theta}_v) = \frac{1}{\|\vec{\tilde{g}}\|^2} \cdot < \vec{\tilde{g}}(t), \vec{\tilde{g}}(0) > , \ \tilde{\Theta}_v > 0$$
$$\vec{\tilde{u}} = \frac{\vec{\tilde{g}}(t) \wedge \vec{\tilde{g}}(0)}{\|\vec{\tilde{g}}(t) \wedge \vec{\tilde{g}}(0)\|}$$

- The angle of the horizontal rotation, about $\vec{g}$ is given by : $\tan(\frac{\Theta_h}{2}) = \frac{<\vec{\eta}, \vec{\nu}>}{r}$, if we note $\tilde{\tilde{\lambda}}((n+1) \cdot \Delta t) = r \oplus \vec{\eta}$, and $\vec{\nu} = \frac{\vec{g}}{\|\vec{g}\|}$. This defines a quaternion $\tilde{\lambda}_h$ as :

$$\tilde{\lambda}_h = \cos(\frac{\Theta_h}{2}) \oplus \sin(\frac{\Theta_h}{2}) \cdot \vec{\nu}$$

- We finally obtain, for the angular orientation :

$$\lambda((n+1) \cdot \Delta t) = \tilde{\lambda}_v \times \tilde{\lambda}_h$$

Number of operations.   Taking into account that $\| \vec{\omega} \|^2$ has been computed previously, if m is the number of multiplications and a the number of additions to compute the $F(t)$ filtering on one sample, this computation requires $96 + m$ multiplications, $52 + a$ additions and 4 square-roots.

5   Estimation of the errors

5.1   A bound for $\vec{\omega}$ and $\vec{\gamma}(0) + \vec{g}$ errors

We are going to compute now, the related experimental errors on $\lambda = \tilde{\lambda}_v \times \tilde{\lambda}_h$, $\vec{\gamma}(0)$ and $\vec{V}$.

The computations being highly non-linear, it does not make sense to calculate, as for linear transformations, the error in term of gain, offset and white, additive, Gaussian noise. Using a different approach we are going to try to estimate a bound for the errors. Let us note :

$$\hat{\vec{\omega}} = \ \ \vec{\omega} + \vec{\epsilon}_\omega$$
$$\hat{\vec{a}} = \ \ \vec{a} + \vec{\epsilon}_a$$

the relation between real value, error, and estimated value, as given elsewhere ([VF89]).

5.2 Expression of the parameters errors

- For the linear acceleration $\vec{\gamma}(O)$:

$$\epsilon_{\vec{\gamma}(O)} = \frac{dG(\vec{a}(t), \vec{\omega}(t))}{d\vec{a}} \cdot \epsilon_{\vec{\omega}} \frac{dG(\vec{a}(t), \vec{\omega}(t))}{d\vec{\omega}} \cdot \epsilon_{\vec{a}}$$

- For the vertical orientation λ_v:

$$\epsilon_{\tilde{\lambda}_v} = \frac{1}{2 \cdot \|\,\tilde{g}\,\|^2} \cdot < \epsilon_{\tilde{g}}, \tilde{g}(0) > \cdot (1 \oplus \tilde{u})$$

$$+ \quad \sin(\frac{\tilde{\Theta}_v}{2}) \cdot \frac{1}{\|\,\tilde{g}(t) \wedge \tilde{g}(0)\,\|} \cdot (0 \oplus \epsilon_{\tilde{g}} \wedge \tilde{g}(0))$$

The norm of $\epsilon_{\tilde{\lambda}_v}$ can be bounded by:

$$\|\,\epsilon_{\tilde{\lambda}_v}\,\| < \frac{2}{\|\,\tilde{g}\,\|^2} \cdot \|\,\epsilon_{\tilde{g}}\,\|$$

- For the linear velocity $\vec{V}$:

$$\epsilon_{\vec{V}(N \cdot \Delta t)} = \sum_{n=0}^{N} \epsilon_{\vec{\gamma}(O)}(n \cdot \Delta t)$$

Since the errors on the linear acceleration $\vec{\gamma}(O)$ and on the vertical orientation λ_v are bounded, this is not the case for the linear velocity $\vec{V}$, which increases linearly with time.
- At last, the horizontal orientation λ_h: is given by the two equations:

$$\begin{aligned} \epsilon_{\tilde{\chi}}(n \cdot \Delta t) &= \quad a(n \cdot \Delta t) + B(n \cdot \Delta t) \cdot \epsilon_{\lambda_h}((n-1) \cdot \Delta t) \\ \epsilon_{\lambda_h}(n \cdot \Delta t) &= \quad c(n \cdot \Delta t) + D(n \cdot \Delta t) \cdot \epsilon_{\tilde{\chi}}(n \cdot \Delta t) \end{aligned}$$

the quaternions being considered as 4-D vectors while [5]:

$$a(n \cdot \Delta t) = \quad (-\frac{\Delta t^2}{4} \cdot \|\,\tilde{\omega}\,\| \cdot \tilde{\epsilon}_{\|\omega\|} \oplus \frac{\Delta t}{2} \cdot \tilde{\epsilon}_\omega) \times \lambda + \Delta \lambda \times \epsilon_{\lambda_v} \times \lambda_h$$

$$B(n \cdot \Delta t) \cdot x = \quad (\Delta \lambda \times \lambda_v) \times x$$

$$c(n \cdot \Delta t) = \quad 0 \oplus (\sin(\frac{\tilde{\Theta}_v}{2}) \cdot \frac{\epsilon_{\tilde{g}}}{\|\,\tilde{g}\,\|} + \frac{< \vec{v}, \epsilon_{\tilde{g}} >}{\|\,\tilde{g}\,\| \cdot \sqrt{r^2 + < \vec{v}, \vec{v} >} \cdot \tilde{g}})$$

$$D(n \cdot \Delta t) \cdot (y \oplus \vec{x}) = \quad \frac{y}{\sqrt{r^2 + < \vec{v}, \vec{v} >}} \oplus < \vec{x}, \frac{\tilde{g}}{\|\,\tilde{g}\,\| \cdot \sqrt{r^2 + < \vec{v}, \vec{v} >}} > \cdot \frac{\tilde{g}}{\|\,\tilde{g}\,\|}$$

We then have, at last:

$$\epsilon_{\lambda_h}(n \cdot \Delta t) = (c(n \cdot \Delta t) + D(n \cdot \Delta t) \cdot a(n \cdot \Delta t)) + (D(n \cdot \Delta t) \cdot B(n \cdot \Delta t)) \cdot \epsilon_{\lambda_h}((n-1) \cdot \Delta t)$$

And if we note:

$$\begin{aligned} \cdot\ e(n \cdot \Delta t) &= \quad c(n \cdot \Delta t) + D(n \cdot \Delta t) \cdot a(n \cdot \Delta t) \\ F(n \cdot \Delta t) &= \quad D(n \cdot \Delta t) \cdot B(n \cdot \Delta t) \end{aligned}$$

the final expression for ϵ_{λ_h} is

$$\epsilon_{\lambda_h}(N \cdot \Delta t) = \sum_{n=0}^{N} (\prod_{p=n+1}^{N} F(p)) \cdot e(n)$$

---

[5] One should note that the product of quaternions is linear in terms of one quaternion coordinates, since:

$$(a \oplus \vec{b}) \times (y \oplus \vec{x}) = (a \cdot y - < \vec{b}, \vec{x} >) \oplus (\vec{b} \wedge \vec{x} + a \cdot \vec{x} + y \cdot \vec{b})$$

This last equation can be used to compute the evolution of the error, if it remains small. Since:

$$\| B(n \cdot \Delta t) \cdot x \| = \quad \| x \|$$

$$\| B(n \cdot \Delta t) \cdot x \| \leq \quad \frac{1}{\sqrt{r^2 + <\vec{v}, \vec{\nu}>^2}}$$

the evolution of ϵ_{λ_h} can be bounded and we are sure to have:

$$\| \epsilon_{\lambda_h}(N \cdot \Delta t) \| \leq \sum_{n=0}^{N} \frac{\| e(n) \|}{\sqrt{r^2 + <\vec{v}, \vec{\nu}>^2}}$$

which increases linearly with time.

6 Conclusion

In this article we have described how to compute angular orientation and linear acceleration from inertial sensors output.

As indicated in the introduction, such computation is the basic process for the use of inertial information in a visual system. Future work will be directed toward the study of the cooperation between vision and inertial sensing, for the estimation of robot motion and orientation.

References

[Aya88] N. Ayache
Construction et fusion de représentations visuelles 3D.
PhD thesis, Université Paris-Sud, Orsay, France, 1988.

[H82] H. Cartan
Cours de calcul différentiel, Chapter II-2, équations différentielles linéaires,
Hermann, Paris, 1982

[Hoc68] G. Hochschild
La structure des groupes de Lie
Dunod, Paris, 1968

[M87] M Le Borgne
Quaternions et contrôle sur l'espace des rotations.
Technical Report 751, Institut National en Informatique et Automatique, Rocquencourt, 1987

[Stu64] J. Stuelpnagel
On the parametrization of the three-dimensional rotation group.
SIAM Review, 6:422–430, 1964

[TNF88] G. Toscani, R. Deriche , O.D. Faugeras
3-D motion estimation using a tocken-tracker.
In: IAPR Workshop on Computer Vision, Special Hardware and Industrial Applications, Tokyo,
pages 257–261, 1988

[VF89] T. Viéville, O.D. Faugeras
Computation of inertial information on a robot.
In: The Fifth International Symposium of Robotic Research, August 1989

A SENSOR PROCESSING MODEL INCORPORATING ERROR DETECTION AND RECOVERY

G.A. Weller, F.C.A. Groen, L.O. Hertzberger
Faculty of Mathematics and Computer Science
Department of Computer Systems
University of Amsterdam
Kruislaan 409
1098 SJ Amsterdam, The Netherlands

ABSTRACT

In this paper we address the problem of providing a sensor-system with the ability to detect measurement errors and to recover from these errors. We propose to equip every sensor module in the sensor system with active and latent tests which verify respectively possible environmental independent and environmental dependent sensor failures. The recovery strategy is based on rules that are local to the different sensor modules. We will show the applicability in an example.

1.   THE PROBLEM

Robot systems that operate autonomously under changing environmental conditions need a flexible and reliable sensor system in order to obtain environmental information. An increase in the flexibility of the robot system and so in the complexity of the tasks of the sensor system should not be accompanied with a decrease of the reliability of the system. Reliability of the sensor system is of great importance since all the actions of the robot will essentially be based on the information that is collected by this system. Equipping the sensor system with a means to maximize its reliability will increase the overall performance of the robot system.

In this paper we address the problem of providing a sensor system with the ability to detect measurement errors and to recover from these errors. Our solution is to break the sensor system up into sensor modules; each of which is equipped with tests to verify certain characteristics in its input data and with tests to verify the internal data and the output data resulting from the execution of the algorithm of the sensor module. A negative test result implies a sensor failure. Two types of tests will be used. The first are tests that verify demands on the data of the sensor module that must be fulfilled at all times. The other type of test verifies demands on the data of the sensor module that only need to be fulfilled under certain conditions. These conditions depend on the environment within which a sensor module must operate. The latter tests are only activated if those specific conditions are valid. A mechanism is introduced that allows activation

NATO ASI Series, Vol. F 63
Traditional and Non-Traditional Robotic Sensors
Edited by T. C. Henderson
© Springer-Verlag Berlin Heidelberg 1990

of these latent tests on basis of the environmental conditions. When a test gives a negative result the sensor module will start a recovery procedure. This procedure is based on adjusting the parameters of the algorithm and/or demanding adjustments of the input to the sensor module. These adjustments are available in the form of rules. Every sensor module is equipped with its own set of rules.

The research described here is done in the Computer Systems group of our faculty and forms part of a research program on intelligent autonomous systems. This research program is focussed on equipping robots with the ability to handle exceptional situations autonomously during the execution of their tasks. The robots that we use are an autonomous vehicle and a robot arm operated by an intelligent robot controller [1]. The intelligent robot controller is capable of detecting an exception during the execution of an assembly task performed by the robot arm. When an exception is detected a diagnosis- and subsequently a recovery stage is entered. The main problem in the recovery strategy is to relate the task knowledge at a high level to elementary robot actions. For this purpose rules are applied that determine the recovery process. The result of the diagnosis stage is used as an input to a set of rules that relate task knowledge to elementary robot actions. The sensor system described here is to be used in these autonomous robots to collect the necessary information to detect failures and perform diagnostics [2]. The strategy of the sensor system to recover from sensor failures is comparable to the strategy that is currently being used by the intelligent robot controller.

2.   PREVIOUS WORK

Obtaining reliable and robust sensor data interpretation is of great importance since erroneous interpretation of the sensor data may lead to unwanted actions in an autonomous system. It is therefore a major area for study as is seen by the interest of many researchers [3],[4],[5]. Techniques that are in use are statistical techniques in which, on basis of more than one measurement, a statistically best measurement is chosen. Within this area the Kalman filter technique is often used.(see for example ref. [6]). Other authors use similar techniques, based on comparing probability distributions of sensor readings from sensors that measure the same object properties [7]. These techniques use the concept of redundancy to refine their measurements or even to reject non-fitting measurements (see for example ref. [8]).

Breaking up a complicated sensor algorithm into parts and applying tests within each part to obtain a measure of its correctness is an obvious technique to determine the exact place within the sensor algorithm where a failure occurs. This idea can be found in a sensor model that has been proposed by Henderson et. al.[9], [10]. Every piece has its own functionality within the total sensor. They called such a part a "logical sensor".

A 'logical' sensor is an extension of the 'regular' sensor. A quantity that a logical sensor could measure is for example, the lines in an image collected with a camera. A 'logical' sensor can use either the output of a transducer, the output of another logical sensor or any combination of these as its input . For convenience the transducer itself is also called a logical sensor. This approach opens up the way to create many different logical sensors through combinations of others, leading to a flexible and modular sensor system structure. Within this logical sensor concept a mechanism has been incorporated to handle an 'erroneous' input to a logical sensor. Every input is judged by an 'acceptance test' which result in the acceptance or rejection of the input . In case of rejection, the logical sensor has a list of alternative logical sensors available that can provide the same input. From this list it picks the next logical sensor which is then activated. A logical sensor fails when its list of alternatives is exhausted.

The former methods may come up with the notion of a sensor failure but they are not able to trace the origin of the failure. Their recovery scheme is based on rejection of erroneous results and/or replacement of the sensor. They don't supply enough measurement data for a recovery scheme based on adjusting the sensor signal and/or parameters.

3.   OUR APPRAOCH

We will adopt the basic idea of the logical sensor system of using building blocks that we call sensors modules (to distinguish it from the logical sensor concept) that each has its own functionality. This will ensure a flexible sensor system. (Figure 1) In our approach we think that before the radical step of replacing a sensor module to obtain 'correct' input is taken, it might prove worthwhile to find the cause for discarding it. After this is determined an attempt can be made to recover from it by adjusting the parameters of the algorithm and/or the input to the sensor. To obtain this we want to express within a tests the expertise knowledge on what to expect from the data in general and in specific situations to ensure the expected functioning of the module. For the recovery stage we also want to use expertise knowledge containing the necessary adjustments of parameters of- or input to the sensor module. The modular sensor system allows to put this knowledge at the appropriate position in the module instead of using an overall knowledge system to check on the consistency of the sensor data interpretation.

Our sensor system is build up hierarchically (figure 1). The transducers are at the bottom level and· at the top level we find the robot specific sensors modules. The higher in this hierarchy the less generally applicable a sensor module will be.

The tests that are incorporated in a sensor module are not used as a mechanism to activate an alternative sensor module in case of a negative test result. We use the tests to pin-point the location where a failure occurs, the nature of the test indicates what went wrong. The scope of the test in a sensor module is not restricted to the input data. In the course of the execution of the algorithm tests will also be activated, increasing the ability to pin-point the exact location and nature of a sensor module error. A negative test result will initiate a recovery stage. The recovery strategy is based on rules that are available locally to the sensor module of which the test is a part. These rules determine the explicit recovery steps.

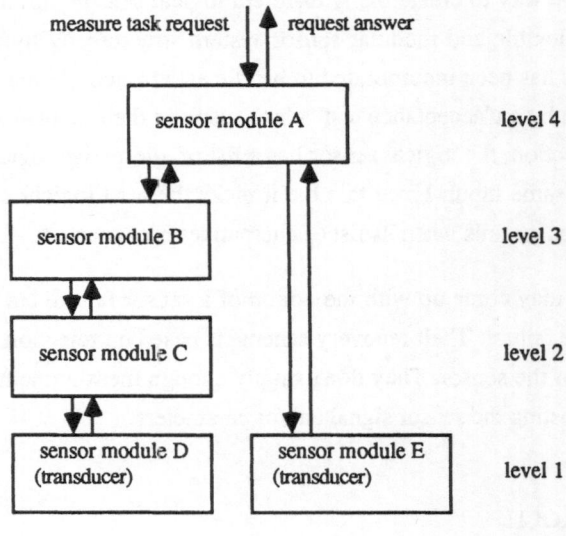

Figure 1

3.1. Determination of an error condition

3.1.1. The environmental information

Tests may be either active or not-active. Only active tests are executed. A not-active test will only become active if certain environmental conditions justify such a test. If, for example, a sensor system contains a sensor module that measures what colours are present in a camera picture, this sensor module may have a latent test available on the distribution ratio of colours in the picture. This test need only be activated when this condition (the distribution ratio) is actually known.

The previous example showed how a test can be initiated based on information concerning the environment within which a measurement is performed. We will call this type of information, such as the distribution ratio, environmental information. Activating latent tests to verify such environmental information increases the number of active tests in a sensor module and therefore the ability to locate and identify a possible failure more precisely.

It will not be possible to obtain such environmental information for every sensor module. This is illustrated in the following example.

A possible sensor module is :

measure-position-part-A

This sensor module may identify an object as object A and if it is so it measures its position. Such a sensor module is used in part assembly tasks. The strategy of this sensor module might be to use a bar-code mark on the object, the code of which can be identified with the name of the object and the orientation of which can be used to determine the orientation of the object. In the hierarchy of activated sensor modules this sensor module will activate the module :

measure-lines-picture

This sensor module determines the algebraic equations and thickness of the straight lines in an image. In this example it is activated indirectly by the measure-position-part-A sensor module but it may be activated in a variety of different sensors modules and is therefore generally applicable.

The strategy of the sensor module in the first example is based on the classification of the bar code pattern of a specific pattern. All the characteristics of the pattern such as the bar-code itself, the sizes of the code mark, the colour of the bar-code mark are actually environmental information. This sensor module uses only part of this environmental information, the bar-code itself, in its measurement. It is however crucial for this module that the picture really contains the bar code image. At the level of the measure-lines sensor module such environmental conditions are not apparent. This module simple extracts any straight line from an input picture. This is actually a consequence of its general applicability. The environmental conditions that specify the environmental information are determined from the functionality of the sensor modules up in the hierarchy of activated sensor modules. The less generally applicable a sensor module, the more specific it may be in its environmental conditions. The next problem is now apparent:

-    Demands derived from environmental conditions that are available on a high sensor module level, can only be verified using data that is exclusively available on a low sensor module level.

Even if the relevant data are still available on a high sensor module level, it is advantageous to verify their validity immediately when they are derived on a lower level. We therefore want the demands formulated from environmental conditions, that are available on a high level sensor module, to be available on the low level sensor module where they can be verified by tests implemented within the sensor module. This way we increase the total number of tests and the ability to measure a sensor module failure in an early stage of data interpretation. The demands on measurable quantities from the data are formulated in the form of restrictions on the values of these quantities.

3.1.2. The relevant environment of a sensor module

A sensor module can be viewed as a black box having an input and an output. Its connections to the outside world ('environment'?) are the sensor modules that it activates (its children) and the sensor module that activated him (the father). All the characteristics of the environment that can have influence on the proper (that is : 'expected') functioning of the sensor module will be incorporated in the input data. Also other characteristics of the environment that will not influence the proper functioning but are easily to be checked from the data available to the sensor module, will (or may) be incorporated in the input data.

3.1.3. The demands on the data

The tests to verify demands on the data logically form part of the sensor module because they are completely characterized by the functionality of the algorithm of the sensor module. We can now distinguish two different sources of demands on the data. These are :

-       Demands that are required by the characteristics of the algorithm that is to work upon the data (Hendersons acceptance test). This is, for example, a minimal number of data points required for an algorithm that solves a set of equations by matrix operations. We will call these demands : Algorithmic demands.

-       Demands that are determined by the environment within which the sensor module is to operate. We will call these demands : Environmental demands.

3.1.4. The environmentally independent tests

The environmental independent tests verify the algorithmic demands. The tests scrutinizes the data to check if the algorithm can accomplish the transformation from input to output . Since the execution of these tests is independent of the specific measurement conditions, they are executed each time the sensor module obtains new input data.

3.1.5. The environmental dependent test

These test verify environmental demands. From the characteristics of the algorithm of the father, demands may be formulated on the output of the child that are only valid for that specific father and not if another father needs the output of the same child. This makes these demands environment dependent in the sense that they are only valid under certain measurement conditions that needed the activation of the specific father.

The difficulty exists to allow every relevant demand that has influence on a low sensor module level to be known at this low level. To keep the modular character of the sensor system we only want a father to be able to formulate demands on the output of a child and not grand- or grandgrand fathers to be able to formulate demands on the output of a grand- or grandgrand child.

3.1.6. Results of the tests

After the execution of a test three modes can be distinguished:

mode 1-   The input data fulfils the demands prescribed by the sensor module
                The internal or output data fulfils the demands prescribed by the sensor.

mode 2-   The input data fulfils the demands prescribed by the logical sensor
                The internal or output data doesn't fulfil the demands prescribed by the sensor module.

mode 3-   The input data doesn't fulfil the demands prescribed by the sensor module

3.2.   Recovering from sensor error.

On basis of a negative test result a sensor error is detected. The sensor is then in mode 2 or 3. The next step is to try to recover from this error. Rules are used that relate a negative test result to changes required in the sensor data such that the test, when executed again, will be positive. These are the strategy rules. Every test must have one or more strategy rules attached to it.

Strategy rules-   Rules attached to a specific test that relate a negative test result to
                        required changes in the data that will lead to a positive test result.

The result of the application of these rules are demanded changes in the data. These changes in the data are used in a next set of rules that translate them to changes in algorithmic parameters or changes in input data .

adaptation rules-   Rules to translate required data changes to algorithmic parameter
                        changes.
propagation rules-   Rules to translate required data changes to demanded input changes.

A depiction of the recovery based on these rules is given in the figure 2.

4.   SENSOR MODULES FOR ROBOT ASSEMBLY APPLICATIONS

The sensor system that we use operates in a assembly robot environment. The tasks of the robot system determine the necessary sensor modules. A sensor module that is needed is the : Recognize part module. The algorithm that we developed for this module basically attempts to match a 2-dimensional graph representing the visible (and detected) edges of an object to a 3-

dimensional graph of an object from the set of possible objects [11]. For this matching process the geometrical features of the objects are used. Environmental knowledge in this case consists, for example, of the set of geometrical features that possibly can be detected in the scene. This can be used to compare geometrical features that are found with this set in order to tests the proper interpretation

Another important sensor module that is needed in our applications is the : Detect edges module. This module is activated by the recognize part module to obtain input consisting of an edge image of the environment. Environmental knowledge at this level can be on the characteristics of arc lines- (for example the maximum curve strength), the maximum length of lines- or the minimum length of lines in the picture. This module needs an image of the environment as its input. It will therefore activate the : Look module. From this module it will obtain an image from the 3-dimensional world. The Look module can be activated by any sensor module that needs an image from the 3-dimensional world. If, for example, an object must be tracked from a sequence of images [12] the Track_object sensor module will activate the Look module to obtain an input image.

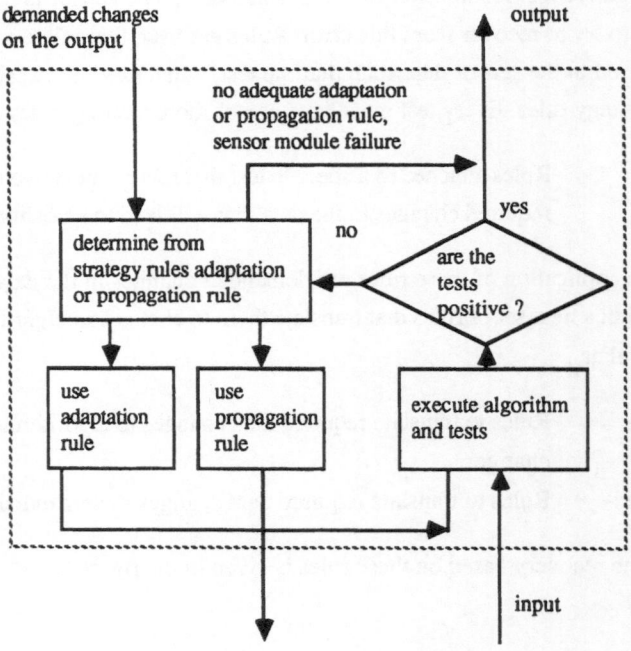

Figure 2

We will show for the Look module how environmental information may cause the activation of tests. For the important Look module errors we will show how the recovery strategy functions.

4.1. Determination of an error condition for the Look module

We will call the sensor module that activates the Look module the father module (figure 3). The output of the Look module is an image of a specific volume in the work space that is to be monitored. The input for this module is a picture of the work space that is recorded by a camera mounted on a robot arm. Of the camera the position, pose, iris, focus and gain factor are adjustable. The number of degree of freedom that influence the input is therefore 9. The output is that part of the input image containing the volume in 3D that is to be monitored.

The Look module activates the Single frame sensor module to obtain a set of coherent measured quantities consisting of : A pointer to the location in memory at which a recorded picture is stored, the position and pose values- and the focal length value of the camera that recorded the picture.

If the Look module has available knowledge about the environment in which the measurement is performed it is able to formulate demands on the data of the Single frame sensor module. The environmental information that is relevant at the level of the Look sensor module is :

Rl :	11•	restrictions on the position and dimensions of the 3D volume.
	12•	restrictions on the resolution of the image of the volume (no. pixels in x and y direction).
	13•	restrictions on the viewing direction.
	14•	restrictions on the dynamic range of the required image.

The first, third and fourth type of demand result from the (by the father) expected characteristics of the world environment at which the measurement is performed. The second type of demand results from the father that needs a certain resolution of the image of the volume to have its algorithm work properly.

From 11 and 12 together with the coherent set of measured quantities that the Look module receives from the Single frame module, it calculates the position and dimensions of the image of the volume in the input picture.

If demands are passed to the Look sensor module, it will form in its turn demands on its input that it obtains from the Single frame module. The environmental information that is relevant at the level of the Single frame are :

Rs :	s1•	restrictions on the focus distance.
	s2•	restrictions on the depth of focus.
	s3•	restriction on the pose of the optical axis of the camera system
	s4•	restrictions on the position of the camera system
	s5•	restriction on the maximum intensity value.

The Look sensor module formulates these demands from the demands that it itself received from the father module. If it hasn't received any, it cannot formulate restrictions on the output of the Single frame sensor module. The demands s1 and s2 are formulated directly from l1 and l2. S3 is determined from l3. Using this pose, l1 and l2, s4 can be determined. For these calculations it also uses the main focal length of the camera and the dimensions of the CCD in the camera. Eventually from l4 the Look module formulates s5.

After formulating, for example, s3, the Look sensor sets a bit in a field that it shares with the Single frame module which indicates that s3 is valid and it deposits the restriction values in predetermined positions in the same field.

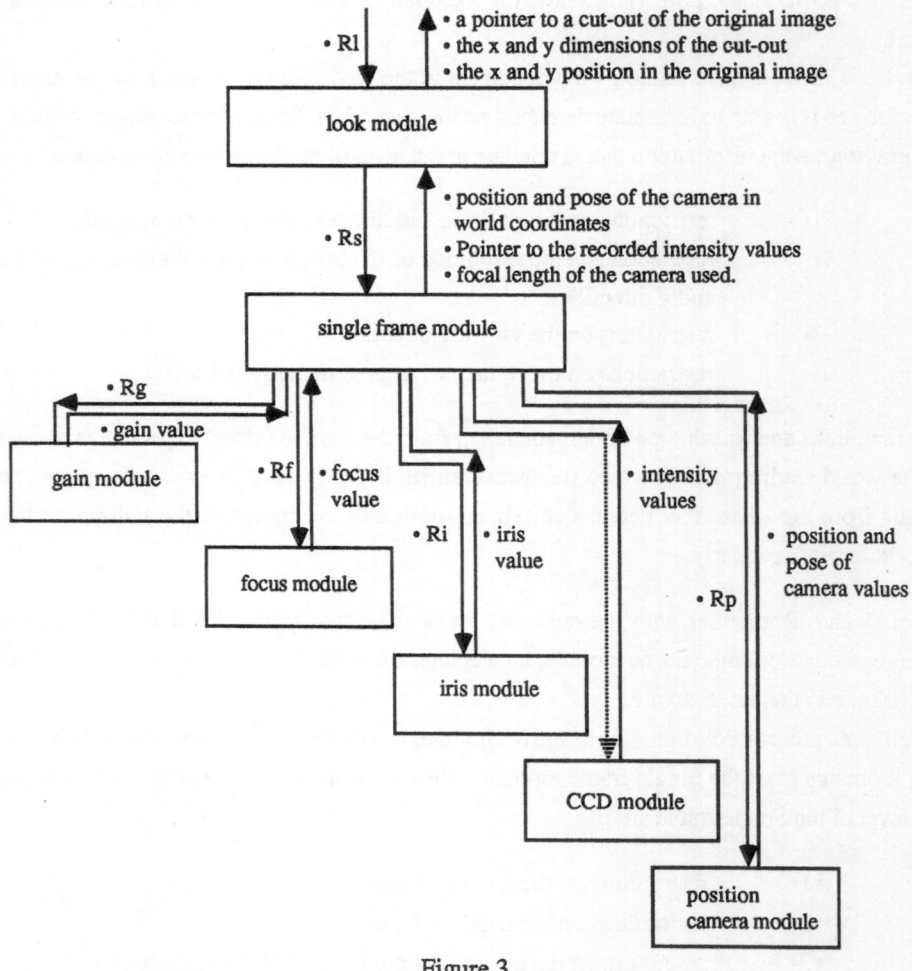

Figure 3

The Single frame module in its turn formulates the demands for the sensor modules that it activates. The restrictions that it outputs to its children are :

Rg :   restrictions on the camera gain value.

Rf :   restrictions on the focus value.

Ri :   restrictions on the iris value.

Rp :   restrictions on the position and pose values.

The iris module receives restrictions on the iris value. This module simply read the contents of a register containing the position of the iris ring of the camera. Because it has available the restrictions on these values, a test can be activated to verify the read value. In figure 3 it is shown that the CCD module cannot be demanded any restrictions (the dotted line) since (currently) it is not possible to adjust parameters in order to fulfil demands on its output (which are, by the way, very difficult to imagine). This sensor module can only read out the contents of the CCD cells.

From the restrictions on the focus distance (s1), the depth of focus (s2) and the maximum intensity value (s5), the valid and allowable values for the gain and iris are obtained. These values determine the restrictions during adjustments of the gain and iris. From s1 and s2, the maximum size of the iris opening is calculated, since the depth of focus is a function of the focus distance and the size of the iris opening. S5 is used to determine the minimum gain factor which, in combination with the maximum iris opening, leads to a number that is identified with an intensity value under moderate illumination conditions. Propagating adjustments on the gain and iris values are restricted in this way. These restrictions are captured in the propagation rules of the Single frame module.

4.2.   Recovering from a sensor error.

When the maximum intensity is too low, the iris opening or the gain should be increased. From the current maximum intensity value and the desired value, the required change is calculated. First the iris opening is increased. When this is at its maximum opening the gain is increased. In this way the conditions for the depth of focus are met and the signal-noise ratio is as high as possible. When the gain is maximal, the error is not recoverable at the level of the Single frame module and the error condition is propagated to the father module.

The father sensor module that activated the look module may demand a change in view direction of the object. The reason for such a demanded change may be to eliminate disturbing highlight in the image of the object. From this demand the look module will formulate, according to a propagation rule, changed restrictions on the pose of the optical axis and a new output from the single frame module. This module in its turn will formulate changed restriction to its children and demand new output from them.

On a very low level a straightforward recovery procedure within the iris module will be activated if the value of the read iris value isn't within the restrictions demanded by the single frame module. The iris sensor module will now adjust its internal parameters, that determine the activation of a motor coupled to the iris ring, which will result in a read iris value that will conform the demanded restrictions.

5.   CONCLUSIONS

We have presented a model for a sensor system that handles the problem of detecting and solving sensor failures. Detection of failures of the sensor modules is based on tests on the data of the modules. We have divided the tests into two groups, environment dependent- and environment independent tests. The first group of tests verifies demands on the data of the sensor module that are determined by the characteristics of the algorithm with respect to the signal. The environment dependent tests verify the demands on the data of the sensor module that are determined by the environment within which a sensor module is to operate. We have shown how the latter group of tests increases the ability of a logical sensor to find a failure in an early stage. We incorporated rules in each sensor module as a means to recover from a sensor failure. The recovery scheme uses these rules to adjust sensor module parameters or input.

Literature

1   Meijer G.R., et. al.. : Monitoring and exception handling for robot control. CIM Europe working conference on production systems. p256, Bremen, 1986.
2   Meijer G.R., Weller G.A. Groen F.C.A. Hertzberger L.O. Sensor based control for autonomous robots. WP-3-4, Proc. IEEE International Conference On Control And Applications, Jerusalem, April 3-6, 1989
3   Pugh: Robot sensors - A personal view, pp. 521-532, ICAR 85.
4   Bogler: Shafer-Dempster reasoning with applications to multi sensor target identification systems.IEEE Transactions on Systems, Man and Cybernetics,vol. SMC-17, no. 6, November/December 1987.
5   Kak A.C., Roberts B.A., Andress K.M. and Cromwell R.L.: Experiments in the integration of world knowledge with sensory information for mobile robots. IEEE Int. Conf. on Robotics and Automation. pp 734-740 1987
6   Ayache N. and Faugeras O.D.: Maintaining representations of the environment of a mobile robot., Int. Symposium on Robotics Research Santa-Cruz, August 1987
7   Luo R.C., Lin M.H. Scherp R.S.: The issues and approaches of a robot multi-sensor integration. IEEE Int. Conf. on Robotics and Automation. pp 1941-1946, 1987
8   Chow E.Y.and Willsky A.S.: Analytical redundancy and the design of robust failure detection systems. IEEE Transactions on Automatic Control. pp 603-614, vol. AC-29, no. 7, July 1984
9 . Henderson and E. Silcrat.: Logical sensor systems. pp. 169 - 193, Journal of Robotic Systems, 1(2) 1984.
10   Henderson, Hansen and Bhamu : The specification of distributes sensing and control. 2(4), 387-396, Journal of Robotic Systems. 1985.

11   J. Buurma, G.A. Weller, F.C.A. Groen, G.R. Meijer, L.O.Hertzberger : Graph-construction and matching for 3D object recognition and pose estimation. To appear in the Proceedings of : Fifth Int. Conf. on Artificial Intelligence and Information-Control Systems of Robots. Strbské Pleso, Czechoslovakia, Nov. 6-10, 1989

12   G.A. Weller, F.C.A. Groen : A gradient comparison method for tracking and focussing purposes. Proc. of the 3rd Int. Workshop on 'Time Varying Image Processing And Moving Object Recognition'. Florence, Italy, May 29-31, 1989

Recognition Strategies for 3-D Objects in Occluded Environments[*]

A. C. Kak, A. J. Vayda, K. D. Smith, and C. H. Chen[†]

Robot Vision Lab
School of Electrical Engineering
Purdue University
W. Lafayette, IN 47907

ABSTRACT

Two different types of approaches will be discussed: one for a model-driven system and the other for generic shape recognition. The model-driven system, called 3D-POLY, has a computational complexity of only $O(n^2)$ for single object recognition, where n is the number of surfaces on the model object. This system achieves its computational efficiency by associating a special *a priori* defined attribute with each object feature and then organizing the object features with respect to this attribute. The generic shape recognition system, called INGEN, is intended for domains where precise models of the objects involved are not available, such as the postal domain; objects in these domains are categorized by their overall shapes with considerable latitude regarding the metrical parameters involved. A unique feature of INGEN, which sets it apart from 3D-POLY, is that object hypotheses are tested by their volumetric consistency, meaning that the hypothesized objects must not violate conditions not only over the space that is visible to the sensor but also over space that may not be visible to the sensor due to occlusions. In other words, while 3D-POLY forms and verifies object hypotheses on the basis of only what is visible to the sensor, INGEN also reasons over the space that is occluded. 3D-POLY is by design limited to drawing all its inferences from the visible data, even in the presence of occlusions, since industrial objects can possess highly complex shape that preclude the kind of volumetric analysis carried out in INGEN.

1. INTRODUCTION

Over the years it has become clear to us that there cannot exist universally applicable strategies for the recognition of 3D objects from range maps — or, for that matter, from sensory data of any type. The reason for this has to do with the fact that the types of distinctions that a system must make between different objects depend upon the uses for which object

* This work was supported by the National Science Foundation under Grant CDR 8803017 to the Engineering Research Center for Intelligent Manufacturing Systems and the US Postal Service Office of Advanced Technology under Contract 104230-86-H-0043.

† C. H. Chen was with the Robot Vision Lab. He is now with SRI International.

recognition is being carried out. For example, the types of distinctions needed for identifying objects in a robotic assembly cell are different from the types of distinctions necessary for sorting parcels in a postal mail stream. For the former, it will usually be necessary to take into account the precise geometric attributes of the various features of objects, since such attributes are important for making distinctions between industrial objects. For the latter, on the other hand, it will usually be sufficient to classify an object as, say, a "box" for large variations in the dimensions of the object. (If one of the dimensions becomes too small, the object might then be called a "flat," again, over large variations in the other two dimensions.)

The two different types of problems outlined above require fundamentally different approaches to object recognition. Where industrial objects are involved, a recognition system must make precise measurements of the various geometric attributes of the object surfaces. On the other hand, for other kinds of objects such as postal objects, a recognition system must be able to ignore the finest level detail, which in most cases would correspond to irrelevant details such as "crumpliness" of the surfaces.

In the rest of this paper, we present two approaches to 3-D object recognition in occluded environments. These two approaches, embodied in systems called 3D-POLY and INGEN, lie at opposite ends of the spectrum of 3-D object recognition. At one end of the spectrum we have feature-based object recognition such as recognition of industrial objects; in this case we have complete 3-D models for all objects that might appear in a scene. Our goal is to match scene features with features in our object database, thus finding the identity and the pose of objects in the scene. Most efforts in object recognition have focused on this domain. At the other end of the spectrum we have generic object recognition such as classification of postal objects. In this case we have incomplete information about objects that might appear in the scene, in the sense that objects are assumed to belong to generic categories based on gross shape where each category allows for large variation in object dimensions. This domain is relatively new and traditional feature-based approaches are not easily adapted to it.

In 3D-POLY, which is a highly computationally efficient system designed for industrial parts, object representation is a most important issue since industrial parts can have complex shapes involving many surfaces and edges and since efficiency of recognition and pose estimation is greatly influenced by the representation used. On the other hand, the most important issue in INGEN is not representation, since the system only has to deal with a small number of generic shapes — flats, parallelepipeds, cylinders, irregulars, etc. — the important issues here deal with topics such as shape characterization of surfaces that may not be smooth at the finest level of detail; being able to cope with surface-irregularity caused breaks in the extracted object surfaces; developing a control structure which allows varied approaches to computing object attributes; and using information derived from the geometrical contacts between objects to find object dimensions.

Another major difference between 3D-POLY and INGEN is based on how the hypothesize-and-verify approaches are executed. Whereas 3D-POLY examines a scene for "landmark" features for hypothesis formation, INGEN, having to deal with only a small number of generic shapes, can form an object hypothesis from any surface segment extracted from the scene. Also, a hypothesis formed in 3D-POLY represents a *particular* object from the CAD library of objects that must be made available to the system. On the other hand, the concept of a CAD library does not exist in INGEN, so when INGEN makes a hypothesis it is about a generic 3D shape, such as a parallelepiped or a cylinder. Since the nature of the hypotheses formed are so different in the two systems, consequently the approaches employed for verification are also different. In 3D-POLY, the particular object whose identity and pose are hypothesized must be verified by examination of additional features from the scene. In INGEN, on the other hand, the hypothesis refers to a generic shape with estimated pose and dimensions, and hypothesis verification becomes more of an exercise in hypothesis refinement where the values of object attributes are estimated with improved accuracy. More precisely, for INGEN we define *refinement* as the process of adding new information to an object hypothesis or improving the accuracy of information in an object hypothesis. To exemplify, the detection of a planar surface fragment in the scene leads to a moment-based computation of the major and minor axes associated with the surface fragment and the formation of the hypothesis that the surface fragment belongs to a parallelepiped. Associated with this hypothesis is a pose transform and initial estimates of the object dimensions as inferred from computations of the extent of the surface along the major, minor, and normal axes; the dimensions involved being the length and width, which lie along the major and minor axes, and the height, which lies along the normal axis. All this information is then fed to the attribute refinement module which recomputes (supposedly, to a higher accuracy) the values for various attributes. The initial estimate of the object height as the extent of the surface in the direction of the surface normal is usually very poor unless the object is actually very thin. The height can be refined (to a maximum value instead of a minimum value) by using the vertical coordinate of the "center of mass" of the surface fragment. This estimate, while not necessarily accurate, is superior to the previous estimate and is useful to the geometric reasoning system. The directions of the major and minor axes can be refined by using the edges associated with the surface fragment. The refinement module looks for edges that are either occluding or convex type and thus not occlusion artifacts. Then, if the longest of these edges has a length that is a significant fraction of the estimated length of the object, the new direction of the major axis is then set to the direction of this edge and the minor axis to a perpendicular direction in the plane of the surface fragment. If no appropriate edges are found then the moment-based major and minor axes are retained. This refinement of the directions for the major and minor axes leads to refinements of attributes such as pose, the various dimensions, postal classification, and confidence in the hypothesis. We have described here only one aspect of the refinement process. In general, this process also includes procedures for upgrading object attributes on the basis of information gleaned from additional object surfaces as they become available by virtue of the action of the model based predictor. More on this later.

Another major difference between 3D-POLY and INGEN lies in how they handle the information that is not visible due to self and mutual occlusion. In 3D-POLY, only the visible features are used for both hypothesis formation and verification. On the other hand in INGEN, while visible features are used for hypothesis formation, verification (refinement) invokes geometrical constraints which involve object interactions in both visible space and the space that may not be visible due to occlusion. To elaborate, the system ensures that an object hypotheses does not intersect the volume of space occupied by other object hypotheses. In the process, it also determines that maximum object dimensions for the object so that it is just touching other objects in the scene, even though the points of contact are not visible in the scene.

Yet another distinction between 3D-POLY and INGEN may be stated in terms of the mathematical relationships between scene objects and their model counterparts. In 3D-POLY, the relationship between a point, p_M, in the model coordinate frame and a point, p_S, in the scene coordinate frame is defined by:

$$p_S = Tp_M$$

where T is a homogeneous transformation matrix for the object pose (position and orientation). On the other hand, in INGEN, the object models do not contain metrical information so a pose transformation alone does not completely specify the relationship between the object and the model. The parallelepiped model is a unit cube and the cylinder model is a right circular cylinder with unit length and radius. These models require three scale factors as well as a pose transformation in order to be matched with objects in the scene. The scale factors define the length of the objects along the three perpendicular axes in the model coordinate frame. The scale factors are incorporated into the transformation between the scene and model coordinate frames as follows:

$$p_S = TSp_M$$

where the scale transformation, S, is defined by:

$$S = \begin{bmatrix} s_x & 0 & 0 & 0 \\ 0 & s_y & 0 & 0 \\ 0 & 0 & s_z & 0 \\ 0 & 0 & 0 & 1 \end{bmatrix}$$

To give the reader an idea of the kinds of scenes that the two systems are capable of interpreting, we will now show some experimental results. Fig. 1a shows a pile of industrial-type objects in random orientations; the scene is made up of the two types of objects shown in Fig. 2. Fig. 1b shows range data derived from a structured light scan of the scene, and Fig. 1c a segmented range map. The sequence of images in Fig. 3 illustrates a robot picking up objects from the scene using 3D-POLY for identifying the objects and estimating their poses. The manipulation does require some grasp planning, since the robot must determine how to approach each object and which two surfaces to make contact with using its two-fingered gripper. Grasp planning for these experiments was done by declaring *a priori* for each object a set of grasps, each characterized by a transformation matrix in the coordinate frame of the object. A product of the transformation matrix associated with the pose of the object and the grasp transformation matrix yielded the actual transform to be used for approaching and grasping the object as it exists in the scene. In the spirit of the Handey system [Lozetal87], these final transforms are then examined from the standpoint of their verticality and the most vertical of the approach directions selected.

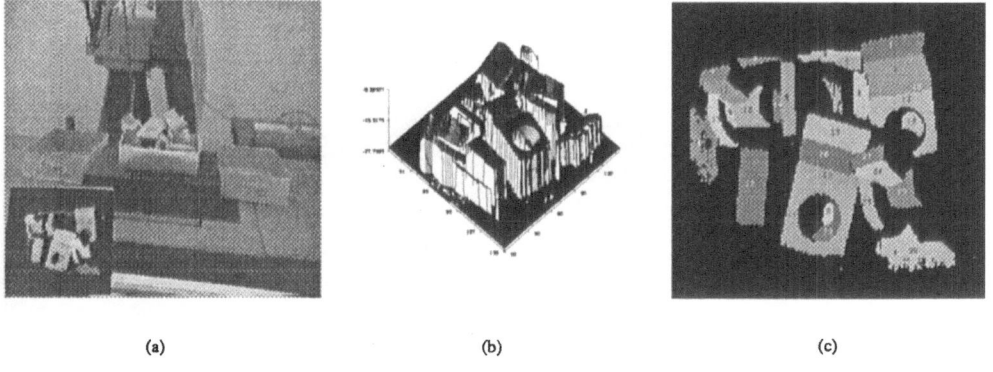

(a)   (b)   (c)

Figure 1. (a) A pile of industrial objects, (b) range data for the scene, and (c) the segmented range map for the scene.

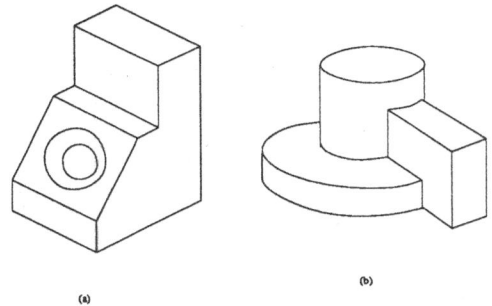

Figure 2. Two object models - (a) square and (b) round.

Figure 3. Robot picking up an object after recognition by 3D-POLY.

Fig. 4a shows a pile of postal objects. Each object in the pile — some of these are shown separately in Fig. 5 — has an overall generic shape, despite the irregularities of the surfaces, that belongs to a small number of categories. Fig. 5 shows typical postal objects from the four major categories. Fig. 4b shows range data derived from a structured light scan of the scene and Fig. 4c and segmented range map. The sequence of images in Fig. 6 illustrates a

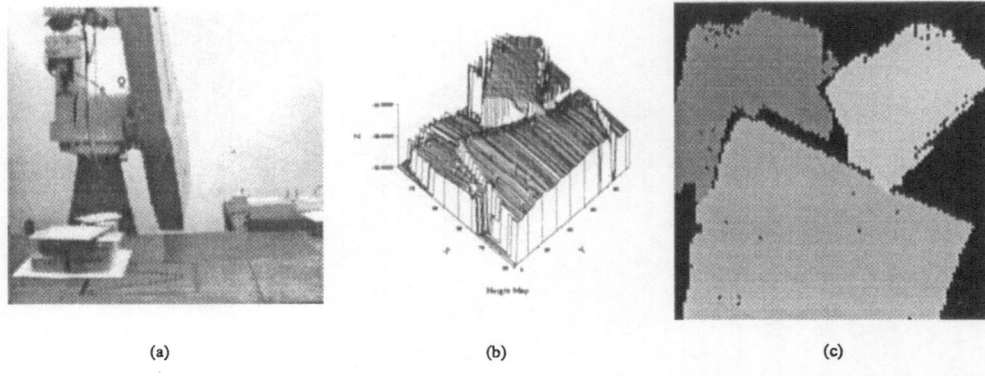

(a)   (b)   (c)

Figure 4. (a) A pile of postal objects, (b) range data for the scene, and (c) the segmented range map for the scene.

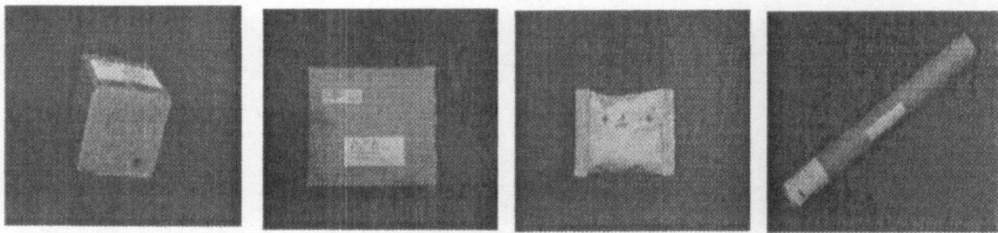

Figure 5. Typical postal objects - Box, Flat, Irregular, and Cylinder.

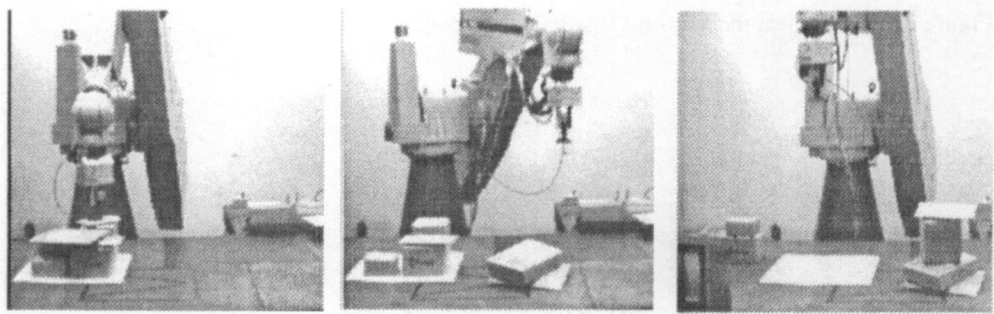

Figure 6. Seven postal objects in a pile are picked up and stacked by a robot using INGEN.

robot picking up objects from the pile and stacking them using INGEN for identifying the objects and estimating their poses. Grasp planning for this system is simplified by the use of a suction gripper. The suction gripper also makes it possible to pick up objects with a wide range of sizes. The grasp point for an object is specified as the center of the largest visible surface. For planar surfaces this is the center of mass of the surface but for cylindrical surfaces it is the projection of the center of mass along the average surface normal until the surface is intersected. The suction gripper moves along the average surface normal as it approaches the object and will be aligned along the major axis of the object. The gripper moves toward the object until a specified force is detected by the force/torque sensor in the wrist of the arm. The suction is turned on and the gripper moves away from the grip point along the average surface normal. The pickup is verified with the force/torque sensor by comparing forces before and after the pickup to determine the weigh of the object. If the weight is too small a error is signaled and error recovery routines are invoked.

In Fig. 6, the photo on the left shows the robot picking the first object from the pile. The center photo shows the results after the first three objects have been stacked. The photo on the right shows the final result where the objects have been moved to two stacks, one to the left of the original pile and the other to the right. Because of the depth of the original pile most of the objects are totally occluded. Data must be collected and INGEN must be run several times as the robot removes the topmost objects and more objects are uncovered. In this example INGEN was run four times. Three objects were found in the first run and two objects were found in the second and third runs. In the final run INGEN determined that the scene was empty. Range data is collected by a structured light scanner mounted above the robot work area.

In the rest of this report is organized as follows. In Section 2 we will discuss the features used in both 3D-POLY and INGEN and in Section 3 we will discuss how we match these features. Section 4 will then focus on the object representations employed in the two systems; as was mentioned before, the approach used in 3D-POLY is the key to its computational efficiency. In Section 5, we will then take up the subject of hypothesis formation; followed by the flow of control, in Section 6, for hypothesis verification and refinement.

3D-POLY was first presented in [ChenKak88, ChenKak89]; the reader is referred to those two references for further details on the system. Further details on INGEN can be found in [VaydaKak89]. Of course, we would be remiss if we did not mention that our work in both 3D-POLY and INGEN would not have been possible had it not been for the earlier pioneering contributions that uncovered many of the problems in hypothesize-and-verify approaches to computer vision [FaugHeb83, GriLoz84, Faug85, HebKan85, BolHor86, FaugHeb86, HutUll87, Low87, Mur87, LamWol88, Umeetal88].

2. FEATURES FOR OBJECT RECOGNITION

In both 3D-POLY and INGEN, a feature can be one of the following three types: primitive surface, primitive edge, and point. Primitive surfaces include planar surfaces, cylindrical surfaces and conic surfaces, which are three special cases of quadric surfaces. In INGEN, an additional surface category, irregular, is used to characterize surfaces that are generally planar but have wide variations in the local surface normals. Primitive edges refer to straight-line features or ellipsoidal-curve features. Point features consist mainly of object vertices and those surface points that have distinctive differential-geometrical properties; surface points falling in the latter category exhibit maximal or minimal curvatures or can be saddle points. These three classes of features are effective in describing the shape of an object, and, what is more, they can be reliably extracted from range images.

A feature is represented by a frame of attribute-value pairs; these frames will be referred to as *feature frames*. To illustrate, 3D-POLY, written in C, uses the following feature frame for the surface feature *s2* of the object shown in Fig. 7.

> <*surface_type*: *cylindrical* >,
> <*radius*: 3>,
> <*axis*: (0.0, 0.0, 1.0)>,
> <*area*: 3>,
> <*point_on_axis*: (0.0, 0.0, 0.0)>,
> <*adjacent_region*: {s1, s3, s4}>, ···

On the other hand, in INGEN, which is written in Prolog, the feature frame is represented by a series of Prolog facts, as shown below for the surface *cube_surface_1* of the generic parallelepiped model:

```
model_surface(cube_surface1).
surface_att_val(cube_surface1, viewpoint, model).
surface_att_val(cube_surface1, type, planar).
surface_att_val(cube_surface1, shape, square).
surface_att_val(cube_surface1, equation, [0.0, 0.0, 1.0, -1.0]).
surface_att_val(cube_surface1, edges,
        [cube_edge1, cube_edge2, cube_edge3, cube_edge4]).
surface_att_val(cube_surface1, match_point_homog,
        [[1.0, 0.0, 0.0, 0.0],
        [0.0, -1.0, 0.0, 0.0],
        [0.0, 0.0, -1.0, 0.0],
        [-0.5, 0.5, 0.0, 1.0]]]).
```

The reader might like to know that in INGEN the feature frames for scene features tend to be more extensive than the corresponding frames for the model features, as illustrated below for a scene surface, *surface4*, which corresponds to the model surface whose feature frame was shown above:

```
surface(surface4).
surface_att_val(surface4, homog,
        [[-0.690338, 0.722662, 0.034514, 0.0],
        [-0.722046, -0.691189, 0.030135, 0.0],
        [0.045633, -0.004117, 0.99895, 0.0],
        [9.39085, 8.52602, 3.11563, 1.0]]).
surface_att_val(surface4, edges, [edge28, edge29, edge30, edge31]).
surface_att_val(surface4, viewpoint, viewpoint1).
surface_att_val(surface4, position, [9.39085, 8.52602, 3.11563]).
surface_att_val(surface4, orientation, [0.045633, -0.004117, 0.99895]).
```

surface_att_val(surface4, type, planar).

surface_att_val(surface4, shape, rectangle).

surface_att_val(surface4, area, 18.7382).

surface_att_val(surface4, num_points, 5211).

surface_att_val(surface4, major_axis, [-0.690338, 0.722662, 0.034514]).

surface_att_val(surface4, minor_axis, [-0.722046, -0.691189, 0.030135]).

surface_att_val(surface4, normal_axis, [0.045633, -0.004117, 0.99895]).

surface_att_val(surface4, cylindrical_axis, [-0.211292, -0.977415, 0.00377]).

surface_att_val(surface4, length, 6.11106).

surface_att_val(surface4, width, 6.12978).

surface_att_val(surface4, height, 0.36800).

surface_att_val(surface4, angle_from_vertical, 2.62584).

surface_att_val(surface4, angle_major_axis, 133.689).

surface_att_val(surface4, plane_equation, [0.045633, -0.004117, 0.99895, -3.50579]).

surface_att_val(surface4, plane_error, 0.012946).

surface_att_val(surface4, xlow, [7.17599, 6.41599, 3.122]).

surface_att_val(surface4, xhigh, [11.612, 9.65199, 2.862]).

surface_att_val(surface4, ylow, [11.466, 6.35799, 3.043]).

surface_att_val(surface4, yhigh, [11.473, 10.786, 2.901]).

surface_att_val(surface4, zlow, [11.612, 9.65199, 2.862]).

surface_att_val(surface4, zhigh, [7.52499, 10.273, 3.23]).

surface_att_val(surface4, highest_point, [7.52499, 10.273, 3.23]).

The scene feature frames are larger because they contain more information which helps to characterize the data. Attributes such as the surface fitting error, major axis, minor axis, cylindrical axis, area are not necessary in the model but are required by the matching process. Size attributes are present in the data but not in the model because a scaling transformation is used to relate scene feature size to model feature size. The viewpoint attribute references the data frame for a viewpoint which includes information such as viewing direction, field of view, and resolution.

As shown above, the value of an attribute can be a number, a symbol, a set of labels of other features, or a list of some of these, depending on the nature of the attribute. The attributes of a feature can be categorized based on their geometric and topological characteristics into three classes: shape, relation and position/orientation.

- Shape Attributes

A shape attribute describes the local geometric shape of the feature, For example, "surface_type," "radius," and "area" are typical shape attributes. We are mostly interested in shape attributes which are transformation (or viewpoint) invariant such as "surface_type" and try to avoid the use of those which are transformation dependent such as "area." However, INGEN does make use of the "area" attribute to help estimate how much of an object hypothesis is visible and to compute a measure of confidence in the object hypothesis.

- Relation Attributes

A relation attribute indicates how a given feature is topologically related to other features. For example, "surface adjacency" is a typical relation attribute. Relation attributes should also be independent of transformation. An attribute such as "on_top_of" is not a proper relation attribute because it depends on the pose of the object.

- Position/Orientation Attributes

A position/orientation (pose) attribute specifies position and/or orientation of a feature with respect to some coordinate system. In general, the pose attributes of a scene feature are measured with respect to a world coordinate system, and those of a model feature are measured with respect to a model-centered coordinate system. As with shape attributes, some pose attributes are transformation dependent, such as the centroid of a surface, or the midpoint of an edge; while others are transformation independent, such as surface normal of a planar surface or the axis of a cylindrical surface. This distinction is important in that transformation independent attributes can be used in the matching process before the transformation is hypothesized.

- A Special Attribute Used by 3D-POLY

For the object representation in 3D-POLY, we associate a special attribute with every feature: *the principal direction*. In an object-centered coordinate system, the principal direction of a feature gives us a fix on the directional position and/or orientation of the feature with respect to the other features on the object. Since one must first establish an object-centered coordinate frame, a principal direction can only be assigned to a model object, or to a scene object whose pose has been estimated during the hypothesis formation stage.

In the next section, we show how a useful data structure — the feature sphere — can be defined using the concept of principal directions. Here, we will define the principal directions for different classes of features, and within each class, for different types of features. For some classes and types of features, the principal direction represents their orientations, for others it represents their directional position on the object.

1. Primitive surfaces:

   For planar surfaces the principal direction is the direction of the outward surface normal. For cylindrical and conic surfaces the principal direction is the direction of the axis. For spherical surfaces the principal direction is the direction of the position vector to the center of the sphere in the object-centered coordinate system. For a quadric surface of the form $x^t Ax + x \cdot B + C = 0$. the principal direction is the direction of the principal eigenvector of the matrix A. In general, the eigenvector associated with the largest eigenvalue is the principal eigenvector of A and the direction associated with this eigenvector usually defines the major axis of a surface.

2. Primitive curves:

   For straight lines the principal direction is the line direction. For circular and ellipsoid curves the principal direction is the normal of the curve's plane.

3. Point features:

For point features the principal direction is the direction of the position vector of the point feature with respect to the object-centered coordinate system.

● A Special Feature Used by INGEN

In 3D-POLY, relations between features are treated as attributes of the features involved in the relation. INGEN, on the other hand, represents relations as separate features. A single relation involves two surfaces which are adjacent in the scene and any number of edges which lie along their common boundary. Relations have attributes which help to characterize them such as the boundary type (concave, convex, occluded, occluding, or unknown), local measurements of the angle and jump discontinuities at the boundary, and global measurements such as the angle between the average normals of the surfaces.

3. FEATURE MATCHING

3.1 3D-POLY

Within the recognition strategy of 3D-POLY, we must be able to efficiently determine the success of a match between a scene feature and an object feature. We do this by applying matching criteria to each of the feature attributes. A successful match is scored if all scene feature attributes agree with all model feature attributes.

We will now provide matching criteria for this matching problem for each of the three attribute classes. Strictly speaking, these conditions are applicable only under the noiseless case. For actual measurements, the comparisons implied by our conditions would have to treated in relation to some user specified thresholds, the magnitudes of the thresholds depending upon the noise and other uncertainties in the system.

● Matching Criteria for Shape Attributes

The reader will recall that we have two different types of shape attributes, those that are viewpoint independent and those that are not. A viewpoint independent shape attribute sa of a scene object feature is said to match the corresponding shape attribute of a model object feature if

$$sa(S_i) = sa(M_{c(i)}).$$

where the function $sa(\cdot)$ returns the value of the attribute sa for the feature that is its argument; S_i is a scene feature, and $M_{c(i)}$ is the candidate model feature that is under test for matching with the scene feature. The above equality must be satisfied for each $sa \in SA(S_i)$, where $SA(S_i)$ represents all shape attributes of the scene feature S_i. For viewpoint dependent shape attributes, clearly we cannot insist upon an equality in the above equation. In general, for such attributes, we require

$$sa(S_i) \le sa(M_{c(i)}).$$

Note that since all the viewpoint dependent shape attributes are numerical in nature, we only have to use numerical inequalities and not, say, subsets, as would be case for symbolic features. For example, we would expect the length of a scene edge be equal to or less than the length of the corresponding model edge due to possible occlusion. Therefore, the matching criterion for this attribute can only be expressed as

$$edge\_length\ (S_i) \leq edge\_length\ (M_{c(i)}).$$

- Matching Criteria for Relation Attributes

Relation attributes are also transformation invariant, thus if a scene feature S_i has relation ra with, say, scene feature S_l, then the model feature $M_{c(i)}$ must have the same relation ra with the model feature $M_{c(l)}$. More precisely, for every $ra \in RA(S_i)$

$$c(ra(S_i)) \subseteq ra(M_{c(i)}).$$

Since all scene features will likely not be visible, we can score a successful match if some subset of them is visible.

- Matching Criteria for Position/Orientation Attributes

For a viewpoint independent position/orientation attribute la, such as the location of a vertex or the direction of an edge, the matching criterion is described by

$$la(S_i) = \mathbf{R} \cdot la(M_{c(i)}) \qquad \text{if } la \text{ is an orientation vector,}$$
$$la(S_i) = \mathbf{R} \cdot la(M_{c(i)}) + t \qquad \text{if } la \text{ is a position vector.}$$

for every $la \in LA(S_i)$. $\mathbf{R}$ and t are the rotation and translation components of the pose transformation.

Many important position attributes are not viewpoint independent. For example, the position attributes *point_on_axis* and *point_on_edge* are both viewpoint dependent since these points can be at arbitrary locations along their respective directions. Despite their arbitrary locations, these points play a vital role during the verification phase of matching. For illustrating this important point, let's say a scene edge is under consideration for matching with a model edge under a given pose transformation. Now, if the directions of the two are identical, that's not a sufficient criterion for the match to be valid, since the identity of directions merely implies that the scene edge is parallel to the model edge. To impose the additional constraint that the two edges be collinear, we need the *point_on_edge* attribute even if the point is arbitrarily located on the edge in the model space.

3.2 INGEN

The major difference between the matching criteria used in INGEN and those used in 3D-POLY is that scale transformations must be used for many shape and position/orientation attributes. We need to divide attributes into two classes based on whether or not they are scale invariant.

Shape attributes such as length, radius, and area are not scale invariant. For example, every surface of the parallelepiped model — which is a unit cube but is used to recognize parallelepipeds of any size — has unit area because all the edges have unit length. The scale transformation must be used to find the actual edge lengths and surface areas for the object. Other shape attributes such as the surface type are scale invariant.

Relation attributes such as surface adjacency are scale invariant because they are purely topological.

Position attributes such as the centroid of a surface, the midpoint of an edge, or the position of a vertex are not scale invariant but orientation attributes such as the surface normal or cylindrical axis typically are scale invariant.

Attributes that are scale invariant are handled as shown above for 3D-POLY. Attributes that are not scale invariant must be scaled appropriately before matching can' take place. The scaling process is different for each attribute because they are affected by scaling differently. Attributes such as the position of a vertex can be scaled by multiplication by the scaling transformation. Other attributes such as the area of a surface are not directly related to the scaling transformation. These attributes must be defined so that they are derivable from other attributes that can be scaled. For example the area attribute for a polygonal planar surface can be computed given the positions of all of its vertices after they have been scaled appropriately.

4. REPRESENTATION OF OBJECTS

4.1 3D-POLY

As discussed in detail in [ChenKak88, ChenKak89], in 3D-POLY an object hypothesis can be verified in linear time with respect to the number of features on an object. This is made possible by a special data structure, the *feature sphere*. In this data structure, each feature is indexed on the surface of a unit sphere by the principal direction associated with the feature. A unit sphere is first tesselated — in the spirit of tesselations employed for Extended Gaussian Images — and all the feature frames associated with an object are then assigned to the tessels on the sphere. A feature frame goes to that tessel whose direction on the unit sphere is closest to the principal direction associated with the feature frame; the direction of a tessel is defined as the direction to the center of the tessel. Fig. 7 shows a simple object and two of its feature spheres; the spheres shown should be interpreted in the sense that there is a pointer located at each marked point on the sphere, the pointer to, in this case, a surface or vertex feature frame for this object. For the same object, different types of features are represented by different feature spheres. So, a complete representation for the object of Fig. 7 would consist of three feature spheres, for the surfaces, for the edges and for the vertices.

Before we proceed further, we should like to mention that the fundamental reason for why the feature sphere data structure leads to highly efficient hypothesis verification strategies is simple and easily accessible to intuition: After the identity and pose of an object are hypothesized from some subset of the scene features, verification consists of computing the

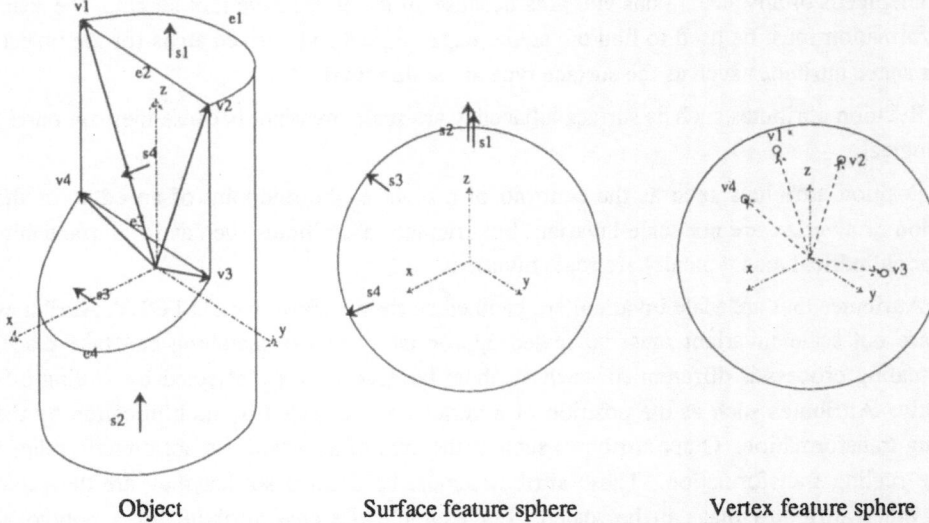

| Object | Surface feature sphere | Vertex feature sphere |

Figure 7. A model object and its surface and vertex feature spheres.

principal directions of the other features; if for a given scene feature the system cannot locate a model feature in (or, in the vicinity of) the tessel corresponding to the computed principal direction of that scene feature, the hypothesis is immediately rejected. This process considerably reduces the computational burden of having to actually match scene features with model features, since a match must be carried out only if a model feature is found in the tessel corresponding to the principal direction of the scene feature; as a consequence, one never has to carry out more than n feature matches if n is the number of features on an object.

Of course, in order to implement the feature sphere idea in a computer one must first devise a strategy for tesselating a sphere and then one must come up with a procedure to index the tessels for efficient storage and retrieval. In 3D-POLY, a unit sphere is tesselated by projecting an inscribed icosahedron onto the unit sphere. The result is a coarse tesselation of the sphere where each tessel is triangular looking. By bisecting each edge of a triangular face and joining the resulting points, it is possible to quadruple the number of tessels; when carried out recursively, this procedure yields even finer tesselations. If such a recursive generation leads to the division of the original triangular edges into Q parts, we associate a *frequency of Q* with the final geodesic tesselation, the number of cells in the tesselation being Q^2 times the number of faces on the icosahedron. Fig. 8 shows an icosahedron on the left and, on the right, a frequency of 4 tesselation of the unit sphere as derived from the icosahedron. In most of our work on hypothesis verification, for scenes like the one shown in Fig. 1, we have used $Q = 4$, which results in 162 tessels on the unit sphere. In actuality, the vertices of the tesselations as shown in Fig. 8 are used as the centers of the cells of a "dual" tesselation which would be obtained by projecting a dodecahedron onto a unit sphere and then subdividing its pentagonal cells.

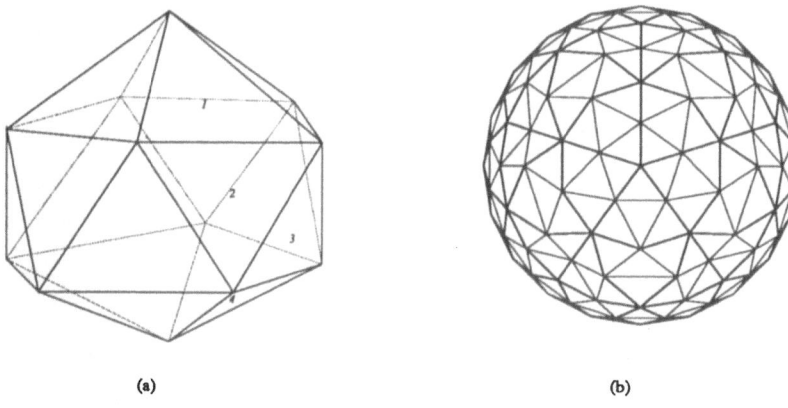

(a) (b)

Figure 8. (a) An icosahedron and (b) a frequency 4 polyhedron derived from the icosahedron.

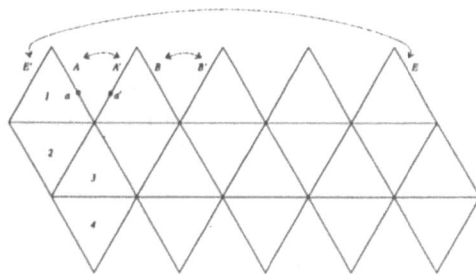

Figure 9. The original icosahedron is flattend out to form five connected parallelograms, each of them consisting of 4 triangular faces.

How the tessels are indexed for computer representation has a great bearing on the overall efficiency of hypothesis verification. In [ChenKak88, ChenKak89], an indexing strategy was presented that results in constant time algorithms both for finding the tessel corresponding to a principal direction and for finding the neighbors of a given tessel; the latter function is important for dealing with noisy data when a model feature may not lie exactly in the tessel whose direction corresponds to the principal direction of the scene feature. The indexing strategy in 3D-POLY is based on the rationale that when we flatten out an icosahedron, we obtain five connected parallelograms, each parallelogram consisting of four triangular faces, as shown in Fig. 9. We have numbered four of the triangular faces in the icosahedron shown in Fig. 8a and their correspondents in the flattened out version shown in Fig. 9 to help the reader visualize the process of flattening out. The flat representation of the tesselated sphere may now be constructed by subdividing each parallelogram of Fig. 9 into $4 \times Q^2$ triangular cells. The vertices thus obtained, as shown in Fig. 10 for the case of $Q=4$, correspond to the vertices of the geodesic polyhedron. In other words, the vertices shown in Fig. 8 correspond to the vertices obtained if we were to divide all the

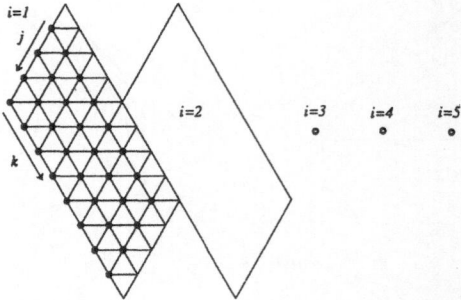

Figure 10. The assignment of the elements of a $Q \times 2Q$ array on a parallelogram.

parallelograms in a manner similar to what was used for the leftmost parallelogram in Fig. 9. Therefore, the sampling points shown in Fig. 8b corresponding to the centers of the cells used for discretizing the space of principal directions. The flattened-out representation, of which Fig. 10 is an example, will be referred to as the *spherical array*.

Each parallelogram in a spherical array consists of $(Q+1) \times (2Q+1)$ vertices. Obviously, the vertices in each parallelogram separately could be represented by a two dimensional array; however, note that the vertices on the border of the parallelograms are shared, meaning, for example, that the vertices a and a' on the edges A and A', respectively, are really the same vertex on the geodesic polyhedron. In other words, before the icosahedron is unfolded to form the spherical array, edge A and edge A' are really the same edge; edge B and edge B' are also the same edge; similarly, with the edges E and E'; and so on (Fig. 9).

The fact that each border vertex should appear only once in an overall indexing scheme for the vertices in a spherical array implies that the size of the index array for representing each parallelogram need only be $Q \times 2Q$. For example, for the case shown in Fig. 10, each parallelogram need only be represented by a 4×8 array. The assignment of array indices for the parallelograms is depicted in Fig. 10 for the $Q = 4$ case. The index i specifies a parallelogram and the indices j and k specify a vertex within the parallelogram. Clearly, we have five $Q \times 2Q$ arrays, for a total of $10 \times Q^2$ indexed points on the spherical array, this number being two less than the total $10 \times Q^2 + 2$ vertices on the geodesic polyhedron. The two missing vertices correspond to the two common vertices of the five parallelograms, one at the top and the other at the bottom. We shall allocate two additional distinguished sets of indices to represent these two vertices and refer to them as the the zenith and the nadir of the tessellated sphere. This indexing scheme implies the following ranges for i, j and k:

$$[i, j, k] \quad 1 \leq i \leq 5, \quad 1 \leq j \leq Q, \quad 1 \leq k \leq 2Q.$$

The zenith and the nadir are assigned the distinguished indices [0, 0, 0] and [−1, 0, 0], respectively.

For lack of space, we will not discuss here the constant time algorithms presented in [ChenKak88, ChenKak89] that use this indexing scheme for finding the neighbors of a given tessel and for finding the tessel corresponding to a given principal direction. Suffice it to say here that in these algorithms the identity of vertices such as a and a' in Fig. 9 is taken care of by applying one of 8 simple mappings, depending on the form of the out of range indices.

4.2 INGEN

As mentioned in the introduction, due to the small number of generic shapes involved, representation is not a major issue in the INGEN system. The hierarchical boundary graph representation which is used is defined by sets of Prolog facts as shown in the previous section on features. Representation graphs for the cylinder and parallelepiped models are shown in Fig. 11 and Fig. 12. For each graph the top node represents the object, the next row of nodes down represents the surfaces, the next row represents edges, and the bottom row represents vertices. Arcs in the graph represent "boundary-of" relations. Each entity in the graph is characterized by a set of attributes. See [VaydaKak89] for more details on the representations used in INGEN.

In INGEN we are not as concerned with the efficiency of accessing features as we are in 3D-POLY. The reasons for this are that INGEN uses only a few generic shape models and there are fewer features per object with the types of objects that INGEN deals with. It is also important to note that 3D-POLY gains efficiency by organizing and accessing features based on the principal direction attribute but no single attribute in INGEN has the same importance. Features must be accessed based on many different attributes during the refinement process.

5. HYPOTHESIS FORMATION

5.1 3D-POLY

Recognition in 3D-POLY employs a traditional two step procedure. First, a small number of features are used to hypothesize the existence of an object with a given pose. Secondly, the remaining object features are used to verify or refute that hypothesis. In the first step, 3D-POLY employs an algorithm that allows it to efficiently select groups of features to hypothesize objects. The scene and model features are grouped into Local Feature Sets (LFS), such that we need only match scene LFS's with model LFS's, as opposed to matching features, one by one. This adds a level of complexity to the hypothesis generation process, but sharply reduces the number of spurious hypotheses.

Ideally, a LFS is a minimal grouping of features that is capable of yielding a unique value for the pose transformation which takes the model object into the scene object. The features in such a minimal grouping could, for example, correspond to one of the rows in Table 1.

More practically, it is desirable that the features in an LFS be in close proximity to one another, so that the probability of their being simultaneously visible from a given viewpoint would be high. In our implementation, we have found useful the following variation on the above idea, which seems to lead to particularly efficient hypothesis generation strategies for objects that are rich in vertices. We allow our LFS's to be larger than minimal groupings

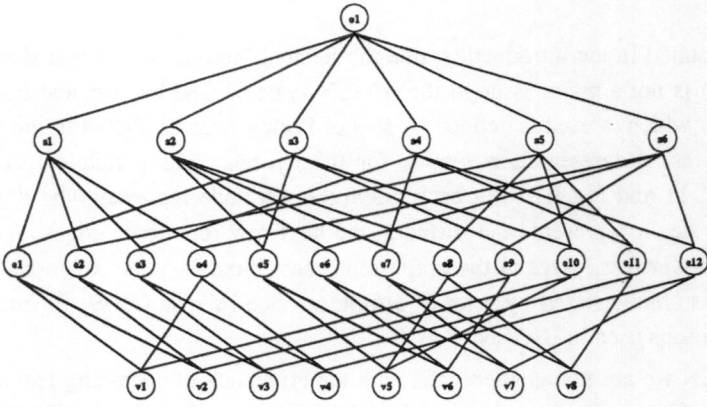

Figure 11. Parallelepiped model graph.

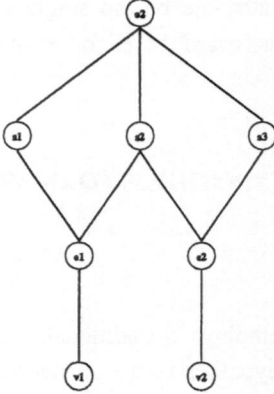

Figure 12. Cylinder model graph.

and insist that each grouping contain a vertex and all the surfaces meeting at that vertex. It would be equally easy to use edges in place of surfaces.

In the scene, a vertex will be called *completely visible* if no occluding edges meet at the vertex. We believe that a completely visible vertex in a scene provides the strongest constraints for calculating the pose transformation, T, associated with an object in a scene. Of course, theoretically, any two of the non-parallel surfaces coming together at the vertex, in conjunction with the vertex itself, are capable of specifying uniquely the T associated with a scene object. Therefore, theoretically at least, any two of the surfaces, together with the coördinates of the vertex, can yield the T. However, in practice, it is difficult to calculate with great precision the position of the vertex itself, primarily because of the nature of discontinuities of some of the spatial derivatives at such a point. Therefore, our approach is that if a completely visible vertex can be found in a scene, it should immediately be used to calculate a T.

Configuration of features	Position/orientation attributes
Three unique, noncollinear points.	The three positions vectors associated with the three points.
One straight edge & one noncollinear point.	The orientation attribute associated with the direction of the edge, the position attribute associated with a point on the edge, and the position attribute associated with the noncollinear point.
One ellipsoidal edge & one noncollinear point.	The orientation attribute associated with the edge, the position attribute associated with a point on the axis of the ellipsoidal edge, and a position attribute associated with the noncollinear point.
Two primitive surfaces & one point.	The two principal directions associated with the two surfaces, and a position attribute associated with the extra point.
Three non-coplanar primitive surfaces.	The orientation attributes associated with any two of the surfaces, and three position attributes associated with some three points, one on each surface.

Table 1. Summary of hypothesis generation feature sets

Of course, it is entirely likely that we may not find any completely visible vertices in a scene, meaning that because of self-occlusion we may be able to see only two of the three surfaces. In such a case, the LFS for the vertex can still be used by assigning appropriate labels to the scene surfaces from the entries in the LFS. In general, if h surfaces meet at a vertex and only k of these are visible in a scene, then there are only h possibilities for matching the k scene surfaces, this happens because of the rotational adjacencies that have to be maintained. Therefore, we can say that in matching k scene features with the h features of an LFS, the overhead is k, which is incurred in matching the k scene features with the potential correspondents from the LFS. Since this can only be done in h ways, the overall complexity associated with matching with an LFS is $O(h \times k)$.

Therefore, the complexity associated with generating hypotheses for an object which has N_{LFS} LFS's is $N_{LFS} \times O(h \times k)$. In practice, $N_{LFS} = O(m)$, where m is the total number of model features. Therefore, the overall complexity associated with generating all the hypotheses is $O(m \times h \times k) = O(m)$. This is significantly better than the $O(n^m)$ results using a brute force method with n scene features.

Before we conclude this section, we would like the reader to note that the gains achieved with the use of LFS's as non-minimal feature groupings is at the cost of more complex flow of control during hypothesis generation. While with minimal groupings, it is possible to institute uniform control, with non-minimal groupings special cases must be handled separately depending upon how many of the features in an LFS can be matched with the scene features.

Also, we have said nothing about the mathematics of how to actually compute a T given that we have a match between some scene features and model features. For lack of space, this mathematics can not be presented here. However, the interested reader might wish to go through what we believe is a very readable derivation in Appendix A of [Chen-Kak88,] which employs quaternions to demonstrate that an optimum T can be computed by solving an eigenvalue problem. This derivation is based on the principles first advanced by Faugeras and Hebert [FaugHeb83].

5.2 INGEN

Recognition in INGEN is based on the hypothesize-and-refine approach rather than the traditional hypothesize-and-verify approach. The relatively simply initial hypotheses required for this approach simplify the hypothesis generation process. At the start of the refinement process we need only the most basic information about an object because during refinement every attribute of the object hypothesis is subject to change. Thus, because of the recomputation that will take place during the refinement process, the accuracy of the attributes in the initial hypothesis is not critical. A purely data-driven approach is used to generate an object hypothesis for every surface in the scene. Other information available for the scene such as edges, vertices, and relations is used for refinement but is not used for hypothesis generation.

The initial object hypothesis consists mostly of information derived from the surface which supports it. The object type is based on the surface type. A cylindrical surface gives rise to a cylindrical object hypothesis. A planar surface gives rise to a parallelepiped object hypothesis. An irregular surface gives rise to an irregular object hypothesis. Irregular objects are handled the same as parallelepiped objects except that special allowances are made during the refinement process to accommodate the irregularity of their surfaces and edges. The average surface normal and the major and minor axes specify the orientation of the object. The center of mass of the surface specifies the position of the object. A homogeneous transform for the object pose is computed from the axes and the position. The lengths along the three axes define the size of the object. The values originally assigned to these attributes will most likely be changed as better estimates are available and values for other attributes that were not included in the initial hypothesis will be computed during attribute refinement processing.

The initial object hypotheses can be inaccurate in a number of ways, all of which can be corrected by the refinement process. For example, the position, orientation, or dimensions of the object may be incorrect but will be recomputed during refinement. An object with multiple surfaces visible in the scene will cause multiple object hypotheses to be formed. During refinement, these hypotheses will be merged into a single object hypothesis. As another example, consider the case where the planar end surface of a cylinder is used as the basis for an object hypothesis. The object will be hypothesized to be a parallelepiped. During refinement, the adjacent cylindrical surface should be merged into the object and the type changed to cylinder. Or, if the surface is found to be circular rather than rectangular then the object type can be changed to cylinder based on that information.

6. FLOW OF CONTROL

6.1 3D-POLY

We have shown an efficient method for hypothesizing objects in our scene through the use of Local Feature Sets. We have also shown how the feature sphere can be used to organize features based on their principal direction. With this facility, it is straightforward to limit matches of scene and model features to those features with the same (within some error tolerance) principal direction. The scope of 3D-POLY, however, is somewhat larger, in that it recognizes objects under occluded conditions, as would be the case when the objects are presented to the sensory system in the form of heaps.

In general, when the range images to be interpreted are of scenes containing piles of overlapping objects, one has to contend with the following two problems: 1) The number of features extracted from a scene will usually be very large; and, 2) since different objects may be made of similar features, it would generally not be possible to set up simple associations between the scene features and the objects. To get around these problems in dealing with multiple object scenes, researchers previously have either performed object segmentation by exploiting range discontinuity information [Fanetal88], or have used a *model-driven*[*] approach to group together scene features belonging to single objects [FaugHeb86], [BolHor86]. However, the former approach usually fails to work especially when the juxta-positions of multiple objects are such that there are no range discontinuities between them and the latter is too inefficient.

We will now present a data-driven approach for aggregating from a complex scene features belonging to single objects. The cornerstone of our approach is the idea that only physically adjacent scene features need be invoked for matching with a candidate object model. For this purpose, the notion of physical adjacency will be applied in the image space as opposed to the object space, implying, for example, that two surface regions sharing a common boundary, even if it is a jump boundary, will be considered adjacent to each other. Using this idea, we will now describe the complete method:

The algorithms uses two sets, UMSFS and MSFS, the former standing for the unmatched scene feature set and the latter for the matched scene feature set. Initially, the algorithm assigns *all* the scene features to the set UMSFS. The process of object recognition starts with a local feature set (LFS) extracted from the UMSFS. The matching of this scene LFS with a model LFS generates a hypothesis about object identity and a pose transforma-tion. The features in the scene LFS are then taken from the UMSFS and assigned to MSFS; note that MSFS keeps a record of all the scene features matched so far with the current can-didate model. Then during the verification stage, only those scene features in the UMSFS

---

[*] To be contrasted with the *data-driven* procedure to be described in this section.

that are adjacent to the features in the MSFS are selected for matching with the candidate model. During the verification stage, if a UMSFS feature does match the candidate model feature, the scene feature is taken out of the UMSFS and added to the MSFS; otherwise the feature is marked as tested under the current hypothesis and left in the UMSFS.

The verification stage terminates when the MSFS stops growing. Once the verification process terminates, the algorithm determines whether or not the features in the MFS constitute enough evidence to support the hypothesis on the basis of some predefined criterion. This criterion may be as simple as requiring a percentage, say 30%, of model features to be seen in the MSFS; or, as complicated as requiring a particular set of model features to appear in the MSFS; or, at a still more complex level, some combination of the two. If a hypothesis is considered verified, the features currently in the MSFS are labeled by the name of the model and taken out of further consideration; otherwise, the hypothesis is rejected and every feature in the MSFS is put back into the UMSFS and the process continued with a new LFS. The entire process terminates after all the LFS's have been examined.

In our current implementation of this algorithm, the recognition criterion requires that at least 33% of a candidate model's features be present in the MSFS for a hypothesis to be considered valid. Note that the acceptance threshold can be no greater than 50% for most objects, especially those that have features distributed all around, since from a single viewpoint only half of an object will be visible. Therefore, 50% is a loose upper limit on the acceptance threshold. On the lower side, the threshold can not be set to be too low, since that would cause misrecognition of objects. We have found 33% to be a good compromise.

6.2 INGEN

The INGEN system consists of five modules as shown in Table 2. Hypothesis generation was discussed in the previous section. In the rest of this section we discuss the other four modules which carry out the hypothesis refinement process. In Table 2 we show some interesting characteristics of these modules. The Focus column shows the extent of the interpretation which the module focuses on: S refers to single objects and M refers to multiple objects. The Driven column shows the basic approach of the module: D refers to a data-driven approach and M refers to a model-driven approach. The Features column shows the data features used by the module: S refers to surfaces, E refers to edges, V refers to vertices, and R refers to relations.

Module	Focus	Driven	Features
Hypothesis Generation	S	D	S
Attribute Refinement	S	D	SEV
Surface Merging	M	D	SER
Model Based Prediction	M	M	SEVR
Geometric Reasoning	M	M	SEV

Table 2. INGEN module characteristics.

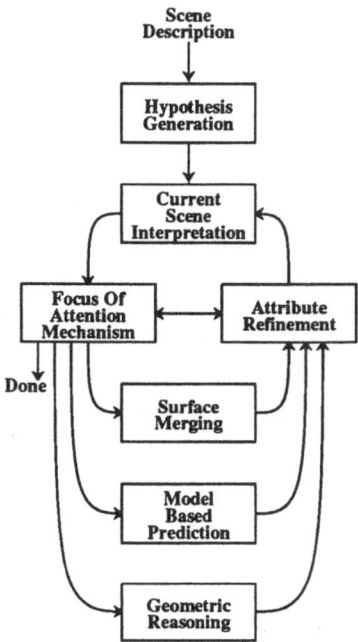

Figure 13. The INGEN system.

The overall flow of control in INGEN is controlled by the Focus of Attention Mechanism as shown in Fig. 13. The first step is hypothesis generation. As described previously, an object hypothesis is created for each surface in the scene and added to the Current Scene Interpretation. As object hypotheses are refined the Current Scene Interpretation database is continuously updated. The next step is Attribute Refinement. For each hypothesis in the Current Scene Interpretation we compute a number of attributes in an entirely data-driven manner. The third step is Surface Merging. The surfaces of all of the object hypotheses (single-surface object hypotheses at this point) are examined and if compatible surfaces (coplanar or cocylindrical) are found then they are merged into a single object hypothesis. The Attribute Refinement Module is then called upon to recompute the attributes of the newly modified hypothesis. The fourth step is Model Based Prediction. For each object hypothesis, the object model is used to predict surfaces which should be visible in the scene. If these surfaces can be found in the scene then they are added to the object hypothesis. The Attribute Refinement Module is then called upon to recompute the attributes of the newly modified hypothesis. Notice that surface merging involves adding surfaces to an object hypothesis based on the relationships between the surfaces to be added and the surfaces in the hypothesis. Model based prediction, on the other hand, adds surfaces to an object hypothesis based on the relationships between the surfaces to be added and the hypothesized object model. The final step is Geometric Reasoning. When uncertainty exists in one of the object's principal dimensions we use geometric reasoning to determine the maximum possible extent of the object in that direction. The geometric reasoning process can be thought of as ''growing'' the object in a particular direction until it comes into contact with another object in the scene. It also helps to ensure that the scene interpretation is geometrically consistent by ensuring that object hypotheses do not intersect each other. As before, the Attribute Refinement Module is then called upon to recompute the attributes of the newly modified hypothesis with one exception. The attributes which have just been computed by the Geometric Reasoning Module will not be changed during attribute refinement.

As we discuss the operation of the Attribute Refinement, Surface Merging, Model Based Prediction, and Geometric Reasoning modules, we will make reference to the data shown in Fig. 14, which was supplied by the GE/RCA - Advanced Technology Labs. This scene contains a letter which is lying on two boxes which in turn are lying on the table. It is important to point out that the INGEN system is still under development. The Attribute Refinement and Surface Merging Modules and the Focus of Attention Mechanism are fully implemented and were used to produce the robot manipulation results discussed in the introduction. The Model Based Prediction and Geometric Reasoning Modules are partially implemented and have not been integrated into the system.

6.2.1 Attribute Refinement

The Attribute Refinement Module uses data-driven processing to determine the values of object attributes for one object at a time. In the course of recognizing the objects in a scene this module may be called upon to process each object several times. Immediately after hypothesis generation, the Attribute Refinement Module processes every hypothesis. Subsequently, the Attribute Refinement Module processes each object whenever it has been modified by the Surface Merging, Model Based Prediction, or Geometric Reasoning modules.

The attribute refinement process is carried out by a group of refinement procedures. Each refinement procedure uses a set of object attributes as its input and produces a set of attributes as its output. A single attribute may be used as input to several procedures and may also be produced as output by several procedures. This arrangement allows for the bootstrapping of attributes. One procedure can find a rough estimate for a particular attribute and another procedure can then use this estimate along with other attributes to determine a superior estimate.

Because refinement procedures are related to each other only by the attributes used as input and output they are independent and can be modified, added to, or removed from the system without the necessity of making changes elsewhere in the system. Dependencies between procedures are determined by the input and output attribute sets. Thus, when an attribute is changed we can use forward chaining through the input and output attribute sets for procedures to determine which procedures must be called to propagate the change and what attributes will be affected.

When performing attribute refinement on a new hypothesis all of the refinement procedures are called for each object. When one of the other modules changes an object hypothesis only the refinement procedures necessary to update the hypothesis are called. The object attributes that have been changed are noted and forward chaining is used to

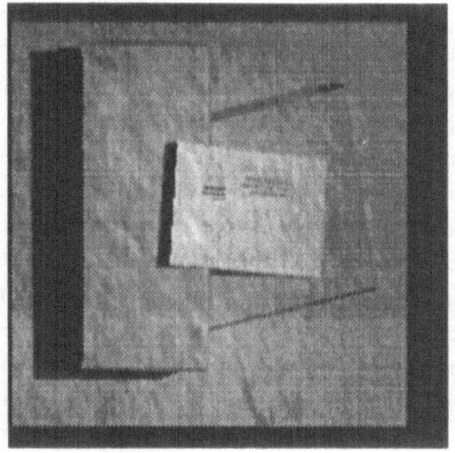 

Reflectance Image   Range Image

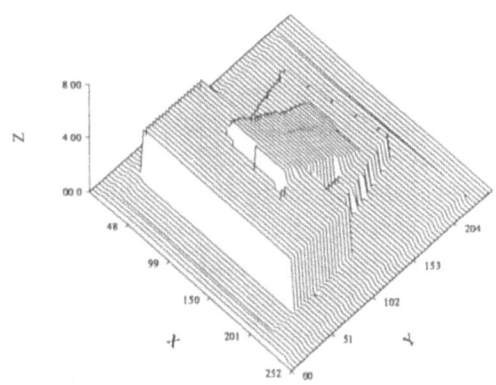

Range Data

Figure 14. Reflectance image, range image, and range data plot of data of a postal scene use for our example.

determine which other procedures depend (directly or indirectly) on those attributes and therefore need to be called to propagate the change.

Here we offer a description of some of the more important refinement procedures. The *general object type* refinement procedure examines the types of the surfaces of the object hypothesis to determine the object type. The *axis* refinement procedure looks at the edges of the object surfaces and defines new object axes if certain conditions are satisfied. The procedure looks for object edges that are either occluding or convex type. Using only these edge types ensures that the system is not fooled by shadows or occlusion effects. If the longest of these edges has a length that is a significant fraction of the initially computed length of the object, then the direction of the major axis is changed to the direction of this edge and the direction of the minor axis is changed to the perpendicular direction in the plane of the surface. If no appropriate edges are found then the initially computed moment-based major and minor axes are retained. In reference to our example, the major axes for the leftmost box and the letter are changed only slightly by the refinement process because the letter is unoccluded and the box is only slightly occluded. The moment-based method provides good estimates of the axes of objects that are unoccluded but can be very inaccurate for occluded objects. The rightmost box in our example scene demonstrates this problem. Three complete edges of the object are visible but due to occlusion only about half of area of the object is visible. The edge-based method produces a much better estimate of the major axis in this case. The *extent* refinement procedure looks at all of the object vertices and finds their distances from the object position. The maximum distances in both directions along each axis are used to define the dimensions of the object. The position is also redefined to be centered on the major and minor axes and to lie on the primary surface of the object. Extent refinement is necessary whenever axis refinement adjusts the orientation of the object. The *height* refinement procedure defines the object height to be the vertical component of the object position. This gives exact results for an object lying flat on the table. For objects that are tilted or lying on other objects the Geometric Reasoning Module will produce better estimates. Thus, for our example the two boxes have their heights estimated accurately but the letter height is overestimated. The *homogeneous transformation* refinement procedure uses the object axes and the object position to define the homogeneous transformation for the object. This transformation is suitable for use by a robot with a suction gripper for picking up the object. It is also used by the model fitting procedure to find the model-to-scene and scene-to-model transformations. The *area* refinement procedure finds the visible area of the object and the area of the object that should be visible. These values can be used to determine the degree to which the object is occluded. The *confidence* refinement procedure produces a measure of the confidence in the object hypothesis. The *postal type* refinement procedure determines the appropriate postal category for the object based on size and shape information. The *model fitting* refinement procedure determines the model-to-scene and scene-to-model transformations for the object which are used by the Geometric Reasoning Module. The object homogeneous transformation and the object dimensions are used along with the model database to find the transformations.

It is important to note that the attribute refinement procedures are data-driven (with the exception of the model fitting procedure) and do not use object models to characterize the object. However, they do make use of object type information to function differently for different types of objects. Also, these procedures are designed to work equally well on object hypotheses with single or multiple surfaces. For details on the various refinement procedures see [VaydaKak89].

6.2.2 Surface Merging

In cluttered scenes object surfaces will frequently be separated into multiple adjacent surface segments due to segmentation errors or non-adjacent surface segments because of occlusion. By merging smaller surfaces into larger ones based on compatibility and continuity criteria we can solve these problems. The surface merging module processes hypotheses in order based on the visible area of their largest surface. For each hypothesis, all single-surface hypotheses with smaller surfaces are tested to determine if merging is appropriate. The order is important because the merging criteria are not symmetrical. We are evaluating whether the hypothesis with the smaller surface should be merged into the multiple surface hypothesis containing the larger surface. When the decision is made to merge two hypotheses the surface of the smaller hypothesis is added to the surface list of the larger hypothesis. The smaller hypothesis is then removed from the Current Scene Interpretation and the larger hypothesis is sent to the Attribute Refinement Module so that its attributes may be updated and then added to the Current Scene Interpretation.

The first criteria for merging is the comparison of the types of the surfaces under consideration. Obviously, merging of surfaces of the same type (parallelepiped, cylinder, or irregular) is allowed but we also allow the merging of irregular surfaces with cylindrical or planar surfaces if other criteria support the merge. The primary criteria for merging is the evaluation of how well the data for one surface fits the equation for the other surface. The fit of the data for the smaller surface to the equation of the larger surface is more important than the reverse fit. For surfaces which are adjacent in the scene, the boundary between them is examined for continuity. Relations between adjacent surfaces contain information about both local and global properties of the adjacency relationship. Global properties include the angle between the average surface normals and the angle between the cylindrical axes. Local properties include measurements of the angle between the surfaces and the jump discontinuity based on examining only data points that are close to the boundary. These measurements allow the categorization of the boundary as one of the following: concave, convex, occluded, occluding, or unknown. For surfaces which are not adjacent in the scene the alignment of the axes of the two objects is checked. The axis alignment check is the primary reason why surface merging is not carried out before the formation of object hypotheses. The axis refinement procedure provides estimates of the object axes that are superior to those provided in the surface description. Currently, object models are not used for surface merging but it is expected that this capability will be added in the future.

6.2.3 Model Based Prediction

One of the attribute refinement procedures fits an object model to the data and finds the model-to-scene and scene-to-model transformations. Using these transformations and information about the viewpoint(s) used to collect the range data it is possible to predict the visibility of the features of the hypothesized object model in the scene. Given the viewpoint direction and the surface normal for a planar surface there are two criteria that must be satisfied for the surface to be visible. First, the normal must point toward the viewpoint. Second, the angle between the normal and the viewpoint direction must be sufficiently small. The actual angle threshold depends on the geometry of the range sensor and the scene but is typically about 45 degrees. For cylindrical surfaces we use the cylindrical axis instead of the surface normal. The scene description is searched to find entities which match the model entities that have been predicted to be visible. Currently only surfaces are used for this purpose. When a surface is found that fits the prediction it is merged into the object hypothesis and its own hypothesis is removed from the current scene interpretation. The Attribute Refinement Module is then called to refine the attributes of the new object hypothesis.

Model based prediction is not always successful in finding other surfaces for the object hypothesis. In many cases no surfaces will be predicted as being visible or those that should be visible will not be found. For these cases the Model Based Predictor performs two more operations. First it considers the possibility that the model used for the object is incorrect and looks at other models. If it can match the object hypothesis to another model and find another surface in the scene which also supports the new model then the object type is changed, the surface is added to the hypothesis, and attribute refinement is carried out for the modified hypothesis. If no supporting surfaces for any object model can be found then it considers the possibility that the object is actually of the type other. Objects of type other consist of any number of convexly adjacent surfaces that do not fit any object model. This object type gives the system the capability of characterizing unknown objects instead of ignoring them. Other type objects do not have object models but a parallelepiped model is used to represent a bounding box for the object. As in the previous cases, the scene is examined for surfaces adjacent to the object hypothesis. The edges between the hypothesis and adjacent surfaces are examined to determine if there is sufficient continuity to declare them as being from the same object. The criteria requires the edge to be convex and have no step discontinuity. With this method, convex objects of any size and shape can be found and characterized.

6.2.4 Geometric Reasoning

The task of the Geometric Reasoning Module is to determine size parameters for object hypotheses that can't be determined from the scene data. It resolves uncertainty in one of the dimensions of an object by ''growing'' it until it contacts another object in the scene. The point of contact need not be visible in the scene data. This process also helps to ensure that the scene interpretation is geometrically consistent by ensuring that object hypotheses do not intersect each other.

With the dimensions of our generic objects being defined by their extent along their three principal axes we can, in general, carry out the growing operation in six directions, two for each axis. In most cases there is uncertainty in only one direction for an object hypothesis, but in the worst case there can be uncertainty in five directions. As an example, consider the simplest case where the top surface of a box sitting on a table is viewed from above. The only uncertainty is in how far the box extends in the direction away from the viewpoint. However, if one of the top edges is occluded then there is also uncertainty in a direction perpendicular to the viewing direction. If all four top edges of the box are occluded then there is uncertainty in four directions perpendicular to the viewing direction as well as in the direction away from the viewpoint. Our current approach is to consider uncertainty in one direction at a time. This approach has both benefits and drawbacks. It is computationally much easier to consider only one direction at a time and in many cases it can be shown that the directions are in fact independent and need not be considered simultaneously.

However, there are some cases where the directions do indeed depend on one another and in these cases the order in which the system considers them affects the outcome, i.e. the algorithm is not commutative. This is an important issue which has yet to be resolved.

The geometric reasoning algorithm used in INGEN is based on a single operation: finding the maximal extent of an object hypothesis in a particular direction. Before we describe our algorithm we must introduce some terminology that will help to clarify the discussion. The *object* is the object hypothesis for which we wish to determine the maximum value of a particular size parameter. The *obstacles* are all of the other object hypotheses in the scene which potentially physically constrain the object. We use this terminology because it is convenient to think of the computation of the maximal extent of an object as the "growing" of the object until it comes in contact with an obstacle.

The algorithm consists of four steps.

1. Find scene-to-model transformation for the object. We assume, without loss of generality, that the parameter to be found is the object's extent along the positive z axis.

2. Use the scene-to-model transformation to transform all obstacles into the model coordinate frame of the object.

3. Find the maximum extent of the object along the positive z axis such that the object is in contact with at least one obstacle in the scene and its volume does not intersect the volume of any obstacle in the scene.

4. Use the model-to-scene transformation to transform the extent measurement back into the scene coordinate frame and thus find its actual value.

The determination of the scene-to-model transformation in step one is carried out during the attribute refinement stage of processing as described previously. Fig. 15 shows four views of the hypothesized objects from our example scene at the conclusion of the attribute refinement stage. Included are orthographic projections for the top (x-y plane), front (y-z plane), and side (x-z plane) views and an axonometric projection. Note that the letter is hypothesized as a box with its estimated height being its height above the table. This results in the letter intersecting both of the other objects and the table.

In step two all of the objects are transformed into the model coordinate frame of the object of interest, in this case the letter as shown in Fig. 16. The letter is transformed into a unit cube, which is upside down with respect to Fig. 15, and the other objects are transformed accordingly.

Clearly, step three of this algorithm is the most difficult one. However, the transformation of the obstacles into the model coordinate frame of the object which was carried out in step two simplifies the computations in step three. Our approach borrows some computational techniques from ray casting algorithms in the computer graphics field and path planning algorithms in the robotics field. The ray casting technique that we borrow is to transform our object and obstacles into a coordinate frame where the object hypothesis is a simple primitive object of fixed size and shape. The path planning technique that we borrow is the use of both parametric and implicit representations of surfaces and edges to aid in the computation of intersections. Roth [Roth82] discusses

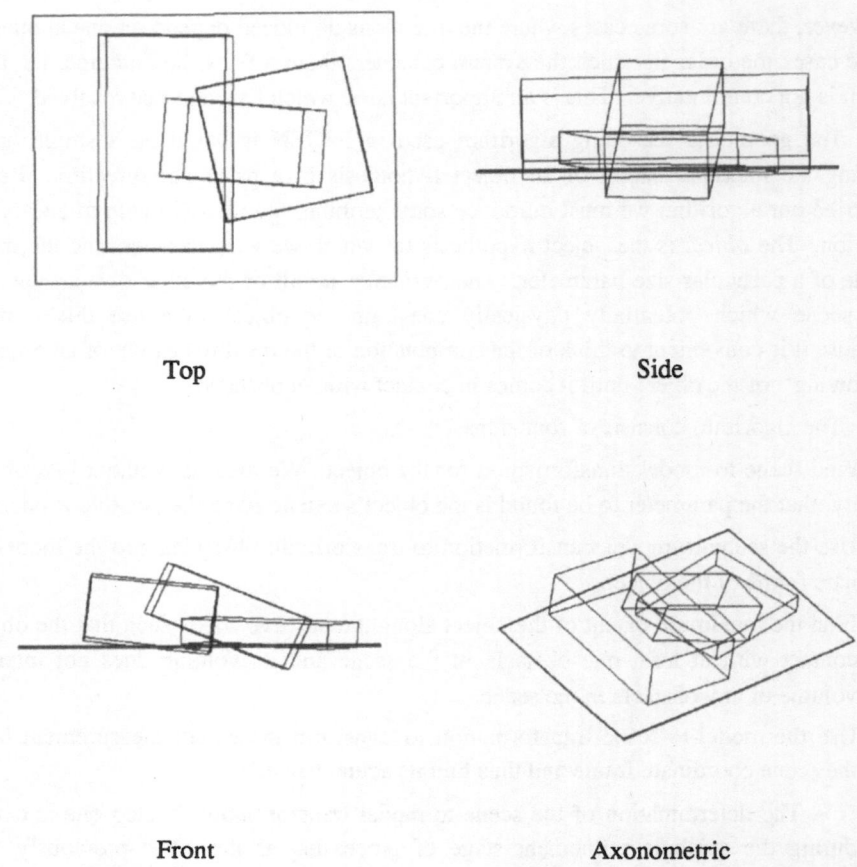

Top   Side

Front   Axonometric

Figure 15. Hypothesized objects.

ray casting techniques and algorithms and Lozano-Pérez [Lozano87] discusses the configuration space approach to robot motion planning.

To make the problem more tractable we restrict ourselves to handling objects which are rectangular parallelepipeds or right elliptical cylinders. For cylindrical objects we assume that the diameter of the cylinder can be determined from the range data for the cylindrical surface so we need only consider the case of a cylinder with unknown length. For obstacles, we consider polyhedra with any number of planar faces and unbounded planar surfaces. To handle cylindrical obstacles we can define a polyhedral model to represent the cylinder to any desired accuracy but in most cases a parallelepiped is accurate enough. Extensions to the algorithm for other types of objects and obstacles are discussed at the end of this section.

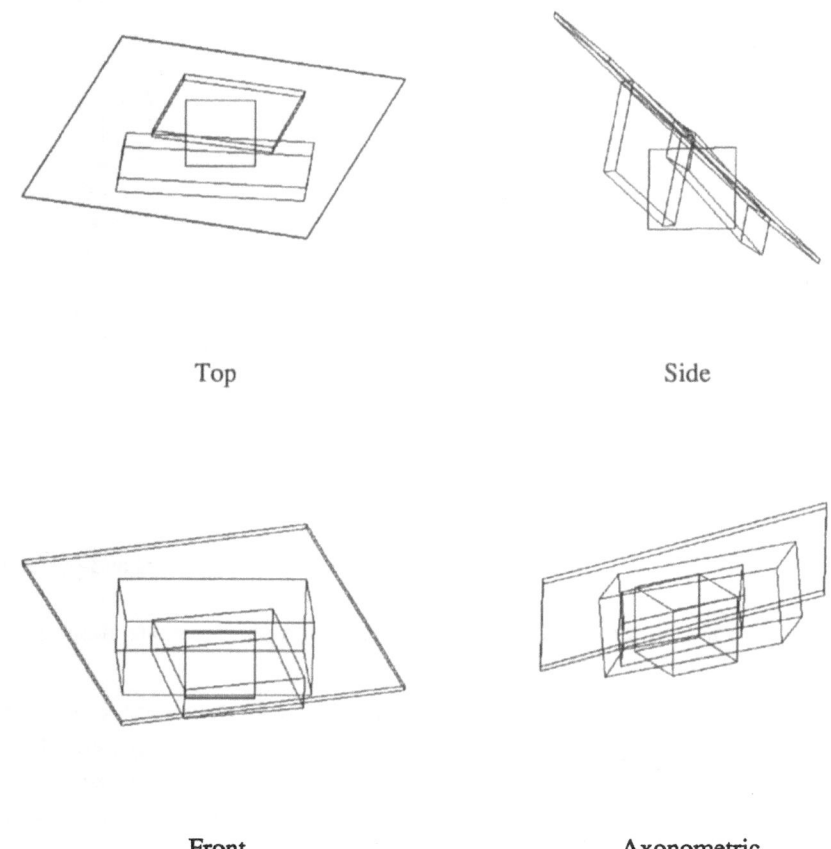

Top   Side

Front   Axonometric

Figure 16. Hypothesized objects after being transformed into the model coordinate frame of the topmost object.

Three types of contact are possible between polyhedral objects and polyhedral obstacles. We use the traditional convention [Lozano87] to name these interactions: *Type A* - an object surface contacting an obstacle vertex, *Type B* - an object vertex contacting an obstacle surface, and *Type C* - an object edge contacting an obstacle edge. The only difference between contacts involving cylindrical objects and those involving polyhedral objects is in the definition of type B contact: *Type B* - an object edge contacting an obstacle surface.

The three types of contact are shown in Fig. 17. At the top are contacts between parallelepiped objects and polyhedral obstacles and at the bottom are contacts between cylindrical objects and polyhedral obstacles. From left to right, type A, B, and C contacts are shown. Note that the objects and the obstacles are positioned in the object's object-centered coordinate frame. Thus, the obstacles always appear above the object.

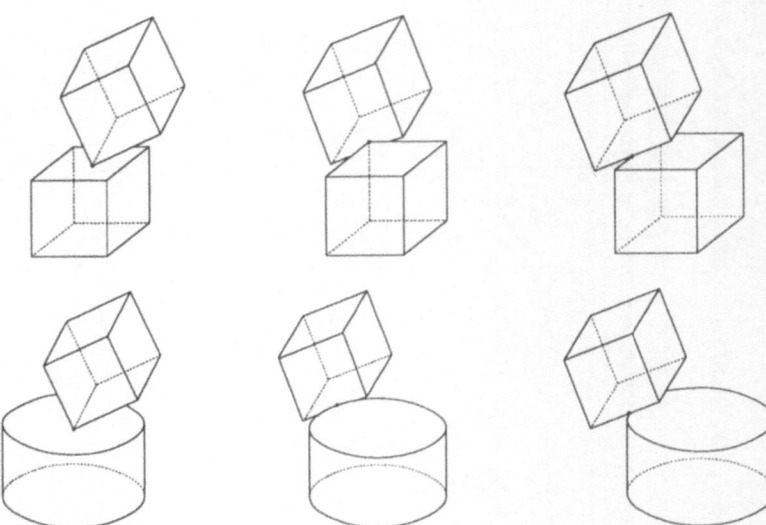

Figure 17. Type A, B, and C (from left to right) contacts between objects and obstacles.

Obstacles that are not above the object in the object-centered coordinate frame do not constrain the size of the object.

For each object-obstacle interaction we need to check for all three types of contact. The maximum object height (in the model coordinate frame) is determined by the minimum height contact with the obstacle. This process is carried out for each obstacle in the scene and the results are then combined to find the maximum height for the object which is determined by the minimum height contact with any obstacle.

In the following sections we give an overview of the computational techniques for dealing with the three types of contact. See [VaydaKak89] for a more complete description. It is important to note that all of the computations are carried out in the model coordinate frame of the object of interest. We will always be finding the maximum extent of the object along the positive z axis, which we refer to as the height of the object, in the model coordinate frame. Thus, when we refer to the top surface of the object we are referring to the surface which has a surface normal that points in the positive z direction in the model coordinate frame regardless of it's actual location and orientation in the scene.

- Type A - An object surface contacting an obstacle vertex

For type A contact we need to find the obstacle vertex which will contact the top surface of the object at the lowest point and thus constrain its height. The procedure is to find all obstacle vertices that could potentially contact the top surface of the object and then find the one with the smallest z component.

For a parallelepiped object the scene-to-model coordinate transformation results in the object being a unit cube in the positive octant with a vertex at the origin. Only obstacle vertices which satisfy $0 \leq x \leq 1$ and $0 \leq y \leq 1$ can potentially contact the top surface of the parallelepiped. For a cylindrical object the scene-to-model coordinate transformation results in the object being a right circular cylinder with unit radius and unit length which lies along the positive z axis and has one end at the origin. Only obstacle vertices which satisfy $x^2 + y^2 \leq 1$ can potentially contact the planar top surface of the cylinder.

- Type B - An object vertex or edge contacting an obstacle surface

For parallelepiped objects we need to find the obstacle surface which will contact an object vertex at the lowest point and thus constrain its height. The procedure is to find the contact points for all obstacle surfaces that could potentially contact the top vertices of the object and then find the one with the smallest z component. Potential contact points are the intersection points of the four lines defined by the vertical edges (parallel to the z axis) of the parallelepiped object and the obstacle surfaces. The four lines are defined by: $\{\, x = 0, y = 0 \,\}$, $\{\, x = 0, y = 1 \,\}$, $\{\, x = 1, y = 0 \,\}$, and $\{\, x = 1, y = 1 \,\}$. Once the intersection points are found they must be tested to find whether or not they are within the boundaries of the obstacle face. The point-in-polygon problem is a common problem in computer graphics and various efficient solutions exist [Pavlidis82].

For cylindrical objects we need to find the obstacle surface which will contact the top edge of the cylinder at the lowest point and thus constrain its height. The procedure is to find the contact points for all obstacle surfaces that could potentially contact the top edge of the object and then find the one with the smallest z component. Thus, we are interested in the extremal points (with respect to the cylindrical axis) for the intersection of a planar surface with a cylindrical surface. These extremal points lie on a line defined by projecting the cylindrical axis, which is the z axis, onto the plane. Thus, the problem reduces to finding the intersection points of a line and a cylinder. Once again, we must only consider contact points which are within the boundaries of the obstacle face.

- Type C - An object edge contacting an obstacle edge

For type C contact we need to find the obstacle edge which will contact the top edge of the object at the lowest point and thus constrain its height. The procedure is to find all points at which obstacle edges intersect the vertical surfaces of the object and then find the one with the smallest z component.

For parallelepiped objects the contact points are the intersection points between obstacle edges and the four vertical faces (parallel to the z axis) defined by the four plane equations: $x = 0$, $x = 1$, $y = 0$, and $y = 1$. For planes $x = 0$ and $x = 1$ we are interested in intersection points such that $0 \leq y \leq 1$. For planes $y = 0$ and $y = 1$ we are interested in intersection points such that $0 \leq x \leq 1$. For cylindrical objects the contact points are the intersection points between obstacle edges and the cylindrical surface defined by the equation: $x^2 + y^2 = 1$.

The result from step three is the maximum extent of the object along the positive z axis in the current model coordinate frame of the object. Note that this coordinate frame depends on the previous dimensions of the object. A value of 1 indicates that the size of

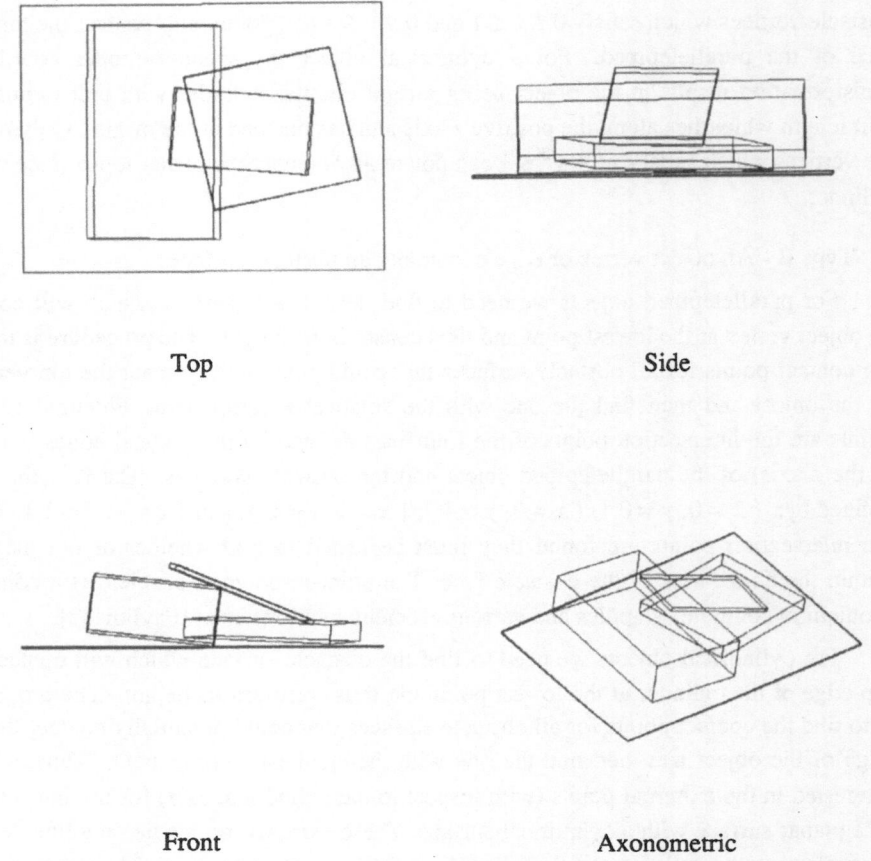

Top   Side

Front   Axonometric

Figure 18. Hypothesized objects after height refinement using the geometric reasoning algorithm.

the object need not be changed. A value less than one indicates that the old dimension was too large and must be reduced and a value greater than one indicates that the old dimension was too small and must be increased. The new object dimension is obtained in step four by simply multiplying the value found in step three by the previous value for the object dimension. This value can then be used to determine new model-to-scene and scene-to-model transformations.

Fig. 18 shows the results after performing several geometric reasoning operations on the example data. The operations performed were intersecting the letter with the two .boxes and the table and intersecting each box with the table. The operation not performed was the intersection of the two boxes with each other. This example illustrates the difficulty involved in considering uncertainty in multiple directions for an object and the difficulty in specifying the order in which to consider objects. The correct order would be to grow the small box toward the table and then to grow (actually shrink) the small box toward the large box. Then we could grow the large box toward the table. If we reverse the order and grow the large box toward the table first it will intersect the small box before the table and will have a height that is too small. Without any extra information the determination of the proper ordering for the geometric reasoning operations for a scene is undecidable. However, any ordering will give results that are geometrically feasible and fit the visible data. Our goal is to develop a control structure which will select the best ordering based on other criteria such as visibility (reasoning about occlusion) or physical stability (reasoning about gravity).

While we currently use only two object models the approach described here can be extended to handle more complex objects in two ways. The first is to define other types of primitive objects besides the parallelepiped and the cylinder. Spheres, trapezoids, tori, generalized cylinders, and polyhedra are candidates that could be considered. However, with the possible exception of polyhedra, they will probably be of limited usefulness in most cases. This approach requires the development of new intersection algorithms for each new type of object. The second path is to consider composite objects which are constructed from combinations of the primitive objects. For the simplest case, primitives are combined by the attachment operation. In this case, the techniques that we are currently using are directly applicable. All that is required is that we consider each of the primitives individually (even if they are part of a more complex object) and then combine the results as if each primitive was a separate obstacle. The only changes necessary would be the overhead required to group the primitives together as objects. Objects would be defined by a single object coordinate frame and a set of primitives with transformations relating them to the object coordinate frame. The more complex, and more general, approach is to use Constructive Solid Geometry (CSG) to define objects. The CSG representation represents objects as sets of primitives combined by regularized volumetric union, intersection, and subtraction operations. As with composite objects, primitives can be considered individually and the results combined for each object. However, the union, intersection, and subtraction operations make the combination of results much more complex. Ray casting programs use this approach successfully [Roth82] for finding the intersections of lines with CSG objects. Our problem of finding the intersections between CSG objects is not quite as simple but is computationally feasible.

7. CONCLUSION

We have discussed two object recognition systems which address recognition problems in two different domains. 3D-POLY uses a feature-based model-driven hypothesize-and-verify approach to efficiently recognize industrial objects. It's efficiency is derived from the use of the feature sphere data structure and local feature sets for hypothesis generation. INGEN uses both data and model-driven processing in its hypothesize-and-refine approach to generic object recognition. It's geometric reasoning capability allows it to make reasonable hypotheses about the parts of an object which are not visible. The differences and similarities between these two systems demonstrate how recognition problems in different domains benefit from different approaches.

ACKNOWLEDGMENTS

We would like to thank Dr. Frederick R. Glickman, Program Director at the US Postal Service, Office of Advanced Technology, for many insightful discussions as the generic object recognition project has progressed from the initial conceptualization through the current implementation. We also thank Mr. Gary P. Herring, Director of the US Postal Service, Office of Advanced Technology, for providing us with global overviews on the relevance of this project to the needs of the postal service.

REFERENCES

[BajSol87]   R. Bajcsy and F. Solina, "Three Dimensional Object Representation Revisited," *Proceedings of the First International Conference on Computer Vision,* IEEE Computer Society, 1987.

[BolHor86]   R. C. Bolles and P. Horaud, "3DPO: A Three-Dimensional Part Orientation System," *The International Journal of Robotics Research,* Vol. 5, No. 3, 1986.

[ChenKak88]   C. H. Chen and A. C. Kak, "3D-POLY: A Robot Vision System for Recognizing Objects in Occluded Environments," Purdue University Technical Report TR-EE 88-48, December 1988.

[ChenKak89]   C. H. Chen and A. C. Kak, "A Robot Vision System for Recognizing 3-D Objects in Low-Order Polynomial Time," *IEEE Transactions on Systems, Man, and Cybernetics,* Vol. 19, No. 6, 1989.

[Fanetal88]   T. J. Fan, F. Medoni, and R. Nevatia, "Matching 3-D Objects Using Surface Descriptions," *Proceedings of the 1986 IEEE International Conference on Robotics and Automation,* IEEE Computer Society, 1987.

[Faug85]   O. D. Faugeras, "New Steps Toward a Flexible 3-D Vision System for Robotics," *Proceedings of the Second International Symposium on Robotics Research,* 1985.

[FaugHeb86]   O. D. Faugeras and M. Hebert, "The Representation, Recognition, and Locating of 3-D Objects," *The International Journal of Robotics Research,* Vol. 5, No. 3, 1986.

[FaugHeb83]   O. D. Faugeras and M. Hebert, "A 3-D Recognition and Positioning Algorithm Using Geometrical Matching Between Primitive Surfaces," *Proceedings of the 8th International Joint Conference on Artificial Intelligence,* 1983.

[GrimLoz84]   W. E. L. Grimson and T. Lozano-Pérez, "Model-Based Recognition and Localization from Sparse Range or Tactile Data," *The International Journal of Robotics Research,* Vol. 3, No. 3, Fall 1984.

[HebKan85]   M. Hebert and T. Kanade, "The 3D-Profile Method for Object Recognition," *IEEE Computer Society Conference on Computer Vision and Pattern Recognition,* 1985.

[HutUll87]   D. P. Huttenlocher and S. Ullman, "Object Recognition Using Alignment," *Proceedings of the First International Conference on Computer Vision,* 1987.

[Kenner76]   H. Kenner, *Geodesic Math And How To Use It,* University of California Press, Berkeley, CA, 1976.

[LamWol88]   Y. Lamdan and H. J. Wolfson, "Geometric Hashing: A General and Efficient Model-Based Reocgnition Scheme," *Proceedings of the Second International Conference on Computer Vision,* 1988.

[Low87]   D. G. Lowe, "Three-Dimensional Object Recognition from Single Two-Dimensional Images," *Artificial Intelligence,* Vol. 31, 1987.

[Lozano87]   T. Lozano-Pérez, "A Simple Motion-Planning Algorithm for General Robot Manipulators," *IEEE Journal of Robotics and Automation,* Vol. RA-3, No. 3, June 1987.

[Lozetal87]   T. Lozano-Pérez, J. L. Jones, E. Mazer, P. O'Donnell, and W. E. L. Grimson, "Handey: A Robot System that Recognizes, Plans, and Manipulates," *Proceedings of the 1987 IEEE International Conference on Robotics and Automation,* IEEE Computer Society, 1987.

[Mur87]   D. W. Murray, "Model-Based Recognition Using 3D Shape Alone," *Computer Vision, Graphics, and Image Processing,* 40, 1987.

[Pavlidis82]   T. Pavlidis, *Algorithms for Graphics and Image Processing,* Computer Science Press, 1982.

[Pugh76]   A. Pugh, *Polyhedra: A Visual Approach,* University of California Press, Berkeley, CA, 1976.

[Roth82]   S. C. Roth, "Ray Casting for Modeling Solids," *Computer Graphics and Image Processing,* 18, 1982.

[Umeetal88]   S. Umeyama, T. Kasvand, and M. Hospital, "Recognition and Positioning of Three-Dimensional Objects by Combining Matchings of Primitive Local Patterns," *Computer Vision, Graphics, and Image Processing,* 44, 1988.,

[VaydaKak89]   A. J. Vayda and A. C. Kak, "INGEN: An Inference Engine for Recognizing Generic Shapes in Occluded Environments," Purdue University Technical Report TR-EE 89-53, August 1989.

Sensor-Based Robot Control Requirements for Space

Ronald Lumia
Intelligent Controls Group
Robot Systems Division
National Institute of Standards and Technology
(formerly the National Bureau of Standards)
Gaithersburg, MD 20817
USA

NASA has begun the development of the Space Station, a
permanently manned facility in space, for a variety of
scientific goals. One part of this project is the Flight
Telerobotic Servicer (FTS) which will help build and maintain
the structure. The FTS is envisioned as a two armed robot
with seven degrees of freedom for each arm. When the FTS is
launched, it is expected to perform several tasks which
include the installation and removal of truss members of the
Space Station structure, changeout of a variety of modular
units, mating a thermal connector, etc. While the FTS will
initially use teleoperation, it is envisioned to become more
autonomous as technology advances. In order for the FTS to
evolve from teleoperation to autonomy, NASA requires that the
NASA/NBS Standard Reference Model (NASREM) be used as the
functional architecture for the control system. The quest for
autonomy inevitably leads to the need for sophisticated
sensors and sensory processing. This paper will explore the
requirements for the tasks envisioned for FTS at first launch
as well as during its evolution phase and show how those tasks
impact research on sensors, sensory processing, and other
parts of the FTS control system. Finally, the current state
of the NASREM implementation at NIST will be presented.

1.   INTRODUCTION

The Flight Telerobotic Servicer (FTS) will be used to
build and maintain the Space Station. It is envisioned to be
used as a teleoperated device initially. However, it is
required to be able to evolve with technology and manifest
this evolution by becoming more and more autonomous. While

This work is partially supported by funding from NASA under
contract S-28187-D and was prepared by U.S. Government
employees as part of their official duties and is therefore a
work of the U.S. Government and not subject to copyright.
Equipment listings in this paper do not imply a recommendation
by NIST nor that the equipment is the best for the purpose.

teleoperation is the first step, there has been a deliberate choice not to pursue telepresence.

There is a basic dichotomy in the evolution of FTS activity from the operator's perspective. The FTS can move toward full telepresence or full autonomy. Reaching either extreme, full telepresence or full autonomy, is a long term, possibly unattainable objective. However, it is instructive to examine the possibilities. A system that consists only of telepresence implies that the human remains in the loop for all task steps. In a telepresence system, the human operator performs the tasks, his "presence" being translated to the remote worksite via the technology. This capability is very useful and even necessary for certain applications, especially where the environment is relatively unknown or unstructured. The development of systems which pursue true "telepresence" where the operator is immersed in sensory feedback will require a great deal of R & D.

The nature of the FTS role on Space Station (i.e., assembly, maintenance, inspection, etc.) does not involve tasks of a scientific or totally unstructured nature where the human operator is actually performing some sort of investigation. Therefore, the FTS is choosing to pursue autonomy for several reasons. First, the repetitive chores performed by the robot are less onerous to the operator if the FTS had some autonomous capabilities. Second, the operator does not give up anything since he may break into any level of the NASREM architecture to take control at will. Third, time delays may preclude telepresence for useful applications. For example, in remote satellite servicing, the time delays incurred between the operator and the FTS preclude the use of force reflection and other sophisticated telepresence techniques. An autonomous mode of operation may be the only viable alternative. There may be only a sparse data return, e.g., one image per minute. A telepresence strategy of move and wait would be too slow to be a realistic option.

This paper explores what is required to have a sensor based robot in space. First the tasks originally envisioned for the FTS are presented along with a set of tasks required

of an evolved version of the FTS. Then, the problems
associated uniquely with space qualification and those
associated with the current state of robot technology are
presented. This is followed by a description of the NASA/NBS
Standard Reference Model for Telerobot Control System
Architecture (NASREM), which is required by NASA for the
control system of the FTS. Finally, the current state of
NASREM development at NIST is described.

2. FTS TASKS -- AT FIRST ELEMENT LAUNCH AND ASSEMBLY COMPLETE

The FTS is required to have certain capabilities at First
Element Launch (FEL) in support of building and maintaining
the Space Station. These tasks are intended to be
representative of the range of work which the FTS can perform
rather than the limit of its capabilities. The tasks are:

o   Installation and removal of Space Station truss
        members
o   Installation of a Structural Interface Adapter
        (SIA) on the truss.
o   Changeout of Space Station Orbital Replaceable Units
        (ORU).
o   Mating of the Space Station thermal utility
        connectors.
o   Performance of inspection tasks.
o   Assembly and maintenance of Space Station Electrical
        Power System Radiator Assembly.

The FTS is required to have, integral to its basic
design, the capability of evolving toward autonomous execution
of the tasks listed previously. In addition, the FTS should
eventually have the capability to perform the following tasks
autonomously:

o   Changeout the Hubble Space Telescope (HST) reaction
        wheel ORU while the HST is secured at the Space

Station or on the Space Shuttle.

o   Refuel the Gamma Ray Observatory (GRO) propellant
     tanks while the GRO is secured at the Space Station
     or on the STS.

o   In-situ servicing and maintenance of Space Station
     platforms and free fliers.

In order to be able to perform these tasks, a significant
amount of sensor and sensor processing technology must be
integrated into the FTS system. Sensor capability must be
available for:

o   Joint control
     For example, the joints of the manipulator could
     command torques and have sensors which measure
     actual torque, position, and velocity.

o   Vision processing
     Image processing algorithms must be able to cope
     with extreme variations in lighting.

o   Contact force measurement
     During operations such as mating a connector, it is
     important to be able to measure the forces generated
     by the robot in contact with the environment.
     Without this information, there is no way for a
     control algorithm to correct for errors. This would
     probably require the use of a force/torque sensor.

o   Safety
     Since safety is extremely important, sensors which
     redundantly measure important parameters must be
     included in the FTS system. Example of these
     sensors include robot joint position, proximity of
     astronauts to the FTS workspace, etc.

The above tasks, as well as the sensors associated with
the tasks, will be performed in space. This implies a fairly

complex FTS, especially in the requirement for sufficient compute power. The next two sections explore some of the problems in sending the FTS into space.

3. SPACE QUALIFICATION PROBLEMS ASSOCIATED WITH ROBOT SYSTEMS

Space qualification is a rigorous process to ensure that the systems which are sent into the harsh space environment will work reliably. It is well beyond the scope of this paper to explore the technical details of any particular problem. However, this section will hopefully provide the reader with an appreciation of the complexity associated with sending a system into space.

The first problem is associated with materials science. The materials which are used to construct robots and the required ancillary equipment must be able to withstand the harsh space environment and function as expected. The materials must be able to resist thermal fatigue caused by the cycles of extreme heat and cold. Furthermore, they must not outgas, i.e., release gases into space.

The space qualification of electronic components, especially computers, presents a formidable challenge. Most electronic circuitry is sensitive to two types of radiation. The first type is the background radiation which is significantly higher in space than on Earth due to the atmosphere. Consequently, all electronic circuitry must be able to withstand a certain level of total dose radiation. This is tested in the qualification process and a level of radiation hardness must be met. The second type of radiation is caused by cosmic rays, with enough energy to change a bit, passing through the circuit. This phenomenon is known as a single event upset (SEU). Clearly, the circuitry must either be insensitive to the SEU or must detect and react to it. Unfortunately, both the radiation-hard and SEU issues often involve the redesign of the manufacturing processes for the integrated circuits.

For mission critical activities or when human safety is at risk, the electronics are required to be "two fault tolerant." If any fault occurs, the system can still operate. If a second fault develops anywhere in the system, then the system fails in a safe manner. This requirement has major implications for redundancy and protocols of switching between subsystems after a fault is detected. This is clearly an area of continuing research.

Thermal considerations play an important role in the total system design since removing heat by convection is not an option in space. Motors, for example, must be capable of either conducting or radiating the heat generated. It is possible to use passive methods for thermal control but often the active methods using fluid loops are more effective. However, the active loops often require hazardous chemicals such as ammonia compounds and are a great deal more complex to handle. Dangerous chemicals invoke safety rules which may preclude servicing in the pressurized unit and extra-vehicular servcing limits the types of possible repairs.

If robots are going to operate in space without umbilical cords for any period of time, batteries are required. The batteries must operate in a vacuum, be insensitive to temperature variations, store significant energy, weigh as little as possible, and take up a small volume.

Simulation has often been used with great success to determine the value of one approach over another. The problem with simulation, of course, is that the simulation may not represent reality sufficiently, i.e., the model may lack fidelity. On Earth, it is possible to simulate first and then empirically test the system to determine the accuracy of the model. For space, the situation is more complex because it is very difficult to predict exactly how the robot system will behave without gravity.

This list of problems for the space qualification of robot systems is by no means complete. However, the number of significant issues should provide some level of appreciation for the complexity associated with putting robots in space.

4. GENERIC ROBOT TECHNOLOGY PROBLEMS

The second class of problems associated with space robotics is also present in ground based automata. The solutions to this class of problems will advance the state-of-the-art in robotics.

Control methods for robots have traditionally centered on position control where a robot is programmed to follow a predefined path. While this approach has proved quite useful for certain applications in industrial automation, it is not entirely satisfactory for factories and therefore is probably also unsuitable for space utilization. The cost associated with errors in space is enormous and more advanced control methods, such as impedance control [1], appear to hold more promise. However, it is not entirely clear which advanced control algorithm is best for a given application. There is a need to study a set of alternative algorithms in the execution of a given battery of tasks. Knowledge of which algorithms work on which tasks and the reasons why is essential to improve robot control technology.

A second problem area in robot technology is associated with modeling the workspace of the robot. A representation of the robot workspace is required to allow the robot to operate with known objects in a sensible fashion. Many alternatives for this representation exist and must be tested in a systematic way to ascertain which approach is most appropriate. A robot in space operates in a reasonably unstructured environment. Although it can be argued that everything sent into space for the Space Station is man-made and a CAD model is available, there will always be slight discrepancies which will make real objects differ from their models. Since it is possible that such a disparity could result in a catastrophe, such as an object which is larger than its model being pushed through a satellite, a highly calibrated workspace is nearly impossible. The workspace representation can be close, but sensors are required to prevent disasters. Sensory processing, therefore, must be an integral part of a robotic system and interact effectively with the model of the robot workspace so that the algorithms controlling robot motion can be as efficacious as possible.

Sensory processing presents other demanding challenges for robot systems in terms of quality and processing rate. It is well known that lighting is crucial for success in industrial implementations of computer vision. Shadows, specular reflection, lens distortion, etc., will have even more impact in space since the environment cannot be controlled as well as in a factory. The speed of sensory processing is also critical because it can limit the rate the control system can move the robot in response to stimuli. The improvements required for sensory processing in space will increase the available knowledge and ultimately benefit all robot systems.

The last area is concerned with how the operator commands a robot, the operator interface. In teleoperation, there are several issues. The first issue is whether the control device, or master, is kinematically similar to or different from the telerobot. The amount of computation required to control the kinematically similar master is much lower than that required for the kinematically dissimilar master. However, the production of a kinematically similar master is required for each new robot.

Another issue in operator control is kinesthetic feedback. Without force reflection, the operator sends commands to the telerobot but cannot perceive the effect of those commands as the telerobot moves in its workspace. With force reflection, the operator can "feel" the reaction of the telerobot. While it is generally desirable to allow the operator more feedback from the telerobot through force reflection, the control is more difficult and could potentially result in instability.

5. NASA/NBS STANDARD REFERENCE MODEL FOR TELEROBOT CONTROL SYSTEM ARCHITECTURE (NASREM)

The fundamental paradigm of the control system is shown in Figure 1. The control system architecture is a three legged hierarchy of computing modules, serviced by a

communications system and a global memory. The task
decomposition modules perform real-time planning and task
monitoring functions; they decompose task goals both
spatially and temporally. The sensory processing modules
filter, correlate, detect, and integrate sensory information
over both space and time in order to recognize and measure
patterns, features, objects, events, and relationships in the
external world. The world modeling modules answer queries,
make predictions, and compute evaluation functions on the
state space defined by the information stored in global
memory. Global memory is a database which contains the
system's best estimate of the state of the external world.
The world modeling modules keep the global memory database
current and consistent.

The first leg of the hierarchy consists of task
decomposition modules which plan and execute the decomposition
of high level goals into low level actions. Task
decomposition involves both a temporal decomposition (into
sequential actions along the time line) and a spatial
decomposition (into concurrent actions by different
subsystems). Each task decomposition module at each level of
the hierarchy consists of a job assignment manager, a set of
planners, and a set of executors. These decompose the input
task into both spatially and temporally distinct subtasks.

The second leg of the hierarchy consists of world
modeling modules which model (i.e. remember, estimate,
predict) and evaluate the state of the world. The "world
model" is the system's best estimate and evaluation of the
history, current state, and possible future states of the
world, including the states of the system being controlled.
The "world model" includes both the world modeling modules and
a knowledge base stored in a global memory database where
state variables, maps, lists of objects and events, and
attributes of objects and events are maintained. The world
model maintains the global memory knowledge base by accepting
information from the sensory system, provides predictions of
expected sensory input to the corresponding sensory system
modules, based on the state of the task and estimates of the

external world, answers "What is?" questions asked by the executors in the corresponding task decomposition modules, and answers "What if?" questions asked by the planners in the corresponding task decomposition modules.

The third leg of the hierarchy consists of sensory processing sensory system modules. These recognize patterns, detect events, and filter and integrate sensory information over space and time. The sensory system modules at each level compare world model predictions with sensory observations and compute correlation and difference functions. These are integrated over time and space so as to fuse sensory information from multiple sources over extended time intervals. Newly detected or recognized events, objects, and relationships are entered by the world modeling modules into the world model global memory database, and objects or relationships perceived to no longer exist are removed. The sensory system modules also contain functions which can compute confidence factors and probabilities of recognized events, and statistical estimates of stochastic state variable values.

The control architecture has an operator interface at each level in the hierarchy. The operator interface provides a means by which human operators, either in the space station or on the ground, can observe and supervise the telerobot. Each level of the task decomposition hierarchy provides an interface where the human operator can assume control. The task commands into any level can be derived either from the higher level task decomposition module, from the operator interface, or from some combination of the two. Using a variety of input devices, a human operator can enter the control hierarchy at any level, at any time of his choosing, to monitor a process, to insert information, to interrupt automatic operation and take control of the task being performed, or to apply human intelligence to sensory processing or world modeling functions.

The sharing of command input between human and autonomous control need not be all or none. It is possible in many cases for the human and the automatic controllers to simultaneously share control of a telerobot system. For example, in an assembly operation, a human might control the position of an end effector while the robot automatically controls its orientation. For a more detailed description of NASREM, see [2].

6.   NIST IMPLEMENTATION OF NASREM

In order to implement a functional architecture, especially one like NASREM which allows evolution with technology, the interfaces must be carefully defined. Although the NASREM functional architecture specifies the purpose of each module in the control system hierarchy, it does not completely specify the interfaces between modules. This section will describe the method by which the interfaces for the SERVO level of the hierarchy have been defined. The method involves gathering all of the algorithms available for SERVO level control, dividing each algorithm into the parts which inherently belong to task decomposition, world modeling, and sensory processing, and then deriving the interfaces which will support these algorithms. Any design, however, must constrain the problem sufficiently so that detailed interfaces can be devised.

With this in mind, the Servo Level design was based on a fundamental control approach which computes a motor command as a function of feedback system state y, desired state (attractor) y_d, and control gains. In this approach, the gains are coefficients of a linear combination of state errors $(y-y_d)$. The system state and its attractor are composed from the physical quantities to be controlled, (i.e. position, force, etc.,) and can be expressed in an arbitrary coordinate system. This type of algorithm is the basis for almost all manipulator control schemes [3]. However, this basic algorithm is inadequate for controlling the gross aspects of manipulator motion, as described in [4]. The algorithm can provide "small" motions so that the dynamics of the servo algorithm itself are not significant. This means that the Primitive Level must generate the gross dynamics of the motion through a sequence of inputs to the Servo Level. This can be achieved through an appropriate sequence of either attractor points [3,5] or gain values [4].

Figure 2 depicts the detailed Servo Level design. The task decomposition module at the Servo Level receives input from Primitive in the form of the command specification parameters. The command parameters include a coordinate system specification C_z which indicates the coordinate system in which the current command is to be executed. C_z can specify joint, end-effector, or Cartesian (world) coordinates. Given with respect to this coordinate system are desired position, velocity, and acceleration vectors (z_d, $\dot{z}_d$, $\ddot{z}_d$) for the manipulator, and the desired force and rate of change of force vectors (f_d, $\dot{f}_d$). These command vectors form the attractor set for the manipulator. The K's are the gain coefficient matrices for error terms in the control equations. The selection matrices (S,S') apply to certain hybrid force/position control algorithms. Finally, the "Algorithm" specifier selects the control algorithm to be executed by the Servo Level.

When the Servo Level planner receives a new command specification, the planner transmits certain information to world modeling. This information includes an attention function which tells world modeling where to concentrate its efforts, i.e. what information to compute for the executor. The executor simply executes the algorithm indicated in the command specification, using data supplied by world modeling as needed.

The world modeling module at the Servo Level computes model- based quantities for the executor, such as Jacobians, inertia matrices, gravity compensations, Coriolis and centrifugal force compensations, and potential field (obstacle) compensations. In addition, world modeling provides its best guess of the state of the manipulator in terms of positions, velocities, end-effector forces and joint torques. To do this, the module may have to resolve conflicts between sensor data, such as between joint position and Cartesian position sensors.

Sensory processing, as shown in Figure 2, reads sensors relevant to Servo and provides the filtered sensor readings to world modeling. In addition, certain information is transmitted up to the Primitive Level of the sensory processing hierarchy. Primitive uses this information, as well as information from Servo Level world modeling, to monitor execution of its trajectory. Based on this data, Primitive computes the stiffness (gains) of the control, or switches control algorithms altogether. For example, when Primitive detects a contact with a surface, it may switch Servo to a control algorithm that accommodates contact forces.

A more complete description of the Servo Level is available in [3] where the vast majority of the existing algorithms in the literature are described. The same process for developing the interfaces based on the literature has also been performed for the Primitive level and is available in [5]. While the procedure is planned for each level in the hierarchy, the amount of literature support tends to decrease as one moves up the NASREM hierarchy.

Once the interfaces are defined, it is possible to choose a computer architecture and begin to realize the system. While every effort is being made to do the job properly, there is no reason to assume that the implementation at NIST is optimal in any way. It is simply illustrates one realistic method to implement the NASREM architecture.

While a functional architecture is technology independent, its implementation obviously depends entirely on the state-of-the-art of technology. The designer must choose existing computers, buses, languages, etc., and, from these tools, produce a computer architecture capable of performing the functions of the functional architecture. The system must adequately meet the real-time aspects of the controller so that adequate performance is achieved through careful consideration of computer choice, multiple processor real-time operating system, inter-processing communication requirements, tasking within certain processors, etc. For a more detailed description of this methodology, see [6].

The NIST implementation considers two aspects of the

software development process: the development environment on which the code is written, debugged, and tested as well as possible, and the target environment where the code for the real-time robot control system is integrated into the system. Figure 3 shows the approach. A network of SUN workstations running UNIX is used for the development environment, sacrificing the speed of the developed code for the ease of development. Once the code is tested as well as possible, it is downloaded to the target system. The target system consists of a VME backplane of several (currently 6) 68020 processors. For rapid iconic image processing, the PIPE system [7] is integrated into the system. The target hardware drives a K-1607 Robotics Research Corp. arm.

From the software side, the multiprocessing operating system used for the target is required to be as simple as possible so that the overhead is minimized. The duties of the operating system are limited to very simple actions such as downloading executable code, starting up the processors, and interprocessor communication. While tasking is not performed at the lower levels of the hierarchy because of the overhead associated with context switches, it is desirable at higher levels in the hierarchy which are not as time critical. NIST researchers are currently investigating three alternatives for tasking: tasking provided by the run-time kernel of the native ADA cross compiler, pSOS tasking, and ADA tasking. Interprocessor communications alternatives including pRISM, sockets, etc., must also be evaluated empirically. The actual application code is written in ADA. Although ADA compilers cannot currently produce code as efficient as other languages such as C, NIST researchers have shown that the gap is steadily decreasing [8].

The application code is developed by programming the processes which achieve the functions associated with the boxes in the functional architecture. The problem then becomes one of assigning each of the processes, such as those shown in Figure 2, to a particular processor. There is a clear trade-off between the cost of the solution and the performance of the system. There are currently no software tools which automatically perform this assignment based on an arbitrary index of performance. The approach at NIST is step-wise refinement of the performance of the system. Given the particular hardware being used, a certain number of processors is chosen arbitrarily. For that configuration, the processes are assigned to the processors. Then, the system is evaluated in terms of its performance. If the performance is unacceptable, the designer has several options. The first option is to add more processors. This alternative is balanced against additional communication required by the processors. Another alternative is to add faster processors or special purpose processors, such as dynamics chips, which optimize particularly compute intensive operations. This trade-off clearly relates to cost. Another alternative is to reassign the processes to the processors in order to balance the workload of each processor. Each of the alternatives can be used by the designer in order to improve the performance of the system. This allows a particular configuration which implements the functional architecture to change with time as improvements in technology are realized.

7. CONCLUSION

The FTS project is the driving force in U.S. space based robots. At first element launch of the Space Station, it will behave as a teleoperated system. However, by using the NASREM architecture, it will be capable of evolving with technology, incorporating greater levels of automation. In order to perform sophisticated autonomous tasks, the FTS must have a significant suite of sensors at first element launch and have the capability to integrate new sensors into its control system as these products become available.

8.   REFERENCES

[1] N. Hogan, "Stable Execution of Contact Tasks Using Impedance Control," IEEE International Conference on Robotics and Automation, 1987.
[2] J.S. Albus, R. Lumia, H.G. McCain, "NASA/NBS Standard

Reference Model For Telerobot Control System Architecture (NASREM)," NBS Technote #1235, also as NASA document SS-GSFC-0027.

[3] J. Fiala, "Manipulator Servo Level Task Decomposition," NIST Technote #1255, April 20, 1988.

[4] J. Fiala, "Generation of Smooth Trajectories without Planning," manuscript in preparation.

[5] A.J. Wavering, "Manipulator Primitive Level Task Decomposition," NIST Technote #1256, January 5, 1988.

[6] T. Wheatley, J. Michaloski, and R. Lumia, "Requirements for Implementing Real Time Control Functional Modules on a Hierarchical Parallel Pipelined System," 1989 Conference on Space Telerobotics, Pasadena, January 30, 1989.

[7] E.W. Kent, M.O. Shneier, and R. Lumia, "PIPE," Journal of Parallel and Distributed Computing, Vol. 2, 1985, pp. 50-78.

[8] S. Leake, "A Comparison of Robot Kinematics in ADA and C on Sun and microVAX," Robotics and Automation Session, IASTED, Santa Barbara, CA., May 25-27,1988.

419

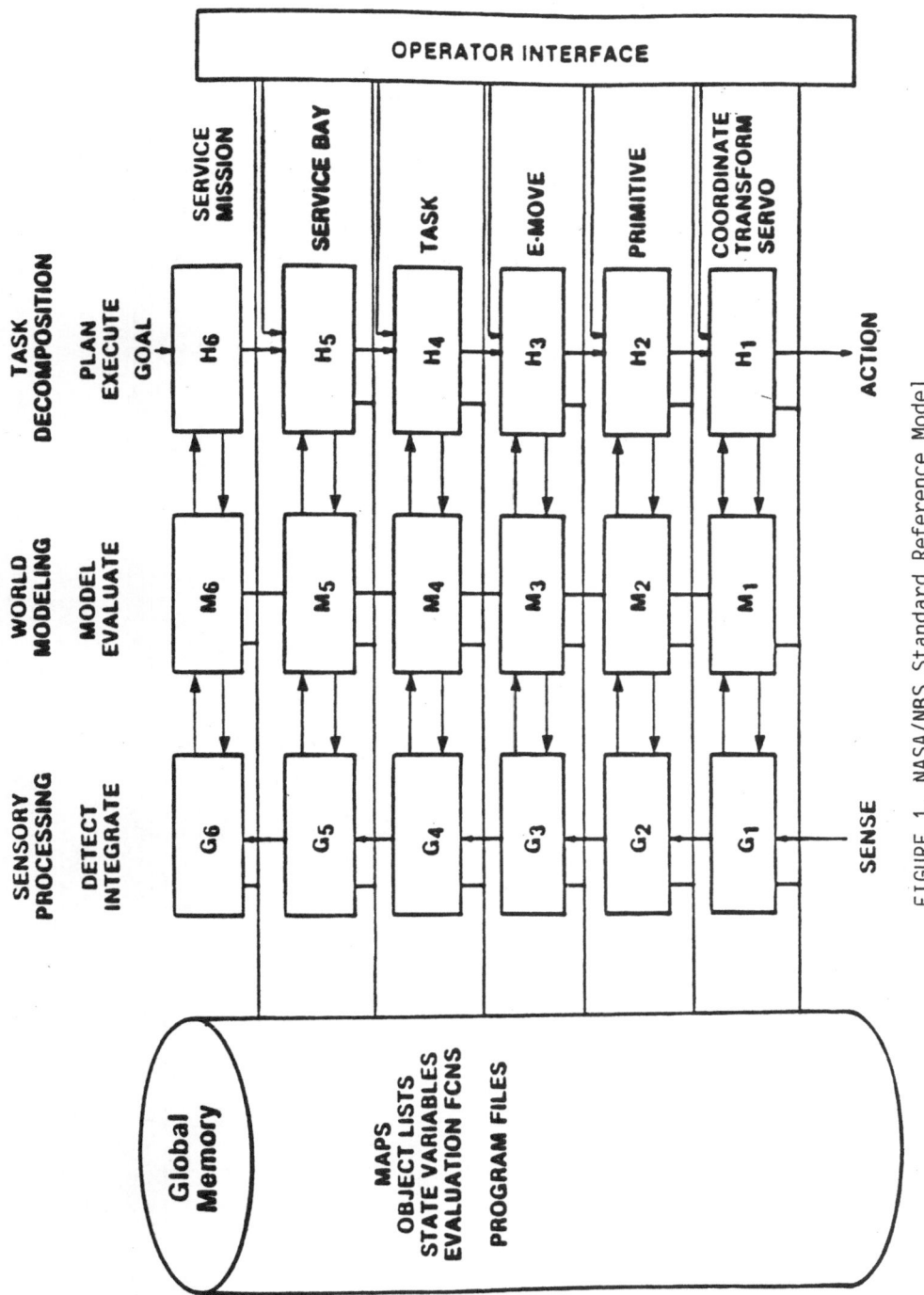

FIGURE 1. NASA/NBS Standard Reference Model
for Telerobot Control System Architecture (NASREM)

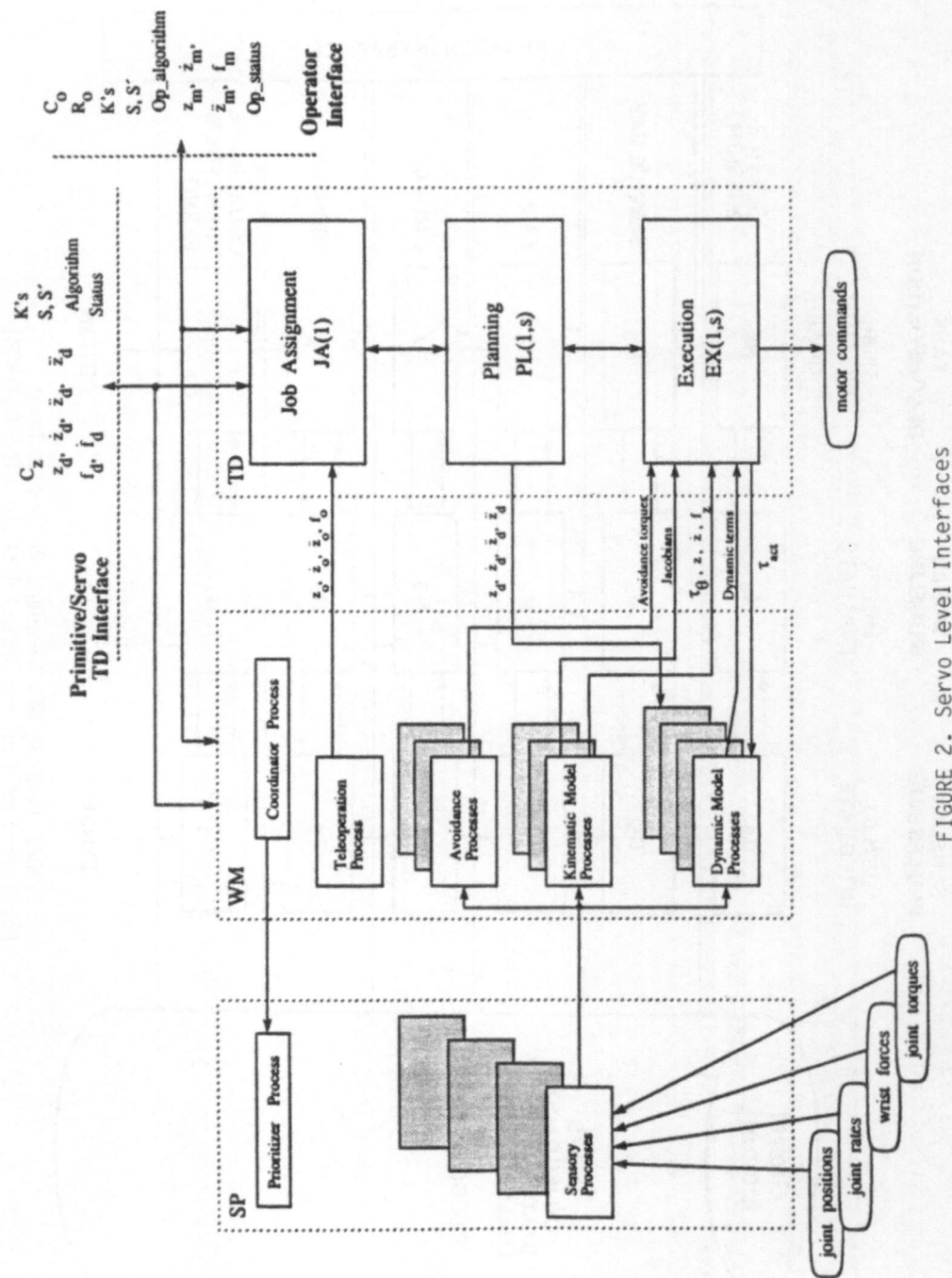

FIGURE 2. Servo Level Interfaces

SYSTEM DEVELOPMENT
(View at Hardware)

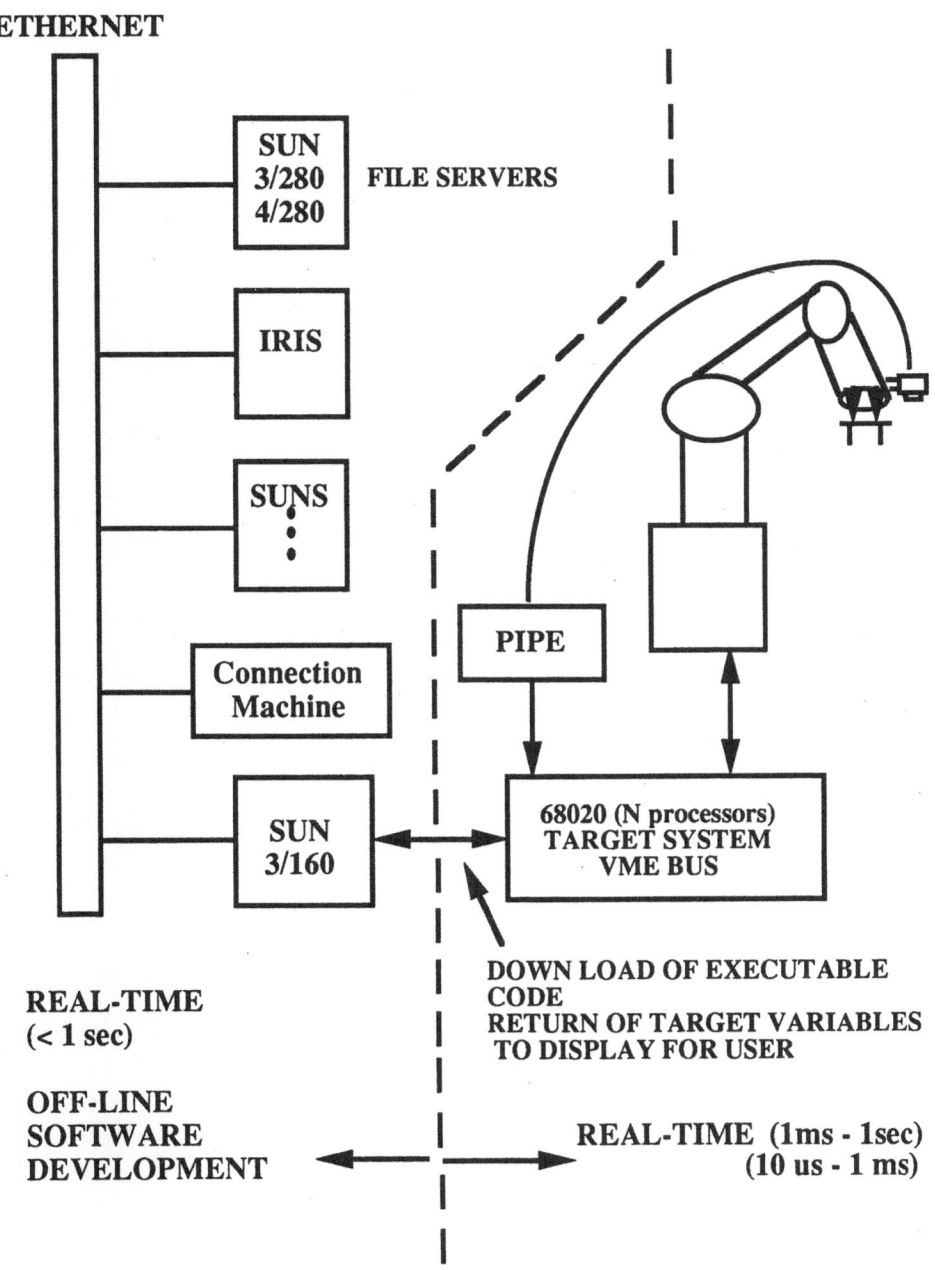

FIGURE 3. NASREM development at NIST

FAST MOBILE ROBOT GUIDANCE

Antonio B. Martínez, Albert Larré
Dept. d'Enginyeria de Sistemes,
Automàtica i Informàtica Industrial
Universitat Politècnica de Catalunya
c/Pau Gargallo,
08028-BARCELONA (Spain)

ABSTRACT

Prior knowledge of the environment constitutes an important help in planning trajectories and avoiding collisions. But sometimes modifications of the workroom due to appearance of possible obstacles are the main drawbacks for getting a high speed of navigation.

The use of multiple sensors provides complementary information that can help to avoid collision.

In order to assure the maintenance of adequate speed of the mobile without producing excessive speed reductions it is nessary to interpret the information acquired from the sensors and focus on the more interesting characteristics of the scene.

The goal of this work is not to develop a navigation system but develop image processors which help us to get the most adequate information needed to carry on the guidance of these mobiles with a response time as low as possible.

The system presented has the function of facilitating the task of fusion of the information obtained from TV cameras and ultrasonic sensors. The facilities provided by the system are oriented to reducing the processing time so that the mobile robot navigation speed is not reduced.

NATO ASI Series, Vol. F 63
Traditional and Non-Traditional Robotic Sensors
Edited by T. C. Henderson
© Springer-Verlag Berlin Heidelberg 1990

1. INTRODUCTION

This paper describes the work already done in the ESAII Department oriented to the fast driving control of a mobile robot.

The goal of this work is not to develop a navigation system but develop the needed image processors which help us to get the most adequate information needed to carry on the guidance of these mobiles with a response time as low as possible.

The research task carried out is in the computer vision field.

To achieve these goals it was decided that a hierarchical information structure was the solution. In a first level, specific problems are solved in parallel with different processors.

The data given by them are more concentrated, which allows the system to work in less time.

The second level performs the fusion of the necessary information acquired from the same area viewed by the utilized sensors.

2. CONFIGURATION OF THE EXPERIMENTAL MOBILE ROBOT "EIXERIT"

In order to evaluate in real conditions different algorithms for image processing and algorithms of control for mobile robots, an experimental platform has been designed. Figure 1 shows a photograph of the "Eixerit" prototype.

The primary consideration behind selection of the hardware components was to provide for the "Eixerit" a structure capable of maneuvering indoors, on road and off road, carrying on board all the power, sensors, and needed compùter support for its autonomous operation.

2.1. Vehicle

The overall vehicle dimensions are 1.8 m long, 70 cm wide, 80 cm high and 400 kg in weight.

The electronics within "Eixerit" are powered by an auxiliary power unit.

The vehicle has six wheels. The two central wheels are the driving ones, each one having its own independent engine in order to attain the desired mobility and maneuverability.

Fig.1   "Eixerit"   prototype

2.2.   Sensors

The "Eixerit" has been endowed with sufficient sensory capability to guarantee the maximun reliability and safety in its movements.

a) Sonar sensor array.

The sonar sensor array consists of a ring of 10 transducers distributed two to each side, three on the front and three on the back.

The sonar devices are Polaroid laboratory grade ultrasonic range transducers. Their measuring range spans distances from 30 cm to 10 m in optimal conditions, and the range accurancy is roughly 1%.

The main lobe of the sensitivity function is contained within a solid angle of 30º.

b) The mobile robot is equipped with two vision cameras which give stereo vision as well as stereomotion.

3. DEVELOPED PREPROCESSORS

The preprocessors developed until now have been oriented to extract some relevant features of scenes in real time. In this aspect we have developed:

- Texture analyzer
- Vertex extraction for 3D information.
- Straight line extractor.
- Recognition and tracking of same preselectable pattern contained in a scene.

Texture analyzer

This preprocessor performs a partition of the image based on the homogeneity of the region.

For instance, after extracting the contour of a scene, the vision system will assign special texture to different things based on previous knowledge, allowing us to distinguish among different areas, for example path, grass, sky, etc.

The desired results obtained with this method were achieved by applying different operators over 8x32 and 5x5 pixels.

Vertex Acquisition

Vertices are relevant points to describe an image because they have the advantage of being invariant on the object. These points are detected by means of special purpose hardware in two steps. During the first step, the contour of the object is obtained by applying a gradient operator of 3x3 pixels, and during a second step, points of the contour with a curvature above a threshold are located by applying a 5x5 pixel operator.

The points corresponding to vertices on the scene are supplied to the "vertex tracker module". The information supplied to the tracker corresponds to the (X, Y) coordinates of each vertex which have to be tracked in every new image.

The chose size of 5x5 pixels gives a rather good approach to vertex detection although in some cases it detects, due to different textures, other points which do not correspond to vertices. This situation is not an inconvenience because the goal of the system is to analyse some points.

If the vertices of a scene can be obtained in real time using a hardware operator and the camera is placed on a mobile, the points corresponding to the vertices detected can be tracked all along their displacement, and the initial and final position of each one will be located and the correspondence problem solved. Since every point has been tracked continuously the homologous points between the initial and final image are matched without difficulty.

In that way, the relative location of these points in space is calculated, thus allowing the position of the objects appearing in the environment of a mobile robot to be obtained (2).

Straight line detection

Straight line identification is performed in two stages by means of a pipeline structure. The first stage performs the contour extraction. The second stage performs straight line identification by matching, for every point in the image, the lines that pass over this point with 38 different patterns corresponding to the set of different orientations in which a 10-pixel length line can be represented. In order to make the straight line detector operator implementable in hardware from its 10x10=100 inputs the window has been segmented into regions of reduced size (Fig. 2). With these reduced regions the operator can be implemented by means of a PALs structure connected in cascade. The input of the PALs corresponds to the pixels of the image that can belong to a straight line in a given orientation, 18 PALs of 14 inputs and 4 outputs, working in parallel over a 10x10 pixel window, are used (Fig.3).

Programming all of these PALs and matching each one of them to the parts of a straight line, 38 patterns are defined and the straight line orientation can be determined.

The minimum length of the lines to be detected can be programmed to be between 6 and 10 pixels by using two of the inputs of the PALs for external control.

In order to increase the minimum length of the straight lines to be detected without increasing the hardware complexity, that is, the number of PALs, a sampling of the pixels of the image can be performed by taking 1 pixel among 2 or 3. The minimum length can be expanded to 20 or 30 pixels. The length can be expanded to 20 or 30 pixels. So the length range to be programmed can be adjusted between 6 and 30 pixels.

Once a straight line has been detected it is necessary to filter the scene information in the sense of keeping only straight lines of lengths greater than a previously defined size. So then the pixels belonging to straight lines have to be validated while the rest of the pixels of the image have to be eliminated.

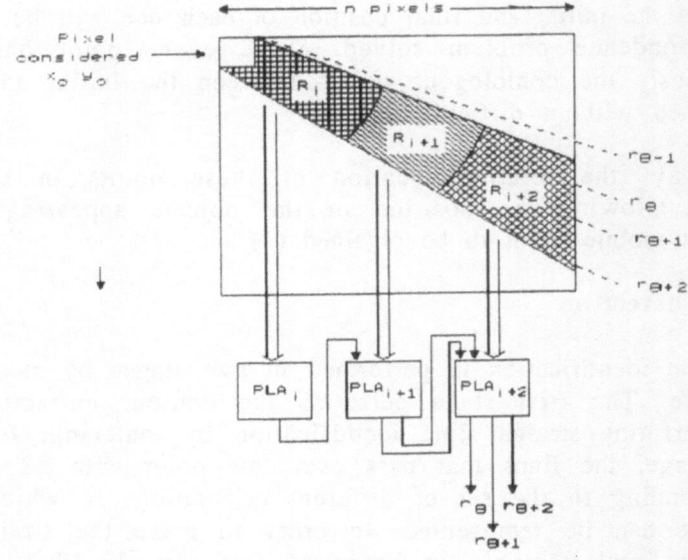

Fig.2 Segmentation of the window

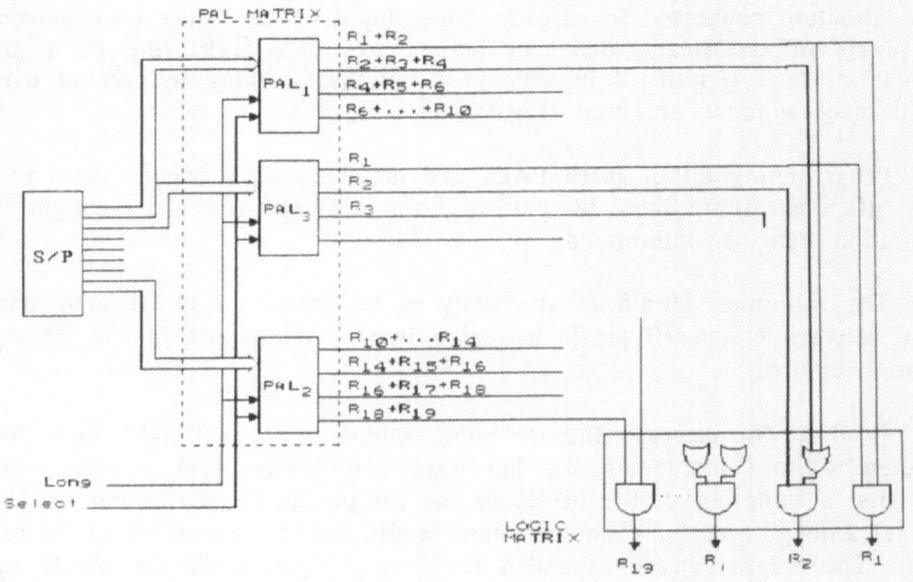

Fig.3 Parallel sampling of the window

Figure 4 shows an image of a road and Figure 5 shows the same image after straight line detection.

Fig.4 View of a road from the "Eixerit"

Fig.5 Straight line detection

Tracking of a preselectable target

The real-time tracking of a mobile target for the longest period of time possible by moving, if necessary, the sensor system has been adopted as a main research line in many research laboratories (3).

The tracking module's main goal is to track a preselectable object of the scene that is sufficiently differentiable from the background and to supply its coordinate continuously.

The recognition and tracking module developed is based on a processor that can determine the point of maximum similitude between the funtion $P_{x_0, y_0}(\alpha)$ Fig.6) obtained at the moment of the acquisition of the image and the function $P_{x,y}(\alpha)$ obtained from the different points of the environment of $(x_1 \; y_1)$ and contained in a predefined window, corresponding to the maximun environment considered as the possible zone of displacement of an object from one image to the following one.

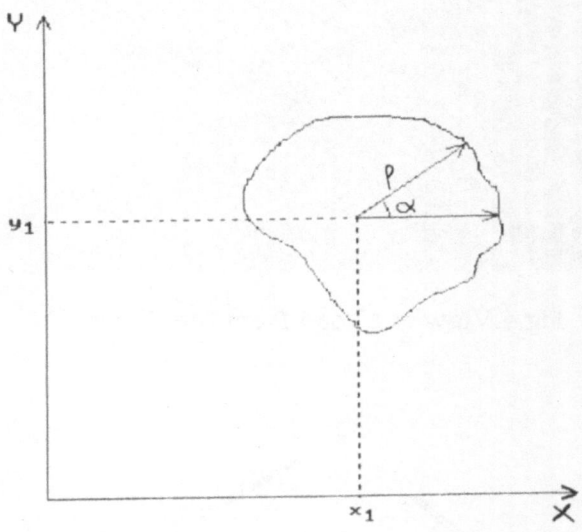

Fig.6 Polar sampling of a scene

This procedure can perform the recognition of a pattern not previously determined, but with some constraints.

The necessary condition to assure the pattern recognition and tracking is that its shape be sufficiently differentiable from the rest of the image of its environment. Due to the fact that the shape of the object to be tracked can apparently change all along its movement, the model is updated with a given time constant, adapting itself to slow variations of shape and size. Figure 7 shows the structure of this module.

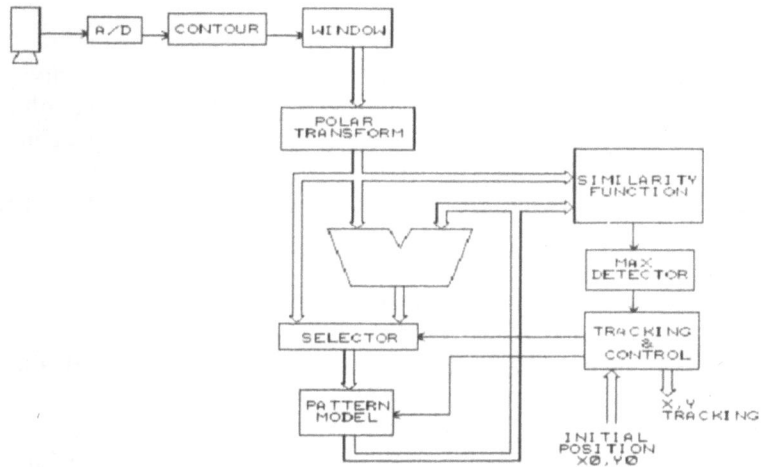

Fig.7 Structure of the tracking module

The size of the window for evaluation of the coincidence function depends on the relative dynamics expected of the target trajectory.

The system is able to recognize and track objects with a wide range of different shapes and can operate in the mode "watch and track". Then its location in space can be calculated by means of two simultaneous observations.

From the tests carried out it has been observed that the precision obtained in the location of the target in every image is not very good, but in spite of this, tracking it all along its course allows extraction of the trajectory followed. The trajectory calculations are based on a polynomial adjustment that allows the system to track it, and even to refind the target after it has been lost due to an occlusion, provided that its trajectory does not vary too much from the foreseen, extrapolated trajectory.

In a second level another processor performs the evaluation of the mobile's position, affecting both the corresponding scale and coordinate changes, and pondering over the weight of the results obtained from each preprocessor as a function of the degree of similitude obtained from previous hypothesis of maximum coincidence.

4.   FUSION

In the first stage the problem that we were trying to solve was to enable the navigation of an obstructed but possible path through the use of a previously programmed trajectory or the tracking of a specific object.

In order to achieve these goals we utilized the following preprocessors:

- Texture analyser for path detection.

- Straight line detector to measure accurately the limits of the path to be followed.

- Vertex detector for description of obstacles with as little time as possible.

- Tracking module (two units) to calculate the position of a significant part of an object.

The four preprocessors used in this experiment supply us with sufficient information about the area to allow us to identify the region of interest with a high level of integration.

The integration of the information is achieved by entering in a blackboard (4) the information coming from the ultrasonic sensor as well as the data acquired by the image preprocessors. By operating in this way during the movement of the robot it is possible to create a model of the probability of found obstacles. The quality of the "passable path" and the quality of the "texture" are acquired through previous knowledge of the model, consultation with the master or perception of the difficulty of movement (power consumption).

The probabilistic map of the path is built primarily by the absence of obstacles detected by the ultrasonic sensor and the uniformity of texture between boundary lines. Figures 8   a,   b,   c show examples of a path with obstacles, without obstacles, and with a probability of found obstacles. A map of velocity can then be made based on this information.

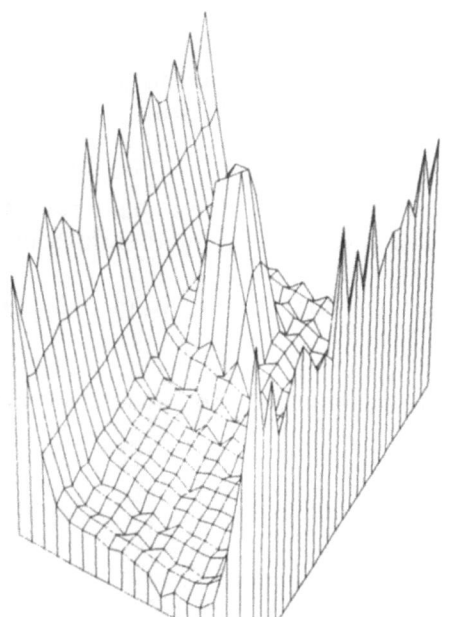

**Fig.8a Path with high
probability of
found obstacle**

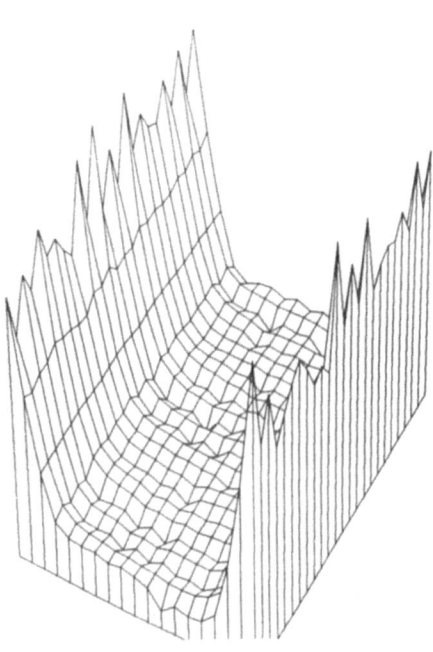

**Fig.8b Path with low
probability of
found obstacle**

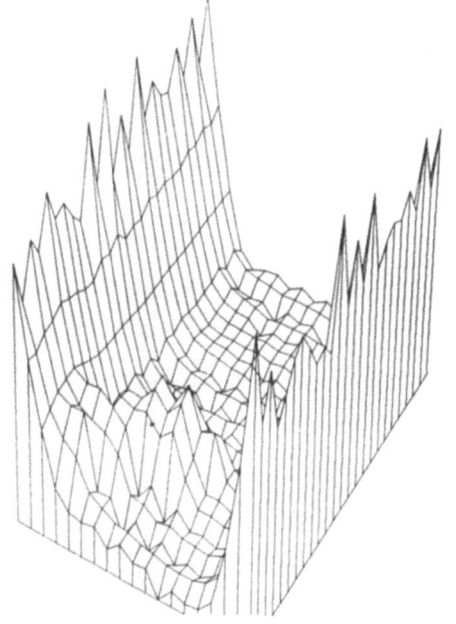

Fig.8c   Path with some probability of found obstacle

RESULTS

Experiments in mobile robot road-following were performed   We succeeded with images of city streets tested. We succeeded with images of country paths.

The uncertainty model probability of found obstacles had several constraints; for example, we realised that shadows are the main drawback.

We also tested the tracking module and we had success if the contour was well defined.

We are presently investigating methods to improve the model   of the possible path.

REFERENCES

(1)   Amat, J., Casals A:
      Visual inspection system with qualitative analysis capabilities.
      NATO ASI Series F, vol.52, Springer-Verlag, Berlin-Heidelberg-
      New York, 1989

(2)   Amat, J., Llario, V:
      A vision system with 3D capabilities.
      IEEE Int. Conference on Robotics and Automation.
      March 1985, St. Louis, USA

(3)   Bird Bhanu and Durga Panda:
      Qualitative reasoning and modeling for robust target tracking
      and recognition from a mobile platform.
      Image Understanding Workshop, 1988.
      Honeywell System & Research Center

(4)   Blackboard Workshop
      Carnegie-Mellon University
      June   1986

BIBLIOGRAPHY

-     Goto, Y., Stentz, A.:
      Mobile robot navigation: the CMU system.
      IEEE Expert (1987)

-     Thorpe, C. et al.:
      Vision and navigation for Carnegie-Mellon
      NAVLAB.
      IEEE Transaction on Pattern Analysis and Machine
      Intelligence,   May 1988

-     Poggio, T. et al.:
      The MIT vision machine.
      Image Understanding Workshop, 1988
      Honeywell System & Research Center

NEURAL SIGNAL UNDERSTANDING FOR INSTRUMENTATION

L.F. Pau, F. Johansen
Bldg. 348/EMI, Technical University of Denmark
DK 2800 Lyngby, Denmark (R 377)

ABSTRACT:
This paper reports about neural signal interpretation theory and techniques, for the purpose of classifying the shapes of a set of instrumentation signals, in order to calibrate the device, diagnose anomalies, generate tuning/settings, and to interpret the measurement results. Neural signal understanding research is surveyed, and the selected implementation is described with its performances in terms of correct classification rates, robustness to noise. Formal results on neural net training time and sensitivity to weights, are given. A theory for neural control is given using functional link nets, and an explanation technique is designed to help neural signal understanding. The results hereof are compared to those of a knowledge based signal interpretation system, within the context of the same specific instrument and data.

Keywords: Neural understanding, Calibration, Signal understanding, Control theory, Neural control, Training time, Sensitivity to noise, Explanation facilities, Knowledge based signal interpretation, Instrumentation, Analytical instrumentation.

1. INTRODUCTION

This paper reports about a project involving neural processing for the purpose of classifying the shapes of a set of instrumentation signals $y_i(t)$ i = 1,..,N, in order to calibrate the device, to diagnose anomalies/wrong settings, and to gen-

NATO ASI Series, Vol. F 63
Traditional and Non-Traditional Robotic Sensors
Edited by T. C. Henderson
© Springer-Verlag Berlin Heidelberg 1990

erate appropriate tuning settings, and to interpret the measurement results, altogether c = 1,.., C classes.

The sensor(s) in the instrument is calibrated by exposing it to a standardized but controlled environment, with control parameters U (t) = {u_i (t), i = 1,..,p} (e.g.: flow, gain, bandwidth, delay, ...). The output of the calibration runs are N 1-D signals y_i (t) i=1,...N depicting the variations over time of a number of heterogeneous responses.

In instrument operations, likewise, signals y_i (t) are recorded for given U (t), except that the sensors, the environment, and the samples can be different or unknown.

Untill recently, a skilled operator was needed to interpret the shapes of the N signals jointly, to relate them to changes to be made in the controls U (t), and to past behaviour/settings of the sensors & electronics. These signal shape changes were supplemented by a diagnosis (hypothesize-then-test) about the functional faults, if any, of the sensor and/or of the set-up.

Three approaches are basically possible [18]:

A   signal feature extraction, and neural processing hereof, with training on known calibration and test data, and later neural classification [2]

B   syntactic pattern recognition of the curve shapes, on the basis of formal symbol grammars, and analysis of the parsed symbol strings by diagnostic knowledge based rules [1] [3].

C   identification of a dynamic model of the (U(t), y_i(t)) relation, in a multi-model framework, with each model corresponding to a separate functional behaviour of the whole instrument [4] [5]; two subclasses co-exist here, one where the in-built control laws in the instrument are known, and one where they are not.

The approach C seemed not to work, so we will report mainly about the approach A (Sections 2-5), and compare it briefly with B, which was implemented as well on exactly the same practical case (see Section 6).

For an introduction to neural processing, the reader is referred to introductory references such as [6] [7] [17].

2. SURVEY OF RELATED NEURAL SIGNAL UNDERSTANDING WORK

Neural signal understanding involves three major issues:

- signal knowledge representation or feature selection, to be presented to the inputs of the neural network

- selection of the neural network architecture

- adaptive control and classification issues, in view of the controls U(t).

2.1 Neural Signal Representation

Four approaches prevail here (Figure 1):

i   signal feature selection, which means defining a battery of feature extraction function φ_j j=1, ..., n applicable to signals:

$$\varphi_j : \{\{y_i (t), t \in [0, T]\}, i = 1, ..., N\} \longrightarrow R$$

Examples are: minimum, maximum, morphomathematical signal features, first spectral component, Karhunen-Loewe expansion components, signal moments, polynomial coefficients, etc. [2] [8].

ii   serial order description [9], consisting in defining a state consisting of two elements:

- a vector of measurements $(Y(t) \triangleq \{y_i (t), i = 1,...N\}$ for any given t.

- a plan P whereby these vectors Y(t) are updated:

$$Y(t+1) = f (Y(t), P)$$

where typically P includes U(t) but also scheduling choices
$(P \ni U(t))$

In this case, there is no explicit representation of temporal order and no explicit representation of action sequences; the input nodes then receive the states (Y(t), P)

iii   an explicit signal representation, where there is one input node attached to each signal sample $y_i(t)$, leading to a total of $n = N \times T$ input nodes [10] [11]

iv   a functional signal expansion, where all signals $y_i(\cdot)$ are series expanded in terms of a basis of known kernel functions h_ℓ $\ell = 1, .., L$

$$y_i(t) = \sum_{l = 1,L} a_{ilt} h_l$$

Examples are: polynomial expansions (Legendre, Hermite), Walsh functions, sinus functions, Fourier functions, etc...[6] [12]. The signal is then represented through $n = L \times N \times T$ coefficients $\{a_{i\ell t}\}$.

Later in this paper, we assume the default signal representation be be iii; changes can readily be made for further developments to accommodate other representations instead.

2.2 Neural Signal Understanding Network Architecture

The selection of the network architecture is, together with the signal representation, paramount for its learning capabilities, convergence, etc.

Most work refers to the back-propagation algorithm [7] [15] [17] operating on a perceptron with hidden layers, z_k, k=1, .., K where the weights $\{w_\ell\}$ control adaptation formulas inserted at each hidden layer node. The input elements are usually kept as linear elements in order to provide for a good dynamic range, while sigmoid elements are used for hidden or output nodes. Sometimes the output nodes are also linear.

Typical adaption formulas inside the hidden nodes are:

i   least mean squares / adaline [11]:

$$z_k = z_{ko} + \sum_{j = 1}^{a} b_j \, y_k \, (t - j \cdot \Delta t)$$

where z_{ko} is the threshold, and $\{b_j\}$ are determined as to minimize in the least square sense the expected error:

$$E = \int \sum_k \left(y_k(t) - \sum_{j = 1}^{a} b_j \, y_k \, (t - j \cdot \Delta t) - z_{ko} \right)^2$$

ii generalized delta rule with self-decay [9], e.g. with a logistic function:

$$z_k = (max - min) / (1 + exp(-y_k(t)) + min$$

where min is the minimum value attained by the logistic function (and therefore the minimum value of activation that any unit can have), and max is the maximum value, or with an explicit decay weight $((\mu/w) < 1$, $\mu < 1)$:

$$z_k(t) = \mu z_k(t-1) + w_{kt} y_k(t)$$

iii functional link network [6], with the gradient minimization operating on the weights applicable to different time-slices $[T_1, T_2]$ and signals y_j $(\cdot)$, thus emphasizing some more or less according to the magnitude of the weights.

3. NEURAL SIGNAL UNDERSTANDING FOR INSTRUMENTATION

For the task defined in Section 1, the following approach was selected in one specific instance of analytical instrumentation [16].

3.1 Signal Classes

The C (C = 9) signal classes were taught by labelling observed evolutions $\{y_i(t),$ $i=1,..,N$ $t=1,...,T)$ by their class labels and U(.) plan/control labels. The issue of the selection of training examples was complicated by the fact that more than one label was often assigned to these samples, enforcing the need for further post-processing of the neural classification outcomes [13].

3.2 Neural Signal Representation

A signal feature selection was selected (see Section 2.1. **i**):

- the N signals $y_i(t)$ were first passed through a second order derivative filter in order to find points of inflexion and curvature changes;

- the ordered lists φ_j of such positions, and their number, were then entered into the network.

More evolved similar procedures for curve feature extraction, involve acquiring the discrete tangent field by taking equidistant curve slices, estimating the tangents and curvature at the slice end points on the curves [16]. Orientation discontinuities appear as multiple tangents at a position, with no curvature estimate. Splines can then be matched to said intersection points.

3.3 Neural Signal Understanding Network Architecture

A layered feed-forward network with back propagation training rule [7] [15] [17] without decay (Section 2.2 ii)) was used and implemented in the self-developed NETSIM package (written in C) [10].

3.4 Neural Classification Performance

Six series of experiments were performed.

a. calibrated signal data, n=20 input nodes, K=20 hidden nodes in one hidden layer, and C=6 output nodes

b. as a., but with one additional hidden layer with K=20 nodes

c. non calibrated signal data, otherwise as a.

d. calibrated signal data, n=15 input nodes, one hidden layer of K=15 nodes, and C=6 output nodes

e. additive noise added to the calibrated signal data a.:

$$y_j'(t) = y_j(t) + \beta \cdot n(t)$$

where β is a constant, and n(t) white noise in the interval [-1, +1]

f. multiplicative noise applied to the calibrated signal data a.:

$$y_j^{''}(t) = (1 + \alpha \cdot n(t)) \, y_j(t)$$

These results were obtained by leave-one-out testing on the entire training set. The results [10] yield the following correct classification results:

a: 65%   d: 40%

b: 60%   e: as **a.** with $\beta < 0.0369$

c: 55%   f: as **a.** with $\alpha < 0.1777$

It also appears that calibrated training signals are less sensitive to multiplicative noise than non-calibrated ones, whereas the latter are less sensitive to additive noise.

3.5 Training Time and Network Sensitivity

A formal study of the convergence of the selected multilayer feed-forward network was established [10], leading to the following result, to be justified in a subsequent paper:

Hypotheses: - $(\lambda + 1)$ is the number of layers

- n_j is the number of neurons in layer j, where $j=0$ is the input layer

- O_j is the output of neuron j

- I_j is the input to neuron j, equal to O_j in layer $j=0$, otherwise equal to 0

- $w_{\ell k}$: is the weight applicable to the link between neurons ℓ and k

- $a_l = \theta_l + \Sigma \; O_j^* \; w_{j\ell}$, $j=1,...,n_\ell - 1$ is the weighted sum of the inputs to neuron ℓ , having threshold value θ_ℓ, when this neuron is in layer ℓ

- the neuron response function is the logistic function $f(x) = 1/(1 + \exp(-x))$ (Fig.3)

- $\phi \triangleq 1/4$ $\delta \triangleq 4/27$ as related to bounds on the logistic neuron response function, and its derivative [10].

Conclusions:

1)   the sensitivity of the network output O_i in layer λ, to changes in the input values I_j in layer $\lambda=0$, is bounded by:

$$\frac{\Delta O_i}{\Delta I_j} \leq \phi^\lambda * \sum_{k_n}^{N_\lambda - 1} \left(w_{k_n i} * \sum_{k_{n-1}}^{N_\lambda - 1} \left(w_{k_{n-1} k_n} * (\dots) \right) \right)$$

2)   the sensitivity of the network output O_i in layer λ, to changes in the weight w_{jk} between nodes at two successive layers $(\ell-1)$ and (ℓ) is bounded by:

$$\frac{\Delta O_i}{\Delta w_{jk}} \leq \phi^{(\lambda - l + 1)} * \left(\sum_{l_n}^{N_\lambda - 1} w_{l_n i} * \left(\sum_{l_{n-1}}^{N_\lambda - 2} w_{l_{n-1} l_n} * \dots \right) \right)$$

3)   the training time τ is proportional to:

$$\tau \sim 1/(\text{incr.} \, \upsilon^{1/2} \cdot D_{min}^{1/\lambda}$$

where
$\upsilon \triangleq E(n_j)$ is the average number of neurons in a layer;
$D \triangleq$ is the minimum euclidian distance between any two training points $y_j(t)$ belonging to two different classes amongst the C possible
incr. $\triangleq E(\Delta w)$ is the average weight change between two iterations.[10] shows a good correlation of this curve with real training time data.

4. NEURAL CONTROL BY FUNCTIONAL LINK NETS

4.1 Introduction

As mentioned in Section 1, the instrument operator will, after calibration, trim some controls U(t) to achieve desired performances, while also complying with specific sensor and electronics behaviour.

Examples of such controls U(t) are: gain, offset, and flow of titrant/solvent.

In this Section 4, we give the formal process whereby to synthesize these controls U(t) by neural processing, assuming the net architecture to be the functional link net [6]. These results are original and of a general nature; therefore, the instrumentation application will not be emphasized in the remains of this Section.

4.2 Functional Link Nets

Functional link nets [6] are essentially characterized by:

- a single layer of input neurons

- a single aggregation node with a non-linear response

- the expansion of the input layer by functional links which are kernel functions of selected components of the input vector X(U,t); as a result the N nodes of the input layer are the N functional links $f_i(X(U,t))$, I=1,..,N   (including evt. $f_i = x_i$ (U,t) ,i-th component of the vector X)

- the back propagation search rule which updates linear weights w_i, i=1, ..., N.

Whereas functional nets allow for supervised learning, unsupervised learning, as well as classification, they provide a formal framework for the synthesis of controls while also encompassing key notions from traditional control theory. This gives the capability of re-using neural processing hardware and firmware, not only for perceptual or learning tasks, but also to replace programmable logic controllers and other control devices (PID controls, etc.).

4.3 Neural Control Theory

Consider a functional net, with inputs X(U,t) subject to controls U(t), and a flat net with N functional links f_i (X) (see Figure 5).

The target z(U,t) and threshold θ require that the linear weights w_i in the net be such that:

$$w_1 \, f_1 \, (X(U, t)) \, + \, .. \, + \, w_N \, f_N \, (X(U, t)) \, = \, z \, (U, t) \, - \, \theta$$

This defines the vector $W(t) = [w_1..w_N \; 1]$ as orthogonal to a manifold subject to the controls $U(t)$; as a result, in line with identification theory, $W \triangleq [w_1..w_N \; 1]$ is a dual control $W(t)$, or point in the w-control space dual to the U-control space.

However, the neural learning schemes whereby $W = [w_1..w_N \; 1]$ is determined, such as a gradient search or in some cases a recursive pseudo-inverse calculation, are of much wider applicability in practice than optimal control schemes, which also suffer from a high numerical complexity.

The neural control functional net will in term operate on functional links $g_1, ..., g_N$ applying to the input $W\,(t)$, with weights $v_1,..,v_N$:

$$v_1 \, g_1 \, (W\,(t)) \, + \, ... \, + \, v_N \, g_N \, (W\,(t)) \, = \, T(t) \, - \, \psi$$

where $T\,(t)$ is the target. Typically this target will deal with global performance criteria of the overall system, such as:

- response time whereby the target $z(U,t)$ is met

- energy level related to the control energy

$\int U(t) \, dt$

Being dual to $W\,(t)$, the weights $V(t) = [v_1,... \; v_N \; 1]$ will belong to a space isomorphic with the original functional space $f(X(U,t))$, of which $V(t)=U(t)$ is a special case.

The above formalism, opens up for three capabilities:

A. by classifying over time the vectors $W(t)$, typically with fixed discriminants trained by a dedicated supervised net, then one can readily select those functional link groups (i.e. subsets of $f_1,..,f_N$) achieving desired properties of the system as chosen via the classification net training on the pairs (W, class (W)), where class (W) is the classification outcome.

B. the dual net V will in fact enforce on the learning net W, long term properties over time, or measured in terms of error on the position $X(U,t)$ itself, by assigning the weights V to belong to fixed domains in the X-space or U-space.

C. temporal properties will be explicitly available by tracking the changes over time of W(t).

5. EXPLANATION FACILITIES TO NEURAL SIGNAL UNDERSTANDING

For the calibration, set-up, diagnostic and measurement interpretation tasks, the instrument operator almost invariably wants a support system to justify its advice through explanation facilities such as: why? how? why not? Unfortunately, such facilities do not exist in neural processing. We therefore describe the formal and practical solution developed to address this issue: ~

0. All inputs X(U,t) are scaled to belong to the range [-1,+1]

1. Our neural explanation facility involves asserting as Prolog facts the weights w_i, i = 1, ..., N (obtained after convergence), as well as the corresponding signal classes N(X(U,t)) = c   c = 1,...,C

2. A sorting utility ranks the weights w_i i=1,...N by decreasing values of $|w_i|$, eventually signed

3. A sorting utility ranks the classes c by decreasing frequencies in both the learning and operational neural signal interpretation modes, for given weights w_i i=1, ..., N.

4. Implement in Prolog standard back chaining or forward chaining explanation facilities operating on the sorted fact bases **2.** and **3.** of hypothesis and goals, respectively.

The result is to allow to select as "why" explanations, those neural input nodes (meaning those inputs x_i (U,(t) or functional links f_i), which have the largest weights and thus contributions to the output neuron responses. Likewise, the "how" explanations will be those classes for which, given the current weights, the classification frequencies will be the highest.Of course, in both cases, text generation in Prolog allows us to relate text to those two explanations described above. By selecting the number of items of highest ranks, the user can request rough meaningful explanations, or comprehensive ones. Furthermore, a knowledge base may filter out the alternatives to eliminate non significant explanations in causality chains, and to generate the corresponding root explanations in such chains.

6. COMPARISON WITH KNOWLEDGE BASED INSTRUMENTATION SIGNAL INTERPRETATION

Fortunately, the neural signal understanding set-up described in Section 3 in a specific case, could be compared with a knowledge based signal interpretation system working with the same instrument and data [16].

The major differences between the two solutions were, in the case of the knowledge based system:

- a signal interpretation inference procedure, based on conceptual graphs, and implemented in Prolog

- an intelligent user interface implemented in Smalltalk, involving logbooks, application notes, sensor behaviour monitoring

- a signal feature extraction procedure (similar to Section 3.2), implemented in Pascal

- an explanation facility, implemented in Smalltalk and Prolog.

The knowledge based signal classification performance, estimated on the same data as in Section 3.4, was a minimum 92% correct classification on calibrated or non-calibrated data.

The knowledge acquisition time was close to 6 man months, compared to a few days in Section 3.4.

The knowledge based signal interpretation sensitivity to additive or multiplicative noise (see Section 3.4) was:

$$e = 74\% \qquad f = 56\%$$

which indicates a better robustness of the neural signal interpretation procedure, for the given knowledge available.

Regarding generating controls U(t), of which two were stationary and a third was a linear time dependent function, the accuracy achieved by the approach of Section 4.3 was higher than that of the knowledge based approach which had a tendency to give class-related settings only (and not measurement related).

In terms of explanation facilities, the knowledge based system was clearly superior in terms of accuracy and depth of the justifications; however, the neural signal explanation facilities of Section 5, executed with the first five ranked candidates, was quite sufficient for all practical purposes.

7. CONCLUSION

Neural signal understanding for instrumentation is a promising technology, esp. when the sensors or processes drift or have memory. It is less performant than the knowledge based signal interpretation approach, which however assumed an extensive knowledge acquisition process (not always practical). The neural training time is still far too high on hardware usually implemented in or next to instruments. Key to better overall performances are improved signal feature extraction procedures, possibly combining knowledge based and neural processing, and high-speed coprocessor boards.

However, depending on operations of the instrument, this may be a minor problem since the neural network only needs to be trained once for each operational setting.

REFERENCES

[1]   K.S. Fu, Syntactic pattern recognition, Academic Press, 1975

[2]   L.F. Pau, Failure diagnosis and performance monitoring, Marcel Dekker, N.Y., 1981

[3]   K.S. Fu (Ed), Applications of pattern recognition, Chapter 6, CRC Press, Boca Raton (FL), 1980

[4]   B. Widrow, S. Stearns, Adaptive signal processing, Prentice Hall Inc., Englewood Cliffs, NJ, 1985

[5]   H.W. Sorensen (Ed), Kalman filtering, IEEE Press, N.Y., 1985

[6]   Y.H. Pao, Adaptive pattern recognition and neural networks, Addison Wesley, Reading (MA), 1989

[7]   R.P. Lippmann, An introduction to computing with neural nets, IEEE ASSP Magazine, April 1987, 4-22

[8]   C.H. Chen (Ed), Digital waveform processing and recognition, CRC Press, Boca Raton (FL), 1982

[9]   M.I. Jordan, Serial order: a parallel distributed processing approach, ICS Report 8604, Institute for cognitive science, Univ. California San Diego, La Jolla, CA 92093, May 1986

[10]   F.S. Johansen, Neural signal tolkning (in Danish), M.S. thesis (EMI) Technical University of Denmark, August. 1988

[11]   A. Lapedes, R. Farber, Non linear signal processing using neural networks: prediction and system modelling, T.R. Los Alamos National Lab., Los Alamos, NM 87545, July 1987

[12]   D. Gabor et al., Proceedings IEE, Vol 1088, July 1960

[13]   S.I. Gallant, Connectionist expert systems, Comm. of the ACM, Vol 31, no 2, Feb. 1988, 152-170

[14]   K. Fukushima, A neural network for visual pattern recognition, IEEE Computer J., vol 21, no 3, March 1988

[15]   W.P. Jones, J. Hoskins, Back propagation: a generalized delta learning rule, BYTE, Oct. 1987, 155-162

[16]   KEMEX, Technical report, Industry and Technology agency, Copenhagen, Nov. 1988

[17]   D.E. Rumelhart, J.L. McClelland (eds.), Parallel Distributed Processing: Exploration in the Microstructure of Cognition, MIT Press, 1986

[18]   L.F. Pau, Intelligent man-machine interfaces in instrumentation, Proc. 1988 Finnish Artificial intelligence conference STEP-88, Publ. Univ. Helsinki, Vol. 1, pp. 12-19.

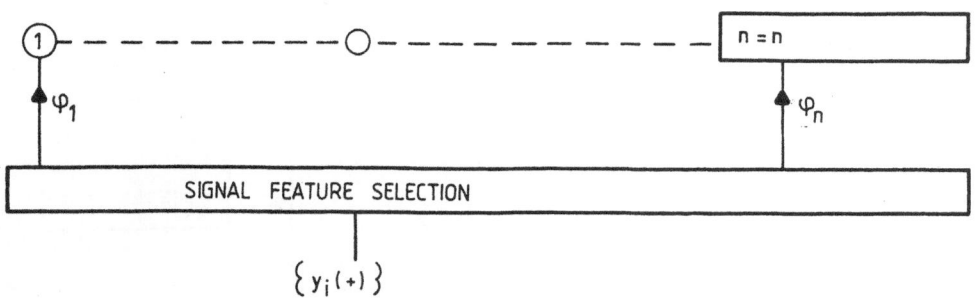

$$\{\, y_i(\cdot)\,\}$$

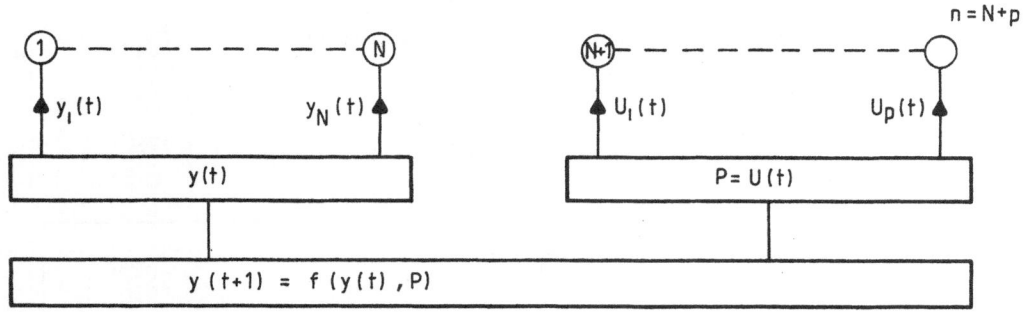

SERIAL ORDER DESCRIPTION

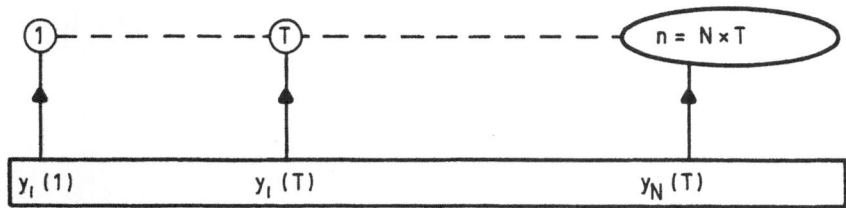

EXPLICIT SIGNAL REPRESENTATION

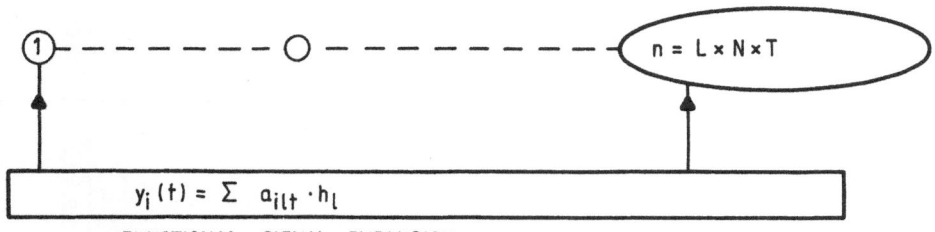

FUNCTIONAL SIGNAL EXPANSION

Figure 1. Neural signal representation

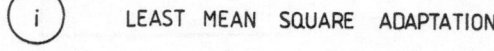

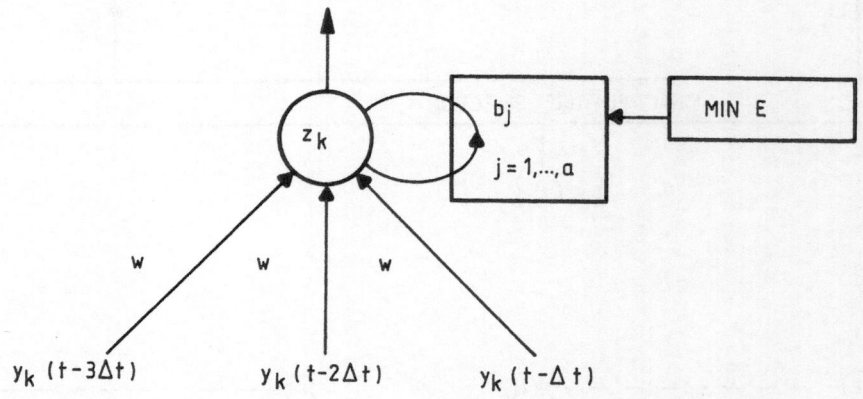

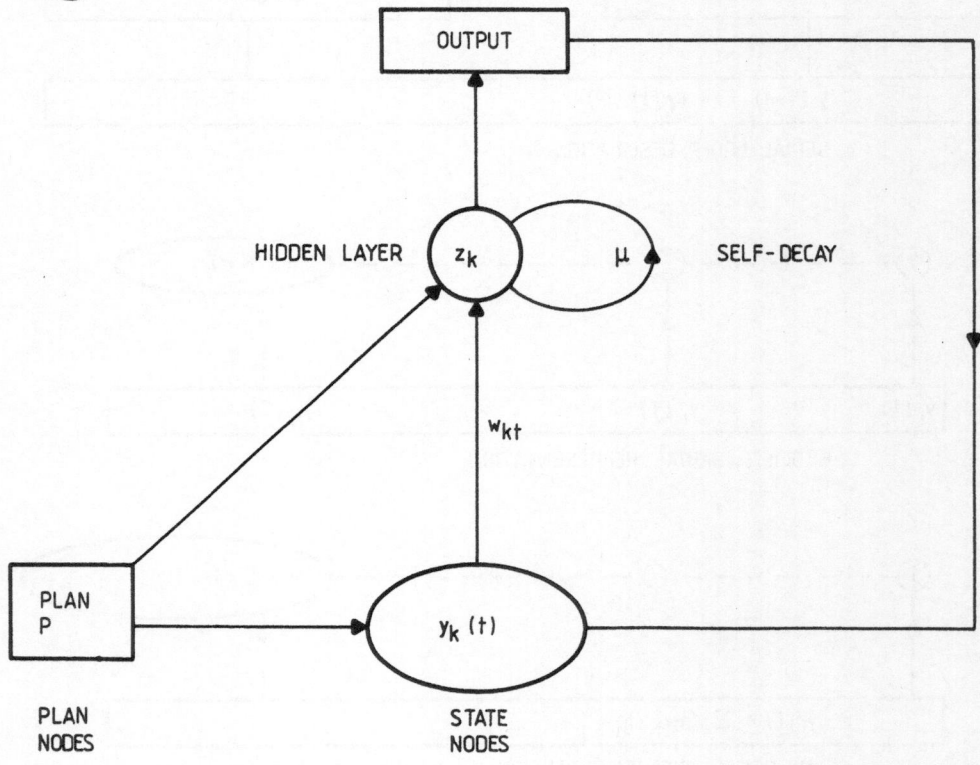

Figure 2. Neural signal understanding network architectures

Figure 3. Neural processing element

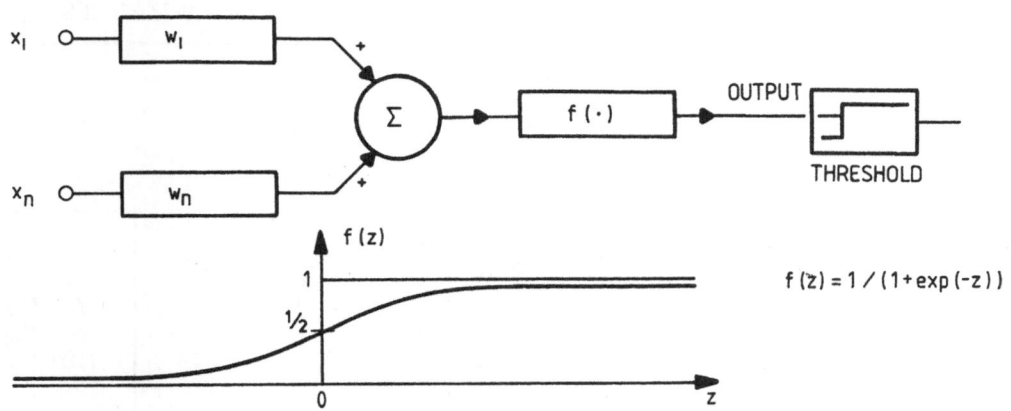

$$f(z) = 1 / (1 + \exp(-z))$$

Figure 4. A $\lambda = 3$ layer feed forward neural network

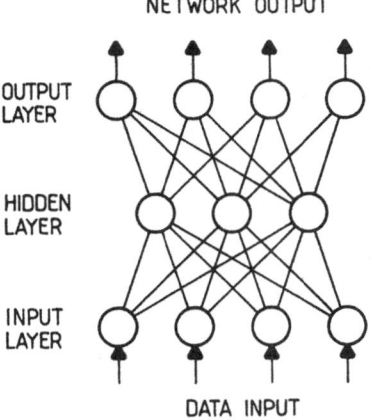

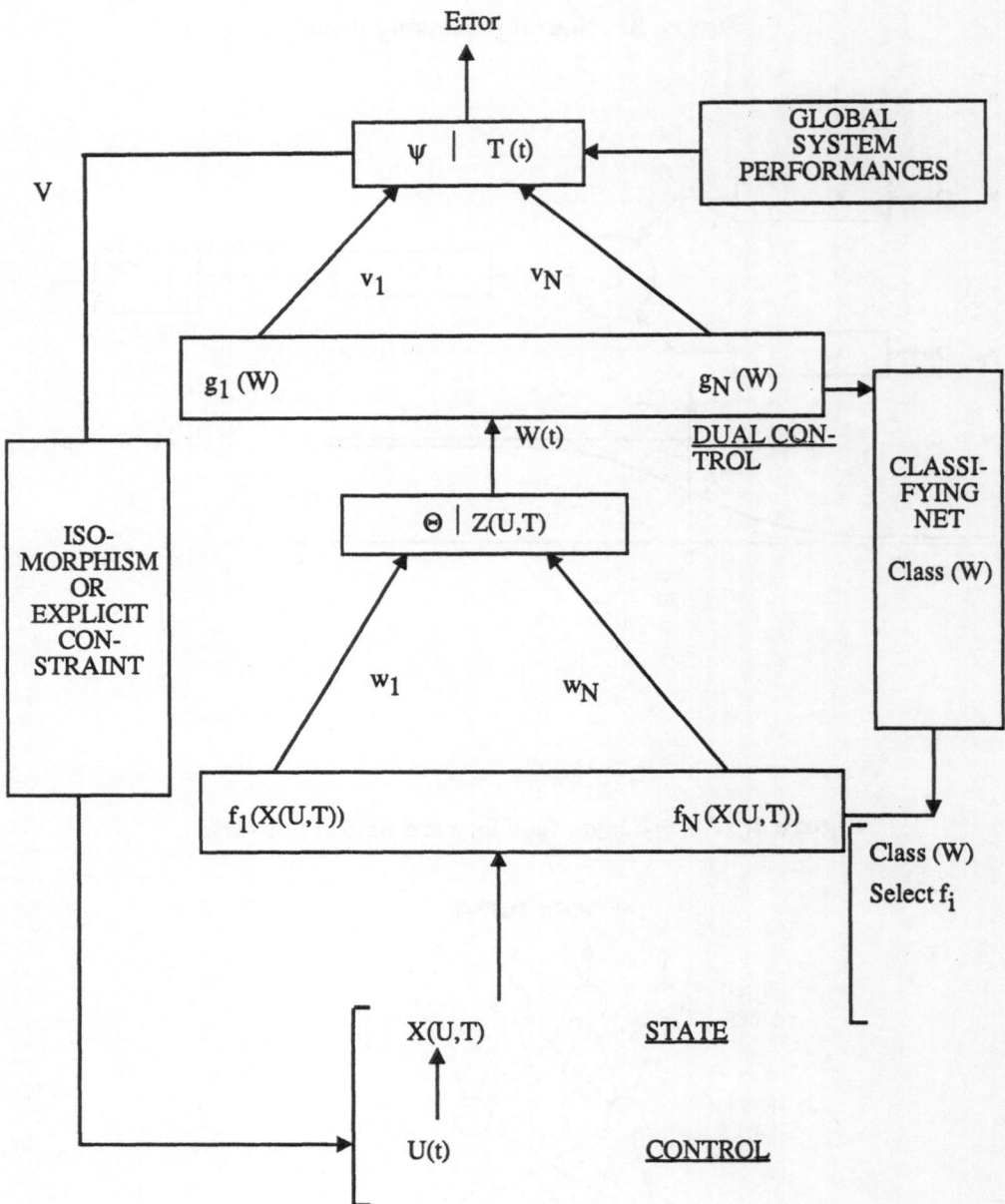

Figure 5.   Neural control by functional link net

AN APPROACH TO REAL-TIME ON LINE VISUAL INSPECTION

Vicenç Llario[1] and Jordi Sanromà[2]

[1] RALUX, S.A. Cai Celi 11, 08014 Barcelona, Spain
[2] EROVI, Enginyeria de Robòtica i Visió, Computer Vision-Image Processing Group, Passeig de Gràcia 11B, 4rt, 08007 Barcelona, Spain

Abstract

In this paper we present some applications of computer vision in well defined industrial environments. In most of those applications the main constraints we have to cope with are, on one hand, real-time response, and on the other hand uncertain illumination conditions intrinsic to this kind of environments. As far as our experience has shown (we suppose everybody has experienced the same), it is very difficult , while developing the vision system in the laboratory, to take into account all the external factors which will influence the performance of the system on the factory floor. Factors such as illumination fluctuations, dirt, noise, vibrations, and high temperatures, among others, will inevitably lead to an unpredictable behavior of the system and as a consequence to a loss of reliability. In order to face those drawbacks we approach every application with a defined methodology which mainly consists of two phases: modelling and design.

During the modelling phase we establish a theoretical model of the process to be inspected as well as the constraints to be considered. This procedure allows us to implement what is commonly called a local analysis of restricted areas of the image by means of special purpose hardware. The principal goal of this process is to attain the best trade-off between speed and reliability.

NATO ASI Series, Vol. F 63
Traditional and Non-Traditional Robotic Sensors
Edited by T. C. Henderson
© Springer-Verlag Berlin Heidelberg 1990

During the design phase we must choose a strategy consistent with the model in order to guarantee the efficient selection of primitives for the final implementation.

1 Introduction

Inspection is a crucial step in most industrial production processes. The main reasons or incentives to apply automatic inspection techniques are, on one hand, to avoid the need for a human to work in dangerous or hazardous environments, and on the other hand, to attain a better mean quality in the inspection process derived from the intrinsic features of an automatic inspection system.

It is assumed that the most difficult task for inspection is the one concerned with visual appearance. The role of visual inspection is double because the system has to be able to discriminate not only functional defects but also cosmetic disparities. While the human visual system is adapted to deal with variety and change, in general the visual inspection process requires analyzing the same type of images repeatedly to detect anomalies. Several studies show that human performance decreases dramatically when dealing with dull, boring and endlessly routine jobs. The direct consequence of this is slow, expensive, heterogeneous and erratic inspection. As a matter of fact, the principal motivation behind automatic visual inspection is increased flexibility and lower cost. What we mean when we say increased flexibility is that it is commonly accepted that those techniques imply increasing productivity and improving product quality. In a lot of applications human inspectors are slow compared to modern production rates and as a consequence the error rate increases dramatically. One of the goals of automatic operation is to match high speed production with high speed inspection and at the same time provide means for monitoring and diagnosis in realtime.[18], [36].

The design, development and implementation of a specific application implies a defined methodology. The role of those methods used to face a real application is to minimize development time, start-up time and consequently overall costs and risks. To attain these goals it is crucial to use efficient,comprehensive and accessible development tools.[27]

We will describe two specific applications which show the competence of vision processing in industrial environments. The characteristics of these applications are:
- 100% inspection
- The inspection process has to be automatic
- Real time response
- Good repeatability
- Including error measurement

2 Calculation of the Growing Parameters of Antibiotic Fermentation Cultures

At this point we will present a system developed to deal with the recognition and characterization of circular geometries which was recently applied to the determination of the growing parameters of fermentation cultures.

It is well known that there are different approaches to deal with the detection of circles and the calculation of their parameters. Some developments are based on the geometrical properties of circular shapes [26],[12]. Another group of researchers has been working with heuristic operators to detect feature regions of the image,[24], [16]. Finally, there is a set of works using the transformation of the image space which is mainly based either on the Radon transform or on the Hough transform.

The Hough transform was introduced in 1969 by Rosenfeld [25]. The transformation, initially conceived to detect straight lines, was later extended to detect curves [10],[34], and arbitrary shapes [2]. There are a lot of authors who have used some kind of Hough transform (in general modified versions to attain a better trade-off between processing time and overall performance in terms of reliability).[14], [4], [5], [31], [22], [6], [21], [7], [38]. There has been some controversy around the origin of the Hough transform, for instance Stanley R. Deans [9] shows how this transformation can be directly derived from a more general one, the Radon Transform [13], [19], [8].

We used a modified version of the H.T. which allowed us a final hardwired implementation to get the best compromise between speed and reliability. The global constraints for the calculation process are the following:

- Samples located on different carriers
- The samples lie on varied color gelatines
- Low contrast between the gelatine and the culture
- Accuracy of the measurement better than 0.1 mm

The proposed solution consists of designing a vision system based on a CCD camera able to move around the inspection area. To provide this motion, an x-y positioning table was designed and built which allows the analysis of a single sample every n frames in order to guarantee the desired accuracy.[Figure 1].

According to the sort of sample-carriers employed, it was decided to use a view against the light to have a better contrast as well as to avoid shadows and reflectance effects over the gelatines. [Figure 2].

As mentioned before, the role of the illumination system is essential. We finally achieved homogeneous light distribution over the whole sample independently of the gelatine by means of what we call an automatic background control which enhances the culture and avoids transparency effects.

The whole process can be divided into five steps as follows:

a: The operator programs the trajectories for the inoculation of the cultures and stores those locations since they correspond also to inspection areas (playback programming mode).
b: Automatic injection following the programmed trajectories.
c: Measurement of the eccentricity of the halos and classification (accept/reject). When the system detects eccentricity in the sample, this corresponds either to a non-uniform culture or to an unnatural growing process.
The system provides the orthogonal diameters as well as the occupied surface of the accepted cultures. [Figure 3].
d: Presentation of the data (display and printer) grouping lots, and storage in the H.D. for further statistics.
e: Calibration
The calibration phase is crucial in order to attain the best reliability of the inspection equipment. The way we proceed is by testing the system with model patterns to detect deviations from the real measurements. The system automatically measures every pattern and corrects the parallax errors derived from the loss of orthogonality between the optical axis and the carrier (parallelism between the image plane and the sample-carrier). This way, the measurements corresponding to circular shapes around the symmetry axis must be corrected with a calculated coefficient. The calculated coefficients are stored

on an array or look-up table which will be accessed during the operating phase.

To automatically adjust the contrast between the sample and the background, the system is able to analyze light transition on the borders in order to effect a rescaling of the interest zones. [Figure 4].

3 Character Recognition

There are several different methodologies to deal with character recognition [20], [32], [35]. The definitions of the major schemes in OCR are the following:

Fixed-font: Recognition of specific fonts (Pica, Elite, Courier, etc.) typewritten characters.

On-line: Generally hand-drawn characters with the provision of pictorial information and timing information of each stroke.

Handwritten character recognition: Single hand-drawn characters (unconnected and not written in calligraphy).

Script recognition: Handwritten characters with no restriction.

The techniques used in character recognition are the same as the ones applied to general image processing and pattern recognition problems. According to Gaillat and Berthod [11] they can be classified as follows:

- Point by point global comparison: This is what is commonly known as template matching pixel by pixel.
- Global transformations: Transformations such as Fourier, Karhunen-Loeve, moments calculation etc.[28]
- Local Properties extraction: endpoints, straight lines, vertices, corners, etc.[37]

- Contour tracking: coding by sequences of curvatures (concavities and convexities detection).
- Structural methods: Topological descriptions and graph matching.[1]

Most commercial machines use template matching techniques which are intrinsically font-dependent. Those machines show a font sensitivity which is commonly emphasized in their specifications, being not so suitable for industrial applications [3].The literature stresses recognition methods based on features rather than on matching since they are less font sensitive and moreover they allow small tilts on the image[15].

The application we present has nothing to do with document processing but with real-time on-line character recognition during the rail manufacturing process,(railway industry), [Figure 5]. The main constraints to deal with this inspection process are as follows:

- Dusty atmosphere
- Maximum speed of displacement for the rail: 1 m/s
- The code to be recognized is formed by: 8 numerals plus one capital letter [Figure 6].
- The letter is always located in the same position with respect to the numerals.
- The characters which must be recognized can be slightly tilted with respect to the camera.
- Character types: Either continuous strokes or sets of spots.

We found some confusion groups as reported in [17] but the simplifications imposed by the codification scheme used in this specific application allowed us to face the problem in a straightforward way. For instance, the probable confusion between 8 and B or between 6 and b was automatically avoided by the fact of the known relative position of the numerals and letters.

We approached the character recognition problem by applying two different techniques. On one hand we used a feature based method which allowed a fast response time, while on the other hand and depending on the results of the pre-classification, we applied a local operator capable of coding the contour image by a sequence of curvatures [23]. So, the final classifier can be viewed as a merging of both techniques to cope with the intrinsic drawbacks of the specific environment in which the system must perform [Figure 7].

The contour analysis is based on the examination of outer pixels of the contour without paying attention to the discrimination capabilities of hole shape and size, although position and size are used in the preliminary classification phase [29]. Before analyzing the contour pixels, the contour image must have been thinned to avoid uncertainty during the tracking process. This uncertainty will lead to a false coding of the sequence of curvatures. Anyway, another important aspect of the whole process is the normalization phase. As a matter of fact, the sequence of curvatures which results from the coding of a numeral has a different length depending on whether it is slightly tilted or not. Moreover, some numerals give rise to sequences longer than others, making difficult the post-processing stages which must distinguish them. The cross-correlation of a sequence obtained from the image with that corresponding to the model pattern allows a fast classification and at the same time provides feedback information about the tilt angle (Shifting between both patterns) [Figure 8].

REFERENCES

[1] Baird,H.S.: Feature identification for hybrid structural/statistical pattern classification. Proc. IEEE Computer Vision and Pattern Recognition Conference, Miami, FL, 1986

[2] Ballard,D.H.: Generalizing the Hough Transform to detect arbitrary shapes. Pattern Recognition Vol. 13 No 2, pp 111-122, 1981

[3] Balm,G.J.: An introduction to optical character reader considerations. Pattern Recognition Vol 2, 1970

[4] Brown,C.M.: Inherent bias and noise in the Hough Transform. IEEE Transactions on Pattern Analysis and Machine Intelligence Vol PAMI-5 No 5, September 1983

[5] Davies,E.R.: Radial histograms as an aid in the inspection of circular shapes. IEE Proceedings Vol 132 No 4, July 1985

[6] Davies,E.R.: A high speed algorithm for circular object location. Pattern Recognition Letters No 6. 1987

[7] Davies,E.R.: A modified Hough scheme for general circle location. Pattern Recognition Letters No 7. 1988

[8] Deans,S.R.: Gegenbauer Transform via the Radon Transform.SIAM Journal of Math. Anal. Vol 10 No 3,1979

[9] Deans,S.R.: Hough Transform from Radon Transform.IEEE Transactions on Pattern Analysis and Machine Intelligence Vol PAMI-3 No 3, 1981

[10] Duda,R.O.;Hart,P.E.:Pattern recognition and scene analysis.New York. Wiley 1973

[11] Gaillat,G.;Berthod,M.: Panorama des techniques d'extraction de traits caracteristiques en lecture optique des caracteres. Revue Technique Thomson-CSF 11. 1979

[12] Haralick,R.M.: Solving camera parameters from the perspective projections of a parameterized curve. Computer Vision, Graphics and Image Processing No 27. 1984

[13] Helgason,S.:The Radon Transform on Euclidean space, compact two point homogeneous space and Grassman manifolds. Acta Mathematica 113, 1965

[14] Ianino,A.;Shapiro,S.D.:A Survey of the Hough Transform and its extensions for curve detection.IEEE Proceedings 1978

[15] Kahan,S. et al.: On the recognition of printed characters of any font and size. IEEE Transactions on Pattern Analysis and Machine Intelligence, Vol PAMI 9, No 2, 1987

[16] Kelley,R.B. et al.:Three vision algorithms for acquiring workpieces from bins.IEEE Proceedings Vol 71 No 7, 1983

[17] Lam,S.; Baird,H.S.: Performance testing of mixed-font, variable size character recognizers. AT&T Bell Labs., Computer Science Technical Report 126, November 1986

[18] Llario,V.: Hardware-software trade-offs in robot vision. NATO Advanced Research Workshop on Real Time Object Measurement and Classification. Maratea, Italy 1987

[19] Ludwig,D.:The Radon Transform on Euclidean space. Communications on Pure and Applied Mathematics. Vol 19. 1966

[20] Mantas,J.:An overview of character recognition methodologies.Pattern Recognition Vol 19 No 6 pp 425-430. 1986

[21] Martinez,A.B.: Preprocesador de imágenes para el reconocimiento y localización de geometrías circulares y elípticas en tiempo real. Tesis Doctoral. Facultat d'Informática de Barcelona. June 1988

[22] Nagao,M.; Nakajima,S.: On the relation between the Hough Transform and the projection curves of a rectangular window. Pattern Recognition Letters. Vol 6 .1987

[23] Perera,Toni: On character recognition. Erovi Technical Report 12/87. December 1987.

[24] Perkins,W.A.: A computer vision system that learns to inspect parts. IEEE Transactions on Pattern Analysis and Machine Intelligence. Vol PAMI 5 No 6. 1983

[25] Rosenfeld,A.: Picture processing by computer. Academic Press, New York 1969.

[26] Rosenfeld,A.: Image analysis: problems, progress and prospects. Pattern Recognition Vol 17, No 1, 1984

[27] Sanromà,J.: Some experiences of computer vision in industrial environments. Erovi Technical Report 7/88. July 1988

[28] Shridhar,M.; Baldreldin,A.: High accuracy character recognition algorithms using Fourier and topological descriptors. Pattern Recognition Vol 17 No 5, 1984

[29] Srihari,S.N.; Computer text recognition and error correction. Silver Spring, MD:IEEE Computer Society Press 1984

[30] Schurmann,J.: Reading machines. Proc. 1982 International Conference on Pattern Recognition. Munich, W.Germany. October 1982

[31] Stockman,G.C.; Agrawala,A.K.: Equivalence of Hough curve detection to template matching. Communications of the ACM Vol 20, No 11. 1977

[32] Suen,C.Y. et al.: Automatic recognition of handprinted characters. The state of the art. Proceedings of IEEE, Vol 68. 1980

[33] Suen,C.Y.: The role of multi-directional loci and clustering in reliable recognition of characters. Proc. 6th International J. Conference on Pattern Recognition. Munich, 1982

[34] Tsuji,S.; Matsumoto,F.: Detection of ellipses by a modified Hough Transformation. IEEE Transactions on Computers. Vol C 27, No 8. 1987

[35] Ullman,J.R.: Picture analysis in pattern recognition. digital picture analysis, A. Rosenfeld Editor. Springer 1976

[36] Venema,W.J.; Sterken,H.P.M.: IdentiVision system. Machine Vision Symposium 1986

[37] Yamamoto,K.; Mori,S.: Recognition of handprinted characters by outermost point method. Proc. Fourth International Joint Conference on Pattern Recognition, Kyoto, Japan.1978

[38] Ye,Q.: A pre-processing method for Hough Transform to detéct circles. 1986 IEEE Conference on Robotics and Automation. San Francisco,CA. 1986

Figure 1.- Imaging System setup with positioning table.

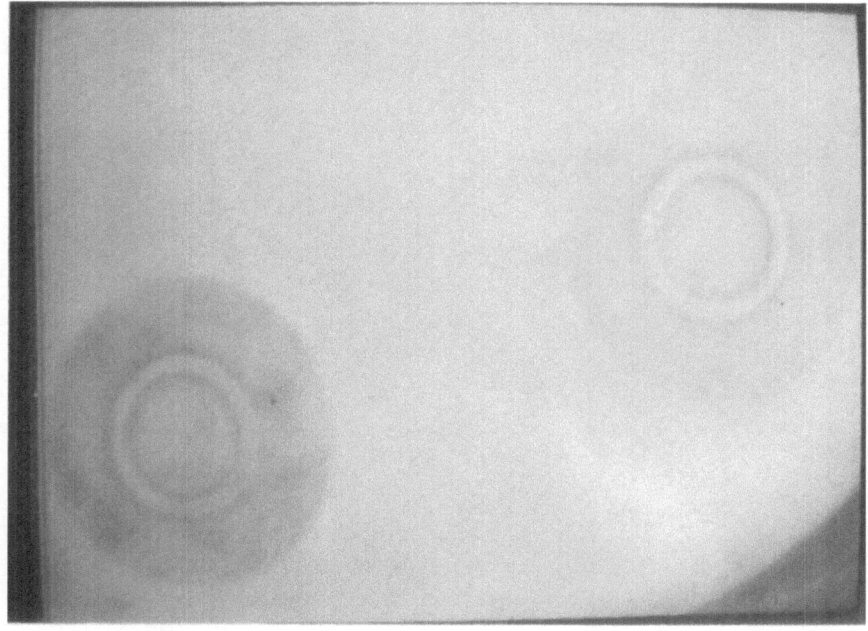

Figure 2.- Low contrast between the biological culture sample and the background gelatine.

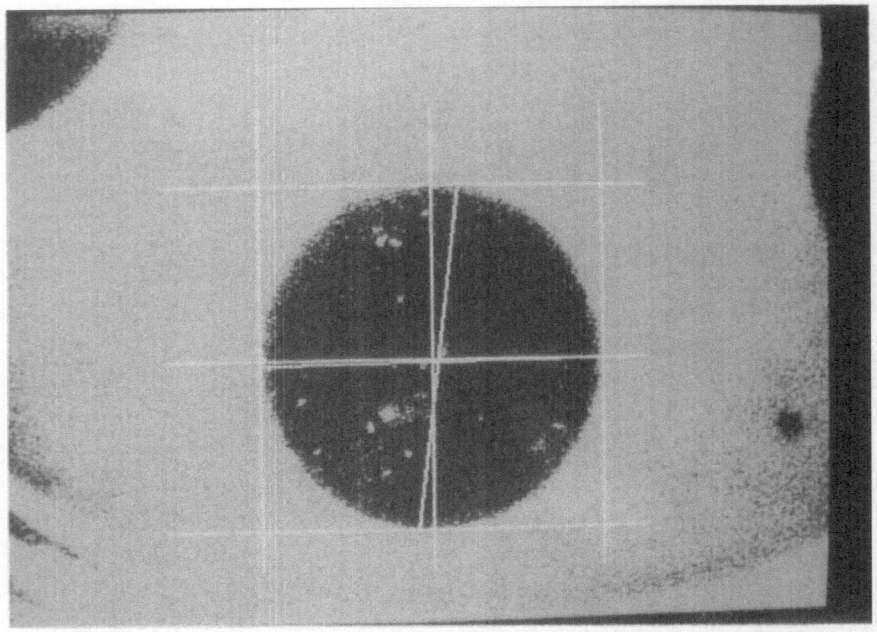

Figure 3.- Measurement of parameters over tha sample.

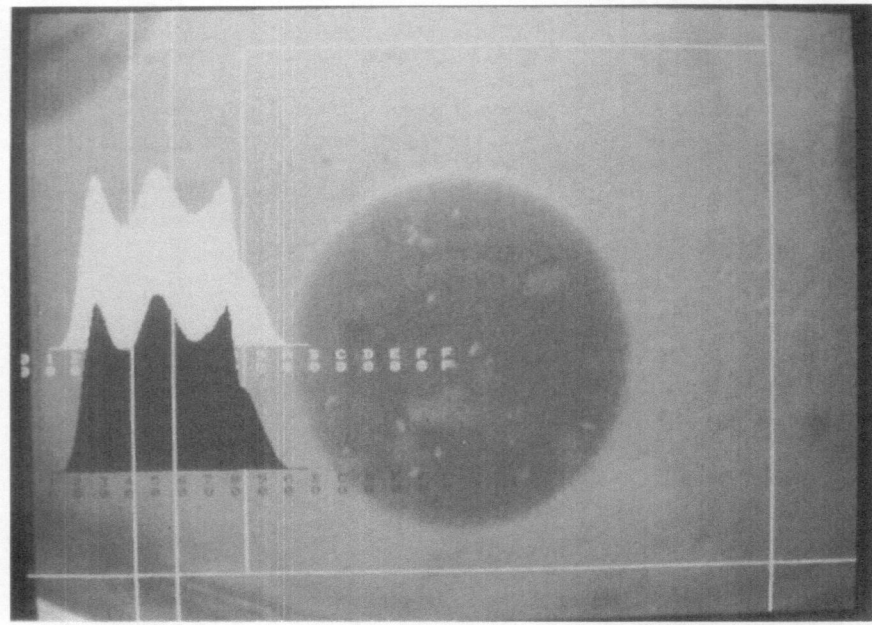

Figure 4 .- Rescaling of interest zones according to the contrast between sample and background.

Figure 5.- Vision System setup at the factory floor.

Figure 6 .- Sample of the format of the code to be recognized.
Characters are formed by spots.

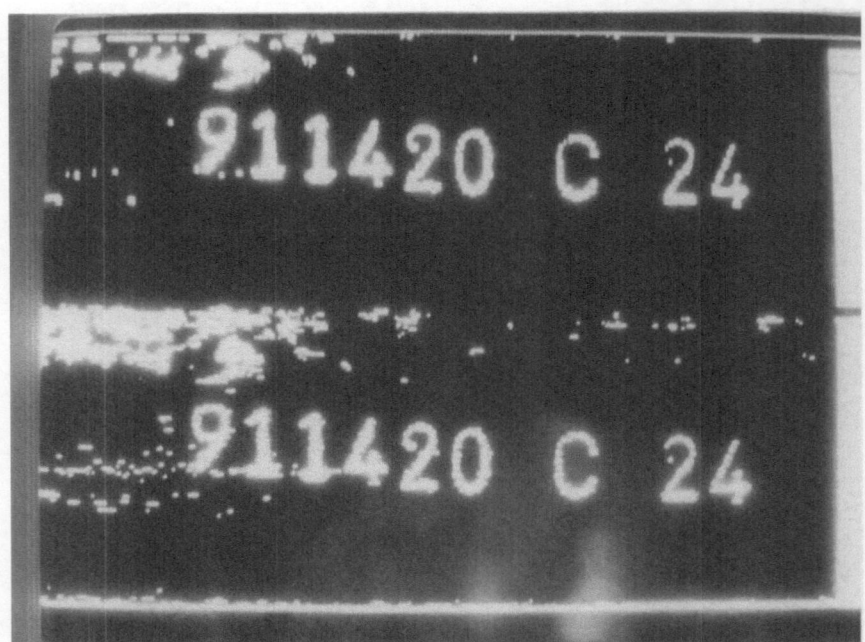

Figure 7.- Pre-processing of the gray scale image.

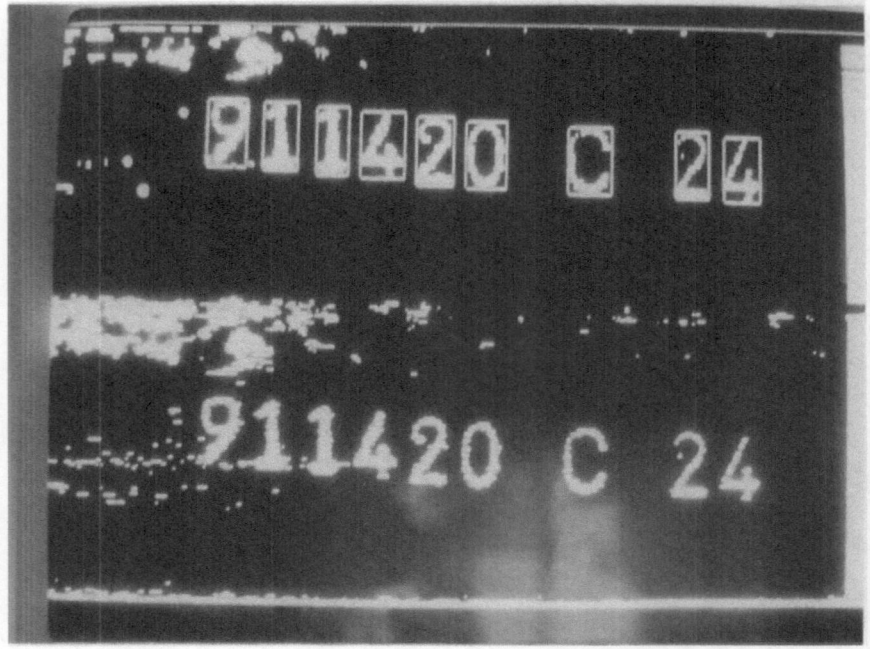

Figure 8 .- Some results of the segmentation stages, the characters not being aligned.

NATO ASI Series F

NATO ASI Series F

NATO ASI Series F

NATO ASI Series F